Veterinary Physiology and Biochemistry

Veterinary Physiology and Biochemistry

Edited by:

Patricia Marques

Delve
PUBLISHING

www.delvepublishing.com

Veterinary Physiology and Biochemistry

Patricia Marques

Delve Publishing
224 Shoreacres Road
Burlington, ON L7L 2H2
Canada
www.delvepublishing.com

Email: orders@arclereducation.com

ISBN: 978-1-77469-172-4 (Hardcover)

DECLARATION

ABOUT THE EDITOR

Patricia obtained her PhD from University College Dublin in 2010. Her interests are on Microbiology and Parasitology. She is currently working as a Postdoc at University of Maryland Baltimore, USA on Chlamydia infections.

TABLE OF CONTENTS

Section 2: Dogs

Section 3: Bovines

Section 6: Animal Models

Section 7: Wild Animals

7.1 Reptiles

Section 7: Wild Animals

7.2. Mammals

Section 7: Wild Animals

7.5. Marsupials

Section 7: Wild Animals

7.6. Rodents

LIST OF CONTRIBUTORS

Emma K. Read
Department of Veterinary Clinical and Diagnostic Sciences, University of Calgary Faculty of Veterinary Medicine, Calgary, Alberta, Canada,

Catriona Bell
Royal (Dick) School of Veterinary Studies, University of Edinburgh, Roslin, Midlothian, Scotland

Susan Rhind
Royal (Dick) School of Veterinary Studies, University of Edinburgh, Roslin, Midlothian, Scotland

Kent G. Hecker
Department of Veterinary Clinical and Diagnostic Sciences, University of Calgary Faculty of Veterinary Medicine, Calgary, Alberta, Canada,

G. Traverso
Division of Gastroenterology, Massachusetts General Hospital, Harvard Medical School, Boston, MA, United States of America,
Department of Chemical Engineering and Koch Institute for Integrative Cancer Research, Massachusetts Institute of Technology, Cambridge, MA, United States of America,

G. Ciccarelli
Bioengineering Systems and Technologies, Massachusetts Institute of Technology Lincoln Laboratory, Lexington, MA, 02420, United States of America

S. Schwartz
Bioengineering Systems and Technologies, Massachusetts Institute of Technology Lincoln Laboratory, Lexington, MA, 02420, United States of America

T. Hughes
Bioengineering Systems and Technologies, Massachusetts Institute of Technology Lincoln Laboratory, Lexington, MA, 02420, United States of America

T. Boettcher
Bioengineering Systems and Technologies, Massachusetts Institute of Technology Lincoln Laboratory, Lexington, MA, 02420, United States of America

R. Barman
Division of Gastroenterology, Massachusetts General Hospital, Harvard Medical School, Boston, MA, United States of America,
Department of Chemical Engineering and Koch Institute for Integrative Cancer Research, Massachusetts Institute of Technology, Cambridge, MA, United States of America,

R. Langer
Department of Chemical Engineering and Koch Institute for Integrative Cancer Research, Massachusetts Institute of Technology, Cambridge, MA, United States of America,

A. Swiston
Bioengineering Systems and Technologies, Massachusetts Institute of Technology Lincoln Laboratory, Lexington, MA, 02420, United States of America

Yu-Mei Chang
Research Office, The Royal Veterinary College, Camden Campus, Royal College Street, London, NW1 OTU, United Kingdom,

Erin Hadox
Department of Clinical Science and Services, Regulatory T Cell Laboratory, The Royal Veterinary College, Camden Campus, Royal College Street, London, NW1 OTU, United Kingdom,

Balazs Szladovits
Department of Pathology and Pathogen Biology, The Royal Veterinary College, Hawkshead Campus, North Mymms, Hatfield, Hertfordshire, AL9 7TA, United Kingdom

Oliver A. Garden
Department of Clinical Science and Services, Regulatory T Cell Laboratory, The Royal Veterinary College, Camden Campus, Royal College Street, London, NW1 OTU, United Kingdom,

Rian C. M. M. Lensen
Department of Veterinary Medicine (IVRU), University of Namur, Namur, Belgium,

Christel P. H. Moons
Department of Nutrition, Genetics and Ethology, Ghent University, Merelbeke, Belgium

Claire DiederichID
Department of Veterinary Medicine (IVRU), University of Namur, Namur, Belgium,

Soo Jin Jeon
Department of Animal Sciences, Institute of Food and Agricultural Sciences, University of Florida, Gainesville, Florida, United States of America,
Emerging Pathogens Institute, University of Florida, Gainesville, Florida, United States of America,

Mauricio Elzo
[1]Department of Animal Sciences, Institute of Food and Agricultural Sciences, University of Florida, Gainesville, Florida, United States of America,

Nicolas DiLorenzo
Department of Animal Sciences, Institute of Food and Agricultural Sciences, University of Florida, Gainesville, Florida, United States of America,
North Florida Research and Education Center, Institute of Food and Agricultural Sciences, University of Florida, Marianna, Florida, United States of America

G. Cliff Lamb
Department of Animal Sciences, Institute of Food and Agricultural Sciences, University of Florida, Gainesville, Florida, United States of America,
North Florida Research and Education Center, Institute of Food and Agricultural Sciences, University of Florida, Marianna, Florida, United States of America

Kwang Cheol Jeong
Department of Animal Sciences, Institute of Food and Agricultural Sciences, University of Florida, Gainesville, Florida, United States of America,
Emerging Pathogens Institute, University of Florida, Gainesville, Florida, United States of America,

Anna C. Davies
School of Clinical Veterinary Science, University of Bristol, Bristol, United Kingdom,

Christine J. Nicol
School of Clinical Veterinary Science, University of Bristol, Bristol, United Kingdom,

Mia E. Persson
Avian Behavioural Genomics and Physiology Group, Linko¨ping University, Linko¨ping, Sweden,

Andrew N. Radford
School of Biological Sciences, University of Bristol, Bristol, United Kingdom

Kate Fenner
School of Animal and Veterinary Sciences, Charles Sturt University, Wagga Wagga, NSW, Australia,

Holly Webb
Faculty of Veterinary Science, University of Sydney, NSW, Australia

Melissa J. Starling
Faculty of Veterinary Science, University of Sydney, NSW, Australia

Rafael Freire
School of Animal and Veterinary Sciences, Charles Sturt University, Wagga Wagga, NSW, Australia,

Petra Buckley
School of Animal and Veterinary Sciences, Charles Sturt University, Wagga Wagga, NSW, Australia,

Paul D. McGreevy
Faculty of Veterinary Science, University of Sydney, NSW, Australia

Yvonne A. Eiby
UQ Centre for Clinical Research, The University of Queensland, Brisbane, Queensland, Australia,

Layne L. Wright
UQ Centre for Clinical Research, The University of Queensland, Brisbane, Queensland, Australia,

Viskasari P. Kalanjati
UQ Centre for Clinical Research, The University of Queensland, Brisbane, Queensland, Australia,

Stephanie M. Miller
UQ Centre for Clinical Research, The University of Queensland, Brisbane, Queensland, Australia,

Stella T. Bjorkman
UQ Centre for Clinical Research, The University of Queensland, Brisbane, Queensland, Australia,

Helen L. Keates
School of Veterinary Science, The University of Queensland, Brisbane, Queensland, Australia,

Eugenie R. Lumbers
UQ Centre for Clinical Research, The University of Queensland, Brisbane, Queensland, Australia,
School of Biomedical Sciences and Pharmacy and the Hunter Medical Research Institute, The University of Newcastle, Newcastle, New South Wales, Australia

Paul B. Colditz
UQ Centre for Clinical Research, The University of Queensland, Brisbane, Queensland, Australia,

Barbara E. Lingwood
UQ Centre for Clinical Research, The University of Queensland, Brisbane, Queensland, Australia,

Fritz Trillmich
Animal Behaviour, Bielefeld University, Morgenbreede 45, 33615 Bielefeld, Germany.

Anja Günther
Animal Behaviour, Bielefeld University, Morgenbreede 45, 33615 Bielefeld, Germany.

Caroline Müller
Chemical Ecology, Bielefeld University, Universitätsstraße 25, 33615 Bielefeld, Germany.

Klaus Reinhold
Evolutionary Biology, Bielefeld University, Morgenbreede 45, 33615 Bielefeld, Germany.

Norbert Sachser
Behavioural Biology, University of Münster, Badestraße 13, 48149 Münster, Germany

Silvia Bolognin
CNR-Institute for Biomedical Technologies, Metalloproteins Unit, Department of Biology, University of Padova, Padova, Italy,

Federica Pasqualetto
CNR-Institute for Biomedical Technologies, Metalloproteins Unit, Department of Biology, University of Padova, Padova, Italy,

Carla Mucignat-Caretta
Department of Human Anatomy and Physiology, University of Padova, Padova, Italy,

Janez Scanca
Department of Environmental Sciences, Jozef Stefan Institute, Ljubljana, Slovenia,

Radmila Milacic
Department of Environmental Sciences, Jozef Stefan Institute, Ljubljana, Slovenia,

Pamela Zambenedetti
Division of Pathology, General Hospital, Dolo (VE), Italy, 5 Department of Experimental Veterinary Science, University of Padova, Legnaro (PD), Italy

Bruno Cozzi[5], Paolo Zatta
CNR-Institute for Biomedical Technologies, Metalloproteins Unit, Department of Biology, University of Padova, Padova, Italy,

Panteleimon D. Mavroudis
Bayer AG, Engineering & Technology- Systems Pharmacology, Leverkusen, Germany,

Helen E. Hermes
Bayer AG, Engineering & Technology- Systems Pharmacology, Leverkusen, Germany,

Donato Teutonico
Bayer AG, Engineering & Technology- Systems Pharmacology, Leverkusen, Germany,

Thomas G. Preuss
Bayer Crop Science, Environmental Safety, Monheim, Germany

Sebastian Schneckener
Bayer AG, Engineering & Technology- Systems Pharmacology, Leverkusen, Germany,

Brian L. Bostrom
Department of Zoology, University of British Columbia, Vancouver, British Columbia, Canada,

T. Todd Jones
Department of Zoology, University of British Columbia, Vancouver, British Columbia, Canada,
Joint Institute for Marine and Atmospheric Research, University of Hawaii at Manoa, National Oceanic and Atmospheric Administration, Kewalo Research Facility, Honolulu, Hawaii, United States of America,

Mervin Hastings
Conservation and Fisheries Department, Ministry of Natural Resources and Labour, Government of the British Virgin Islands, Road Town, Tortola, British Virgin Islands

David R. Jones
Department of Zoology, University of British Columbia, Vancouver, British Columbia, Canada,

Gregory A. Lewbart
North Carolina State University, College of Veterinary Medicine, Raleigh, North Carolina, United States of America,

Maximilian Hirschfeld
University San Francisco de Quito, Galapagos Science Center, Puerto Baquerizo Moreno, Galapagos, Ecuador,

Judith Denkinger
University San Francisco de Quito, Galapagos Science Center, Puerto Baquerizo Moreno, Galapagos, Ecuador,

Karla Vasco
University San Francisco de Quito, Galapagos Science Center, Puerto Baquerizo Moreno, Galapagos, Ecuador,

Nataly Guevara
University San Francisco de Quito, Galapagos Science Center, Puerto Baquerizo Moreno, Galapagos, Ecuador,

Juan García
Galapagos National Park Service, Puerto Ayora, Galapagos, Ecuador,

Juanpablo Muñoz
University San Francisco de Quito, Galapagos Science Center, Puerto Baquerizo Moreno, Galapagos, Ecuador,

Kenneth J. Lohmann
Department of Biology, University of North Carolina, Chapel Hill, North Carolina, United States of America

Shawna J. Cikanek
Department of Clinical Sciences, College of Veterinary Medicine, Kansas State University, Manhattan, Kansas, United States of America,

Simon Nockold
Panama Amphibian Rescue and Conservation Project, Smithsonian Tropical Research Institute, Gamboa, Republic of Panama,

Janine L. Brown
Smithsonian Conservation Biology Institute, National Zoological Park, Front Royal, Virginia, United States of America

James W. Carpente
Department of Clinical Sciences, College of Veterinary Medicine, Kansas State University, Manhattan, Kansas, United States of America,

Angie Estrada
Panama Amphibian Rescue and Conservation Project, Smithsonian Tropical Research Institute, Gamboa, Republic of Panama,

Jorge Guerrel
Panama Amphibian Rescue and Conservation Project, Smithsonian Tropical Research Institute, Gamboa, Republic of Panama,

Katharine Hope
Smithsonian Conservation Biology Institute, National Zoological Park, Front Royal, Virginia, United States of America

Roberto Ibáñez
Panama Amphibian Rescue and Conservation Project, Smithsonian Tropical Research Institute, Gamboa, Republic of Panama,

Sarah B. Putman
Smithsonian Conservation Biology Institute, National Zoological Park, Front Royal, Virginia, United States of America

Brian Gratwicke
Smithsonian Conservation Biology Institute, National Zoological Park, Front Royal, Virginia, United States of America

Erin E. Wiedower
Grazingland Animal Nutrition Lab, Blackland Research and Extension Center, Texas A&M University, Temple, Texas, United States of America,

Andrew J. Kouba
Department of Conservation and Research, Memphis Zoological Society, Memphis, Tennessee, United States of America,

Carrie K. Vance
Department of Biochemistry and Molecular Biology, Mississippi State University, Mississippi, United States of America,

Rachel L. Hansen
Department of Conservation and Research, Memphis Zoological Society, Memphis, Tennessee, United States of America,

Jerry W. Stuth
Department of Rangeland Ecology and Management, Texas A&M University, College Station, Texas, United States of America,

Douglas R. Tolleson
School of Natural Resources and the Environment, V Bar V Ranch, The University of Arizona, Camp Verde, Arizona, United States of America

David S. Jachowski
Department of Fisheries and Wildlife Sciences, University of Missouri, Columbia, Missouri, United States of America,

Rob Slotow
Amarula Elephant Research Programme, School of Life Sciences, University of KwaZulu-Natal, Durban, South Africa

Joshua J. Millspaugh
Department of Fisheries and Wildlife Sciences, University of Missouri, Columbia, Missouri, United States of America,

Lee McMichael
School of Veterinary Science, The University of Queensland, Gatton, Qld 4343, Australia
Biosecurity Queensland, Department of Agriculture, Fisheries and Forestry, Brisbane, Qld 4108, Australia

Daniel Edson
Biosecurity Queensland, Department of Agriculture, Fisheries and Forestry, Brisbane, Qld 4108, Australia

Amanda McLaughlin
Biosecurity Queensland, Department of Agriculture, Fisheries and Forestry, Brisbane, Qld 4108, Australia

David Mayer
Department of Agriculture Fisheries and Forestry, Brisbane, Qld 4103, Australia

Steven Kopp
School of Veterinary Science, The University of Queensland, Gatton, Qld 4343, Australia

Joanne Meers
School of Veterinary Science, The University of Queensland, Gatton, Qld 4343, Australia

Hume Field
EcoHealth Alliance, New York, NY 10001, United States of America

Claire E. Couch
Oregon State University, Department of Integrative Biology, Corvallis, Oregon, United States of America

Morgan A. Movius
Oregon State University, College of Veterinary Medicine, Corvallis, Oregon, United States of America

Anna E. Jolles
Oregon State University, Department of Integrative Biology, Corvallis, Oregon, United States of America
Oregon State University, College of Veterinary Medicine, Corvallis, Oregon, United States of America

M. Elena Gorman
Oregon State University, College of Veterinary Medicine, Corvallis, Oregon, United States of America

Johanna D. Rigas
Utah State University, School of Veterinary Medicine, Logan, Utah, United States of America

Brianna R. Beechler
Oregon State University, College of Veterinary Medicine, Corvallis, Oregon, United States of America

Geraldine A. Wright
Centre for Behaviour and Evolution, Institute of Neuroscience, Newcastle University, Newcastle upon Tyne, United Kingdom

Joshua L. Lillvis
Neuroscience Institute, Georgia State University, Atlanta, Georgia, United States of America

Helen J. Bray
School of Biology, Newcastle University, Newcastle upon Tyne, United Kingdom

Julie A. Mustard
School of Life Sciences, Arizona State University, Tempe, Arizona, United States of America

Jared R. Eckroth
Department of Molecular Biosciences, University of Oslo, Oslo, Norway

Øyvind **Aas-Hansen**
Norwegian Institute of Food, Fisheries and Aquaculture Research, Nofima, Tromsø, Norway

Lynne U. Sneddon
Institute of Integrative Biology, University of Liverpool, Liverpool, United Kingdom

Helena Bichão
Norwegian Institute of Food, Fisheries and Aquaculture Research, Nofima, Tromsø, Norway

Kjell B. Døving
Institute of Integrative Biology, University of Liverpool, Liverpool, United Kingdom

Nicole Ashley Davies
School of Biomedical Sciences, The University of Queensland, St Lucia, Queensland, Australia

Galina Gramotnev
Landscape Ecology and Conservation Group, School of Geography, Planning and Environmental Management, The University of Queensland, St Lucia, Queensland, Australia

Clive McAlpine
Landscape Ecology and Conservation Group, School of Geography, Planning and Environmental Management, The University of Queensland, St Lucia, Queensland, Australia

Leonie Seabrook
Landscape Ecology and Conservation Group, School of Geography, Planning and Environmental Management, The University of Queensland, St Lucia, Queensland, Australia

Greg Baxter
Landscape Ecology and Conservation Group, School of Geography, Planning and Environmental Management, The University of Queensland, St Lucia, Queensland, Australia

Daniel Lunney
Office of Environment and Heritage New South Wales, Hurstville, New South Wales, Australia
School of Veterinary and Life Sciences, Murdoch University, Perth, Western Australia, Australia

Jonathan R. Rhodes
Landscape Ecology and Conservation Group, School of Geography, Planning and Environmental Management, The University of Queensland, St Lucia, Queensland, Australia

Adrian Bradley
School of Biomedical Sciences, The University of Queensland, St Lucia, Queensland, Australia

Simon J. Girling
Veterinary Department, Living Collection, Royal Zoological Society of Scotland, Edinburgh, United Kingdom

Roisin Campbell-Palmer
Faculty of Arts and Sciences, Department of Environmental and Health Studies, Telemark University College, Bø i Telemark, Norway
Conservation Programmes, Royal Zoological Society of Scotland, Edinburgh, United Kingdom

Romain Pizzi
Veterinary Department, Living Collection, Royal Zoological Society of Scotland, Edinburgh, United Kingdom

Mary A. Fraser
School of Life, Sport and Social Science, Edinburgh Napier University, Sighthill Campus, Edinburgh, United Kingdom

Jonathan Cracknell
Longleat Safari and Adventure Park, Longleat, Warminster, Wiltshire, United Kingdom

Jon Arnemo
Department of Forestry and Wildlife Management, Faculty of Applied Ecology and Agricultural Sciences, Hedmark University College, Campus Evenstad, Koppang, Norway
Department of Wildlife, Fish and Environmental Studies, Faculty of Forest Sciences, Swedish University of Agricultural Sciences, Umeå, Sweden

Frank Rosell
Conservation Programmes, Royal Zoological Society of Scotland, Edinburgh, United Kingdom,

Ayelen T. Eberhardt
Laboratorio de Ecología de Enfermedades, Instituto de Ciencias Veterinarias del Litoral, Universidad Nacional del Litoral - Consejo de Investigaciones Científicas y Técnicas (UNL - CONICET). Esperanza, Santa Fe, Argentina

Sebastián A. Costa
Laboratorio de Ecología de Enfermedades, Instituto de Ciencias Veterinarias del Litoral, Universidad Nacional del Litoral - Consejo de Investigaciones Científicas y Técnicas (UNL - CONICET). Esperanza, Santa Fe, Argentina

M. Rocío Marini
Cátedra de Patología Básica, Facultad de Ciencias Veterinarias, Universidad Nacional del Litoral. Esperanza, Santa Fe, Argentina

Andrea Racca
Laboratorio de Ecología de Enfermedades, Instituto de Ciencias Veterinarias del Litoral, Universidad Nacional del Litoral - Consejo de Investigaciones Científicas y Técnicas (UNL - CONICET). Esperanza, Santa Fe, Argentina

Cecilia J. Baldi
Laboratorio de Ecología de Enfermedades, Instituto de Ciencias Veterinarias del Litoral, Universidad Nacional del Litoral - Consejo de Investigaciones Científicas y Técnicas (UNL - CONICET). Esperanza, Santa Fe, Argentina

M. Rosario Robles
Centro de Estudios Parasitológicos y de Vectores (CONICET) La Plata, Buenos Aires, Argentina

Pablo G. Moreno
Laboratorio de Ecología de Enfermedades, Instituto de Ciencias Veterinarias del Litoral, Universidad Nacional del Litoral - Consejo de Investigaciones Científicas y Técnicas (UNL - CONICET). Esperanza, Santa Fe, Argentina

Pablo M. Beldomenico
Laboratorio de Ecología de Enfermedades, Instituto de Ciencias Veterinarias del Litoral, Universidad Nacional del Litoral - Consejo de Investigaciones Científicas y Técnicas (UNL - CONICET). Esperanza, Santa Fe, Argentina

LIST OF ABBREVIATIONS

AAP	Academy of Pediatrics
AD	Alzheimer's disease
AIC	Akaike's information criterion
ALKP	Alkaline phosphatase
ALP	Alkaline phosphatase
ALT	Alanine aminotransferase
AMDF	Average magnitude difference function
ANOVA	Analysis of Variance
APP	Amyloid precursor protein
AS	Alternative splice
AST	Aspartate aminotransferase
CAMs	Cell adhesion molecules
CCL	Curved carapace lengths
CDA	Canonical discriminant analysis
CFU	colony forming unit
CK	Creatine kinase
CLs	Corpora lutea
CLSI	Clinical and Laboratory Standards Institute
CNS	Central nervous system
CSF	Cerebrospinal fluid
CVMA	Chinese Veterinary Medical Association
CZA	Committee of Zoos and Aquariums
DenDF	Denominator degrees of freedom
DP	Dry period
DS	Dissolution shapes
ECG	Electrocardiogram
ESM	Electronic supplementary materials
ESU	Evolutionary significant units

ETT	Endotracheal tube
FCM	Faecal cortisol metabolite
FMDV	Foot-and-mouth disease virus
FNIRS	Fecal near infrared reflectance spectroscopy
FVP	Facial vein phlebotomy
GA	Global assessment
GFR	Glomerular filtration rate
GIT	Gastrointestinal tract
GLM	General linear model
GLMM	General linear mixed models
GO	Gene Ontology
GRS	Global rating scales
GxA	Gene x Age
HC	Hemorrhagic colitis
HCCs	Hair cortisol concentrations
HD	Hydroxybutyrate dehydrogenase
HPA	Hypothalamic-pituitary-adrenal axis
HR	Heart rate
HRV	Heart-rate variability
IQC	intelligent quality control
IQR	Interquartile range
IRIS	International Renal Interest Society
IUCN	International Union for Conservation of Nature
KS	Kolmogorov-Smirnov
LMM	Linear mixed effects model
LP	Lactation period
LSD	least significant difference
MAP	Mean arterial pressure
MCHC	Mean cell haemoglobin concentration
MCV	Mean cell volumes
MXEs	mutually exclusive exons
NEC	Necrotising enterocolitis
NGS	Next-generation sequencing
NumDF	Numerator degrees of freedom

PBPK	Physiologically based pharmacokinetic
PBS	Phosphate saline buffer
PCA	Principal component analysis
PCV	Packed cell volumes
PFGE	Pulsed-Field Gel Electrophoresis
PMN	Polymorphonuclear
PPG	Photoplethysmography
QCEID	Queensland Centre for Emerging Infectious Diseases
QML	Queensland Medical Laboratories
RC	Resistor-capacitor
RCV	Reference change value
RI	Reference interval
RSP	Retro-bulbar sinus puncture
RVC	Royal Veterinary College
SAM	Sympatho-adrenal-medullary
SD	Standard deviation
SEQ	South East Queensland
SI	Standard international
SNPs	Single nucleotide polymorphisms
SNR	Signal-to-noise ratio
SPM	Strokes per minute
SSR	Sum of squared residuals
TH	Tyrosine hydroxylase
THI	Temperature–humidity index
UCVM	University of Calgary Faculty of Veterinary Medicine
UDs	Utilization distributions
WBC	White blood cell
WHWT	West Highland White Terrier

PREFACE

Veterinary Physiology focuses on the scientific study of the functional dynamics of animal biological systems and their relationship to the diagnosis and treatment of disease and injury.

Veterinary physiology deals with the investigation of animal systems and the functioning of these biological systems. It includes the study of study of how animals work and the physical and chemical processes that occur within animals. ... Veterinary physiology includes both anatomy and physiology of an animal.

Biochemistry is one of the few basic sciences where animal and plant kingdoms meet. In veterinary education and research, biochemistry is highly relevant to the metabolism and function of animals in health and disease and forms the basis for an intelligent understanding of major aspects of veterinary science and animal husbandry. Hence the need for adequate recognition of the place of biochemistry in veterinary education cannot be overemphasized. It is only the possession of a high level of knowledge of biochemistry that will enable one to study, or to pursue a line of research, intelligently in applied sciences such as animal nutrition, dairy science, pharmacology, animal physiology, animal genetics, clinical pathology, medicine and so on. In the absence of a grounding in biochemistry, it is difficult to obtain an insight into these subject areas.

In veterinary education and research, biochemistry is highly relevant to the metabolism and function of animals in health and disease and forms the basis for an intelligent understanding of major aspects of veterinary science and animal husbandry.

This book is organized in 7 sections. Section 1 is introduction with 2 different chapters, one about global rating physiological scale and the other discussing physiologic status monitoring via the gastrointestinal tract.

The second section is about physiological and biochemistry studies in dogs.

Sections 3 to 5, concentrate on livestock animals (bovines, chickens and horses, respectively).

Section 6, is dedicated to studies in animal models, including pigs (used as models), guinea pigs, mice, macaques and rabbits.

Section 7 (wild animals) is divided in subsections, Reptiles (turtles and frogs), Mammals (pandas, elephants, lynx, flying-fox, yak and antelope), Insects (honeybees), Fish (cod), Marsupials (koala) and Rodents (castor, and capybara).

SECTION 1: INTRODUCTION

CHAPTER 1

The Use of Global Rating Scales for OSCEs in Veterinary Medicine

Emma K. Read[1], Catriona Bell[2], Susan Rhind[2], Kent G. Hecker[1]

[1]Department of Veterinary Clinical and Diagnostic Sciences, University of Calgary Faculty of Veterinary Medicine, Calgary, Alberta, Canada,
[2]Royal (Dick) School of Veterinary Studies, University of Edinburgh, Roslin, Midlothian, Scotland

ABSTRACT

OSCEs (Objective Structured Clinical Examinations) are widely used in health professions to assess clinical skills competence. Raters use standardized binary checklists (CL) or multi-dimensional global rating scales (GRS) to score candidates performing specific tasks. This study assessed the reliability of CL and GRS scores in the assessment of veterinary students, and is the first study to demonstrate the reliability of GRS within veterinary medical education. Twelve raters from two different schools (6 from University of Calgary [UCVM] and 6 from Royal (Dick) School of

Citation: Read EK, Bell C, Rhind S, Hecker KG (2015) The Use of Global Rating Scales for OSCEs in Veterinary Medicine. PLoS ONE 10(3): e0121000. https://doi.org/10.1371/journal.pone.0121000

Veterinary Studies [R(D)SVS] were asked to score 12 students (6 from each school). All raters assessed all students (video recordings) during 4 OSCE stations (bovine haltering, gowning and gloving, equine bandaging and skin suturing). Raters scored students using a CL, followed by the GRS. Novice raters (6 R(D)SVS) were assessed independently of expert raters (6 UCVM). Generalizability theory (G theory), analysis of variance (ANOVA) and t-tests were used to determine the reliability of rater scores, assess any between school differences (by student, by rater), and determine if there were differences between CL and GRS scores. There was no significant difference in rater performance with use of the CL or the GRS. Scores from the CL were significantly higher than scores from the GRS. The reliability of checklist scores were .42 and .76 for novice and expert raters respectively. The reliability of the global rating scale scores were .7 and .86 for novice and expert raters respectively. A decision study (D-study) showed that once trained using CL, GRS could be utilized to reliably score clinical skills in veterinary medicine with both novice and experienced raters.

INTRODUCTION

Educating veterinary students to become competent, autonomous practitioners requires ongoing assessment of students' abilities and performance using methods that provide reliable and valid scores. For professional skills globally, and veterinary clinical skills specifically, the Objective Structured Clinical Examination (OSCE) has become one of the key performance based methods of assessment. [1,2] While the OSCE was first reported over 30 years ago in human medicine [3] it has only relatively recently been adopted for use in veterinary medicine but is gaining widespread acceptance. [2,4,5] Advantages of the OSCE over previously used methods of assessment of clinical skills include standardization of the tasks performed by all students, the ability to use trained non-subject matter experts as raters, and the reliability of judgments made between raters. [6] It is well documented that for OSCE scores to be reliable and valid, the OSCE must have a series of timed stations, the exam should be blueprinted with a specified set of tasks that are performed in the presence of trained raters and that tasks, timing and raters are standardized over all students.[6]

Scoring methods for OSCEs are typically either analytic, meaning they incorporate binary checklists (CLs) that quantify small elements of performance as yes/no, or holistic, meaning they utilize global rating scales (GRS) that consider skill performance across several domains using a

Likert–type scale. [7] The use of CLs is generally accepted in veterinary medical OSCEs and affords high inter-rater reliability if they are well written, revised after pilot testing and involves adequate rater training. [1,8] However, CLs may not be suitable in all situations and they may reward thoroughness without consideration of timeliness or proficiency of action. [9] Some CLs incorporate a global assessment (GA) at the end to allow raters to provide a subjective score of the student performance and this may or may not correlate well to the overall CL score. [4] More advanced clinical trainees have been shown to skip steps and more rapidly proceed to an endpoint while taking short cuts which could penalize them when being graded using a binary checklist. [9]

Global rating scales (GRS) have also been used in medicine to assess OSCE performance. [8, 10] Almost twenty years ago, GRS were first compared to CLs for scoring technical skills performance. [11, 12] GRS are reported to assess more qualitative performance values, such as overall preparation for the task or efficiency of performance, which can be useful, but does bring up concerns about the need for highly trained raters and inter-rater reliability to avoid subjectivity. [8,10] Despite these concerns, it is reported that assessors are able to discriminate performance better when not tied to reducing clinical performance into a number of individual steps (process-level observation). [13,14] Recently, GRS scores were shown to yield reliable data when compared with CL scores for assessment of more advanced learners (residents) in the field of medicine. [8, 10,15] To date, only one recent report has discussed the use of GRS for the assessment of clinical skills proficiency in veterinary students, however, these authors only addressed issues of pre and post score student satisfaction with the tool and did not demonstrate whether scores from the GRS tool were reliable or valid. [5]

To explore the use of GRS for assessment of veterinary clinical skills, student and rater performance in the context of two different veterinary programs were compared using a fully crossed design in which all students were rated on all stations, by all raters, from standardized video recordings of four different OSCE stations.

The purpose of this study was twofold: First, to assess novice and expert rater performance differences on CL and GRS scores, and second, to assess the reliability of CL and GRS scores from both types of raters. While previous studies have compared CL and GRS scores, this study adds to the literature by providing a direct comparison of novice and expert raters and demonstrates that CLs can be used to inform the use of GRS with both types of raters.

MATERIALS AND METHODS

Institutions

The University of Calgary Faculty of Veterinary Medicine (UCVM) in Canada, was established in 2005, accepted its first cohort of students in 2008. This program has a heavy emphasis on clinical skills training with approximately 20% of each year's curriculum devoted to formalized training beginning in first year, with regular assessment using summative OSCEs. [16] The Royal (Dick) School of Veterinary Studies (R(D)SVS) at the University of Edinburgh was established in 1823, and has clinical skills training throughout the 4 and 5-year programs that it offers. Clinical and practical skills are assessed using different summative practical examinations aligned to specific courses, often as a short series of stations, but there is no large multiple station OSCE format in place.

Participants

Video recordings of student performances on all four OSCE stations were completed in April and May 2013. The six students from UCVM (5 females and 1 male) had recently completed their first year of the four-year veterinary program resulting in 120 hours each of dedicated clinical skills training, and 3 summative multi-station OSCEs. The six students from R(D)SVS (5 females and 1 male) had partially completed their third year of the five-year veterinary program and had 15 hours of dedicated clinical skills training (with no summative OSCE's) plus additional practical skills training in two other courses (with associated summative assessments). Twelve raters were identified from UCVM and R(D)SVS (6 from each school). The raters from UCVM (four females, 2 males) had 1–6 years experience rating OSCEs and were clinicians who had been trained on other occasions using the CL assessment method by 2 of the authors (EKR and KGH). The R(D)SVS raters (5 females, 1 male) had no previous experience assessing OSCEs. All ratings were completed between July and September 2013.

Ethics Statement

The study was approved by conjoint health research ethics board at the University of Calgary, and the Medicine and Veterinary Medicine Education Research (EREC) Project ethics board at the University of Edinburgh. Informed written consent was obtained from all study participants prior to participation in the study. The consent procedure was approved by the ethics board of each institution.

Measures

The OSCE stations used in this study were previously developed at UCVM and included bovine haltering, skin suturing, equine bandaging, and surgeon preparation (gowning and gloving). [4] The original OSCE checklists were modified, for the purpose of this study, by EKR and CB. Each checklist was binary, comprising 8 to 40 items. A student's CL score for a given station was the sum of the number of items that were scored as yes. Station scores were converted to mean percent scores.

The GRS were developed using the checklists as a reference to allow grouping of similar skills into dimensions that were assessed using a 5-point scale, with points 1, 3, and 5 being anchored by explicit descriptors (1 being the lowest level of performance and 5 being the highest level of performance). Each GRS had between 6 and 8 dimensions corresponding to specific groupings of checklist items into broader categories. Other dimensions such as time and motion, that were not related directly to the checklist items, were also developed. A student's GRS score for a given station was the sum of each 5 point item over the entire tool. Station scores were converted to mean percent scores. See Fig. 1 for a CL example and Fig. 2 for a GRS example.

UNIVERSITY OF CALGARY
FACULTY OF VETERINARY MEDICINE

Clinical Skills OSCE

Station Number: **Skin Suturing**

The candidate should do the following (insert more lines if needed):	YES	NO
1. Selects an instrument tray and proceeds to unwrap it using proper aseptic technique (including holds on flat hand turned palm up or places on counter top. opens tabs in turn being careful not to touch the inner lining of the wrapper or the tray of instruments with their bare hands. opens the far tabs first to avoid leaning across the sterile field).	O	O
2. Selects appropriate sized surgical gloves.	O	O
3. Applies the gloves in an aseptic manner using open gloving technique and taking care not to break sterility.	O	O
4. Selects a pair of needle drivers to suture with (not any other instrument).	O	O
5. Selects a pair of sharp-blunt scissors to cut suture material with (not tissue scissors).	O	O
6. Selects a pair of rat-toothed thumb forceps to hold skin and to receive suture needle from needle drivers (does not use Russian forceps or fingers).	O	O
7. Selects the 0-Prolene suture material to suture the skin with and asks the examiner to open it using proper aseptic technique.	O	O
8. Uses rat-toothed thumb forceps for retracting the skin and receiving the needle when suturing. Holds the forceps in a pencil grip (does not palm grip).	O	O
9. Holds the needle drivers using appropriate tripod grip.	O	O
10. Ties a surgeon's knot at the start of the incision (not past the end of the incision so that they leave a gap at the start of the incision) (must be a surgeon's knot – they are starting a tension suture pattern).	O	O
11. Correctly places Suture #1 - Correctly ties interrupted vertical mattress suture in the wound.	O	O
12. Correctly places Suture #2 - Correctly ties interrupted vertical mattress suture in the wound.	O	O

Figure 1. Example checklist instrument for the skin suturing station.

Global Rating Scale Assessment Tool

UNIVERSITY OF CALGARY
FACULTY OF VETERINARY MEDICINE

Skin Suturing (Equine)

Please circle the numbers below that best describe the candidate's actions. Choose only one per dimension.

	Clearly below expectations		Acceptable performance		Performs above expectations
	1	2	3	4	5
Safety	Fails to respect safety aspects - places self, assistant, or horse in repeated circumstances of compromised safety.		Occasionally tentative in approach to horse or advice to handler and may result in a position of compromised safety.		Always conscious of safety and demonstrates confidence. Never places self, assistant or horse in a position of compromised safety.
	1	2	3	4	5
Organisation and selection of materials	Fails to select correct materials and/or instruments for task. Does not organize appropriately prior to starting task.		Slowly selects appropriate aterials. Organizes them but not a manner that necessarily helps prove efficiency with performing sk or has some materials but not organized before performance of the task begins.		Quickly and confidently selects correct materials and instruments for task. Organizes in a deliberate manner that facilitates task. Materials are ready before performance of the task begins.
	1	2	3	4	5
Appropriate aseptic technique	Demonstrates deficient knowledge or execution of aseptic technique without recognising and correcting it.		Competent performance that indicates working knowledge of aseptic technique but committed some minor errors that were recognised and corrected.		Superior attention to aseptic technique. Demonstrates good understanding of sterile versus clean technique. No mistakes made.
	1	2	3	4	5
Correct use of instruments	Student may use an instrument incorrectly and may or may not correct mistake but is still able to get the job done.		Student uses instrumentation correctly for most part. Occasional errors are made but awareness and correction is demonstrated.		Student holds and uses instruments correctly throughout using proficient and smooth technique. No errors in use or handling occur.
	1	2	3	4	5
Correct suture pattern and technique	Student uses inappropriate pattern, bite sizes or spacing.		Student uses appropriate pattern, bite sizes and spacing for the majority of the task but not all.		Student consistently places sutures using appropriate pattern, bite sizes and spacing in a proficient manner.
	1	2	3	4	5
Appropriate disposal of equipment	Improper disposal or failure to dispose of materials.		Delayed or partial disposal of materials.		Prompt and proper disposal of materials.

Figure 2. Example global rating scale instrument for the skin suturing station.

Evaluations

Each student was video recorded performing each of the 4 stations in the same order. Raters were not trained in the use of the assessment tools (beyond their prior experience) but were given a brief explanation by the authors (EKR, CB) as to the purpose of each tool prior to their use if requested. All raters independently rated all students from all schools. Each rater was asked to perform the CL rating first and then the GRS second for each student.

Analyses

Generalizability theory was used to calculate the reliability of the OSCE scores and to perform a decision (D) study. We ran two generalizability analyses, one for the UCVM (expert) rater data and one for the R(D)SVS (novice) rater data. The generalizability (G) studies were fully crossed designs with the following facets, participants (students) (12), stations (4), and raters (6). A decision (D) study was then run to identify the optimal number of stations given one or two raters per station. A two-way Analysis of Variance (ANOVA) was used to assess rater and student school differences in CL and GRS scores. Dependent *t* tests were used to assess differences in CL and GRS scores by the same rater.

RESULTS

Rater CL and GRS scores (mean percent and standard deviation) for each of the OSCE stations and all stations combined are presented in Table 1. The two–way ANOVAs revealed a significant difference in student performance by school in both the CL ($F_{(1,140)} = 78.53$, $p <. 001$) and GRS scores ($F_{(1,140)} = 105.84$, $p <. 001$) with the R(D)SVS students performing significantly lower than UCVM students. There was no difference between rater scores from the two schools and there was no interaction effect between rater school and school that the students attended. There were significant differences in rater scores using CL versus GRS regardless of rater school. Specifically, total CL scores were significantly higher than total GRS scores (UCVM $t(71) = 9.17$, $p<0.001$; RDSVS $t(71) = 15.11$, $p<0.001$).

Table 1. Mean percent scores (and standard deviations) from UCVM and R(D) SVS raters for OSCE Checklist and Global rating scores.

	Checklist		**Global Rating Score**	
	CL—UCVM	**CL—R(D)SVS**	**GRS—UCVM**	**GRS—R(D)SVS**
Station1	90.59 (8.16)	91.60 (8.35)	68.89 (18.99)	71.11(14.90)
Station 2	84.03(16.28)	77.60 (17.16)	70.79(14.91)	68.06(16.21)
Station 3	72.98 (16.64)	73.48 (17.91	72.61(16.51)	72.11(17.71)
Station 4	60.46 (16.15)	66.76(14.82)	58.06(18.39)	56.30(16.67)
Total Score	77.02 (10.26)	77.36 (8.77)	67.59 (13.12)	66.89 (10.86)

CL—checklist; GRS—global rating scale; UCVM—UCVM raters; RDSVS—R(D)SVS raters

The reliability of the CL and GRS scores were. 76 and. 86 respectively for UCVM raters suggesting that scores for students across raters and stations were generally consistent (see Table 2). The reliability of the CL and GRS from R(D)SVS raters were. 42 and. 70 suggesting scores weren't as consistent across raters and stations for CL scores as they were for the GRS scores.

Table 2. Generalizability (G) study (for participant (student) (12), rater (6) and station (4)) and Decision (D) study with reliability coefficient (G-coef) calculated using the variance components.

	CL—UCVM		**CL—R(D)SVS**		**GRS—UCVM**		**GRS—R(D)SVS**	
	σ^2	**%**	**σ^2**	**%**	**σ^2**	**%**	**σ^2**	**%**
p	47.23	11.94	17.79	5.23	88.11	24.92	54.49	16.77
r	15.31	3.87	15.91	4.68	55.20	15.61	12.23	3.76
s	167.02	42.22	97.44	28.65	33.83	9.57	42.33	13.03
p|r	11.05	2.79	0.00	0.00	1.49	0.42	6.64	2.04
p|s	36.75	9.29	83.16	24.45	36.62	10.36	68.42	21.06
r|s	18.87	4.77	26.14	7.69	25.37	7.17	19.00	5.85
p|r|s, error	99.36	25.12	99.65	29.30	112.95	31.95	121.80	37.49
G-coef	0.76		0.42		0.86		0.70	
D- study values								
raters	**stations**	**CL—UCVM**	**CL—R(D)SVS**	**GRS—UCVM**	**GRS—R(D)SVS**			
2	4	.64	.35	.79	.60			
1	4	.51	.28	.69	.50			
2	5	.67	.40	.82	.65			
1	5	.55	.33	. 74	.55			
2	6	.70	.45	.84	.69			
1	6	.58	.37	.77	.59			
2	7	.73	.48	.86	.71			
1	7	.61	.41	.79	.62			
2	8	.74	.52	.88	.74			
1	8	.63	.44	.81	.64			
2	9	.76	.55	.89	.75			
1	9	.64	.47	.83	.66			
2	10	.77	.57	.90	.77			
1	10	.66	.49	.84	.68			
2	11	.78	.60	.91	.78			
1	11	.67	.52	.85	.69			
2	12	.79	.62	.91	.79			
1	12	.68	.54	.86	.79			

CL—checklist; GRS—global rating scale; UCVM—UCVM raters; R(D)SVS—R(D)SVS raters p- participant (student); r—rater; s- station; **σ^{2-}** variance; %—percentage.

The amount of variance for students (participants; p) was higher in the GRS scores compared to the CL scores for both groups of raters. Interestingly the amount of variance accounted for by raters was relatively low (3.76%, 3.87%, 4.68%) except for the GRS-UCVM value at 15.61%. Variability due to rater, especially when variance due to rater has been higher than variance due to student/participant, has often been identified as a source of error and has caused much discussion regarding rater training and selection for OSCEs and clinical observations [17,18]. The station facet (s) accounted for considerably more variance in checklist scores from both groups of raters (42.22% UCVM, 28.65% R(D)SVS) than in GRS scores (9.57% UCVM, 13.03% R(D)SVS). The participant (student) by station (p|s) facet for the R(D)SVS raters was 24.45% and lower (9.29%) for the UCVM raters for CL scores and there was a similar trend in the GRS scores (21.06% R(D) SVS and 10.36% UCVM respectively). While the trend in the GRS was for a higher G-coefficient and higher variance due to participant, unsystematic error and varying scores of students across raters and stations accounted for 31.95% and 37.49% of the variation respectively in the UCVM and R(D) SVS scores.

The D-study determined that more than 12 stations (using the benchmark of a reliability coefficient of 0.7) are required for OSCE scores to be reliable when using one rater per station for both UCVM and R(D)SVS using CLs. For GRS the D study determined that at least 5 stations are required for UCVM raters but you would need at least 12 for the R(D)SVS raters for their scores to achieve an acceptable level of reliability when using one rater per station.

DISCUSSION

The findings from this study were the following. First, there was no difference in rater scores of student performance between schools using CL and GRS; second, scores from CLs were significantly higher than scores from GRS; and third, scores from GRS regardless of rater type, demonstrated greater reliability than CL, where the use of CL preceded the use of GRS.

CL scores were greater than GRS scores, which is not surprising given that the GRS allow raters greater opportunity to assess other dimensions not represented in CL such as time, efficiency, motion and safety. This has also been identified in previous work. [10, 19] Two different students can perform the same task and score "yes" for the same CL items but have differing performance. A student may quickly and efficiently perform the items ending up with a similar score to a student who slowly and repeatedly performs the items until they eventually obtain a yes for each one. This implies that, at least for some tasks, it may not be possible to sufficiently discriminate between candidates using a simple binary CL. [10, 15,19] GRS provides the opportunity to score additional dimensions that separate a superior performance from an average or poor one, and provide the opportunity for more qualitative feedback. [19] It is also important to consider that GRS may be subject to more inherent rater biases of student performance, these biases could range from personal preference of particular technique/method to be used in a station to age/appearance or gender biases. [10, 19] However, in this study we did not explore these biases with the participating raters. Conversely, CLs focus raters on the parts of the task because they are based on directly observable behaviors, e.g. the student either performed the items or did not. The findings from this study are not surprising and similar to those reported elsewhere. [10]

Scores from raters who had not been trained (R(D)SVS) demonstrated lower reliability because they had not been exposed to rater training, were not familiar with the assessment tool or testing process in advance, and did not have an opportunity to practice scoring, thus supporting the widely recognized importance of rater training prior to the use of any new assessment tool (including CLs). [1,6,9] However, while the reliability coefficients were lower, the same general trend existed between novice and expert raters with no significant differences in student performance by rater type.

Scores from the GRS were more reliable for UCVM and R(D)SVS raters. This might be due to (1) the way in which the scale was constructed and utilized, perhaps making it easier to use than the CL, (2) the possibility that raters developed familiarity with the items while using the CL that in turn made the GRS easier to use, or (3) the possibility that raters developed familiarity with student performances while using the CL that improved the precision of their judgments while using the GRS. Regardless, the reliability was found to be adequate for both the expert and the novice raters. Furthermore, the D studies performed demonstrated that 5 stations with one expert rater per station had a reliability of. 74. For novice raters the D study showed that 12 stations with one rater per station would be required to achieve a similar reliability with GRS. Conversely, greater than 12 stations would be required to achieve a reliability of greater than. 7 when using CL alone no matter whether expert (G = .68) or novice (G = .54) as a rater.

It is important to recognize that there are limitations to our study. First of all, the order of tool use was not randomized and the CL was always scored before the GRS. CL's were used first because this follows the logical order associated with development of these tools. Typically, the CL is created using a cognitive task analysis or some other rigorous development approach, and then the GRS is created, by grouping these individual items into dimensions. Missing factors are then included such as safety, attention to sterile technique, time and motion. Our results may indicate that the CL does inform the consistency of the measurement of the GRS. Interestingly, Cunnington et al. compared CL and GRS scores from experienced raters, whereby two independent raters assessed the same student concurrently with different tools. They found that the tools performed similarly, with GRS being at least as reliable as CL scoring. [19] Their study did not evaluate the effect of rater experience or training as was carried out here. The results from their study and our own indicates that scores from the GRS are at least as reliable as CL scores; [19]

Another limitation may be the perceived smaller sample size of both raters and students. Due to the fully crossed study design, raters reviewed 96 recorded OSCE interactions (12 students x 4 stations x 2 tools). This design allowed for all raters to rate all students across all stations yielding variance components for all facets, as opposed to typical OSCE data where raters are nested within station. In this study, we were able to identify the variance due to type of rater (novice, expert), participant (student), and station thus providing a clear understanding of where the error variance occurs.

The raters were asked to assess the students following observation of video recordings. One could criticize that raters might have been unable to see all of the student's performance depending on camera angles or image quality compared to a live experience where the examiner is free to move around to shift perspective. Based on a previous study documenting that video recording is a valuable assessment tool for OSCE grading with both CL and GRS, we felt justified to use this method to compare two different school populations of raters and students. [20]

CONCLUSIONS

This study has demonstrated for the first time within veterinary medical education that scores from GRS demonstrate an acceptable level of reliability and may allow for better discrimination than CL between students of varying qualities by both novice and experienced raters. This determination of a good versus average performance becomes more critical as a student advances in their training. The use of CLs may be helpful to inform raters and allow training for GRS use, but this study also confirmed that the use of CL without training of novice raters did not result in a reliable assessment methodology.

ACKNOWLEDGMENTS

Thank you to our colleagues and students for participating in the data collection. Thank you to our technical staff for assistance in setting up the trials.

FUNDING STATEMENT

This study was funded by the University of Calgary Faculty of Veterinary Medicine (EKR, KGH) and the Higher Education Academy, UK (CB, SR). The funders had no role in the study design, data collection and analysis, decision to publish, or preparation of the manuscript.

REFERENCES

1. May SA, Head SD. (2010) Assessment of Technical Skills: Best Practices. JVME 37(3): 258–265. 10.3138/jvme.37.3.258
2. Hecker KG, Adams CL, and Coe JB. Assessment of first-year veterinary students' communication skills using objective structured clinical examination: The importance of context. 2012; JVME 39(3): 304–310. 10.3138/jvme.0312.022R
3. Harden RM, Stevenson M, Downie W, Wilson GM. Assessment of clinical competence using objective structured examination. 1975; Br Med J 1(5955): 447–451.
4. Hecker K, Read EK, Vallevand A, Krebs G, Donszelmann D, Muelling CK, et al. Assessment of first-year students' clinical skills using objective structured clinical examinations. 2010; JVME 37 (4): 395–402. 10.3138/jvme.37.4.395
5. Schnabel LV, Maza PS, Williams KM, Irby NL, McDaniel CM, Collins BG. Use of a formal assessment instrument for evaluation of veterinary student surgical skills. 2013; Vet Surg 42(4): 488–496. 10.1111/j.1532-950X.2013.12006.x
6. Hodges BD. The objective structured clinical examination: Three decades of development. 2006; JVME 33(4): 571–577.
7. Hodges B, McIlroy JH. Analytic global OSCE ratings are sensitive to level of training. Med Educ. 2003; 37: 1012–1016.
8. Adler MD, Vozenilek JA, Trainor JL, Eppich WJ, Wang EE, Beaumont JL, et al. Comparison of checklist and anchored global rating instruments for performance rating of simulated pediatric emergencies. Sim in healthcare. 2011; 6: 18–24.
9. Hodges B, Regehr G, McNaughton N, Tiberius R, Hanson M. OSCE checklists do not capture increasing levels of expertise. 1999; Acad Med. 74: 1229–1134.
10. Ma IWY, Zalunardo N, Pachev G, Beran T, Brown M, Hatala R, et al. Comparing the use of global rating scale with checklists for the assessment of central venous catheterization skills using simulation. Adv in Health Sci Educ 2012; 17: 457–470.
11. Regehr G, MacRae H, Reznick RK, Szalay D Comparing the psychometric properties of checklists and global rating scales for assessing performance on an OSCE-format examination. Acad Med. 1998; 73(9): 993–997.

12. Reznick R, Regehr G, MacRae H, Martin J, McCulloch W. Testing technical skill via an innovative "bench station" examination. Am J Surg. 1997; 173(3): 226–230.
13. Van Luijk SJ, vander Vleuten CPM. A comparison of checklists and rating scales in performance based testing In Hart IR, Harden RM, Des Marchais J, editors. Current development in assessing clinical competence, Can-Heal Publications: Montreal: 1992; pp. 357–382.
14. Tavares W, Boet S, Theriault R, Mallette T, Eva KW. Global rating scale for the assessment of paramedic clinical competence. Prehosp Emerg Care 2013; 17(1): 57–67. 10.3109/10903127.2012.702194
15. Crossley J, Jolly B. Making sense of work-based assessment: ask the right questions, in the right way, about the right things, of the right people. Med Educ 2012; 46(1): 28–37. 10.1111/j.1365-2923.2011.04166.x
16. Read EK, Hecker KG. The development and delivery of a systematic veterinary clinical skills education program at the University of Calgary. J Vet Sci Technol 2013; Available at: 10.4172/2157-7579.S4-004. Accessed 2014 August 14.
17. Harasym PH, Woloschuk W, Cunning L. Undesired variance due to examiner stringency/leniency effect in communication skill scores assessed in OSCEs. Adv Health Sci Educ Theory Pract. 2008; 13;617–632.
18. McGill DA, van der Vleuten CP, Clarke MY. Supervisor assessment of clinical and professional competence of medical trainees" A reliability study using workplace data and a focused analytical literature review. Adv Health Sci Educ Theory Pract. 2011; 16; 405–425. 10.1007/s10459-011-9296-1
19. Cunnington J, Neville A, Norman G. The risks of thoroughness: reliability and validity of global ratings and checklists in an OSCE. Adv Health Sci Educ, 1996; 1(3), pp.227—233. 10.1007/BF00162920
20. Friedman Z, Katznelson R, Devito I, Siddiqui M, Chan V. Objective assessment of manual skills and proficiency in performing epidural anesthesia—video-assisted validation. Reg Anes Pain Mgmt. 2006; 31(4): 304–310.

CHAPTER 2

Physiologic Status Monitoring via the Gastrointestinal Tract

G. Traverso[1,2], G. Ciccarelli[3], S. Schwartz[3], T. Hughes[3], T. Boettcher[3], R. Barman[1,2], R. Langer[2], A. Swiston[3]

[1]Division of Gastroenterology, Massachusetts General Hospital, Harvard Medical School, Boston, MA, United States of America,

[2]Department of Chemical Engineering and Koch Institute for Integrative Cancer Research, Massachusetts Institute of Technology, Cambridge, MA, United States of America,

[3]Bioengineering Systems and Technologies, Massachusetts Institute of Technology Lincoln Laboratory, Lexington, MA, 02420, United States of America

INTRODUCTION

Heart rate (HR) and respiratory rate (RR) are essential vital signs for evaluating the physiologic status of children and adults in clinical and non-clinical settings. These two vital signs constitute the initial measurements in acutely ill patients and provide the basis for clinical severity stratification

Citation: Traverso G, Ciccarelli G, Schwartz S, Hughes T, Boettcher T, Barman R, et al. (2015) Physiologic Status Monitoring via the Gastrointestinal Tract. PLoS ONE 10(11): e0141666. https://doi.org/10.1371/journal.pone.0141666

[1, 2] as well as markers of response to life-saving cardiopulmonary resuscitation [3]. Additionally, HR and RR serve as non-diagnostic indicators of performance status in service members [4, 5] and in performance athletes.

There are numerous methods for monitoring HR and RR, but most require the attachment of superficial sensors to the body. HR can be monitored using electrical methods such as electrocardiogram (ECG), optical methods such as photoplethysmography (PPG, pulse oximetry), or mechanical methods such as ballistocardiography. RR can be monitored directly using trans-thoracic plethysmography and expired gas analysis approaches, or indirectly using advanced processing methods applied to PPG [6]. All of these methods have some limitations, as they may cause patient discomfort by being obtrusive or irritating the skin[7, 8], and many methods cannot be used reliably in high physical activity contexts where motion may corrupt the signal. Furthermore, some key vital signs, namely core temperature, must be measured internally, and thus an ingestible device can provide the best signal quality. Over the last decade, ingestible medical devices have gained broad acceptance; for example, ingestible devices can measure temperature [9, 10], and video capsule endoscopy is widely used for diagnosis of gastrointestinal (GI) pathology [11]. We hypothesized that vital sign monitoring from within the GI tract would be a safe and effective alternative to existing superficial clinical monitoring systems, while overcoming some of their limitations. Moreover, we selected miniature components in our design to ensure that the ultimate size of ingestible PSM devices is even smaller than video capsule endoscopes while maintaining other safety standards.

We present initial proof-of-concept experiments in a porcine model to show that HR and RR can be measured simultaneously and with high fidelity from within the GI tract using a single acoustic waveform. Using an endoscopically-guided miniature electret microphone, we measured acoustic data along the GI tract from the mouth to the colon. We evaluated the impact device contact with GI tissue and previously ingested food had on acoustic data quality. We then developed a robust signal processing algorithm to analyze these raw waveforms. Our results support that an ingestible, ultra-miniature acoustic monitoring device could accurately measure vital signs. This technology is likely to be adaptable to a wide range of clinical and non-clinical uses.

RESULTS

Data Collection and Signal Processing

We performed physiological monitoring experiments in six sedated Yorkshire pigs using an endoscopically-guided electret microphone to collect acoustic waveforms along the GI tract. We concurrently recorded physiological waveforms using a standard veterinary vital signs monitor, including external 3-lead ECG, PPG, capnography (via expired breath CO_2 analysis), and a superficial microphone positioned directly above the heart. A total of ~407 minutes each of HR and RR data were collected from all segments of GI tract over the course of four experimental days in six Yorkshire pigs. Specifically, 80.3, 66.7, 149.7, 54.3, and 60.3 minutes of raw audio data were collected from the oral cavity, esophagus, stomach, proximal duodenum and rectum respectively. A schematic of our experimental setup and a representative dataset is shown in Fig 1.

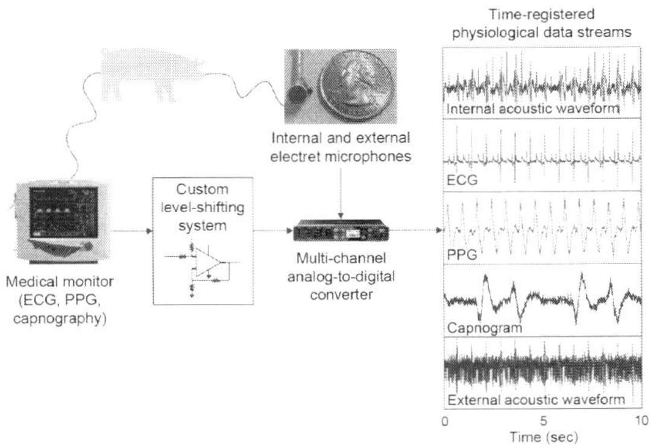

Figure 1. Schematic of the experimental setup and representative dataset.

The animal model is anesthetized and attached to a medical monitoring system that measures ECG, capnography, and photoplethysmography (PPG) waveforms simultaneously. We built a custom voltage level-shifting circuit for each waveform, which outputs to a commercial A/D converter. Two electret microphones, one controlled internally by the endoscope, the other attached superficially on the pig's chest just above the heart, are also sending data to the A/D converter. The final result is perfectly time-registered data streams for heart and lung function as well as the acoustic waveforms. Example concurrent physiological data measurements were taken from the proximal third of a porcine stomach including the acoustic waveform from our internal electret microphone, ECG, PPG (which indicates systemic oxygen perfusion levels and heart rate), capnography from expired CO_2 content, and the acoustic waveform from the external microphone positioned above the heart. (Note that the raw PPG data from the SurgiVet system seems to be inverted, and the raw capnography data appears to be the flow *rate* of CO_2, thus giving a first derivative of the more familiar capnogram waveform.).

Raw waveforms were processed using a phonocardiogram HR estimation algorithm [12] modified to enable simultaneous extraction of HR and RR from a single raw waveform and with processing steps amenable to implementation on a microcontroller (such as a Texas Instruments MSP430). The raw waveform is split and copied into a HR and RR track, and each is processed with parameters characteristic of each signal [13–15] (see Fig 2). The first processing stage consists of an emulated analog front end, namely resistor-capacitor (RC) based bandpass filters. A 10–40 Hz anti-aliasing filter is applied to capture the majority of signal energy yet avoid significant aliasing from downsampling (see spectrogram in Fig 3).

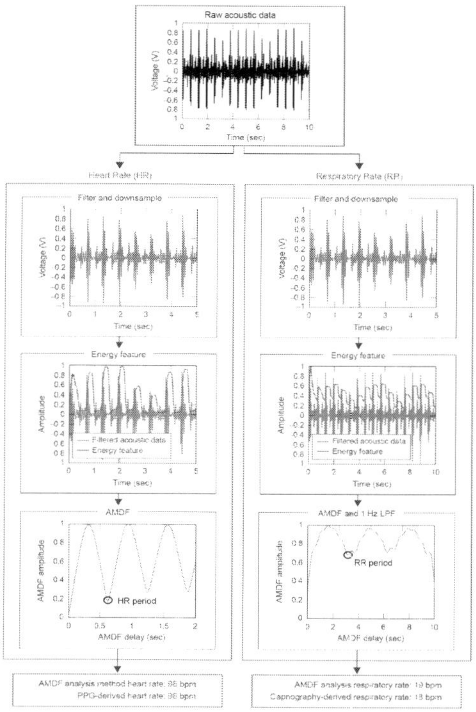

Figure 2. Schematic processing flow chart for HR and RR estimation from internal microphone data.

The second and third stages increase the signal-to-noise ratio (SNR) by successively computing the signal's energy and its average magnitude difference function (AMDF). The final stage uses a robust valley-detection algorithm for estimating HR and the RR from the AMDF. All processing in this study was performed on non-overlapping 20s data frames during which a single average HR and RR is reported. A 20s frame duration was chosen as a compromise between reduced latency and a sufficient window length to encompass more than one breath. There were 1228 and 1219 total HR and RR frames, respectively. Additional signal processing details may be found in the Methods section.

The signal is copied into a HR track and RR track and then analog filtered and down-sampled. A sliding window computes an energy feature (see Methods) that is input to the average magnitude difference function (AMDF). The RR is further low pass filtered with a 1 second Hamming window. The first valley of the AMDF is the estimated vital sign.

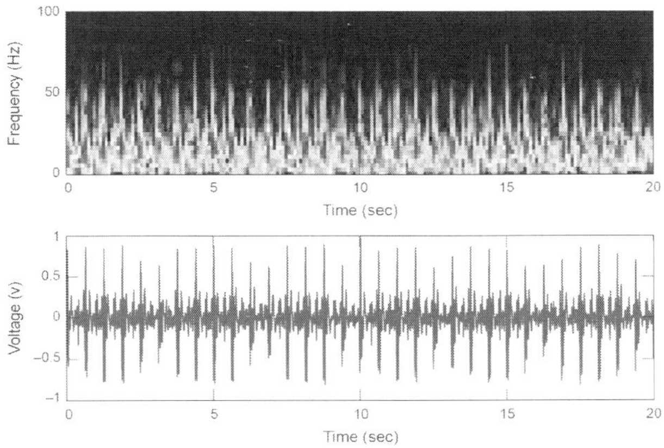

Figure 3. Example spectrogram and corresponding time course data of the internal acoustic signal measured in the proximal third of the stomach. A majority of the signal energy is <50Hz.

HR concordance with external PPG is very strong in the esophagus, stomach (including with and without food and tissue contact), and duodenum: the HR is detected within 5 bpm 97%, 95%, 98% of the time, respectively. The RR is detected within 5 breaths per minute 84%, 82%, 88% of the time in these locations, though many errors in these locations are due to the AMDF algorithm's valley-finding which very predictably doubles the respiratory rate (see Discussion below). Resting respiratory and heart rate in swine ranges are noted between 32–58 breaths/min and 70–120 beats/min respectively[16]. Our 10%:90% quantiles for respiratory and heart rate were 15:43 and 57:107 bpm. While we did measure an acoustic waveform in the mouth and colon, agreement with standard vital sign monitoring was poor (see detailed analysis in Fig 4 and Discussion below). We measured ambient operating room noise during data collection which averaged ~70dB, and peaked at ~80dB.

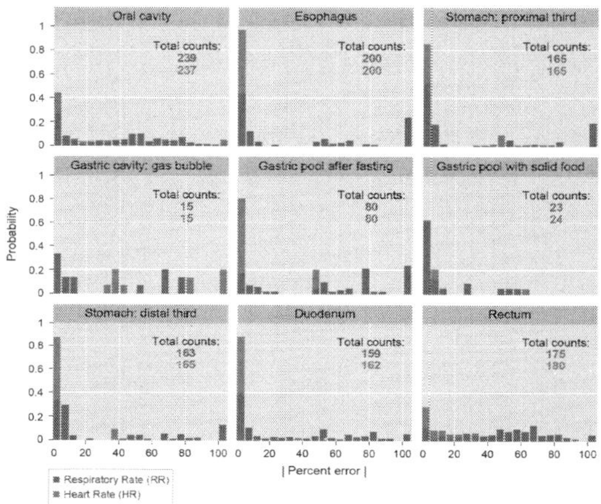

Figure 4. HR and RR estimation performance histograms for all data collected as a function of anatomical location.

The x-axis represents the absolute value of the percent error in a given 20s frame; the y-axis is the number of such frames normalized by the "Total Counts" (for each location and vital sign) used to build the histogram. The AMDF valley-finding algorithm can trigger upon higher-order harmonics of the fundamental period, on noise, or on the incompletely removed heart rate period, giving percent errors concentrated at 50% and 100%; these errors can be easily addressed with more sophisticated AMDF algorithms or running median filters (see Discussion).

Vital sign monitoring in heterogeneous GI environments

An ingestible monitoring device must be able to function under a variety of common GI environmental conditions, including fasted and fed states and device contact with tissue. To demonstrate our system's applicability to these heterogeneous GI environments, we measured waveforms in the gastric content (either solid, liquid, or in a gas bubble, for a total of 39.7

minutes of data collected from the six animals), and in contact with the gastric wall (approximately 110 minutes of data collected from the six animals, both in the proximal and distal thirds of the stomach). All these areas demonstrated HR and RR values in good agreement with external vital sign monitor values. The HR median absolute percent error was noted to be 0% in the proximal third, gastric pool, and distal third. For RR, the median percent error was 4.3%, 10%, and 6.7% in the proximal third, gas bubble, and distal third. Additional median percent errors are noted in Fig 5.

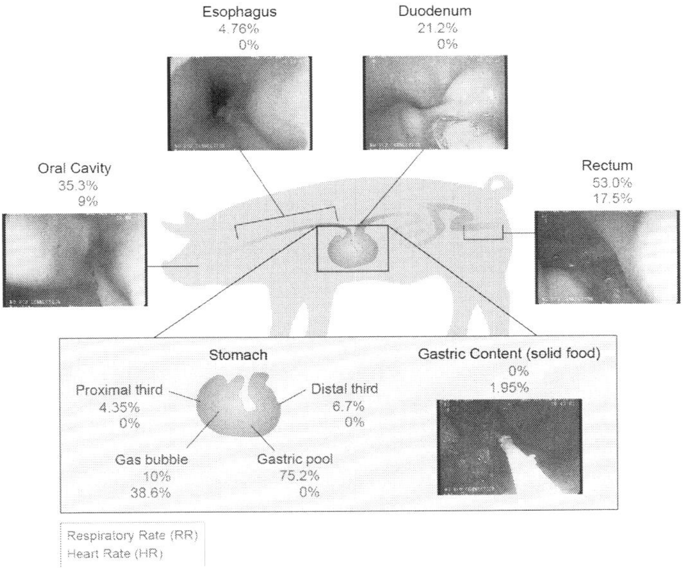

Figure 5. Vital sign algorithm performance as a function of anatomical site.

Median absolute percentage error between PPG and capnography derived HR and RR and acoustically derived HR and RR via our average magnitude difference function (AMDF)-based analysis is reported at each measurement site.

DISCUSSION

Physiological status monitoring is central to the clinical evaluation of patients and increasingly used in non-clinical settings for safety (for example, military service members and first-responders) and performance monitoring (such as professional athletes). Though significant development has focused on low-profile external vital sign monitoring systems, extended monitoring in *non-wearable* formats has seen little development. Wearable systems can be associated with skin irritation from allergic responses or repetitive abrasion during extended use in high physical activity settings, as well as from constriction, and all are subject to the logistical burdens of user compliance. Furthermore, wearable systems are not always capable of directly measuring some key physiological variables externally, namely core temperature.

To address these limitations, we have established a new technique for physiologic status monitoring using technologies and methods compatible with ingestion. Our method is capable of sensing primary vital signs (HR and RR) in a heterogeneous set of environments in the GI tract using a single sensing modality. We have demonstrated measuring HR and RR using this technique *in vivo* in a large animal model at various locations along the GI tract. Signal fidelity was maintained with and without microphone tissue contact, as well as within solid and liquid food material. Our signal processing algorithm robustly detected HR and RR in excellent agreement with measurement from standard external vital signs monitors (PPG and capnography) in the majority of the GI tract.

Ingestible devices have a number of advantages over wearable systems. The ergonomic profile of an ingestible-sized device is minimal; unless a patient has a specific contra-indication for such a device (bowel obstruction, etc.), the user will remain asymptomatic for the duration of monitoring. Such small devices will be similar in size to a bolus of food or pharmaceutical capsule (the largest commonly used size, 000, is ~2.5cm long x 1cm diameter) and will remain free-floating in the GI tract; with an appropriate material choice for the device capsule, this will minimize any abrasive irritation of the GI mucosa. There is also the possibility of implementing this sensor concept as a dissolvable electronic device, thereby eliminating contra-indications. Furthermore, ingestible electronic systems offer the possibility for *in vivo*, yet *non-implantable* medical devices able to monitor difficult-to-measure variables (such as core temperature) simultaneously with HR and RR. Burdens associated with implantable devices, namely the need for a surgeon to introduce the device and potential negative health outcomes (infection, fibrosis), have prevented their wide adoption, and which ingestible devices overcome.

Our signal processing algorithm produced accurate HR estimates for the majority of the GI tract. Overall HR performance is very strong in the esophagus, stomach, and duodenum: concordance with external PPG is detected within 5 bpm 97%, 95%, 98% of the time, respectively. While we did measure an acoustic waveform in the mouth and colon, agreement with standard vital sign monitoring was poor. Because of the excellent performance of the sensor and algorithm proximal to the heart and lungs, we suspect these sites were too distant from the heart and lungs for the sensitivity of the particular microphone chosen (-45dB ±4dB). Higher sensitivity electret or MEMS microphones may be chosen, but must also be small enough for ingestion and have sufficient frequency sensitivity (from 10 to 40 Hz for both HR and RR).

Percent error histograms (Fig 4) show similar empirical distributions among anatomical locations when there are sufficient samples. Namely, measurement error is concentrated near 0%, 50%, and 100%. This consistent and repeatable distribution results from the AMDF valley-finding algorithm triggering on incorrect valleys. With no noise, the first AMDF function valley represents the fundamental period of the heart rate or breathing rate; however, there are additional valleys at multiples of this fundamental period due to the periodicity of the signal, small valleys due to incomplete removal of heart rate modulation in the respiratory track, and spurious smaller valleys due to noise. For example, a doubled pitch period yields errors of 50%. Any detections on spurious valleys before the first true valley result in over estimations of the parameter of interest; if these estimates occur at less than half the true period, the error value becomes > 100%, which are clustered together in Fig 4. The harmonic nature of the AMDF function and its corresponding error modes are a recognized failing of parameter estimation from a periodic waveform such as the AMDF. Potential improvements in future work include implementing more robust period estimation methods such as Extended AMDF [17], normalized autocorrelation, Comb transformation or cepstral estimation algorithms with a period tracker (such as a Kalman filter), or a combination of such methods.

To statistically measure our algorithm's performance, we performed an equivalency test between HR and RR derived from our acoustic approach and the gold-standard value based upon the median difference between matched pairs. We similarly performed a test for correlation significance between acoustic estimates and gold-standard values. (See Methods for details on both.) For HR, the median error difference is equivalent to 0 (within +/- 5bpm at the 95% confidence level) for the esophagus, proximal third of stomach, distal third of stomach, gastric pool, stomach with food, and duodenum. Furthermore, our acoustic estimates are significantly positively correlated (at the 95% level) for each of these locations except the stomach with food and duodenum. Median significant correlations for HR are 1.0, 0.53, 0.79, and 0.61 for the esophagus, proximal third of stomach, distal third of stomach, and gastric pool, respectively.

For RR, the median error difference is equivalent to 0 (within +/- 5bpm at the 95% confidence level) for the esophagus, proximal stomach, and stomach with food, but we were unable to demonstrate significant correlation. As mentioned above, the full histograms (see Fig 4) imply particular susceptibility of our algorithm to pitch period doubling and low signal quality which can be mitigated with filtering above the 20s frame level as well as implementing signal-to-noise measures on the raw acoustic data and AMDF output. To assess algorithm output performance improvements if these issues were addressed, we removed respiration rate pairs with absolute percent errors greater than 35% (that is, data likely to be excluded using filtering and signal-to-noise measures) and performed our equivalency and correlation tests again. With this new limited dataset, all locations have a median difference of 0 and have significantly correlated acoustic and gold standard estimates (median correlations of at least 0.90 for mouth, esophagus, proximal and distal third of stomach, gastric pool, duodenum, and rectum, and 0.59 for stomach with food).

One considerable limitation to our proposed system is that GI transit time varies considerably among healthy individuals [18]. Our initial data suggests that our chosen components and processing algorithm yields best results in the upper GI tract (from the esophagus to the small intestine), and for some individuals, this residence time may be 12h or even shorter. However, this duration of monitoring is similar with many other ambulatory physiological monitoring systems, and this one of the same limitations of the existing "gold standard" ingestible core temperature monitoring solution (the VitalSense™ capsule system [9]). Another potential limitation is the effect of ambient noise pollution on acoustic signal fidelity. Reassuringly, data collected here appeared robust in spite of room noise contributions ranging from 70 to 80 dB. Noise pollution, either from ambient or internal sources, may also be addressed using more robust signal processing algorithms, such as match filtering for known physiological acoustic signatures. Further experimentation in various simulated and environmental contexts is required to fully characterize signal fidelity using this technology as well as inter-individual variability to inform adaptive on-board algorithms of future devices.

Extended vital sign monitoring via ingestible devices could be applied in emergency triage settings in the field, for post-operative patients, outpatient telemedicine monitoring, and performance measurements (such as in professional athletes and military service members). In this study we measured respiratory and heart rates ranging from 6–56 breaths/min and 48–128 beats/min with high correlation to standard monitoring devices. Further testing to the extremes encountered in active (i.e., exercising) or pathophysiologic states will be required to fully delineate the limitations of this technology for monitoring, but are the focus of current hardware and algorithm development efforts. Continuous auscultation, when coupled with more advanced signal processing algorithms focusing on anomaly detection or machine learning-based classifiers, could provide advanced diagnostic tools for pulmonary (chronic obstructive pulmonary disease, asthma, etc.) or cardiac (arrhythmias, stenosis, etc.) pathologies. Future vital sign device designs may even include coupling with *in situ* treatment delivery devices to provide remote, automated systems for rapid diagnosis and treatment of high-risk patients.

METHODS

Porcine Model Vital Sign Monitoring Experiments

In vivo porcine studies were performed in 6 Yorkshire pigs weighing between 50 and 65kg. All animals were female and between 6–7 months of age. For evaluations free of food material, the pigs received a liquid diet for 48 hours prior to the procedure. Otherwise animals were fasted overnight. On the day of the procedure, the morning feed was held and the animal was sedated and intubated. Following induction of anesthesia with intramuscular injection of Telazol (tiletamine/zolazepam) 5mg/kg, xylazine 2mg/kg, and atropine 0.05 mg/kg, the pigs were intubated and maintained on isoflurane 1–3%. An endoscope guiding a miniature electret microphone (PUI Audio, part # POM-2245L-C10-R) was introduced in the esophagus and recordings taken from the mouth, esophagus, stomach, and duodenum with and without tissue contact. Additionally an enema was administered and recordings were taken from the colon. All procedures were conducted in accordance with protocols approved by the Massachusetts Institute of Technology Committee on Animal Care Protocol #0113-009-16. After data collection, all animals were returned to the animal colony and used in other experimental protocols.

Raw acoustic data was sampled at 44.1 kHz. We built a custom level-shifting device (shifting from 0-5V to 0–1.6V) capable of interfacing with existing medical monitoring equipment attached to the animal (Surgivet Advisor™, Smiths Medical) and with a multi-channel A/D converter (Roland Octa-Capture™) capable of handling all 5 input streams (ECG, PPG, capnography, internal and external electret microphones) and outputting via a USB connection to a laptop running Audacity audio collection software.

Signal Processing Algorithm

Our algorithm was specifically designed for implementation in a low size, weight, and power device. First, the 44.1 kHz signal is segmented into 20s segments and copied into two parallel processing tracks; one track ultimately estimates HR, and the other RR. All acoustic data is highpass filtered with a 100dB attenuation Chebyshev Type II filter with a transition band from 5-10Hz to eliminate motion artifacts caused by mechanical compressions rather than acoustic vibrations. This filter is digitally implemented and would not necessarily be part of the final design. The signal in both HR and RR tracks is filtered with an emulated analog RC bandpass filter constructed from a low pass filter with a 3 dB cutoff frequency at 40 Hz and a high pass filter with a 3 dB cutoff frequency at 10 Hz. Finally a 60Hz comb filter is applied to reduce line noise and its harmonics. Both signals are then decimated by a factor of 450 to a sampling rate of 98 Hz to emulate minimal data collection, storage, and processing capabilities.

The analog filtered signal, $x[n]$, is normalized to a maximum amplitude of -1 or 1 and is denoted $x[n]_{norm}$. Then, a sliding window over the 20 second frame computes the energy, $E[n]$, according to Eq 1 below. The HR track uses a window of N_{window} = 25 samples or approximately 0.25 seconds, and the RR track uses a window of N_{window} = 98 samples or 1 second.

If $1 \leq n < N_{window}$

$$E[n] = \frac{1}{n} \cdot \sum_{k=1}^{n} (x_{norm}[k])^2$$

else if $N_{window} \leq n$

$$E[n] = \frac{1}{N_{window}} \cdot \sum_{k=1}^{N_{window}} (x_{norm}[n - N_{window} + k])^2$$

Eq (1). Energy feature calculation

The average magnitude difference function (AMDF) computes a waveform $D[n]$ similar to the output from an autocorrelation operation, but without using any multiplications [19] (again, chosen in anticipation of implementing in ultra-low power and size hardware). The AMDF slides the signal over itself and computes the average difference between the overlapped segments (Eq 2). When the two segments are similar, the AMDF outputs a low value, and when they are dissimilar, the AMDF outputs a high value. The AMDF increases the signal to noise ratio by exploiting the periodicity of the HR and the RR waveforms over the frame. N_E is the number of samples in the frame of the energy signal ($N_E = 1960$, corresponding to 20 seconds). d varies from 0 to 196 (2 seconds) or 0 to 980 (10 seconds) for HR or RR estimation, respectively, because normal cardiac and respiration cycles have periods less than these upper bounds.

$$N = N_E - d_{max}$$

$$D[d+1] = \frac{1}{N} \cdot \sum_{k=1}^{N} |E[k] - E[k+d]|$$

$$\text{where } d = 0, 1, 2, \ldots, d_{max}.$$

Eq (2). Average magnitude difference function

For the RR track only, the AMDF output is further low pass filtered with a 1 second hamming window. This reduces the likelihood that high frequency heart rate modulation will be detected instead of the low frequency respiratory modulation. The final stage in each track is the HR and RR estimation from the AMDF function. Vital sign estimation reduces to a valley-detection-in-noise problem because the true HR or RR period will be the first significant dip at which the AMDF becomes close to zero (though not counting the AMDF value at zero lag, as this has an AMDF value of exactly zero since the two segments being subtracted are identical.) As the segments slide over each, the heart beat or breath that was well aligned in each segment becomes misaligned, and the AMDF value increases. However, eventually the segments will have slid far enough apart that the original heart beat or breath will overlap with the second heart beat or breath in the frame. Because of the consistency in heart rate and breath morphology as an output from the energy stage, the AMDF function will compute a small, non-zero value before increasing once again. If several heart beats are within a frame, the AMDF function will appear to have a ripple or sinusoidal pattern as the first heart beat overlaps each of the successive heart beats. The ripple period is the period of the vital sign to be estimated.

Data Analysis & Statistical Measures

We compared the results from the signal processing algorithm above with the "gold-standard" PPG and capnography signals to determine algorithm performance. Data from these gold-standard methods was resampled to 98 Hz using MATLAB's (MathWorks, Natick MA) resample function, which applies an anti-aliasing filter, and then processed through the AMDF and valley detection stages as described earlier. Only HR values between 40 and 200 beats per minute and RR values below 60 breaths per minute were accepted as physiologically reasonable. In the rare event (10/1229 frames for RR and 6/1234 frames for HR) that the valley finding function did not return an estimate from the acoustic data but the gold standard estimate was valid, the frame was excluded from analysis. For each 20s segment, an absolute percent error is calculated, defined as

$$Absolute\ Percent\ Error = 100 \times \left| \frac{Rate_{algorithm} - Rate_{gold-standard}}{Rate_{gold-standard}} \right|$$

where $Rate_{algorithm}$ is the result of the method above, and $Rate_{gold\text{-}standard}$ is the rate derived from PPG or capnography for HR and RR, respectively. Histograms of these values, normalized by the total counts for each anatomic site, are reported in Fig 4 and the median presented in Fig 5.

We used two methods to statistically determine output similarity between the gold-standard methods (PPG and capnography) and our acoustic approach. The first was a median difference between matched pairs for repeated samples equivalency test, which checks for bias while still being robust to outlier failures from period doubling; the second was a test for correlation significance, which quantitatively assesses whether our acoustic based estimates track the corresponding gold-standard vital signs.

For the equivalency test, the null hypothesis is that a non-tolerable difference exists between the acoustic and gold standard estimates, and the alternative hypothesis is that the median difference is equivalent to 0 within the tolerance limits (at the 95% confidence level). We chose a non-tolerable difference of +/- 5 bpm for both the HR and RR median differences. We performed a bootstrap procedure for the equivalency test given that our data is non-normal and has few independent samples (at most 6 independent pigs for a given anatomical location), though we have repeated samples for each pig at a given location. For each location, we randomly sampled one measurement pair (acoustic estimate and gold-standard value) from each pig that had data for that location. We computed the difference for the pair and took the median of the differences for all the pigs to create one bootstrapped datum. We repeated this procedure 10000 times to create a distribution of median difference errors, and then computed the 5 percentile and 95 percentile points. If both the 5 percentile and 95 percentile points lie within the tolerable difference region of +/- 5 bpm, we reject the null hypothesis (that the median is significantly different than 0 within the tolerance bounds) in favor of the alternative. Note that 5 and 95 percentile points are used for a 95% confidence level instead of 2.5 and 97.5 percentiles because both ends of the confidence interval must be within the tolerance bounds to reject the null hypothesis.

A correlation test was also performed with a similar bootstrapped approach. For each location, one measurement pair was taken for each pig. The Pearson correlation among pairs was computed to create one bootstrapped datum. We repeated this procedure 10000 times to create a distribution of correlations, and then computed the 2.5 and 97.5 percentile points. If the 2.5 to 97.5 percentile range did not include 0, then we reject the null hypothesis (that our acoustic measurements are uncorrelated with the gold standard) in favor of the alternative that data are correlated at the 95% confidence level.

In the case of only one pig being available for the stomach with food condition, the samples were treated as independent and half the samples were randomly picked to create a bootstrap datum. In one case, the gold-standard vital sign was constant, and consequently, we can only report on median difference error and not report a correlation value.

ACKNOWLEDGMENTS

The MIT Lincoln Laboratory portion of this work was sponsored by the Department of the Air Force under Air Force Contract #FA8721-05-C-0002. Opinions, interpretations, conclusions and recommendations are those of the author and are not necessarily endorsed by the United States Government. The MIT Campus portion of this work was sponsored by the Office of the Assistant Secretary of Defense for Research and Engineering. This work was funded in part by NIH grant EB000244 (to R.L.) and NIH grant T32DK7191-38-S1 (to G.T.). We would like to thank Pentax for providing the endoscopic equipment used for this research, and in particular Mr. Mike Fina for facilitating access to the equipment. We also thank Kari Siegle for her excellent graphics support, and Monica Siddalls and Robert Marini for their excellent veterinary support.

FUNDING STATEMENT

The MIT Lincoln Laboratory portion of this work was sponsored by the Department of the Air Force under Air Force Contract #FA8721-05-C-0002. Opinions, interpretations, conclusions and recommendations are those of the author and are not necessarily endorsed by the United States Government. The MIT Campus portion of this work was sponsored by the Office of the Assistant Secretary of Defense for Research and Engineering. This work was funded in part by NIH grant EB000244 (to R.L.) and NIH grant T32DK7191-38-S1 (to G.T.).

DATA AVAILABILITY

Data cannot be made publicly available due to legal restrictions implemented by the United States Air Force, MIT Lincoln Laboratory's overseeing government entity. Data are available upon request to Dr. Albert Swiston (ude.tim.ll@notsiws.trebla).

REFERENCES

1. Egdell P, Finlay L, Pedley DK. The PAWS score: validation of an early warning scoring system for the initial assessment of children in the emergency department. Emerg Med J. 2008;25(11):745–9. Epub 2008/10/29. 10.1136/emj.2007.054965 .
2. Knaus WA, Draper EA, Wagner DP, Zimmerman JE. APACHE II: a severity of disease classification system. Crit Care Med. 1985;13(10):818–29. Epub 1985/10/01. .
3. Lopez-Herce J, Fernandez B, Urbano J, Mencia S, Solana MJ, Rodriguez-Nunez A, et al. Hemodynamic, respiratory, and perfusion parameters during asphyxia, resuscitation, and post-resuscitation in a pediatric model of cardiac arrest. Intensive Care Med. 2011;37(1):147–55. Epub 2010/09/15. 10.1007/s00134-010-2006-2 .
4. Welles AP, Buller MJ, Margolis L, Economos D, Hoyt RW, Richter MW. Thermal-work strain during Marine rifle squad operations in Afghanistan. Mil Med. 2013;178(10):1141–8. Epub 2013/10/03. 10.7205/milmed-d-12-00538 .
5. Beekley MD, Alt J, Buckley CM, Duffey M, Crowder TA. Effects of heavy load carriage during constant-speed, simulated, road marching. Military medicine. 2007;172(6):592–5. Epub 2007/07/10. .
6. Lee J, Florian JP, Chon KH. Respiratory rate extraction from pulse oximeter and electrocardiographic recordings. Physiol Meas. 2011;32(11):1763–73. Epub 2011/10/27. 10.1088/0967-3334/32/11/s04 .
7. Deswysen AC, Zimerson E, Goossens A, Bruze M, Baeck M. Allergic contact dermatitis caused by self-adhesive electrocardiography electrodes in an infant. Contact Dermatitis. 2013;69(6):379–81. Epub 2013/11/13. 10.1111/cod.12137 .
8. Rafiei H, Amiri M, Moghaddasi J. Skin irritation because of electrocardiograph lead in patients in intensive care unit. Int Wound J. 2013;10(1):116 Epub 2012/01/10. 10.1111/j.1742-481X.2011.00915.x .
9. McKenzie JE, Osgood DW. Validation of a new telemetric core temperature monitor. Journal of Thermal Biology. 2004;29(7,Äì8):605–11. 10.1016/j.jtherbio.2004.08.020
10. O'Brien C, Hoyt RW, Buller MJ, Castellani JW, Young AJ. Telemetry pill measurement of core temperature in humans during active

heating and cooling. Medicine and science in sports and exercise. 1998;30(3):468–72.

11. Wang A, Banerjee S, Barth BA, Bhat YM, Chauhan S, Gottlieb KT, et al. Wireless capsule endoscopy. Gastrointest Endosc. 2013;78(6):805–15. Epub 2013/10/15. 10.1016/j.gie.2013.06.026 .
12. Chen YH, Chen HH, Chen TC, Chen LG. Robust heart rate measurement with phonocardiogram by on-line template extraction and matching. Conf Proc IEEE Eng Med Biol Soc. 2011;2011:1957–60. Epub 2012/01/19. 10.1109/iembs.2011.6090552 .
13. Emmanuel BS. A review of signal processing techniques for heart sound analysis in clinical diagnosis. J Med Eng Technol. 2012;36(6):303–7. Epub 2012/07/17. 10.3109/03091902.2012.684831 .
14. Rangayyan RM, Lehner RJ. Phonocardiogram signal analysis: a review. Crit Rev Biomed Eng. 1987;15(3):211–36. Epub 1987/01/01. .
15. Sakai A, Feigen LP, Luisada AA. Frequency distribution of the heart sounds in normal man. Cardiovascular Research. 1971;5(3):358–63. 10.1093/cvr/5.3.358
16. Merck&Co. The Merck Veterinary Manual. 10th ed. Kahn CM, Line S, editors. Whitehouse Station, NJ: Merck & Co., Inc.; 2010.
17. Muhammad G, editor Noise Robust Pitch Detection Based on Extended AMDF. Signal Processing and Information Technology, 2008 ISSPIT 2008 IEEE International Symposium on; 2008 16–19 Dec. 2008.
18. Maqbool S, Parkman HP, Friedenberg FK. Wireless Capsule Motility: Comparison of the SmartPill GI Monitoring System with Scintigraphy for Measuring Whole Gut Transit. Dig Dis Sci. 2009;54(10):2167–74. 10.1007/s10620-009-0899-9
19. Rabiner LR, Schafer RW. Theory and applications of digital speech processing. 1st ed. Upper Saddle River: Pearson; 2011. xiv, 1042 p. p.

SECTION 2: DOGS

CHAPTER 3

Serum Biochemical Phenotypes in the Domestic Dog

Yu-Mei Chang[1], Erin Hadox[2], Balazs Szladovits[3], Oliver A. Garden[2]

[1]Research Office, The Royal Veterinary College, Camden Campus, Royal College Street, London, NW1 OTU, United Kingdom,

[2]Department of Clinical Science and Services, Regulatory T Cell Laboratory, The Royal Veterinary College, Camden Campus, Royal College Street, London, NW1 OTU, United Kingdom,

[3]Department of Pathology and Pathogen Biology, The Royal Veterinary College, Hawkshead Campus, North Mymms, Hatfield, Hertfordshire, AL9 7TA, United Kingdom

ABSTRACT

The serum or plasma biochemical profile is essential in the diagnosis and monitoring of systemic disease in veterinary medicine, but current reference intervals typically take no account of breed-specific differences. Breed-

Citation: Chang Y-M, Hadox E, Szladovits B, Garden OA (2016) Serum Biochemical Phenotypes in the Domestic Dog. PLoS ONE 11(2): e0149650. https://doi.org/10.1371/journal.pone.0149650

specific hematological phenotypes have been documented in the domestic dog, but little has been published on serum biochemical phenotypes in this species. Serum biochemical profiles of dogs in which all measurements fell within the existing reference intervals were retrieved from a large veterinary database. Serum biochemical profiles from 3045 dogs were retrieved, of which 1495 had an accompanying normal glucose concentration. Sixty pure breeds plus a mixed breed control group were represented by at least 10 individuals. All analytes, except for sodium, chloride and glucose, showed variation with age. Total protein, globulin, potassium, chloride, creatinine, cholesterol, total bilirubin, ALT, CK, amylase, and lipase varied between sexes. Neutering status significantly impacted all analytes except albumin, sodium, calcium, urea, and glucose. Principal component analysis of serum biochemical data revealed 36 pure breeds with distinctive phenotypes. Furthermore, comparative analysis identified 23 breeds with significant differences from the mixed breed group in all biochemical analytes except urea and glucose. Eighteen breeds were identified by both principal component and comparative analysis. Tentative reference intervals were generated for breeds with a distinctive phenotype identified by comparative analysis and represented by at least 120 individuals. This is the first large-scale analysis of breed-specific serum biochemical phenotypes in the domestic dog and highlights potential genetic components of biochemical traits in this species.

INTRODUCTION

The serum or plasma biochemical profile is a crucial tool in the diagnosis of systemic disease in veterinary medicine. Abnormal analytes may focus subsequent diagnostic efforts, such as imaging or serological tests, and monitoring of dogs may involve the submission of repeated biochemical profiles to assess the impact of treatment [1, 2]. Specific analytes may be of particular importance in certain disease setting. For example, the sodium: potassium ratio may be used as a screening tool for hypoadrenocorticism [3] and creatinine concentration provides the basis for the International Renal Interest Society (IRIS) staging of chronic kidney disease [4].

The biochemical profile requires optimal reference intervals for accurate interpretation, since clinical decision making is often influenced by the extent to which values depart from expected. However, most current reference intervals are based on reference values across several different types of breed of varying shape, size, age, sex, and neutering status. Age-related changes in biochemical analytes such as alkaline phosphatase (ALP) activity and concentrations of albumin, globulin, calcium, and phosphate have been recognized for some time [5–8]. In contrast, breed-specific phenotypes remain under-investigated, although certain breeds have more widely recognized biochemical characteristics that call into question the validity of common reference intervals, including greyhounds [9–11], Bernese mountain dogs [12] and Dogues de Bordeaux [13].

The dog is the most polymorphic terrestrial species on the planet. The Kennel Club recognizes over 200 different breeds in the United Kingdom, each defined by particular behavioral and physical characteristics that have been artificially selected by humans over the past 500 years [14]. The result is a high degree of interbreed heterogeneity, but remarkable intra-breed homogeneity [15]. Recent human genome-wide association studies have identified multiple loci associated with liver enzyme activity [16], and calcium [17] and magnesium [18] concentrations. Similar loci are likely to influence these and other biochemical analytes in dogs. Given the extent of inter-breed heterogeneity and intra-breed homogeneity of genetic diversity in dogs, breed-specific biochemical phenotypes are not only predictable but may also offer a powerful new approach to the genetic dissection of biochemical traits in health, with concurrent implications for a variety of diseases.

The objective of this study was therefore to perform the first large-scale retrospective analysis of phenotypic diversity of biochemical analytes by age, sex and breed in the domestic dog, both to assess the need for breed-specific reference intervals and to lay the foundations for the genetic dissection of these traits in the future, ultimately triangulating phenotype, breed and genetic predisposition.

MATERIALS AND METHODS

Inclusion and exclusion criteria

Serum biochemical profiles were collected from the database of the Royal Veterinary College (RVC) Diagnostic Laboratory (DL) from January 2004 to October 2013, comparing each dog's profile to the generic reference intervals used by the DL for all dogs. Serum biochemical profiles were included in the study if every analyte fell within the reference interval, excluding glucose concentrations which were considered separately. Dogs without available glucose or if the glucose concentrations falling outside the reference interval were not considered as an exclusion criterion for the interpretation of the other biochemical analytes, but only those glucose concentrations falling within the reference interval were considered in the analysis of glucose. Data were collected from both pure breed dogs and a control group of mixed breed dogs, anonymizing all patient details in accordance with general guidelines laid down by the Royal Veterinary College Ethics and Welfare Committee.

Specimen collection, analysis, and quality control

All blood samples were collected by licensed veterinarians, veterinary technicians, or veterinary students under supervision, for routine diagnostic purposes under the Veterinary Surgeons Act (1966), following written informed consent by the owners of the dogs. Venepuncture was performed in routine fashion, with minimal discomfort to individual animals. All samples were obtained for veterinary diagnostic, rather than experimental, purposes. The vast majority of blood samples were collected into plain serum tubes (99.8%) and analyzed using an IL600 Clinical Chemistry Analyzer (Instrumentation Laboratory Ltd; Cheshire, UK). Reported analytes included total protein, albumin, globulin, sodium, potassium, chloride, calcium, inorganic phosphorus, urea, creatinine, cholesterol, total bilirubin, alanine aminotransferase (ALT), ALP, creatinine kinase (CK), amylase, lipase and glucose.

Statistical analysis

Age of the dogs was classified into eight mutually exclusive categories: less than or equal to one year of age (designated ≤ 1), greater than one year to less than or equal to two years of age (designated $>1:\leq 2$), and $>2:\leq 4$, $>4:\leq 6$, $>6:\leq 8$, $>8:\leq 10$, and >10 years of age. (For clarity of expression in the text of the Results and Discussion sections, these age categories are respectively labeled ≤ 1, 1–2, 2–4, 4–6, 6–8, 8–10, and >10 years.) A linear mixed effects model (LMM) was used to assess the effect of age, sex, neutering status, and all two-way and three-way interactions on each analyte, taking breed as a random effect. Type I error rate was set at 5%. Adjusted means ± standard errors were presented. Residuals were defined as the observed values minus the estimated fixed effects of age, sex, and neutering status. For those breeds represented by at least 10 dogs, principal component analysis (PCA) was undertaken on all predicted random breed effects of the 18 biochemical analytes from the LMMs but the mixed breed dogs. Principal components with Eigenvalues greater than 1 were presented. Those breeds with at least one principal component of less than -2 or greater than +2 were considered to have distinctive phenotypes. A complementary analytical approach was then adopted to verify whether those breeds with distinctive phenotype based on PCA have significantly different distributions of biochemical analytes from outbred dogs with a heterogeneous genetic background. For each biochemical analyte the distributions of residuals for all breeds represented by at least 10 dogs were compared with those of the mixed breed dogs by two-sample Kolmogorov-Smirnov (KS) tests. To facilitate the pictorial representation of the data analyzed by KS tests, all breed codes submitted through the RVC Diagnostic Laboratory were classified into groups previously defined by haplotype analysis of single nucleotide polymorphisms (SNPs) [19], or to their closest matches, namely Ancient, Toy, Working, Sight Hound, Mastiff-like, Retriever/other Mastiff-like, Herding, Terrier, Scent Hound and Spaniel/Pointer groups. Those breeds not classified in this haplotype grouping and without an obvious nearest match were considered in a miscellaneous group labeled 'Other'. Where significant differences were identified for breeds with at least 120 individuals, tentative breed-specific reference intervals were calculated based on ASVCP guidelines [20]. Normality of the analytes were assessed by the Kolmogorov-Smirnov tests and outliers assessed using Horn's algorithm. Non-parametric methods were employed to establish the 95% reference intervals and the 90% confidence intervals of the lower and upper reference limits. The lower (upper) limit for the 95% breed-specific reference interval per analyte may be overestimated (underestimated) if more than 2% of the recorded data were equal to the existing RVC DL lower (upper) reference limit, thus suggesting 'truncation' of the data. This truncation was not applicable to the lower limit for total bilirubin since its value was already 0. All analyses were carried out using statistical software packages SAS 9.2 and R 3.1.2.

RESULTS

Study population

Of 3045 normal serum biochemical profiles identified in the study period, 1495 had glucose concentrations within the reference interval. The study population included 311 intact females, 961 neutered females, 687 intact males and 1086 neutered males. Dogs ranged in age from four weeks to 18 years. Sixty breeds plus the mixed breed control group included 10 or more individuals, for which descriptive statistics. Five breeds plus the mixed breed group included 120 or more individuals, including—in descending popularity—the Labrador Retriever (n = 327), Cavalier King Charles Spaniel (CKCS; n = 174), German Shepherd Dog (GSD; n = 160), Boxer (n = 146), and Golden Retriever (n = 121).

Effects of age, sex and neutering status

All measured analytes except sodium ($p = 0.4376$), chloride ($p = 0.0730$) and glucose ($p = 0.8430$) varied with age ($p<0.05$; Fig 1), though in some cases the differences were marginal despite being statistically significant. Total protein, globulin, potassium, ALT and lipase all generally increased with age (all analytes: $p<0.0001$), while other analytes showed a more variable pattern: albumin concentration ($p<0.0001$) was similar among younger dogs and decreased from 4–6 years of age; calcium ($p<0.0001$) and phosphorus ($p<0.0001$) concentrations were both higher among dogs ≤ 1 year of age and then decreased to a plateau from 2–4 to >10 years of age; urea concentration ($p = 0.0053$) decreased from ≤ 1 to 6–8 years of age in all but the intact male dogs and then increased to >10 years of age in all gender groups; creatinine concentration ($p = 0.0004$) increased to 1–2 years of age and then decreased from 4–6 or, in the case of neutered female dogs, 8–10 to >10 years of age, except in the intact male dogs, which showed an increase from 8–10 to >10 years of age; and ALP activity ($p<0.0001$) decreased from ≤ 1 to 2–4 years of age and then gradually increased to levels similar to those ≤ 1 year of age. The effects of age on phosphorus, cholesterol, total bilirubin, CK and amylase depended on the sex and/or neutering status of the dogs, and for cholesterol and total bilirubin were marginal. Cholesterol concentration decreased with age in neutered female dogs, but in intact male dogs decreased from ≤ 1 to 1–2 years of age and then increased from 4–6 to >10 years of age; total bilirubin concentration increased from ≤ 1 to 2–4 years of age, remained at a plateau until 6–8 years of age, and then decreased again to >10 years of age in intact female dogs; CK activity decreased from ≤ 1 year of age to 2–4 years of age and then increased to >10 years of age in intact male dogs; and amylase activity decreased from ≤ 1 year of age to 4–6 years of age and then increased again to >10 years of age in intact female dogs, but in neutered female dogs increased from ≤ 1 year of age to >10 years of age.

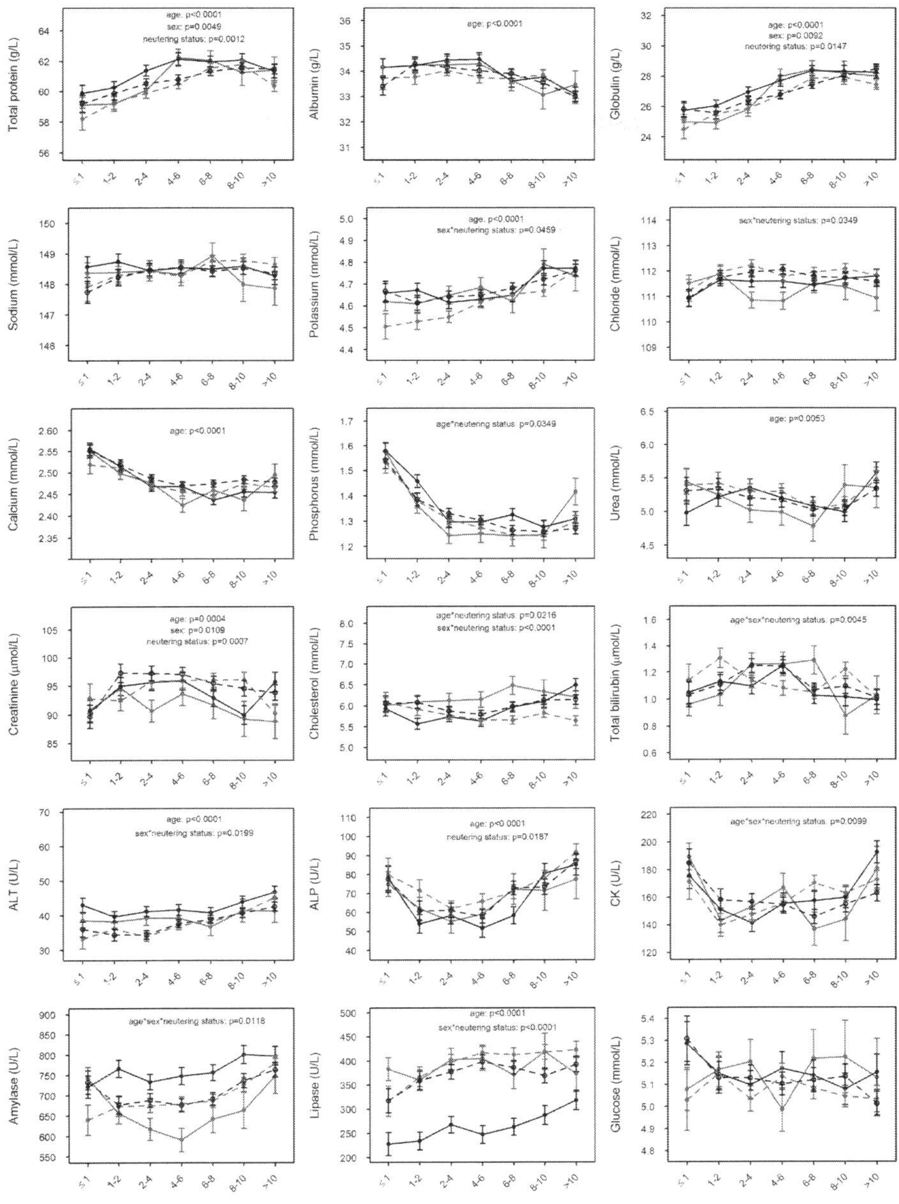

Figure 1. Effects of age, sex and neutering status on serum biochemical analytes.

The adjusted mean values ± 1 standard error bars—accounting for breed—for each of the 18 biochemical parameters are represented on the respective *y* axes, showing age in years on the *x* axes. Intact male and female dogs are represented by the solid black and red lines, respectively; neutered male and female dogs are represented by the interrupted black and red lines, respectively. When present, significant differences (in age, sex, neutering status, or age*sex, age*neutering status, sex*neutering status or age*sex*neutering status interactions) are shown at the top of the figure; there was no significant difference in glucose between age, sex or neutering status. These analyses were all undertaken on the complete dataset of dogs, including both mixed and pure breeds.

There were no differences in albumin ($p = 0.4579$), sodium ($p = 0.7232$), calcium ($p = 0.0722$), phosphorus ($p = 0.0692$), urea ($p = 0.6335$), ALP ($p = 0.2325$) and glucose ($p = 0.2833$) between sexes, but concentrations of total protein (female 60.5±0.2 vs male 61.0±0.2 g/L, $p = 0.0049$), globulin (female 26.7±0.2 vs male 27.1±0.2 g/L, $p = 0.0092$) and creatinine (female 93±1 vs male 94±1 μmol/L, $p = 0.0109$) were all lower in female than male dogs, though the differences were generally marginal (Fig 1). There were significant differences in total protein (neutered 60.5±0.2 vs intact 61.0±0.2 g/L, $p = 0.0012$), globulin (neutered 26.8±0.2 vs intact 27.1±0.2 g/L, $p = 0.0147$), creatinine (neutered 95±1 vs intact 93±1 μmol/L, $p = 0.0007$) and ALP (neutered 72±2 vs intact 67±2U/L, $p = 0.0187$) between neutered and intact dogs: in general, total protein and globulin concentrations were higher in intact than neutered dogs of each respective sex, while creatinine concentration and ALP activity were higher in neutered than intact dogs of each respective sex (Fig 1).

There were significant interactions between sex and neutering status for potassium ($p = 0.0459$), chloride ($p = 0.0349$), cholesterol ($p<0.0001$), ALT ($p = 0.0199$) and lipase ($p<0.0001$), although in all cases except lipase the differences were marginal. In general, potassium concentrations were higher in intact (4.67±0.02 mmol/L) than neutered female (4.61±0.02 mmol/L), but not male, dogs; chloride concentrations were lower in intact (111.3±0.2 mmol/L) than neutered female (111.9±0.1 mmol/L), but not male, dogs; cholesterol concentrations were higher in intact (6.2±0.1 mmol/L) than neutered female (5.8±0.1 mmol/L), but not male, dogs; and ALT activities were higher in intact (42±1 U/L) than neutered male (38±1 U/L), but not female, dogs. However, lipase activities were notably lower in intact (264±11 U/L) than neutered male dogs (371±10 U/L), but there were no differences in values for intact and neutered female dogs.

There were significant three-way interactions between age, sex and neutering status for total bilirubin ($p = 0.0045$), CK ($p = 0.0099$) and amylase ($p = 0.0188$), although once again the differences were marginal in all but amylase: intact male dogs had notably higher amylase activity than intact female dogs from 1–2 to 6–8 years of age, while values for neutered male and female dogs, which could not be distinguished, were intermediate in magnitude.

The observations recorded in this section relate to the complete dataset of dogs, including both mixed and pure breeds. However, broadly similar trends were observed when the most populous breed—the Labrador Retriever (n = 327)–was examined independently.

Principal component analysis identifies breed-specific serum biochemical phenotypes

Eighteen principal components were identified in the current dataset, of which the first six had Eigenvalues of greater than 1 (Fig 2A), cumulatively accounting for 63% of the total variance of the dataset. The first two principal components identified 26 breeds with distinctive phenotypes; the inclusion of principal components 3 to 6 added an additional 10 breeds to this list. Thus, a total of 36 of 60 breeds showed distinctive phenotypes on the basis of the first six principal components (Fig 2B). There were 10 breeds with distinctive phenotypes on the basis of principal component 1 only, six on the basis of principal component 2 only, one on the basis of principal component 3 only, two on the basis of principal component 4 only, and three each on the basis of principal component 5 or 6 (Fig 2B). Eleven breeds had distinctive phenotypes on the basis of multiple principal components. In particular, the boxer had a distinctive phenotype on the basis of principal components 1, 4, 5 and 6; the pug, on the basis of principal components 2, 3 and 4; the West Highland White Terrier (WHWT), on the basis of principal components 2, 4 and 6; the CKCS, on the basis of principal components 2, 5 and 6; the Shih Tzu, on the basis of principal components 1 and 4; and the Golden Retriever, on the basis of principal components 4 and 5; Airedale on the basis of principal components 1 and 2; both Flat-coated Retriever and the GSD on the basis of principal components 2 and 4; Boston Terrier and Dogue de Bordeaux on the basis of principal components 1 and 3.

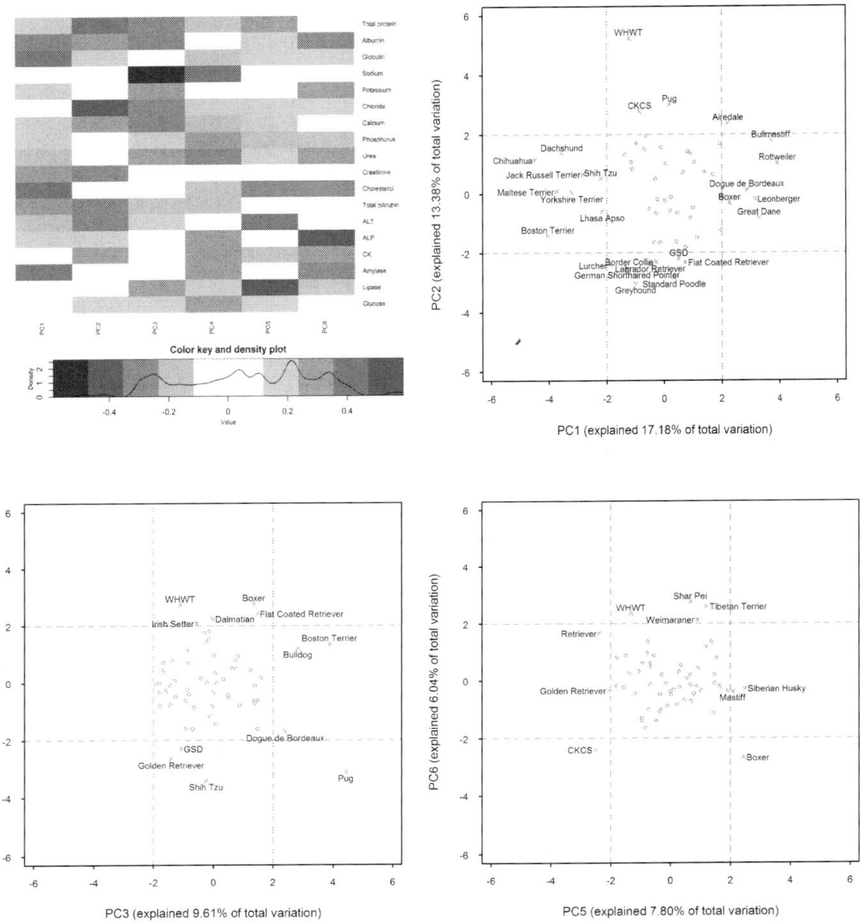

Figure 2. Principal component analysis identifies breed-specific serum biochemical phenotypes.

Of the 18 principal components identified in the current dataset, the first six had Eigenvalues of over 1 and cumulatively accounted for 63% of the total variance. A heat map showing correlations between the serum biochemical parameters and the first six principal components (PC1 to PC6) is shown in part (a). Biplots (PC1 *vs* PC2; PC3 *vs* PC4; PC5 *vs* PC6) are shown in part (b), revealing outlying breeds with principal components of greater than +2 and less than -2, representing distinctive phenotypes. These outlying breeds are labeled in the biplots.

Comparisons with a control group of mixed breed dogs confirm breed-specific phenotypes

All biochemical analytes except urea and glucose showed significant differences between specific individual breeds and the mixed breed group (Fig 3). Notable observations included the differences between pure and mixed breed dogs in total protein concentration in the Mastiff-like group (Boston Terrier); differences in albumin concentration in the Terrier (WHWT) and Spaniel/Pointer (Weimaraner) groups; differences in globulin concentration in the Ancient (Shar Pei) and Spaniel/Pointer (CKCS) groups; differences in sodium concentration in the Toy group (Pug); differences in potassium concentration in the Toy (Pug) and Mastiff-like (Bulldog) groups; differences in chloride concentration in the Toy (Pug), Terrier (Border Terrier) and Spaniel/Pointer (CKCS) groups; difference in calcium concentration in the Mastiff-like group (Boston Terrier); differences in inorganic phosphorus concentration in the Mastiff-like (Boxer, Bulldog), Retriever/other Mastiff-like (Rottweiler), Terrier (WHWT) and Spaniel/Pointer (CKCS) groups; differences in creatinine concentration in the Toy (Chihuahua), Working (GSD), Mastiff-like (Boxer), Retriever/other Mastiff-like (Labrador Retriever), Terrier (WHWT, Yorkshire Terrier), Scent hound (Dachshund), Spaniel/Pointer (Cocker Spaniel) and Other (Jack Russell Terrier) groups; differences in cholesterol concentration in the Mastiff-like (Staffordshire Bull Terrier (SBT)) and Retriever/other Mastiff-like (Rottweiler, Golden Retriever) groups; differences in total bilirubin concentration in the Scent hound (Dachshund) and Spaniel/Pointer (CKCS) groups; differences in ALT activity in the Mastiff-like (Boxer), Terrier (Border Terrier), Scent hound (Basset Hound) and Spaniel/Pointer (CKCS) groups; differences in ALP activity in the Working (GSD), Terrier (WHWT) and Spaniel/Pointer (CKCS) groups; differences in CK activity in the Terrier group (WHWT); differences in amylase activity in the Toy (Shih Tzu), Mastiff-like (Boxer, Bulldog), Retriever/other Mastiff-like (Rottweiler, Golden Retriever) and Terrier (WHWT) groups; and differences in lipase activity in the Mastiff-like (Boxer), Retriever/other Mastiff-like (Labrador Retriever, Golden Retriever and Retriever), Terrier (WHWT) and Spaniel/Pointer (Cocker and Springer Spaniel, Weimaraner) groups (Fig 4).

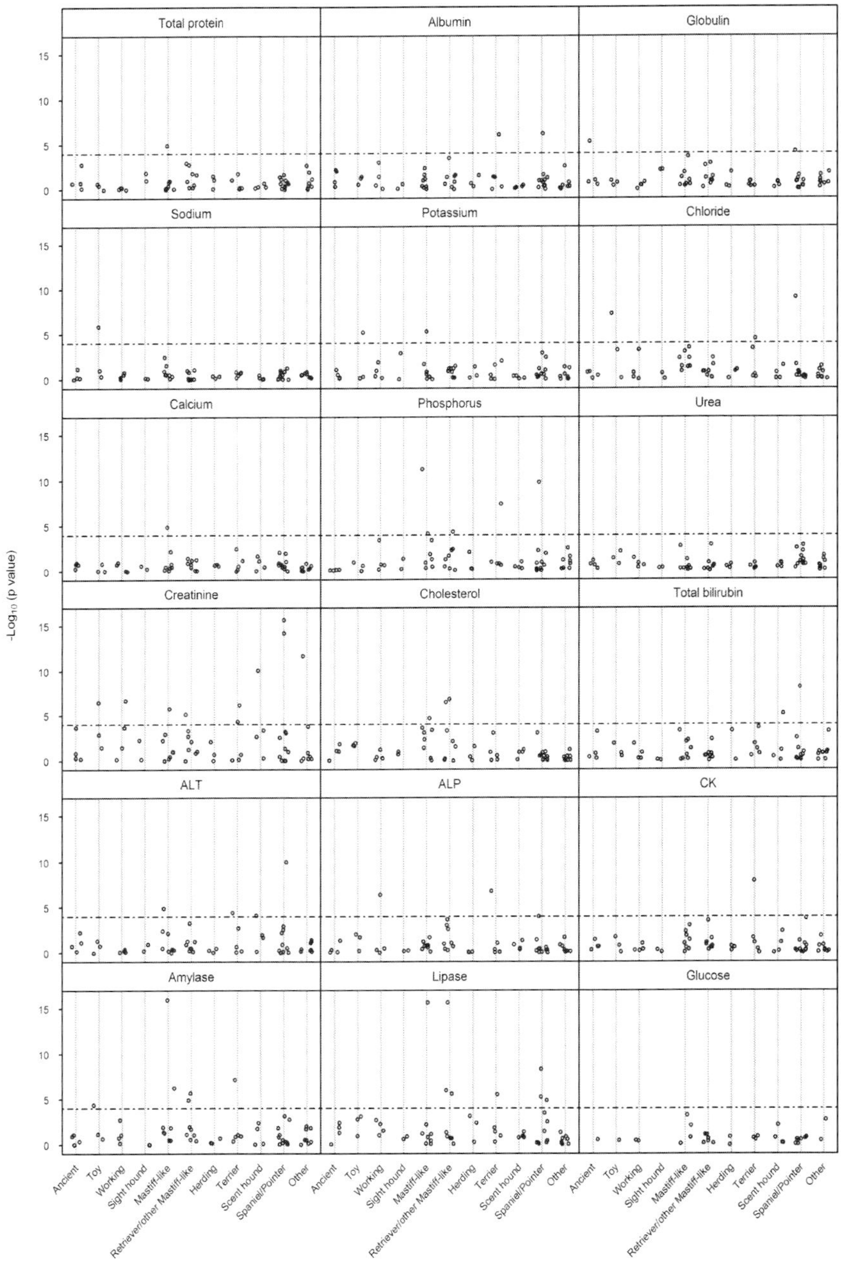

Figure 3. Pairwise comparisons of serum biochemical analytes between pure breed and mixed breed dogs confirm breed-specific phenotypes.

The results of two-sample Kolmogorov-Smirnov tests to compare the distributions of residuals for the pure breed *versus* mixed breed dogs (Residuals were defined as the observed values minus the estimated fixed effects of age, sex and neutering status). Respective *y* axes show -$\log_{10}$ (p value) of specific individual breeds assigned to one of 11 genetically-related groups, shown on the *x* axes. A Bonferroni correction yielded a threshold for significance of 10^{-4}, represented by the interrupted line. All breeds with residual values significantly different from those of the mixed breed population are shown above the line.

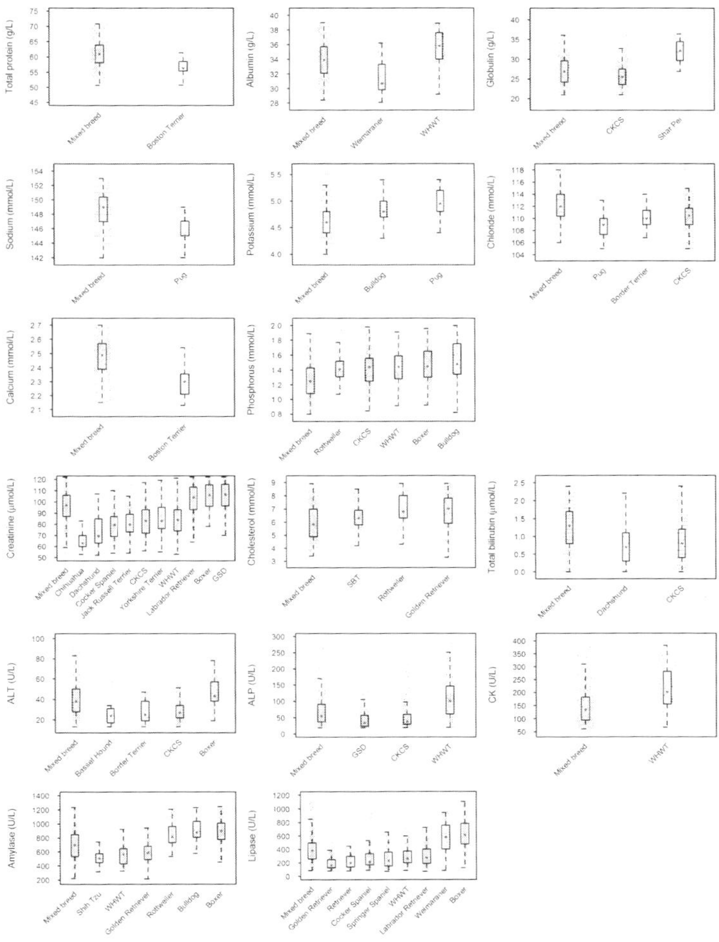

Figure 4. Combined box-and-whisker / dot plots representing serum biochemical data from breeds showing a significant difference from the mixed breed group for each of the 16 parameters for which differences were documented.

Each small dot represents an individual dog, the boxes show the respective 25th and 75th percentiles, the larger red dots median values, and the whiskers the lowest and highest data points still within 1.5 times the interquartile range of the respective lower and upper quartiles.

Breeds with three or more biochemical differences from the mixed breed group were the Pug (sodium, potassium, chloride), Bulldog (potassium, phosphorus, amylase), Boxer (phosphorus, creatinine, ALT, amylase, lipase), Golden Retriever (cholesterol, amylase, lipase), Rottweiler (phosphorus, cholesterol, amylase), WHWT (albumin, phosphorus, creatinine, ALP, CK, amylase, lipase), and CKCS (globulin, chloride, phosphorus, creatinine, total bilirubin, ALT, ALP).

Towards breed-specific serum biochemical reference intervals

Of the 23 breeds identified by pairwise comparisons with the mixed breed group, five were represented by at least 120 dogs (Labrador Retriever, CKCS, GSD, Boxer, Golden Retriever). Histograms of the serum biochemical profiles. Normality of the analytes were rejected for globulin (Labrador Retriever), sodium (Labrador Retriever, GSD, CKCS), potassium (Labrador Retriever, CKCS), chloride (Labrador Retriever, Boxer), urea (Labrador Retriever, Boxer), creatinine (Labrador Retriever, GSD), total bilirubin (Labrador Retriever), amylase (Golden Retriever), lipase (Labrador Retriever, Golden Retriever); ALT, CK and ALP did not followed Gaussian distributions for all 5 breeds. Multiple outliers were detected for chloride (Labrador Retriever, CKCS, GSD, Boxer) and amylase (CKCS, GSD, Golden Retriever). One or two outliers were detected for globulin, potassium, ALT and ALP. Tentative breed-specific reference intervals were calculated for these breeds.

DISCUSSION

To the best of the authors' knowledge, the current study represents the largest retrospective review of normal serum biochemical profiles in the dog. Data from over 3000 individuals representing 60 pure breeds plus a mixed breed control population were examined, revealing effects attributable to age, sex, neutering status and breed. Two complementary analytical strategies, PCA and pairwise comparisons of every pure breed with the control population, demonstrated distinct breed-specific serum biochemical phenotypes, adding to the growing body of literature highlighting the physical, behavioral and physiological diversity of this polymorphic species [21, 22].

Age, sex and neutering status variably impacted many of the serum biochemical analytes. Serum calcium, phosphorus and ALP were higher in younger dogs, as previously published in a number of species [13, 23–28], thought to reflect the impact of increased concentrations of growth hormone on bone turnover and renal phosphate reabsorption [29], an effect that was apparent until 2–4 years of age in this study. A number of novel observations were also made. Serum globulin concentration showed a sustained increase up to 6–8 years of age, coinciding with a decrease in serum albumin from 4–6 years of age. Comprising both acute phase proteins and gamma globulins, we speculate that the rising globulin concentration reflected a progressively increasing inflammatory burden with age, coined 'inflammaging' [30, 31]. A concurrent blunting of hepatic synthesis of albumin, a negative acute phase protein, alongside subclinical renal and/or intestinal losses of this protein could have accounted for the decreasing concentration of this analyte, an observation also made in Dogues de Bordeaux [32] and elderly people in whom albumin concentrations less than 40g/l, but still reported as normal, are associated with increased risk of mortality [33, 34]. Though small in magnitude, serum urea and creatinine showed variable changes, creatinine generally increasing and then decreasing with age, thought to reflect growth of muscles to maturity and then age-related sarcopenia; and urea decreasing to 6–8 years of age and then increasing to >10 years of age, thought to reflect changes in dietary protein intake and declining renal function or subclinical dehydration with age. The lack of an increase in creatinine with advancing age in all but the intact male dogs was intriguing and at odds with changes in serum creatinine in the aging human population [35, 36], but concordant with recent data in aging healthy dogs [32, 37]. From 4–6 years of age, serum ALT and ALP activities both increased, possibly also reflecting a rising subclinical inflammatory burden of the liver with age, one source of which could have been the aging gastrointestinal tract [38]. However, this observation was at odds with a study of Dogues de Bordeaux, which revealed a decrease in serum ALT activity with age [32]. Furthermore, whilst we were careful to include only biochemical profiles from dogs in which every analyte was within existing reference intervals, unhealthy individuals would inevitably have contributed to the sample pool. Unfortunately, not all blood samples were associated with clear information on why the samples were submitted, precluding a systematic analysis of this possibility. Pre-existing disease and the administration of medications, including corticosteroids, could therefore have contributed to the rise in activity of these analytes. Additional findings included the higher CK activity in intact male dogs

<1 and >10 years of age, the former observed as a general phenomenon in other species and attributed to growth and heightened activity of young animals [23, 27], and the latter thought to reflect low-level inflammation and the increasingly recumbent lifestyle of older dogs. The rising potassium concentration was also notable and likely to reflect the combined influence of age on absorption, excretion and transcellular shifts; similar observations have been made in human patients [36, 39].

Sex and neutering status were additional confounding factors that we considered in our analysis. Subtle differences between the sexes were seen in total protein, globulin and creatinine, all of which were higher in male than female dogs. In the case of creatinine, this was thought to reflect the generally larger size and muscle bulk of male dogs, but in the case of globulin and total protein this was a novel finding that we speculate represented the impact of sex hormones, especially androgens, on protein turnover. In contrast to total protein and globulin, creatinine and ALP were higher in neutered than intact dogs of either sex, presumably reflecting diminished clearance of these analytes in the neutered state. More striking differences between the sexes were seen in amylase and lipase, which appeared to be reciprocally influenced by male sex hormones, intact male dogs displaying the highest amylase but the lowest lipase activity. The lower lipase activity in intact male dogs was at odds with a study in silver foxes that showed males had higher serum lipase activity than females [40], suggesting species differences in sex hormone modulation of this enzyme. In contrast to lipase, amylase also appeared to be influenced by female sex hormones, intact female dogs displaying the lowest activity of this enzyme. Reconciling with this observation, estradiol was shown to decrease pancreatic amylase concentration in guinea pigs [41]. However, in a more recent study in rats estradiol was necessary to maintain amylase content in the pancreas [42], again suggesting species-specific hormonal influences. To the best of our knowledge, this is the first time that gender influences on serum activities of these enzymes have been documented in dogs, reconciling with known gender differences of several analytes in a number of species [43, 44]. Such observations underline the need to account for sex as well as breed in assessing what is normal.

Adopting a similar approach to that of our recent study of canine hematological phenotypes [45], two complementary analytical techniques were employed to identify distinctive serum biochemical phenotypes with breed, namely PCA of all of the data from pure breed dogs and pairwise comparisons of data from the pure breed and mixed breed dogs.

Observations made between the two techniques were generally concordant, both revealing 18 breeds with distinctive phenotypes. However, in common with our work on hematological phenotypes PCA was ultimately more sensitive, yielding an additional 18 unique breeds that were not identified by pairwise comparisons; in contrast, pairwise comparisons revealed only five breeds that were not identified by PCA. While the sensitivity of PCA to detect unique breeds would clearly have been determined in part on the principal component thresholds that were employed to define 'uniqueness', the inferred multidimensionality of this analytical technique, which incorporates data from the whole profile rather than individual analytes, was thought to account for its greater ability to distinguish outlying breeds.

Notwithstanding the statistical nuances of these two approaches, several interesting observations were made. Analytes for which there were multiple inter-breed differences included creatinine, lipase, amylase, phosphorus and ALT. Serum creatinine concentration is known to vary with muscle mass [37, 46, 47], predicated in large part on body size, and is the basis for important diagnostic inferences and therapeutic decisions in the assessment of renal function. In our opinion, the current work emphasizes the need for breed-specific reference intervals for serum creatinine, normalized for sex and neutering status, to refine the diagnostic value of this analyte in canine medicine, as also suggested by a recent study of different sighthound breeds [48]; moreover, discrepant serum creatinine reference intervals between laboratories seem to reflect different reference populations rather than analytical variation [49]. Canine breed-specific differences in lipase, amylase and ALT activity presumably reflect global differences in the synthesis, catabolism and excretion of these enzymes, which we speculate are determined in part by genetic 'set-points'. According with our data, recent work has suggested that serum amylase activity in Bernese mountain dogs [12] and ALT activity in Italian greyhounds differ from generic reference intervals [48], but the possibility of subclinical disease influencing these analytes could not be ruled out, with a similar caveat applying in our study. Moreover, the differences in serum ALT activity observed between breeds, while significant, were small and unlikely to impact clinical decision-making, and the diagnostic importance of serum amylase and lipase activity in dogs is questionable: the argument for breed-specific reference intervals for these analytes is therefore less compelling. We speculated that breed-specific differences in serum phosphorus concentration were also referable to genetically determined differences in its absorption, mobilization, deposition and/or excretion, perhaps predicated on varying concentrations of—or receptor sensitivity to—hormones important in its regulation, consistent with genetic loci impacting serum phosphorus concentration in humans [50]. A recent study revealed a departure from generic reference intervals for serum phosphorus in various sighthound breeds [48], again speaking to a breed-specific biochemical phenotype. Nevertheless, serum electrolyte concentrations tend to be maintained within tight reference intervals within a species owing to their critical physiological importance and factors such as dietary phosphorus intake, transcellular shifts with respiratory alkalosis, subclinical renal disease [51], or even diurnal variations [52], which could all differ between breeds, remained credible alternative explanations for this observation.

A number of breeds showed distinctive serum biochemical phenotypes on the basis of both PCA and pairwise comparisons with the control population, but in many cases the departure of values from those of the control population was small and of questionable clinical significance. Furthermore, relatively few sighthounds were recruited into our study, probably reflecting the known departure of serum biochemical analytes from generic reference intervals in these breeds and underlining the likelihood of under-estimation of breed-specific differences in this work. Nevertheless, several interesting, novel observations were made. Of particular note were pugs and CKCS. Pugs demonstrated notably lower serum sodium and chloride, and higher serum potassium concentrations than the mixed breed group, raising the possibility of differences in renal and / or gastrointestinal absorption and excretion of these electrolytes, perhaps reflecting differences in their mineralocorticoid regulation that are at least in part genetically determined. Of all the breeds in our study, the CKCS showed the greatest number of analytes with significant differences from the control group, reconciling with findings in our recent study of haematological phenotypes in dogs [45] and emphasizing the unique nature of this breed in both conformational and physiological traits.

In summary, this study has revealed the rich phenotypic diversity of serum biochemical analytes in the domestic dog, building on a number of smaller studies in this polymorphic species [12, 13, 53, 54]. While many of the absolute differences were small and of questionable clinical significance, some—in particular, those for serum creatinine—were notable and would warrant breed-specific reference intervals adjusted for age, sex and neutering status. While the optimal method to develop reliable reference intervals is to sample (a priori) a large number of rigorously assessed healthy dogs of varying age, sex, neutering status and breed, thus controlling for a number of pre-analytical variables [55–57], the practicalities of this approach in the real world present a formidable challenge. While data of this nature should be gathered in the fullness of time, more practicable approaches, such as indirect sampling (a posteriori) presents an alternative method. While our sampling method for breed specific reference intervals is a limitation of the study, the low number of instances in which truncation was considered suggests that most of the analytes of the presented breeds truly fell within the currently used intervals (further suggesting that generic canine intervals tend to be too wide). A strategy towards subject-based reference values or personalization of reference intervals for serum biochemical analytes has been suggested in human medicine, to account for age, gender and ethnicity [58, 59]In addition, the genetic homogeneity of canine breeds [15] offers a truly unique resource to investigate the genetic basis for homeostatic control of serum biochemical analytes in health, thus ultimately promising to disclose novel therapeutic targets in disease.

ACKNOWLEDGMENTS

We would like to thank Mr Roger Bishop for his help in accessing the Diagnostic Laboratory database in the course of this study and the many clinicians, both past and present, who managed canine patients from which blood samples were submitted for serum biochemical analysis.

Funding Statement

The authors have no support or funding to report.

REFERENCES

1. Thrall MA. Veterinary hematology and clinical chemistry. 2nd ed ed. Ames, Iowa: Wiley-Blackwell; 2012.
2. Stockham SL, Scott MA. Fundamentals of Veterinary Clinical Pathology: Wiley-Blackwell; 2013.
3. Adler JA, Drobatz KJ, Hess RS, Adler JA, Drobatz KJ, Hess RS. Abnormalities of Serum Electrolyte Concentrations in Dogs with Hypoadrenocorticism. Journal of Veterinary Internal Medicine. 2007;21(6):1168–73. 10.1111/j.1939-1676.2007.tb01933.x
4. Elliot J. Staging chronic kidney disease In: Elliot J, Grauer GF, editors. BSAVA manual of canine and feline nephrology and urology. 2nd ed Quedgeley: BSAVA; 2007. p. 159–66.
5. Ikeuchi J, Yoshizaki T, Hirata M, Ikeuchi J, Yoshizaki T, Hirata M. Plasma biochemistry values of young beagle dogs. The Journal of toxicological sciences. 1991;16(2):49
6. Hilppo M. Some haematological and clinical-chemical parameters of sight hounds (Afghan hound, saluki and whippet). Nordisk veterinaermedicin. 1986;38(3):148–55. .
7. Harper JE H R, Wilkinson J, Heaton PR. Age-related variations in hematologic and plasma biochemical test results in Beagles and Labrador Retrievers. JAVMA-J Am Vet Med Assoc. 2003;223(10):1435–42.
8. Rosset E, Rannou B, Casseleux G, Chalvet-Monfray K, Buff S. Age-related changes in biochemical and hematologic variables in Borzoi and Beagle puppies from birth to 8 weeks. Veterinary clinical pathology / American Society for Veterinary Clinical Pathology. 2012;41(2):272–82. Epub 2012/03/17. 10.1111/j.1939-165X.2012.00415.x.
9. Dunlop MM, Sanchez-Vazquez MJ, Freeman KP, Gibson G, Sacchini F, Lewis F, et al. Determination of serum biochemistry reference intervals in a large sample of adult greyhounds. J Small Anim Pract. 2011;52(1):4–10. 10.1111/j.1748-5827.2010.01008.x
10. Watts M. Haematology and blood chemistry values in greyhounds. Vet Rec. 2006;159(4):128-. WOS:000239677600022.
11. Feeman WE, Couto CG, Gray TL, Feeman WE, Couto CG, Gray TL. Serum Creatinine Concentrations in Retired Racing Greyhounds. Veterinary Clinical Pathology. 2003;32(1):40–2. 10.1111/j.1939-165X.2003.tb00312.x

12. Nielsen L, Kjelgaard-Hansen M, Jensen AL, Kristensen AT. Breed-specific variation of hematologic and biochemical analytes in healthy adult Bernese Mountain dogs. Veterinary clinical pathology / American Society for Veterinary Clinical Pathology. 2010;39(1):20–8. Epub 2009/10/02. 10.1111/j.1939-165X.2009.00186.x .
13. Lavoué R, Geffré A, Braun JP, Peeters D, Trumel C, Lavoué R, et al. Breed-specific biochemical reference intervals for the adult Dogue de Bordeaux. Veterinary Clinical Pathology. 2013;42(3):346–59. 10.1111/vcp.12067
14. Karlsson EK, Lindblad-Toh K. Leader of the pack: gene mapping in dogs and other model organisms. Nature reviews Genetics. 2008;9(9):713–25. Epub 2008/08/21. 10.1038/nrg2382 .
15. Jones P, Chase K, Martin A, Davern P, Ostrander EA, Lark KG. Single-nucleotide-polymorphism-based association mapping of dog stereotypes. Genetics. 2008;179(2):1033–44. 10.1534/genetics.108.087866
16. Chambers JC, Zhang WH, Sehmi J, Li XZ, Wass MN, Van der Harst P, et al. Genome-wide association study identifies loci influencing concentrations of liver enzymes in plasma. Nature Genet. 2011;43(11):1131–U129. 10.1038/ng.970
17. O'Seaghdha CM, Wu H, Yang Q, Kapur K, Guessous I, Zuber AM, et al. Meta-analysis of genome-wide association studies identifies six new Loci for serum calcium concentrations. PLoS genetics. 2013;9(9):e1003796 10.1371/journal.pgen.1003796
18. Meyer TE, Verwoert GC, Hwang SJ, Glazer NL, Smith AV, van Rooij FJA, et al. Genome-Wide Association Studies of Serum Magnesium, Potassium, and Sodium Concentrations Identify Six Loci Influencing Serum Magnesium Levels. PLoS Genet. 2010;6(8). 10.1371/journal.pgen.1001045
19. Vonholdt BM, Pollinger JP, Lohmueller KE, Han E, Parker HG, Quignon P, et al. Genome-wide SNP and haplotype analyses reveal a rich history underlying dog domestication. Nature. 2010;464(7290):898–902. Epub 2010/03/20. 10.1038/nature08837
20. Friedrichs KR, Harr KE, Freeman KP, Szladovits B, Walton RM, Barnhart KF, et al. ASVCP reference interval guidelines: determination of de novo reference intervals in veterinary species and other related topics. Veterinary Clinical Pathology. 2012;41(4):441–53. 10.1111/vcp.12006 WOS:000312555200002.

21. Lindblad-Toh K, Wade CM, Mikkelsen TS, Karlsson EK, Jaffe DB, Kamal M, et al. Genome sequence, comparative analysis and haplotype structure of the domestic dog. Nature. 2005;438(7069):803–19. 10.1038/nature04338 .

22. Boyko AR. The domestic dog: man's best friend in the genomic era. Genome biology. 2011;12(2):216 Epub 2011/02/23. 10.1186/gb-2011-12-2-216 ; PubMed Central PMCID: PMCPmc3188790.

23. Ikeuchi J, Yoshizaki T, Hirata M. Plasma biochemistry values of young beagle dogs. J Toxicol Sci. 1991;16(2):49–59. Epub 1991/05/01. .

24. Comazzi S, Pieralisi C, Bertazzolo W. Haematological and biochemical abnormalities in canine blood: frequency and associations in 1022 samples. The Journal of small animal practice. 2004;45(7):343–9. Epub 2004/07/23. .

25. Rubio AV, Hidalgo-Hermoso E, Bonacic C. Hematology and serum biochemistry values of Culpeo foxes (Lycalopex culpaeus) from central Chile. Journal of zoo and wildlife medicine: official publication of the American Association of Zoo Veterinarians. 2014;45(3):589–93. Epub 2014/10/16. 10.1638/2013-0280r2.1 .

26. Aroch I, King R, Baneth G. Hematology and serum biochemistry values of trapped, healthy, free-ranging rock hyraxes (Procavia capensis) and their association with age, sex, and gestational status. Veterinary clinical pathology / American Society for Veterinary Clinical Pathology. 2007;36(1):40–8. Epub 2007/02/22. .

27. Levy JK, Crawford PC, Werner LL. Effect of age on reference intervals of serum biochemical values in kittens. J Am Vet Med Assoc. 2006;228(7):1033–7. 10.2460/javma.228.7.1033 .

28. Rortveit R, Saevik BK, Eggertsdottir AV, Skancke E, Lingaas F, Thoresen SI, et al. Age-related changes in hematologic and serum biochemical variables in dogs aged 16–60 days. Veterinary clinical pathology / American Society for Veterinary Clinical Pathology. 2015;44(1):47–57. Epub 2015/01/07. 10.1111/vcp.12220 .

29. Nap RC, Hazewinkel HA. Growth and skeletal development in the dog in relation to nutrition; a review. The Veterinary quarterly. 1994;16(1):50–9. Epub 1994/03/01. 10.1080/01652176.1994.9694488

30. De Martinis M, Franceschi C, Monti D, Ginaldi L. Inflamm-ageing and lifelong antigenic load as major determinants of ageing rate and longevity. FEBS Lett. 2005;579(10):2035–9. 10.1016/j.febslet.2005.02.055 .

31. Franceschi C. Inflammaging as a major characteristic of old people: can it be prevented or cured? Nutrition reviews. 2007;65(12 Pt 2):S173–6. Epub 2008/02/05. .
32. Lequarre AS, Andersson L, Andre C, Fredholm M, Hitte C, Leeb T, et al. LUPA: a European initiative taking advantage of the canine genome architecture for unravelling complex disorders in both human and dogs. Veterinary journal (London, England: 1997). 2011;189(2):155–9. Epub 2011/07/15. 10.1016/j.tvjl.2011.06.013 .
33. Okamura T, Hayakawa T, Hozawa A, Kadowaki T, Murakami Y, Kita Y, et al. Lower levels of serum albumin and total cholesterol associated with decline in activities of daily living and excess mortality in a 12-year cohort study of elderly Japanese. J Am Geriatr Soc. 2008;56(3):529–35. 10.1111/j.1532-5415.2007.01549.x .
34. Visser M, Kritchevsky SB, Newman AB, Goodpaster BH, Tylavsky FA, Nevitt MC, et al. Lower serum albumin concentration and change in muscle mass: the Health, Aging and Body Composition Study. Am J Clin Nutr. 2005;82(3):531–7. .
35. Tiao JY, Semmens JB, Masarei JR, Lawrence-Brown MM. The effect of age on serum creatinine levels in an aging population: relevance to vascular surgery. Cardiovascular surgery (London, England). 2002;10(5):445–51. Epub 2002/10/16. .
36. Wilding P, Rollason JG, Robinson D. Patterns of change for various biochemical constituents detected in well population screening. Clin Chim Acta. 1972;41:375–87. .
37. Hall JA, Yerramilli M, Obare E, Yerramilli M, Melendez LD, Jewell DE. Relationship between lean body mass and serum renal biomarkers in healthy dogs. Journal of veterinary internal medicine / American College of Veterinary Internal Medicine. 2015;29(3):808–14. Epub 2015/04/29. 10.1111/jvim.12607 .
38. Steele AK, Lee EJ, Vestal B, Hecht D, Dong Z, Rapaport E, et al. Contribution of intestinal barrier damage, microbial translocation and HIV-1 infection status to an inflammaging signature. PloS one. 2014;9(5):e97171 Epub 2014/05/14. 10.1371/journal.pone.0097171 ; PubMed Central PMCID: PMCPmc4018269.
39. Bourdel-Marchasson I, Laksir H, Puget E. Interpreting routine biochemistry in those aged over 65 years: a time for change. Maturitas. 2010;66(1):39–45. Epub 2010/03/04. 10.1016/j.maturitas.2010.02.004 .

40. Zhan YM, Yasuda J, Too K. Reference data on the anatomy and serum biochemistry of the silver fox. The Japanese journal of veterinary research. 1991;39(1):39–50. Epub 1991/05/01. .
41. Guo YS, Singh P. Effect of estradiol on pancreatic amylase and cholecystokinin binding in ovariectomized guinea pigs. J Steroid Biochem. 1989;33(3):459–64. .
42. Park HS, Kim SH, Kim DK, Kim KS, Park HJ. Effects of gonadal steroid hormones on amylase secretion in pancreatic lobules of rats. Pancreas. 2007;34(3):373–7. .
43. Choi SY, Hwang JS, Kim IH, Hwang DY, Kang HG. Basic data on the hematology, serum biochemistry, urology, and organ weights of beagle dogs. Laboratory animal research. 2011;27(4):283–91. Epub 2012/01/11. 10.5625/lar.2011.27.4.283 ; PubMed Central PMCID: PMCPmc3251758.
44. Griffin PM, Elliott SL, Manton KJ. Fasting increases serum bilirubin levels in clinically normal, healthy males but not females: a retrospective study from phase I clinical trial participants. Journal of clinical pathology. 2014;67(6):529–34. Epub 2014/03/07. 10.1136/jclinpath-2013-202155 .
45. Lawrence J, Chang YM, Szladovits B, Davison LJ, Garden OA. Breed-specific hematological phenotypes in the dog: a natural resource for the genetic dissection of hematological parameters in a mammalian species. PloS one. 2013;8(11):e81288 10.1371/journal.pone.0081288
46. Miyagawa Y, Takemura N, Hirose H. Assessments of factors that affect glomerular filtration rate and indirect markers of renal function in dogs and cats. The Journal of veterinary medical science / the Japanese Society of Veterinary Science. 2010;72(9):1129–36. Epub 2010/04/23. .
47. Concordet D, Vergez F, Trumel C, Diquelou A, Lanore D, Le Garreres A, et al. A multicentric retrospective study of serum/plasma urea and creatinine concentrations in dogs using univariate and multivariate decision rules to evaluate diagnostic efficiency. Veterinary clinical pathology / American Society for Veterinary Clinical Pathology. 2008;37(1):96–103. 10.1111/j.1939-165X.2008.00007.x .
48. Uhrikova I, Lacnakova A, Tandlerova K, Kucharova V, Rehakova K, Janova E, et al. Haematological and biochemical variations among eight sighthound breeds. Australian veterinary journal. 2013;91(11):452–9. Epub 2014/02/28. 10.1111/avj.12117 .

49. Ulleberg T, Robben J, Nordahl KM, Ulleberg T, Heiene R. Plasma creatinine in dogs: intra- and inter-laboratory variation in 10 European veterinary laboratories. Acta veterinaria Scandinavica. 2011;53:25 Epub 2011/04/12. 10.1186/1751-0147-53-25 ; PubMed Central PMCID: PMCPmc3087694.

50. Kestenbaum B, Glazer NL, Kottgen A, Felix JF, Hwang SJ, Liu Y, et al. Common genetic variants associate with serum phosphorus concentration. J Am Soc Nephrol. 2010;21(7):1223–32. 10.1681/ASN.2009111104

51. Cortadellas O, Fernandez-del Palacio MJ, Talavera J, Bayon A. Serum phosphorus concentrations in dogs with leishmaniosis at different stages of chronic kidney disease. Vet Rec. 2009;164(16):487–90. Epub 2009/04/21. .

52. Lopez I, Aguilera-Tejero E, Estepa JC, Bas S, Mayer-Valor R, Jimenez A, et al. Diurnal variations in the plasma concentration of parathyroid hormone in dogs. Vet Rec. 2005;157(12):344–7. Epub 2005/09/20. .

53. Sharkey L, Gjevre K, Hegstad-Davies R, Torres S, Munoz-Zanzi C. Breed-associated variability in serum biochemical analytes in four large-breed dogs. Veterinary clinical pathology / American Society for Veterinary Clinical Pathology. 2009;38(3):375–80. 10.1111/j.1939-165X.2009.00134.x .

54. Zaldivar-Lopez S, Marin LM, Iazbik MC, Westendorf-Stingle N, Hensley S, Couto CG. Clinical pathology of Greyhounds and other sighthounds. Veterinary clinical pathology / American Society for Veterinary Clinical Pathology. 2011;40(4):414–25. 10.1111/j.1939-165X.2011.00360.x

55. Jensen AL, Wenck A, Koch J, Poulsen JSD, Jensen AL, Wenck A, et al. Comparison of results of haematological and clinical chemical analyses of blood samples obtained from the cephalic and external jugular veins in dogs. Research in Veterinary Science. 1994;56(1):24–9. 10.1016/0034-5288(94)90191-0

56. Jones G, Barker A, Jones G, Barker A. Reference intervals. The Clinical biochemist Reviews / Australian Association of Clinical Biochemists. 2008;29 Suppl 1:S93

57. Walton RM. Subject-based reference values: biological variation, individuality, and reference change values. Veterinary clinical pathology / American Society for Veterinary Clinical Pathology. 2012;41(2):175–81. 10.1111/j.1939-165X.2012.00414.x .

58. Gomez Roman VR, Jensen SS, Leo-Hansen C, Jensen KJ, Janitzek CM, Rodrigues CM, et al. Assessment of HIV-1 patient recruitability in the Republic of Guinea-Bissau using African versus North American hematology and biochemistry reference intervals. Clin Vaccine Immunol. 2012;19(8):1322–5. 10.1128/CVI.00170-12
59. Ichihara K, Itoh Y, Lam CW, Poon PM, Kim JH, Kyono H, et al. Sources of variation of commonly measured serum analytes in 6 Asian cities and consideration of common reference intervals. Clinical chemistry. 2008;54(2):356–65. Epub 2007/12/20. 10.1373/clinchem.2007.091843

Physiological Stress Reactivity and Recovery Related to Behavioral Traits in Dogs (Canis Familiaris)

Rian C. M. M. Lensen[1], Christel P. H. Moons[2], Claire DiederichID[1]

[1]Department of Veterinary Medicine (IVRU), University of Namur, Namur, Belgium,

[2]Department of Nutrition, Genetics and Ethology, Ghent University, Merelbeke, Belgium

ABSTRACT

This study investigated whether stress responsiveness (in one context) can be used to predict dog behavior in daily life. On two occasions (NT1 = 32 puppies; NT2 = 16 young adults), dogs' physiological stress response after a behavioral test at home was measured in terms of reactivity (10 min post-test) and recovery (40 min post-test) for three salivary markers: cortisol, chromogranin A (CgA) and secretory immunoglobulin A (sIgA). For

Citation: Lensen RCMM, Moons CPH, Diederich C (2019) Physiological stress reactivity and recovery related to behavioral traits in dogs (Canis familiaris). PLoS ONE 14(9): e0222581. https://doi.org/10.1371/journal.pone.0222581

each marker, it was determined whether dogs with a strong physiological response displayed different behavior in daily life compared to dogs with a weaker physiological response. The results revealed three main findings: first, for CgA and cortisol, different patterns were identified according to sample time. High reactivity related to desirable traits, whereas slow recovery after the behavioral test related to undesirable traits. The findings suggest that increased levels of CgA and cortisol 10 minutes after the behavioral test reflected an adaptive stress response, whereas elevated levels 40 minutes after the test reflected unsuccessful coping. Second, patterns for sIgA differed from CgA and cortisol: significant associations were only found with behavioral traits at T2, mostly considered desirable and related to Trainability. Possibly, the delayed reaction pattern of sIgA caused this difference between markers, as sIgA reflects the (secondary) immune response to stress, due to immunosuppressive effects of cortisol. Third, predictive capacity of puppies' physiological stress response (T1) was inconclusive, and contrary relations were found with behavioral traits at T2, suggesting that developmental factors play an important role. This study provides new insights about the relation between stress physiology and behavioral traits, and methodological advice is given to study these patterns further. In conclusion, physiological markers could provide additional insights in dogs' tendencies to display certain behaviors, especially at the young adult stage. Further studies are needed to confirm these patterns.

INTRODUCTION

Consistent individual-specific patterns in behavior and stress physiology have been reported in various animal species, but the relation between both types of parameters remains inconclusive (reviewed in [1,2]). It has been suggested that reactivity (i.e., an animal's responsiveness to environmental stimuli [3,4]) is a fundamental factor shaping personality [5,6] and plays a role in various behavior problems in dogs (*Canis familiaris*) [7]. In theory, a reactive animal responds more quickly and intensely to environmental stimuli, which can be expressed by different behavioral manifestations (e.g., excitability, fear or aggression [2,7]). Accordingly, one might expect that assessment of a dog's stress responsiveness could be used to predict its behavior in daily life. If this were true, an objective, quantifiable assessment of stress responsiveness could be a valuable indicator for the potential to develop behavior problems. An early detection of individuals "at risk" allows for timely measures to be taken to prevent behavioral problems.

To test whether stress responsiveness can be used to predict general behavior in daily life, the physiological stress response of dogs in one context should be compared to their behavioral responses in another unrelated context (e.g., [8]). However, studies on dog personality often focus on behavioral parameters (e.g., [9,10]), whereas physiological measures are more commonly used for assessing levels of stress or stress responsiveness in specific contexts (e.g., [11,12]). Furthermore, physiological parameters that are included in personality studies (i.e., assessing consistent patterns over contexts and time [13]) are generally taken on the same occasion as the behavioral parameters (e.g., [14,15]). In the current study, we assessed whether phyiological measures of stress reactivity in one context related to behavioral traits in daily life, as assessed by dog owners.

Saliva sampling is a relatively easy, non-invasive method that has often been used in dogs to monitor short-term physiological changes [16]. Cortisol is a well-known stress marker in dogs and reflects the activity of the hypothalamic-pituitary-adrenal (HPA) axis [17]. Saliva concentrations of this marker increase in response to sudden non-social stressors [18], fear-inducing events [19], and novel environments [12]. However, physiological markers are regulated through complex pathways [20,21]; and their expression could be influenced by factors like circadian fluctuations [22], physical health [23], activity [24], as well as stimulus intensity [18] and experience [25]. Hence, the measurement of multiple phsyiological parameters is recommended for a more accurate estimation of stress levels [17,26].

A more immediate response ("fight or flight") to a perceived or actual threat is regulated through the sympatho-adrenal-medullary (SAM) axis through the release of catecholamines (adrenalin and noradrenalin). However, these markers are difficult to measure in saliva due to their low concentrations, rapid degradation, and instability in the sample [27] and are found to be poor indicators of acute sympathetic activity in humans [28]. Instead, Chromogranin A (CgA) can be measured, as it is co-released with adrenalin from the adrenal medulla and more stable in the circulatory system than catecholamines [27,29]. In dogs, plasma levels of CgA increased after physiological stress (insulin-induced hypoglycemia [29]); and increases in salivary CgA were found after psychogenic stress in humans (arithmetic task [30]), pigs (immobilization with nose snare [31]), and cows (social isolation [32]). However, little is known about this analyte as potential stress marker in dogs [33], so it was included in our study to explore its value for future studies.

Components of the immune system can also provide insights into the physiological stress response, due to the immunosuppressive effects of cortisol [23,27]. As such, markers of immune system activity provide an indirect (secondary) indication of an animal's stress response. Decreased levels of secretory immonoglobulin A (sIgA) were found in dog saliva after acute stress [22] and 10 days after introduction into a novel kennel [34]. Therefore, this analyte was included to assess its added value to the abovementioned markers for assessing the physiological stress response.

The aim of this exploratory study was to examine if physiological indicators of stress responsiveness are related to behavioral traits in dogs as reported by the owners. For this purpose, three endocrine and immune biomarkers (cortisol, CgA, sIgA) were measured in saliva collected before and at two time points (10 and 40 min) after a behavioral test and compared to owner ratings of their dog's behavior in daily life (by means of the validated Canine Behavioral Assessment & Research Questionnaire: C-BARQ [35]). This was done at two test stages, for the same dogs: puppy (T1) and young adult (T2). We hypothesized that a strong physiological response would be indicative of undesirable behavioral traits, based on findings that saliva levels of cortisol increased [18,19] and levels of sIgA decreased [22] after dogs were exposed to (potentially) aversive stimuli. Furthermore, we hypothesized that dogs with a longer recovery time (i.e., time before levels of physiological markers returned to baseline) would be less successful in coping with stressful situations and thus might be more prone to display undesirable ("maladaptive") behaviors [7,36].

MATERIAL AND METHODS

Subjects

Owners were recruited by participating veterinarians (N = 20) during their puppies' first vaccination visit at the veterinary practice over a 21-month period (Table 1). The owners received two months of free dog food from a well-known brand upon completion of their participation, and the veterinarians received a gift voucher for an online general goods store.

Table 1. Dogs in this study and the time at which they were tested at home.

Dogs tested at T1				Dogs tested at T1 & T2			
Breed	**Sex**	**Age**	**Time**	**Breed**	**Sex (neut.)**	**Age**	**Time**
BSD	M	13.43	14:00	Border Collie	M (-)	16.14; 61.14	14:00
BSD*	M	22.01	14:00	Border Collie	M (-)	22.15; 60.15	10:00
X BSD / BMD	M	16.57	14:00	Border Collie	F (N)	16.72; 59.72	14:00
GSD	F	17.43	10:00	Border Collie	F (N)	18.14; 64.42	10:00
Dogo Argentino	F	18.01	14:00	X B. Collie / Husky	F (N)	13.29; 56.29	14:00
Great Dane	M	19.01	14:00	BSD	M (-)	14.72; 58.72	10:00
Rottweiler	F	15.85	14:00	BMD	F (-)	15.01; 60.00	10:00
AmStaff	F	13.57	14:00	St. Bernard	F (-)	19.58; 61:58	10:00
AmStaff	F	15.85	14:00	Dachshund	M (-)	20.99; 61.00	10:00
Engl. Cocker Sp.	M	15.43	14:00	Engl. Cocker Sp.	M (-)	14.42; 57.42	14:00
Labrador Retriever	M	21.28	10:00	Am. Cocker Sp.	F (N)	18.15; 64.29	10:00
Port. Water Dog	F	14.14	10:00	Fl. C. Retriever	F (-)	14.43; 61.43	10:00
Shih Tzu	F	21.14	14:00	Boston Terrier	M (N)	16.43; 57.43	10:00
Shih Tzu	F	14.99	14:00	French Bulldog	M (-)	18.15; 61.15	14:00
Whippet	F	19.14	10:00	X Tibetan Terrier	F (N)	21.60; 66.57	10:00
American Bulldog	F	20.14	14:00	X	M (N)	24.29; 62.29	10:00

The left panel shows dogs only tested at T1 (N = 16 puppies), the right panel shows dogs also tested at T2 (N = 16 young adults). Neuter status is only shown for T2 (puppies were not neutered). Dog age is shown for each test stage, and time of testing was the same on both occasions.

BSD: Belgian Shepherd dog, Malinois; BMD: Bernese Mountain Dog; GSD: German Shepherd dog; AmStaff: American Staffordshire Terrier; Engl. Cocker Sp.: English Cocker Spaniel; Port. Water Dog: Portuguese Water Dog; Husky: Siberian husky; Am. Cocker Sp.: American Cocker Spaniel; Fl. C. Retriever: Flat Coated Retriever; X: crossbred

(N): neutered

* No owner ratings available for this dog

At test stage 1 (T1), saliva was collected from 32 puppies (mean age ± SD: 17.57 ± 2.94 weeks) subjected to a behavioral test at home, and owner ratings were available for 31 of these puppies (procedures described below). At test stage 2 (T2), 9.91 (± 0.61) months later, 16 dogs were tested a second time at home (age: 60.85 ± 2.64 weeks), and owner ratings were available for all 16 dogs.

Owner ratings

At both stages (T1 and T2), owners were asked to rate the behavior of their dog in daily life by means of an online questionnaire (C-BARQ: updated version of PennBARQ [35]) in Dutch or French. The C-BARQ describes 14 different behavior categories, each represented by the mean of a subset of questions (presented in [37,38]): Stranger-directed aggression (ten questions), Owner-directed aggression (eight questions), Dog-directed aggression (four questions), Stranger-directed fear (four questions), Dog directed fear (four questions), Non-social fear (six questions), Separation-related behavior (eight questions), Attachment and attention-seeking (six questions), Trainability (eight questions), Chasing (four questions), Excitability (six questions), Touch sensitivity (four questions), Energy level (two questions) and Dog rivalry (four questions). The higher the score, the more present a particular trait is in a dog according to the owner.

Owners completed the questionnaire before the test to ensure their dog's behavior during the test would not influence their ratings. They were instructed to answer all questions but could skip a question pertaining to a situation if their dog had not experienced it yet. For any individual dog, if more than 25% of the questions for a category were missing, it was considered a missing value.

Saliva collection & analysis

At both test stages (T1 and T2), saliva samples were collected before (pre-test), as well as 10 (post10) and 40 minutes (post40) after a behavioral test. In short, the first 16 minutes of this test consisted of different social situations (e.g., entrance of a stranger, being ignored by persons in the room, being left alone), and in the second part (27 min), the dogs were exposed in a random order to several stimuli with varying visual and acoustic properties (e.g., sound of a barking dog, squeaky toy, weasel ball, vacuum cleaner). Two startling stimuli were presented at the end (umbrella opened suddenly and shaking of a metal cylinder containing keys), so that the overall test remained as little stressful as possible to the dogs.

The test was conducted in a familiar room at home to minimize influences of unfamiliar settings on the stress response [19,39]. To minimize influences from circadian fluctuations, testing only took place at one of two predetermined time points (10:00 or 14:00), and, on both occasions (T1 & T2), the dogs were tested on the same day (i.e., with a similar routine).

Saliva samples were collected according to a standardized protocol that was refined over several pilot trials [16]. Owners habituated their dogs to the sample procedure, starting four days before the test, and collected the saliva samples to prevent influences of unfamiliar persons [19,40]. On the test day, owners withheld food from the dogs at least one hour before sampling and did not let the animal drink or engage in strenuous activities at least half an hour beforehand to prevent contamination of saliva samples and influence of physical activity on the physiological measures [41,42]. At each sampling occasion (pre-test, post10, post40), the owners put on clean latex gloves and collected saliva samples with four swabs. They put one swab in each cheek pouch and gently held the dog's muzzle for one minute, after which they placed the two swabs back in their centrifugation tubes. Immediately after, they took samples with the two remaining swabs. The owners were instructed not to pet the dog during sampling, as this could influence the physiological response [43]. The dogs were not restrained, and sampling was stopped if an animal appeared overly stressed by the procedure (e.g., low posture, avoidance, resistance). Sampling occurred within four minutes to prevent handling from influencing marker concentrations [44]. Owners noted sampling time, sampling duration, last eating/drinking time, activity before sampling, and any additional remarks. They then placed the four centrifugation tubes, containing one swab each, in the freezer until the end of the behavioral test.

After the test, the samples were transported on ice to the tester's home and stored within two hours (-20°C). Within three months [42], samples were transported on ice to the University of Namur, where they were centrifuged at 3000 rpm (1851.41 x g for 15 min at 10°C) and samples containing blood or other visible contamination were discarded (N = 61 or 11.30%; from an additional 32 samples [5.92%] no saliva could be extracted). The extracted saliva volume per swab was noted, then saliva was pooled per sampling occasion (four swabs), resulting in three samples per test day: pre-test, post10, post40. The samples were assayed for the following three analytes, given sufficient saliva was available: CgA, cortisol, sIgA (respective kits: Human Chromogranin A EIA kit, Yanaihara Institute, Shizuoka, Japan; Salivary cortisol enzyme immunoassay kit, Salimetrics LLC, State College, Pennsylvania; Dog IgA ELISA Quantitation set, Bethyl Laboratories, Montgomery, Texas). Samples had to be diluted to assay CgA (1:8) and sIgA (1:3000), so the required sample volumes for duplicate measurements were 8 µl, 50 µl and 3 µl for CgA, cortisol and sIgA, respectively [16].

To minimize assay influences, all samples from a dog (collected on one test day) were assayed on the same ELISA plate, and at both test stages the same dogs were assayed together on a plate. For example, if at T1 the ELISA plate contained samples from dogs 1–4 then the ELISA plate at T2 also contained samples from dogs 1–4. Intra-assay coefficients of variation (CV) were calculated by dividing the standard deviation of duplicate measurements of a sample by the mean of these measurements. Samples with an intra-assay CV > 15% were excluded from further analysis (N = 5, 0 and 16 for CgA, cortisol, and sIgA, respectively), so that analytical imprecision was less than half of the within-subject variation [45]. Mean intra-assay CVs for the final samples were 4.56% (CgA), 5.16% (cortisol) and 4.58% (sIgA). Inter-assay CVs, based on repeated measurements of two control samples (aliquoted before freezing) were relatively high: 67.90% (CgA), 11.60% (cortisol), and 41.06% (sIgA). However, over the whole assay period (almost two years), the control samples were frozen longer than the recommended 6 months [42], so instability of the analytes might have contributed to this high variance. Acceptable inter-assay CVs of 9.31% (CgA), 4.98% (cortisol), and 6.38% (sIgA) were found for the same assay kits during pilot trials [16]. Each dog served as its own control, and all samples from that dog (one test day) were assayed on the same plate, so inter-assay variation is expected to minimally influence the measured values for each dog (expressed as *changes* in value over time, see "statistics").

Ethics considerations

The described procedures (behavioral test, saliva sampling) were approved by the Ethical Commission of the University of Namur, which did not consider them animal experiments based on the non-invasive nature (as attested by seven international researchers, experienced with these types of measurement). The protocol was agreed upon by the Belgian State authority for Directive 2010/63/EU.

The owners were informed about the study aims and procedures and signed a letter of informed consent. Their participation was voluntary, and they were told that they could interrupt the behavioral test at any moment. Also, endpoints were formulated for the behavioral test to protect the welfare of the dogs.

Statistics

As high inter-individual variation in marker concentrations may confound interpretation of reactivity and recovery parameters [46], we analyzed differences between time points, rather than absolute concentration values. Pre-test values were subtracted from post10 and post40 values [47] to create two variables: Δ10, Δ40.

Given the low sample size, and in consultation with a statistician, non-parametric analyses were used to identify potential patterns between physiological and behavioral parameters. Per marker and for each variable, the dogs were median-split to distinguish those with the largest physiological response from those with the smallest response. Individuals were considered to have a "strong response" if their saliva concentrations *increased* above median for cortisol and CgA or *decreased* below median for sIgA [22]. Scatter plots were used to verify if the median split the sample into two distinctive groups. For three variables, a dog was moved from the "weak response" to the "strong response" group (T1: CgA Δ40; sIgA Δ10) or conversely (T2: CgA Δ10) because their values resembled more closely to the newly assigned group. The first author then used a Mann-Whitney U Test (IBM® SPSS® Statistics, Version 20) to determine whether these groups differed significantly ($\alpha < 0.050$) for the different C-BARQ scores at T1, T2 and when comparing physiological responses at T1 with behavioral scores at T2.

RESULTS

The following sections describe significant associations. To minimize reporting bias [48], all tested associations and ranges in marker concentrations.

Associations between physiological and behavioral variables at the same test stage

For CgA and cortisol, significant associations were found with behavioral traits at both test stages (Table 2). Puppies with a strong physiological response 10 minutes after the behavioral test (Δ10) received lower scores for Separation-related behavior (CgA), whereas those with a slow recovery (Δ40) received lower scores for Trainability (CgA) and higher for Dog Rivalry and Stranger-directed fear (cortisol). These results suggest that dogs with a strong physiological response shortly after the test were less likely to display undesirable behaviors upon separation. This finding is in contrast with our hypothesis, as we expected *more* undesirable behaviors for dogs with a strong physiological response (both for Δ10 and Δ40). Indeed, we found that a strong physiological response *40 minutes* after the test was associated with *more* undesirable behaviors in daily life, as reported by the dog owners.

Table 2. Significant differences in C-BARQ scores (owner ratings) for dogs with small/large physiological changes in response to the behavioral test (Mann-Whitney U test).

Marker	Physiol. change	C-BARQ	Weak response (median ± IQR)	Strong response (median ± IQR)	U	P	N	ES
CgA	$\Delta 10_{T1}$	SEP_{T1}	1.07 ± 0.39	0.63 ± 0.25	17.5	0.02	19	0.83
	$\Delta 40_{T1}$	TRA_{T1}	2.57 ± 0.82	2.38 ± 0.25	18.0	0.04	18	0.78
Cortisol	$\Delta 40_{T1}$	RIV_{T1}	0.00 ± 0.00	1.00 ± 1.00	3.5	0.04	10	0.86
	$\Delta 40_{T1}$	SDF_{T1}	0.00 ± 0.00	0.25 ± 0.25	14.5	0.03	17	0.80
CgA	$\Delta 10_{T2}$	SDF_{T2}	0.75 ± 1.00	0.00 ± 0.00	2.0	0.02	10	0.76
	$\Delta 40_{T2}$	EXC_{T2}	2.58 ± 0.92	1.42 ± 0.59	0.0	0.02	8	1.00
Cortisol	$\Delta 10_{T2}$	TRA_{T2}	2.38 ± 0.63	3.13 ± 0.44	0.5	0.03	8	0.97
	$\Delta 40_{T2}$	NRG_{T2}	1.50 ± 1.75	3.00 ± 0.50	0.0	0.02	8	1.00
sIgA	$\Delta 10_{T2}$	TRA_{T2}	2.63 ± 0.38	3.13 ± 0.38	0.0	0.03	7	1.00
	$\Delta 40_{T2}$	TRA_{T2}	2.50 ± 1.00	3.13 ± 0.38	0.0	0.05	6	1.00
	$\Delta 40_{T2}$	NSF_{T2}	1.50 ± 0.50	0.67 ± 0.67	0.0	0.05	6	1.00
	$\Delta 40_{T2}$	TCH_{T2}	1.50 ± 0.75	0.25 ± 0.75	0.0	0.05	6	1.00

The upper panel presents data from T1 (puppy stage), the lower panel from T2 (young adult stage). "Strong response" dogs are those with the largest physiological change (above median), as expected for that marker (increase for CgA and cortisol, decrease for sIgA).

Δ10 / Δ40: change in salivary marker concentration 10/40 min after the behavioral test, compared to pre-test

SEP: Separation-related behavior; TRA: Trainability; RIV: Dog rivalry; SDF: Stranger-directed fear; EXC: Excitability; NRG: Energy level; NSF: Non-social fear; TCH: Touch sensitivity

ES: effect size (probabilistic index)

At T2, similar patterns emerged for CgA and cortisol. High reactivity (Δ10) was associated with desirable behaviors in daily life: low Stranger-directed fear (CgA) and high Trainability (cortisol). In contrast, a slow recovery (Δ40) was associated with low Excitability (CgA) and high Energy level (cortisol), which might (arguably) be considered undesirable by owners (see discussion).

In contrast to the primary stress markers cortisol and CgA, significant associations between sIgA levels and behavior were only found at the young adult stage (T2), and only for desirable traits. A strong physiological change in sIgA levels, both at 10 and 40 minutes after the test, was associated with the desirable trait Trainability. In addition, slow recovery after the test (Δ40) was associated with low levels of Non-social fear and Touch sensitivity.

Associations between physiological variables at T1 and behavior at T2

Ambiguous associations were found between puppies' stress response, measured by levels of cortisol and CgA at T1, and their future behavioral traits (T2). Puppies with high physiological reactivity (Δ10) at T1 received higher scores at T2 for Stranger-directed fear, and Excitability (CgA) and lower scores for Attachment and attention-seeking (cortisol; Table 3). Those puppies with sustaining levels of CgA after the test (Δ40) received higher scores for Excitability and Energy level later in their life.

Table 3. Significant differences in C-BARQ scores (owner ratings at T2) for puppies with small/large physiological changes in response to the behavioral test (Mann-Whitney U test).

Bio-marker	Physiol. change	C-BARQ	Weak response (median ± IQR)	Strong response (median ± IQR)	U	P	N	ES
CgA	$\Delta 10_{T1}$	SDF_{T2}	0.00 ± 0.50	1.75 ± 2.25	4.0	0.01	13	0.91
	$\Delta 10_{T1}$	EXC_{T2}	2.17 ± 1.66	3.25 ± 1.33	4.0	0.02	13	0.91
	$\Delta 40_{T1}$	EXC_{T2}	1.67 ± 0.50	2.83 ± 1.16	5.0	0.04	12	0.71
	$\Delta 40_{T1}$	NRG_{T2}	2.00 ± 0.50	3.00 ± 1.00	4.5	0.03	12	0.71
Cortisol	$\Delta 10_{T1}$	ATT_{T2}	2.67 ± 1.00	2.17 ± 0.17	5.0	0.04	12	0.85
sIgA	$\Delta 10_{T1}$	TRA_{T2}	2.38 ± 1.00	3.07 ± 0.50	3.0	0.03	11	0.83
	$\Delta 40_{T1}$	DDF_{T2}	0.00 ± 0.25	1.25 ± 0.50	0.0	0.02	8	1.00

"Strong response" dogs are those with the largest physiological change (above median) as expected for that marker (increase for CgA and cortisol, decrease for sIgA).

Δ10 / Δ40: change in salivary marker concentration 10/40 min after the behavioral test, compared to pre-test

SDF: Stranger-directed fear; EXC: Excitability; NRG: Energy level; ATT: Attachment and attention-seeking; TRA: Trainability; DDF: Dog-directed fear

For sIgA, again a positive association was found between large physiological reactivity following the test (Δ10) and the desirable trait Trainability. In contrast, a slow recovery (Δ40) at the puppy stage (T1) was associated with more Dog-directed fear later in life (T2).

DISCUSSION

This study investigated whether stress responsiveness, as measured via physiological markers in one context were related to dogs' general behavior in daily life (as rated by their owners). At two stages (T1: puppy, T2: young adult), the dogs' physiological stress response after a behavioral test at home was measured in terms of reactivity (10 min post-test) and recovery (40 min post-test) for three salivary markers: cortisol, chromogranin A (CgA) and secretory immunoglobulin A (sIgA). Based on these values, it was determined whether dogs with a strong physiological response (for each marker separately) displayed different behavior in daily life, compared to dogs with a weaker physiological response. We hypothesized that a strong physiological response 10 minutes after the test would be associated with undesirable behavioral traits, and we expected the same for a longer recovery time (i.e., time before physiological levels return to baseline). The results revealed three main findings: 1) for CgA and cortisol, different patterns were identified according to sample time post-test, 2) patterns for sIgA differed from CgA and cortisol, and 3) predictive capacity of puppies' physiological stress response (T1) was inconclusive, and ambiguous relations were found with behavioral traits at T2.

First, for CgA and cortisol, high reactivity at 10 minutes after the behavioral test related to desirable traits, whereas slow recovery 40 minutes after the test related to undesirable traits. The former result suggests that dogs with a relatively strong physiological response following the test respond to stressors in an adaptive manner [23,49], reflected in daily life by less fear and higher trainability. In contrast, the latter result might indicate unsuccessful coping (slow recovery), as reflected by the associations with undesirable traits. This could also explain the high Energy (cortisol) and low Excitability (CgA) scores at T2. Increased activity can be a mechanism to cope with prolonged stress [24,50], and, if the dogs were in a negative emotional state, they may have responded less excitedly to the situations described in the C-BARQ score Excitability (e.g., going for walks, arrival of visitors, owner coming home).

The finding that associations of physiological responses with behavioral traits vary according to the sample times indicates the importance of measuring *changes* in marker concentrations, at *multiple* time points. Collection of baseline samples before the onset of a specific event allows for a correction of inter-individual variation in marker concentrations [46] and thus provides a more accurate indication of each dog's physiological response to that event. This is supported by findings from Wormald et al. [8], who only found a significant difference between dogs that either passed or failed an aggression test, when considering the *change* in salivary cortisol levels (pre-post venipuncture). Likewise, Sherman et al. [15] found only significant associations between "emotional reactivity" (measured in a behavioral test) and cortisol levels in saliva and plasma when considering the *change* compared to baseline but not when using absolute post-test values.

Furthermore, the collection of multiple samples is necessary to monitor changes over time, thereby distinguishing dogs with a high (initial) reactivity from dogs with a more blunted (prolonged) stress response. In fact, we hypothesized that a strong physiological response (in itself) would be indicative of undesirable behavioral traits [2,19]. Instead, our findings suggest that the recovery time tells us more about a dog's overall stress responsiveness. Possibly, dogs that are more prone to stress in daily life have frequent elevations of stress markers, which might have led to an attenuated stress response due to negative feedback [21]. In line with this, Beerda et al. [51] found that dogs (N = 8) subjected to 5 weeks of social and spatial restriction, showed an increase over time in daily cortisol levels but had an attenuated cortisol response when exposed to a noise stressor. Dreschel and Granger [19] found that dogs living with other dogs had a lower increase in cortisol levels 20 and 40 minutes after exposure to a fearful event (thunderstorm recording), and their baseline cortisol levels tended to be higher. They argued these findings might reflect stress in daily life from living with another dog. Likewise, Rosado et al. [52] found higher plasma levels of cortisol in dogs diagnosed as aggressive, which they attributed to stress most likely resulting from an inconsistent and unpredictable living environment. Translating these findings to our study, it is possible that relatively small changes in marker concentration shortly after the behavioral test reflected an attenuated stress response to that specific event.

Second, this study revealed differences among physiological markers. Contrary to findings for cortisol and CgA, changes in sIgA levels only related to behavioral traits at T2, most of which were considered desirable. In particular, Trainability was positively associated with reactivity and recovery at T2 and also predicted by high reactivity at T1. Likewise, Kikkawa et al. [53] found sIgA levels to differ significantly between dogs (11–14 months old) selected (N = 25) or rejected (N = 44) for guide dog training. During a 2-week assessment period, the dogs were kenneled and their sIgA levels were measured on days 1, 2, 3, 7 and 14. The sIgA levels of subsequently selected guide dogs gradually increased, whereas sIgA levels of rejected dogs remained relatively low, despite similar housing and caring conditions. On day 14 the selected dogs had significantly higher sIgA levels than rejected dogs, which the authors interpreted as a difference in "adaptive ability". These findings seem to reveal the same pattern as described above for cortisol and CgA (Δ40): consistently low sIgA levels suggest that the rejected dogs had low adaptive capacity, with slow recovery during a period of prolonged stress (kenneling). In our study, however, a strong response (i.e., relatively low levels of sIgA) was associated with *high* Trainability scores, which appears to be more in line with the Δ10 results for CgA and cortisol. These contrary findings are likely a result of different protocols: in our study the samples were collected after startling stimuli (two at the end of the test), whereas Kikkawa et al. [53] took samples at predetermined time points, not associated with a particular stimulus. Consequently, sIgA measures in our study appear to reflect the more immediate response to a stressor, in contrast to the long-term measures of Kikkawa et al. Indeed, in an extensive review on stress and secretory immunity in humans, Bosch et al. [54] identified an overall distinction between acute and chronic stress: chronic stress generally induces a *decrease* in sIgA levels, acute stressors mostly induce an *increase*. One would then expect the dogs from our "strong response" group to be more prone to chronic stress, and thus have a *lower* trainability (as found by Kikkawa et al. [53]). On the other hand, if we take into account a possible attenuation in cortisol response for dogs under chronic stress (as described above), it is possible that this results in less immunosuppression, hence relatively *higher* sIgA levels, in these dogs after an acute stressor. If this were true, then this would explain why the "weak response" group had relatively low Trainability scores. However, many factors affect the findings (including type of stressor, timing of sampling [22], and the balance of sympathetic-parasympathetic control [27,54]. Consequently these complex interactions need to be further explored and elucidated. Nevertheless, it is noteworthy that we found three associations with the Trainability trait that were consistent over time, and these findings along with those from Kikkawa et al. [53] suggest that sIgA holds potential as a marker of (long-term) coping ability. Further investigation is needed to unravel individual differences in sIgA levels, both in response to short-term and long-term stressors.

The differences between CgA and cortisol on one hand and sIgA on the other hand are in line with findings from other studies, and distinguishes them as primary and secondary markers of stress. Similar reaction patterns were found for CgA and cortisol in dog plasma [29], pig saliva [31] and human saliva [55]. Svobodová et al. [56] reported no significant association between salivary cortisol and sIgA values after a stressful condition, though Skandakumar et al. [34] reported a negative (semi-logarithmic) relation between cortisol and sIgA levels in repeated saliva samples from six dogs. Possibly these contradictory findings reflect the delayed reaction pattern for sIgA, in comparison to cortisol: in response to a stressor cortisol is first released, which subsequently inhibits components of the immune system [27], thereby decreasing sIgA levels. So, strictly speaking sIgA is a marker of immune response, which responds to stress, instead of a primary stress marker. The delayed reaction pattern for sIgA could explain the correlations with desirable behavioral traits and sIgA measured 40 minutes after the behavioral test (as opposed to cortisol and CgA). After a stressor sIgA levels return to baseline only after 30–60 min [22]. Consequently, 40 minutes after the test might have been too soon to detect differences between dogs in coping capacity (recovery), as we found for CgA and cortisol. Unfortunately, combined analyses with all markers (e.g., pairwise comparisons; Principal Component Analysis) were not possible due to incomplete datasets for the individual markers. So, more research is needed to further elucidate the interaction between different stress markers.

Third, predictive capacity of stress physiology at the puppy stage (T1) remains inconclusive. Apart from Trainability, sIgA responsiveness in puppyhood only related to an *undesirable* behavioral trait at T2 (Dog-Directed fear), in contrast with the findings described above. Only one association was found for cortisol, and contradictory associations were found for CgA: Stranger-directed fear at T2 was associated with *high* reactivity (Δ10) at T1 but with *low* reactivity at T2. Likewise, Excitability at T2 was associated with *slow* recovery (Δ40) at T1 but with *fast* recovery at T2. These contradictions could partly be explained by the different datasets at both test stages, as the findings describe *relative* patterns for each test population [57]. For instance, a dog's stress response may have been relatively strong compared to the other dogs at T1, but the same response may have been relatively weak at T2. Nevertheless, it is interesting that associations with the same behavioral traits were found at both test stages, even though inverted. Probably the relation with behavioral traits is shaped by developmental factors [43,58], and further studies should elucidate the predictive capacity of the described salivary markers.

Limitations

One important limitation of this study is the small sample size. Data were inevitably lost during laboratory analyses (insufficient or contaminated saliva samples) and by excluding samples with an intra-assay CV > 15%. Effect sizes were relatively high, but this is partially caused by the small sample size (i.e., the denominator of the probabilistic index). Furthermore, potential influences like breed, sex, or neuter status [40,59] could not be analyzed, so the results should be interpreted with caution. Though the patterns identified in this study should be confirmed, they highlight important considerations for future studies.

Another limitation of this study is the high dependence on dog owners. Participation rate was low for this longitudinal and multidisciplinary study, despite proactive recruitment by veterinarians and offering incentives. Furthermore, not all owners adhered equally well to the specific test and sampling procedures. This may have caused inter-individual variation and possibly also some intra-individual variation. By median-splitting the group, the influence of such variation was expected to be minor: individuals with a large stress response would fall into the "strong response" group regardless of the exact change in marker concentration and vice versa. However, the use of the median was an arbitrary choice, and the created groups might not reflect 'true' differences in stress response. Therefore, the data were visually explored to ensure they were split into two distinctive groups.

Furthermore, dogs that would normally cope by means of owner interaction, activity or play had limited options for recovery, as interaction and activity were limited for standardization. Nevertheless, the saliva samples still provided valuable information about their coping ability when owner support or outlet options are limited. For interpretation of the results, standardization of the test situation was thus preferred. For future studies, it might be useful to note signs of stress at the time of sampling, instead of merely activity, and to question owners how their dog usually copes in a stressful situation. This would allow retrospective inspection whether or not the dogs recovered, and if not, whether this might have been caused by the absence of relevant maintenance stimuli [60]. Alternatively, the potential value of hair samples to determine dogs' general stress responsiveness could be further explored. Although hairs contain a mixture of analytes accumulated over weeks, it might give an estimation of dogs' stress reactivity, when taking into account activity levels and coat color [61,62].

Another potential influence concerns the baseline levels before the test. The owners met the tester outside 20–30 min before the test but the dogs' reactions to their owners' return could not be controlled. Excited dogs might have had higher baseline levels of the measured markers, which would hypothetically result in lower differences between post-test and pre-test marker concentrations. However, no such patterns were identified in the dataset, and the findings described above appear to reflect plausible differences in reactivity. This suggests that the influence of different baseline levels was minor.

CONCLUSIONS

The findings of this exploratory study suggest that individual canine differences in physiological stress response relate to behavioral traits, mainly describing fearfulness, reactivity or responsiveness to training. Different patterns were found for physiological stress reactivity and recovery, suggesting that both are important for determining dog stress response and coping ability. High reactivity in response to a stressor is ambiguous to interpret, as it could indicate an adaptive or a maladaptive response. Slow recovery, on the other hand, reflects a maladaptive response. Differences in markers were also found, confirming that a combined analysis of multiple endocrine and immune biomarkers would give a more accurate indication of stress levels [17,26]. Finally, predictive capacity of puppies' physiological stress response was inconclusive, and more research is needed to assess the influence of developmental factors. Though these tenuous conclusions are based on a relatively small dataset, they shed new light on the interpretation of physiological markers and their application in future studies. Hopefully, our findings will inspire other researchers to measure both reactivity and recovery and to use multiple markers for a more accurate indication of the physiological stress response. Further research is needed to confirm these patterns in a larger population.

ACKNOWLEDGMENTS

The authors thank all dog owners who voluntarily participated to this study. The assistance and insights of IVRU staff members (Veterinary Department, University of Namur), in particular Laetitia Wiggers, Prof. Nathalie Kirschvink, Prof. Benoît Muylkens, and Dr. Damien Coupeau were much appreciated. Daniela Späder (Master in Ethology, Paris 13, France) and Alice Depiereux (Bachelor in Animal Technics, HELHA, Belgium) helped with filming saliva sampling instruction videos. The authors also thank Prof. James Serpell (University of Pennsylvania School of Veterinary Medicine) for providing translated versions of the C-BARQ and valuable insights for its interpretation. Finally, the statistical advice from Dr. Etienne Hanon was much appreciated.

FUNDING STATEMENT

This study (author Rian Lensen) was supported by the Federal Public Service of Health, Food Chain Safety and Environment (https://www.health.belgium.be/en), and the Public Service of Wallonia (https://spw.wallonie.be/); contract RT 10/12 CONDOG. The funders had no role in study design, data collection and analysis, decision to publish, or preparation of the manuscript.

REFERENCES

1. Carere C, Caramaschi D, Fawcett TW. Covariation between personalities and individual differences in coping with stress: Converging evidence and hypotheses. Curr Zool. 2010;56: 728–740.
2. Rayment DJ, De Groef B, Peters RA, Marston LC. Applied personality assessment in domestic dogs: Limitations and caveats. Appl Anim Behav Sci. 2015;163: 1–18. 10.1016/j.applanim.2014.11.020
3. Sforzini E, Michelazzi M, Spada E, Ricci C, Carenzi C, Milani S, et al. Evaluation of young and adult dogs' reactivity. J Vet Behav. 2009;4: 3–10. 10.1016/j.jveb.2008.09.035
4. Jones AC, Gosling SD. Temperament and personality in dogs (Canis familiaris): A review and evaluation of past research. Appl Anim Behav Sci. 2005;95: 1–53.
5. Wolf M, van Doorn GS, Weissing FJ. Evolutionary emergence of responsive and unresponsive personalities. Proc Natl Acad Sci U S A. 2008;105: 15825–15830. 10.1073/pnas.0805473105
6. Koolhaas JM, de Boer SF, Coppens CM, Buwalda B. Neuroendocrinology of coping styles: Towards understanding the biology of individual variation. Front Neuroendocrinol. 2010;31: 307–321. 10.1016/j.yfrne.2010.04.001
7. Overall KL. Manual of Clinical Behavioral Medicine for Dogs and Cats. St. Louis, Missouri: Elsevier; 2013.
8. Wormald D, Lawrence AJ, Carter G, Fisher AD. Physiological stress coping and anxiety in greyhounds displaying inter-dog aggression. Appl Anim Behav Sci. 2016;180: 93–99. 10.1016/j.applanim.2016.04.007
9. Barnard S, Marshall-Pescini S, Passalacqua C, Beghelli V, Capra A, Normando S, et al. Does Subjective Rating Reflect Behavioural Coding? Personality in 2 Month-Old Dog Puppies: An Open-Field Test and Adjective-Based Questionnaire. PLoS One. 2016;11: e0149831 10.1371/journal.pone.0149831
10. Riemer S, Müller C, Virányi Z, Huber L, Range F. The predictive value of early behavioural assessments in pet dogs—A longitudinal study from neonates to adults. PLoS One. 2014;9: e101237 10.1371/journal.pone.0101237
11. Ley J, Coleman GJ, Holmes R, Hemsworth PH. Assessing fear of novel and startling stimuli in domestic dogs. Appl Anim Behav Sci. 2007;104: 71–84. 10.1016/j.applanim.2006.03.021
12. Ng ZY, Pierce BJ, Otto CM, Buechner-Maxwell VA, Siracusa C, Werre SR. The effect of dog–human interaction on cortisol and behavior in reg-

istered animal-assisted activity dogs. Appl Anim Behav Sci. 2014;159: 69–81. 10.1016/j.applanim.2014.07.009
13. Stamps J, Groothuis TGG. The development of animal personality: Relevance, concepts and perspectives. Biol Rev. 2010;85: 301–325. 10.1111/j.1469-185X.2009.00103.x
14. Horváth Z, Igyártó B-Z, Magyar A, Miklósi Á. Three different coping styles in police dogs exposed to a short-term challenge. Horm Behav. 2007;52: 621–630. 10.1016/j.yhbeh.2007.08.001
15. Sherman BL, Gruen ME, Case BC, Foster ML, Fish RE, Lazarowski L, et al. A test for the evaluation of emotional reactivity in Labrador retrievers used for explosives detection. J Vet Behav. 2015;10: 94–102. 10.1016/j.jveb.2014.12.007
16. Lensen CMM, Moons CPH, Diederich C. Saliva sampling in dogs: How to select the most appropriate procedure for your study. J Vet Behav. 2015;10: 504–512. 10.1016/j.jveb.2015.08.006
17. Hekman JP, Karas AZ, Sharp CR. Psychogenic stress in hospitalized dogs: Cross species comparisons, implications for health care, and the challenges of evaluation. Animals. 2014;4: 331–347. 10.3390/ani4020331
18. Beerda B, Schilder MBH, van Hooff JARAM, de Vries HW, Mol JA. Behavioural, saliva cortisol and heart rate responses to different types of stimuli in dogs. Appl Anim Behav Sci. 1998;58: 365–381. 10.1016/S0168-1591(97)00145-7
19. Dreschel NA, Granger DA. Physiological and behavioral reactivity to stress in thunderstorm-phobic dogs and their caregivers. Appl Anim Behav Sci. 2005;95: 153–168. 10.1016/j.applanim.2005.04.009
20. Mormède P, Andanson S, Aupérin B, Beerda B, Guémené D, Malmkvist J, et al. Exploration of the hypothalamic-pituitary-adrenal function as a tool to evaluate animal welfare. Physiol Behav. 2007;92: 317–339. 10.1016/j.physbeh.2006.12.003
21. Protopopova A. Effects of sheltering on physiology, immune function, behavior, and the welfare of dogs. Physiol Behav. 2016;159: 95–103. 10.1016/j.physbeh.2016.03.020
22. Kikkawa A, Uchida Y, Nakade T, Taguchi K. Salivary secretory IgA concentrations in beagle dogs. J Vet Med Sci. 2003;65: 689–693. 10.1292/jvms.65.689
23. Lucassen PJ, Pruessner J, Sousa N, Almeida OFX, Van Dam AM, Rajkowska G, et al. Neuropathology of stress. Acta Neuropathol. 2014;127: 109–135. 10.1007/s00401-013-1223-5

24. Hiby EF, Rooney NJ, Bradshaw JWS. Behavioural and physiological responses of dogs entering re-homing kennels. Physiol Behav. United States; 2006;89: 385–391. 10.1016/j.physbeh.2006.07.012
25. Haverbeke A, Diederich C, Depiereux E, Giffroy JM. Cortisol and behavioral responses of working dogs to environmental challenges. Physiol Behav. 2008;93: 59–67. 10.1016/j.physbeh.2007.07.014
26. Ott S, Soler L, Moons CPH, Kashiha MA, Bahr C, Vandermeulen J, et al. Different stressors elicit different responses in the salivary biomarkers cortisol, haptoglobin, and chromogranin A in pigs. Res Vet Sci. 2014;97: 124–128. 10.1016/j.rvsc.2014.06.002
27. Ivković N, Božović Đ, Račić M, Popović-Grubač D, Davidović B. Biomarkers of Stress in Saliva / Biomarkeri stresa u pljuvački. Acta Fac Medicae Naissensis. 2015;32: 91–99. 10.1515/afmnai-2015-0010
28. Kennedy B, Dillon E, Mills PJ, Ziegler MG. Catecholamines in human saliva. Life Sci. 2001;69: 87–99. 10.1016/s0024-3205(01)01111-0
29. Akiyoshi H, Aoki M, Shimada T, Noda K, Kumagai D, Saleh N, et al. Measurement of plasma chromogranin A concentrations for assessment of stress responses in dogs with insulin-induced hypoglycemia. Am J Vet Res. 2005;66: 1830–1835.
30. Toda M, Morimoto K. Effect of lavender aroma on salivary endocrinological stress markers. Arch Oral Biol. 2008;53: 964–968. 10.1016/j.archoralbio.2008.04.002
31. Escribano D, Soler L, Gutiérrez AM, Martínez-Subiela S, Cerón JJ. Measurement of chromogranin A in porcine saliva: validation of a time-resolved immunofluorometric assay and evaluation of its application as a marker of acute stress. Animal. 2013;7: 640–647. 10.1017/S1751731112002005
32. Ninomiya S, Sato S. The assessment of the effect of presenting a companion's face picture on social isolation stress using saliva sampling in cows. Anim Sci J. 2011;82: 787–791. 10.1111/j.1740-0929.2011.00896.x
33. Kanai K, Hino M, Hori Y, Nakao R, Hoshi F, Itoh N, et al. Circadian variations in salivary chromogranin a concentrations during a 24-hour period in dogs. J Vet Sci. 2008;9: 421–423. 10.4142/jvs.2008.9.4.421
34. Skandakumar S, Stodulski G, Hau J. Salivary IgA: a Possible Stress Marker In Dogs. Anim Welf. 1995;4: 339–350.
35. Hsu Y, Serpell JA. Development and validation of a questionnaire for measuring behavior and temperament traits in pet dogs. J Am Vet Med Assoc. 2003;223: 1293–1300. 10.2460/javma.2003.223.1293

36. Mills DS. Medical paradigms for the study of problem behaviour: a critical review. Appl Anim Behav Sci. 2003;81: 265–277. 10.1016/S0168-1591(02)00286-1
37. Duffy DL, Serpell JA. Predictive validity of a method for evaluating temperament in young guide and service dogs. Appl Anim Behav Sci. 2012;138: 99–109. 10.1016/j.applanim.2012.02.011
38. Foyer P, Bjällerhag N, Wilsson E, Jensen P. Behaviour and experiences of dogs during the first year of life predict the outcome in a later temperament test. Appl Anim Behav Sci. 2014;155: 93–100. 10.1016/j.applanim.2014.03.006
39. van den Berg L, Schilder MB, de Vries H, Leegwater PA, van Oost BA. Phenotyping of aggressive behavior in golden retriever dogs with a questionnaire. Behav Genet. 2006;36: 882–902. 10.1007/s10519-006-9089-0
40. Sandri M, Colussi A, Perrotta MG, Stefanon B. Salivary cortisol concentration in healthy dogs is affected by size, sex, and housing context. J Vet Behav. 2015;10: 302–306.
41. Glenk LM, Kothgassner OD, Stetina BU, Palme R, Kepplinger B, Baran H. Salivary cortisol and behavior in therapy dogs during animal-assisted interventions: A pilot study. J Vet Behav. 2014;9: 98–106. 10.1016/j.jveb.2014.02.005
42. Salimetrics LLC, SalivaBio. Saliva collection and handling advice, 3rd edition [Internet]. 2013 [cited 10 Feb 2014]. Available: https://www.salimetrics.com/assets/documents/Saliva_Collection_Handbook.
43. Hennessy MB T. Williams M, Miller DD, Douglas CW, Voith VL. Influence of male and female petters on plasma cortisol and behaviour: can human interaction reduce the stress of dogs in a public animal shelter? Appl Anim Behav Sci. 1998;61: 63–77. 10.1016/s0168-1591(98)00179-8
44. Kobelt AJ, Hemsworth PH, Barnett JL, Butler KL. Sources of sampling variation in saliva cortisol in dogs. Res Vet Sci. 2003;75: 157–161. 10.1016/s0034-5288(03)00080-8
45. Ricos C. Biologic variation and desirable specifications for QC [Internet]. [cited 9 Feb 2015]. Available: https://www.westgard.com/guest17.htm
46. Schalke E, Stichnoth J, Ott S, Jones-Baade R. Clinical signs caused by the use of electric training collars on dogs in everyday life situations. Appl Anim Behav Sci. 2007;105: 369–380. 10.1016/j.applanim.2006.11.002
47. King T, Hemsworth PH, Coleman GJ. Fear of novel and startling stimuli in domestic dogs. Appl Anim Behav Sci. 2003;82: 45–64. 10.1016/s0168-1591(03)00040-6

48. Forstmeier W, Wagenmakers E-J, Parker TH. Detecting and avoiding likely false-positive findings–a practical guide. Biol Rev. 2016;92: 1941–1968. 10.1111/brv.12315
49. Hennessy MB. Using hypothalamic-pituitary-adrenal measures for assessing and reducing the stress of dogs in shelters: A review. Appl Anim Behav Sci. 2013;149: 1–12. 10.1016/j.applanim.2013.09.004
50. Beerda B, Schilder MBH, Van Hooff JARAM, De Vries HW, Mol JA. Behavioural and hormonal indicators of enduring environmental stress in dogs. Anim Welf. 2000;9: 49–62.
51. Beerda B, Schilder MBH, Bernadina W, Van Hooff JARAM, De Vries HW, Mol JA. Chronic Stress in Dogs Subjected to Social and Spatial Restriction. II. Hormonal and Immunological Responses. Physiol Behav. 1999;66: 243–254. 10.1016/s0031-9384(98)00290-x
52. Rosado B, García-Belenguer S, León M, Chacón G, Villegas A, Palacio J. Blood concentrations of serotonin, cortisol and dehydroepiandrosterone in aggressive dogs. Appl Anim Behav Sci. 2010;123: 124–130. 10.1016/j.applanim.2010.01.009
53. Kikkawa A, Uchida Y, Suwa Y, Taguchi K. A novel method for estimating the adaptive ability of guide dogs using salivary sIgA. J Vet Med Sci. 2005;67: 707–712. 10.1292/jvms.67.707
54. Bosch JA, Ring C, de Geus EJC, Veerman ECI, Nieuw Amerongen AV. Stress and secretory immunity. Int Rev Neurobiol. 2002; 52: 213–253. 10.1016/S0074-7742(02)52011-0
55. Nakane H, Asami O, Yamada Y, Harada T, Matsui N, Kanno T, et al. Salivary Chromogranin A as an index of psychosomatic stress response. Biomed Res. 1998;19: 401–406. 10.2220/biomedres.19.401
56. Svobodová I, Chaloupková H, Končel R, Bartoš L, Hradecká L, Jebavý L. Cortisol and Secretory Immunoglobulin A Response to Stress in German Shepherd Dogs. PLoS One. 2014;9: e90820 10.1371/journal.pone.0090820
57. Uher J. Individual behavioral phenotypes: An integrative meta-theoretical framework. Why "behavioral syndromes" are not analogs of "personality." Dev Psychobiol. 2011;53: 521–548. 10.1002/dev.20544
58. Day MJ. Immune System Development in the Dog and Cat. J Comp Pathol. 2007;137: S10–S15. 10.1016/j.jcpa.2007.04.005
59. Cobb ML, Iskandarani K, Chinchilli VM, Dreschel NA. A systematic review and meta-analysis of salivary cortisol measurement in domestic canines. Domest Anim Endocrinol. 2016;57: 31–42. 10.1016/j.domaniend.2016.04.003

60. Appleby D, Pluijmakers J. Separation anxiety in dogs. The function of homeostasis in its development and treatment. Vet Clin North Am Small Anim Pract. 2003;33: 321–344. 10.1016/S0195-5616(02)00101-8
61. Bennett A, Hayssen V. Measuring cortisol in hair and saliva from dogs: coat color and pigment differences. Domest Anim Endocrinol. 2010/08/14. 2010;39: 171–180. 10.1016/j.domaniend.2010.04.003
62. Siniscalchi M, McFarlane JR, Kauter KG, Quaranta A, Rogers LJ. Cortisol levels in hair reflect behavioural reactivity of dogs to acoustic stimuli. Res Vet Sci. 2013;94: 49–54. 10.1016/j.rvsc.2012.02.017

SECTION 3: BOVINES

CHAPTER 5

Evaluation of Animal Genetic and Physiological Factors That Affect the Prevalence of Escherichia coli O157 in Cattle

Soo Jin Jeon[1,2], Mauricio Elzo[1], Nicolas DiLorenzo[1,3], G. Cliff Lamb[1,3], Kwang Cheol Jeong[1,2]

[1]Department of Animal Sciences, Institute of Food and Agricultural Sciences, University of Florida, Gainesville, Florida, United States of America,
[2]Emerging Pathogens Institute, University of Florida, Gainesville, Florida, United States of America,
[3]North Florida Research and Education Center, Institute of Food and Agricultural Sciences, University of Florida, Marianna, Florida, United States of America

Citation: Jeon SJ, Elzo M, DiLorenzo N, Lamb GC, Jeong KC (2013) Evaluation of Animal Genetic and Physiological Factors That Affect the Prevalence of Escherichia coli O157 in Cattle. PLoS ONE 8(2): e55728. https://doi.org/10.1371/journal.pone.0055728

ABSTRACT

Controlling the prevalence of Escherichia coli O157 in cattle at the pre-harvest level is critical to reduce outbreaks of this pathogen in humans. Multilayers of factors including the environmental and bacterial factors modulate the colonization and persistence of E. coli O157 in cattle that serve as a reservoir of this pathogen. Here, we report animal factors contributing to the prevalence of E. coli O157 in cattle. We observe the lowest number of E. coli O157 in Brahman breed when compared with other crosses in an Angus-Brahman multibreed herd, and bulls excrete more E. coli O157 than steers in the pens where cattle were housed together. The presence of super-shedders, cattle excreting .105 CFU/rectal anal swab, increases the concentration of E. coli O157 in the pens; thereby super-shedders enhance transmission of this pathogen among cattle. Molecular subtyping analysis reveal only one subtype of E. coli O157 in the multibreed herd, indicating the variance in the levels of E. coli O157 in cattle is influenced by animal factors. Furthermore, strain tracking after relocation of the cattle to a commercial feedlot reveals farm-to-farm transmission of E. coli O157, likely via super-shedders. Our results reveal high risk factors in the prevalence of E. coli O157 in cattle whereby animal genetic and physiological factors influence whether this pathogen can persist in cattle at high concentration, providing insights to intervene this pathogen at the pre-harvest level.

INTRODUCTION

The prevalence of Escherichia coli O157 in cattle herds (ranging 0–61%) [1], [2], [3], [4] is positively correlated to outbreaks of this pathogen causing severe diseases including hemorrhagic colitis (HC) and hemolytic uremic syndrome (HUS), which can cause kidney failure and be fatal [3], [5]. Despite the implementation of government regulations and development of process interventions, food recalls and human illness related to E. coli O157 remain concerns around the world. Reducing the prevalence of this pathogen in cattle at the pre-harvest level has been highlighted recently as a critical control point to decrease the number of E. coli O157 entering the food chain [6], [7], [8], [9], [10]. Awareness of the risk factors that may increase the prevalence of E. coli O157 at the pre-harvest level can provide insights to develop intervention technologies to reduce its prevalence. Although risk factors on farms have been extensively studied from the bacterial perspectives, information regarding animal factors that may contribute to the prevalence of this pathogen is lacking. Here we identify high risk factors that significantly affect the prevalence of this pathogen in cattle.

Cattle are the primary reservoir of E. coli O157, and ground beef remains a significant source of foodborne transmission with other sources such as fresh vegetables [11]. Cattle that excrete more than 10^4 colony forming unit (CFU)/g of cattle feces have been defined as super-shedders [9], [12]. The super-shedders are responsible for about 90% of the total number of bacteria in the cattle herd [9], [12] and raise the prevalence of cattle infected with this pathogen on farms, making them a high risk factor at the pre-harvest level [6], [7], [13]. However, colonization of this pathogen in cattle is usually asymptomatic due to the lack of Shiga toxin receptor, globotriaosylceramide (Gb3), in cattle endothelial cells [14] that prevents elimination of super-shedding cattle contaminated with this pathogen at farms.

Several aspects influence the prevalence of E. coli O157 in cattle. The colonization of E. coli O157 at the rectal anal junction (RAJ) likely allows this pathogen to persist and shed high levels of bacteria for weeks or months [15], [16]. E. coli O157 primarily colonizes the mucosal epithelium at the RAJ [17], although it can be isolated from the gall bladder and along the gastrointestinal tract [18], [19]. More than 100 genes are involved in the colonization of the bovine intestine identified by genetic and biochemical analyses [20], including the E. coli O157 type III secretion system that enables the translocation of effector proteins into host cells and is required for colonization of cattle [21], [22], [23], [24], [25]. Besides the bacterial factors, environmental factors are believed to contribute to the prevalence of E. coli O157 in cattle. The prevalence of E. coli O157 in feces fluctuates by season with the peak between late spring and early fall [26], [27], [28]. Water, soil, wild animals, insects, and dirty equipment are important vectors for spreading and transmission of E. coli O157 [29], [30], [31], [32]. Transmission of E. coli O157 by environmental sources is one of the challenges to the development of pre-harvest interventions.

Even though similar E. coli O157 strains are flourishing in the same herds with identical cattle husbandry practices applied, a portion of animals are considered super-shedders [9], [12], suggesting that the phenomenon of super-shedders are controlled by multilayers of factors. However, underling mechanisms by which certain animals (2–5% in herds) become super-shedders are not clearly understood. In early studies of animal inoculation with E. coli O157, inoculated orally with water, some animals were never colonized with this pathogen, indicating that some animals were resistant to E. coli O157 [10], [32]. On the basis of these observations, we hypothesized that, in addition to bacterial factors, animals play a critical role in modulating the colonization of this pathogen in animals. This study was designed to address the impact of animal genetic factors on E. coli O157 prevalence, as well as animal husbandry. Here, we present our findings that genetic and physiological factors of animals and animal husbandry practices significantly affect the prevalence of E. coli O157, suggesting potential intervention practices to reduce this pathogen entering to the food production chain.

MATERIALS AND METHODS

Ethics Statement

Standard practices of animal care and use were applied to animals used in this project. Research protocols were approved by the University of Florida Institutional Animal Care and Use Committee (IACUC number 201003744).

Animal Genetic Background

Cattle belonged to the Angus-Brahman multibreed herd at the University of Florida. The herd was established in 1988 to conduct long-term genetic studies in beef cattle under subtropical environmental conditions. Cattle were assigned to six breed groups according to the following breed composition ranges: calf breed group 1=100% to 80% of Angus, 0% to 20% of Brahman; calf breed group 2=79% to 60% of Angus, 21% to 40% of Brahman; calf breed group 3=62.5% of Angus, 37.5% of Brahman, calf breed group 4=59% to 40% of Angus, 41% to 60% of Brahman, calf breed group 5=39% to 20% of Angus, 61% to 80% of Brahman, and calf breed group 6:19% to 0% of Angus, 81% to 100% of Brahman. Mating was diallel, i.e., sires from the 6 breed groups defined above were mated to dams of these same 6 breed groups [33]. Calves (n=91) were kept at the Beef Research Unit of the University of Florida before weaning and were moved to the University of Florida Feed Efficiency Facility (UFEF) after weaning.

Cattle Maintenance

Angus, Brahman, and Angus × Brahman crossbred steers (n=80, 253±38 kg) and bulls (n=11, 345±29 kg) were housed in the UFEF at the North Florida Research and Education Center in Marianna, Florida for 97 days

(From October 19, 2011 until January 24, 2012). Upon arrival (day 0) to the UFEF, cattle were fitted with electronic identification tags (Allflex USA Inc., Dallas-Fort Worth, TX) and were randomly allocated to 5 concrete floor pens of 108 m^2 each with wood shavings bedding for a total of 18 animals per pen with the exception of one pen which contained 19 animals. The 11 bulls were allocated to only 2 of the 5 pens (one pen with 5 and another pen with 6 bulls). Cattle had ad libitum access to feed and water at all times and intake was monitored continuously via a GrowSafe system (GrowSafe Systems Ltd., Airdrie, Alberta, Canada). Each pen contained two GrowSafe nods, thus the mean stocking rate per node was 9 cattle. Cattle received a diet comprised of (DM basis) 34% ground bahiagrass hay, 31% corn gluten feed pellets, 31% soybean hulls pellets, 4% of supplement containing molasses, urea, vitamin and minerals. The diet was formulated to have 14.7% crude protein and 0.97 Mcal of Net Energy of gain/kg of diet dry matter.

Detection and Enumeration of E. coli O157

The presence or absence of E. coli O157 in swabs of the RAJ was determined as previously described with minor modifications [10]. Rectal anal junction swab samples were collected from 91 cattle at the UFEF and a commercial feedlot. Samples were taken to the lab on ice within 4 h of collection to minimize bacterial growth and further tested for microbiological identification. Swab samples were resuspended in 2 ml of Tryptic soy broth (Difco) and serially diluted in 0.1% (w/v) peptone. Two hundred microliters of diluted samples (neat, 10^{-1}, 10^{-2}, 10^{-3}, and 10^{-4}) were then plated on MacConkey sorbitol agar (Difco) supplemented with cefixime (50 μg/liter; Lederle Labs, Pearl River, N.Y.) and potassium tellurite (2.5 mg/liter; Sigma) (CT-SMAC) to determine the number of E. coli O157 [34]. Plates were incubated at 37°C for 18–24 h and typical E. coli O157 colonies (i.e., sorbitol-negative colonies and multiplex PCR positive, described below) were enumerated. The minimum detection limit of the direct plating procedure was approximately 10 CFU/swab. Enrichment was used for the presence or absence determinations on samples that did not yield E. coli O157 by direct plating. For this purpose, samples were enriched in TSB supplemented with novobiocin (20 μg/ml; Sigma) for 18 to 24 h at 37°C with shaking, and E. coli O157 was detected by direct plating after serial dilution (neat, 10^{-1}, 10^{-2}, 10^{-3}, and 10^{-4}) with 0.1% (w/v) peptone. Sorbitol-negative colonies were confirmed by multiplex PCR.

Multiplex PCR

Multiplex PCR was conducted to confirm E. coli O157. Bacterial strains used in this paper are listed (Table 1). E. coli O157:H7 EDL933 and DH5α were used as a positive and negative control of multiplex PCR, respectively. Primers were designed to detect stx1, stx2, hly, and rbfE (Table 2). stx1 and stx2 primers detected subunit A of Stx1 and Stx2, respectively. Each PCR wells contained 25 μl of reaction mix, comprised of 2.5 μl of 10X buffer, 0.5 μl of dNTP, 1 μl of Taq polymerase and a mixture of the 8 primers. PCR cycling conditions were as follows: 94°C for 5 min for pre-denature, 94°C for 30 sec, 55°C for 30 sec, 72°C for 1 min each 30 cycles, and 72°C for 10 min for a final extension. PCR products were visualized on 1.5% agarose gel in Tris-EDTA buffer after electrophoresis.

Table 1. Strains used in this study.

Strain name	Description	references
E. coli O157:H7 EDL933	ATCC43895	[37]
E. coli DH5α	Lab collection	
KCJ1220	Isolated from medium-shedding animal in pen 17	This study
KCJ1225	Isolated from medium-shedding animal in pen 13	This study
KCJ1231	Isolated from super-shedding animal in pen 17	This study
KCJ1237	Isolated from medium-shedding animal in pen 15	This study
KCJ1238	Isolated from medium-shedding animal in pen 16	This study
KCJ1242	Isolated from low-shedding animal in pen 15	This study
KCJ1244	Isolated from low-shedding animal in pen 16	This study
KCJ1252	Isolated from super-shedding animal in pen 14	This study
KCJ1254	Isolated from medium-shedding animal in pen 14	This study
KCJ1265	Isolated from low-shedding animal in pen 13	This study
KCJ1266	Isolated from super-shedding animal in pen 13	This study
KCJ1268	Isolated from low-shedding animal in pen 14	This study
KCJ1430	Isolated from a commercial feedlot	This study
KCJ1432	Isolated from a commercial feedlot	This study

Table 2. Oligonucleotides used in this study.

Target	Name	Sequence (5′- 3′)	Orientation	Size (bp)
rfbE [a]	KCP57	CGGACATCCA TGTGATATGG	F	259
	KCP58	TTGCCTATGTA CAGCTAATCC	R	
stx1 [b]	KCP11	TGTCGCATAGTGGAACCTCA	F	655
	KCP12	TGCGCACTGAGAAGAAGAGA	R	
stx2 [b]	KCP13	CCATGACAACG GACAGCAGTT	F	477
	KCP14	TGTCGCCAGTTA TCTGACATTC	R	
hlyA [b]	KCP19	GCGAGCTAAGCAGCTTGAAT	F	199
	KCP20	CTGGAGGCTGCACTAACTCC	R	

[a]Referenced by Valadez et al. [38].

[b]Referenced by Bai et al. [39].

Subtyping of E. coli O157 Isolates Using Pulsed-field Gel Electrophoresis

Pulsed-Field Gel Electrophoresis (PFGE) was performed to subtype farm isolates in accordance with PulseNet standardized laboratory protocol. A colony purified on CT-SMAC was grown overnight in a shaker in Luria Broth (LB) at 37°C. Concentration of cell suspension was adjusted to an optical density of 1.0 at 600 nm. Cells (400 μl) were mixed with 20 μl of proteinase K (20 mg/ml stock, Fisher Scientific) and 1% Sekem Gold agarose (Lonza). The mixture was placed into a well of disposable plug molds (Bio-Rad Laboratories). Agarose plugs were lysed in cell lysis buffer (50 mM Tris, 50 mM EDTA, pH 8.0 and 1% Sarcosyl) for 2 h at 55°C with constant shaking at 170 rpm. Lysed plugs were washed one time with sterile distilled-water, followed by TE buffer (10 mM Tris, 1 mM EDTA, pH 8.0) twice at 55°C. Plugs were digested with 10 units of XbaI (New England BioLabs), and then electrophoresed using a CHEF Mapper (Bio-Rad Laboratories) under the following condition: 0.5 X Tris-Borate EDTA buffer at 14°C, 6 V/cm for 19 h, and an initial switch time from 2.16 s to 54.17 s. Lambda Ladder PFG Marker (New England BioLabs) was run as a size marker. The PFGE patterns were visualized by using GelDoc™ XR+ with Image Lab™ software (Bio-Rad Laboratories). The banding patterns were analyzed with GelCompar II software (Applied Maths, Kortrijk, Belgium).

Statistical Analysis

Statistical analyses were performed by GraphPad InStat version 3.10. Differences among pens were analyzed by the one-way analysis of variance (ANOVA) test followed by Tukey's test. Proportions of the super-shedders between bulls and steers were compared by Fisher's exact test. All data were expressed as mean ± standard error. A P value of <0.05 was considered significant.

RESULTS

Prevalence of E. coli O157 in the Multibreed Cattle

As a first step to understand the animal factors that may affect the prevalence of E. coli O157 in cattle, we enumerated this pathogen at the RAJ. A total of 91 animals were produced by the diallel mating system described previously [33] and assigned to six groups according to their estimated breed composition. As shown in figure 1A, calf breed group 1 contained the largest expected portion of genetic traits from Angus (100–80%) and the smallest expected portion from Brahman (0–20%). As the calf breed group numbers increase, the Angus portion decreases and Brahman portion increases. Thus, breed group 1 is more closely related to Angus while breed group 6 is more closely related to Brahman. Six different breed groups were randomly housed in 5 pens, but bulls were housed only in pen 1 and 2 to test if castration may affect the levels of this pathogen in the groups (Fig. 1B).

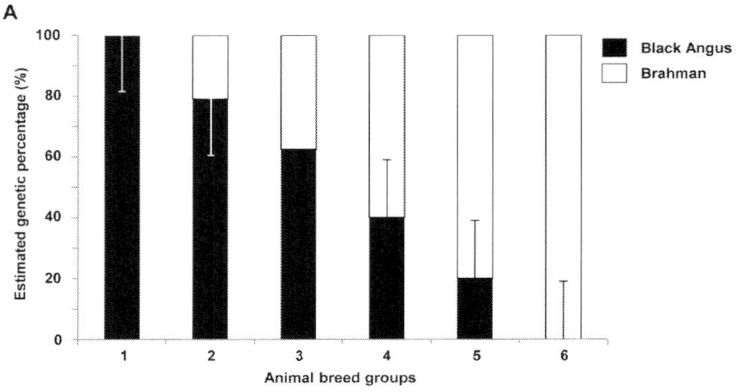

B

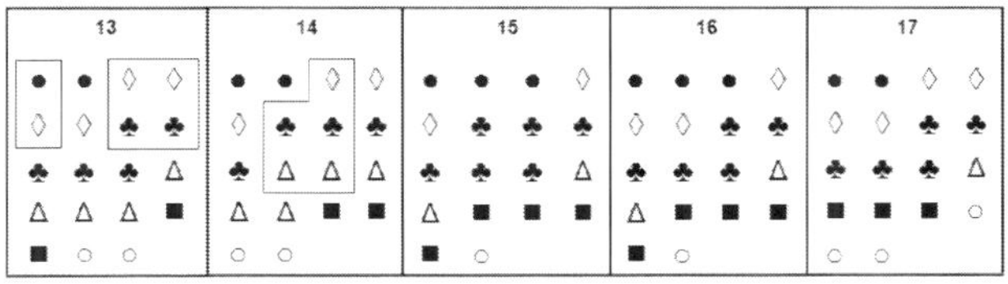

Figure 1. Estimated genetic backgrounds of animals used and housing in five pens. (A) Cattle consisted of six Angus-Brahman multibred groups. Group 1=100% to 80% Angus, 0% to 20% Brahman; Group 2=79% to 60% Angus, 21% to 40% Brahman; Group 3=62.5% Angus, 37.5% Brahman, Group 4=59% to 40% Angus, 41% to 60% Brahman, Group 5=39% to 20% Angus, 61% to 80% Brahman, Group 6:19% to 0% Angus, 81% to 100% Brahman. Bars indicate the expected portion of Angus and Brahman genetic traits in each breed group. (B) Animals were systemically housed in five pens according to their genetic background and sex. • is Calf breed group 1, ◊ is Calf breed group 2, ♣ is Calf breed group 3, Δ is Calf breed group 4, ▪ is Calf breed group 5, and ○ is Calf breed group 6. Bulls in pen 13 and 14 are boxed.

The total number of E. coli O157 from the RAJ swab samples was enumerated by a direct plating method without enrichment to monitor the real number of bacteria colonized on the RAJ. Swab samples were serial diluted before plated on CT-SMAC plate and incubated for 24 hours. Colonies with characteristic sorbitol negative color were picked and confirmed by using multiplex PCR. Out of 91 samples, 37 samples (40.66%) were found to be positive for E. coli O157 with a detection limit of 10 CFU/swab (Fig. 2). Samples that were negative for the direct plating method were enriched overnight followed by direct plating on CT-SMAC after serial dilution, but O157 was not detected from these samples (data not shown). The total number of E. coli O157 from the RAJ varied between animals, ranging from 10^1 to more than 10^6 (Fig. 2). The majority of positive samples contained this pathogen at 10^2–10^5 CFU/swab (n=31; 83.78% of positive samples) and 6 cattle (16.21% of positive samples) contained more than 10^5 CFU/swab (Table 3). Cattle were categorized into four groups depending on the number of E. coli O157 shedding, non-shedder ($<10^1$/swab; n=54; 59.34% of cattle), low-shedder (10^1–10^3 CFU/swab; n=16; 17.59% of cattle), medium shedder (10^3–10^5 CFU/swab, n=15; 16.48% of cattle), and super-shedder ($>10^5$ CFU, n=6; 6.6% of cattle). Previously, super-shedder was defined as an animal that excretes E. coli O157 at more than 10^4 CFU/g of feces. However, in this study we defined a super shedder at excretions of more than 10^5 CFU/swab because previous results showed that RAJ swab samples were 10 fold higher than fecal samples [10].

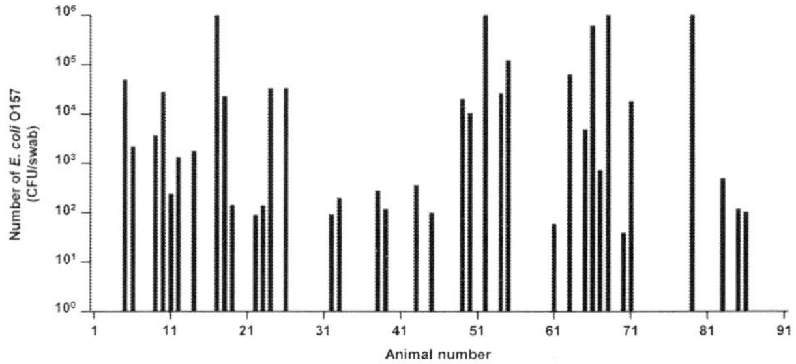

Figure 2. The number of E. coli O157 isolated from rectal anal junction in cattle.

Rectal swabs were enumerated by direct plating on CT-SMAC and further identified by multiplex PCR detecting the stx1, stx2, rbfE, and hlyA genes. Cattle were designate as low-shedders (10^1–10^2 CFU), medium-shedders (10^3–10^4 CFU), and super-shedders (>10^5 CFU). Six out of 91 cattle were super-shedders, accounting for 6.6% of all calves. Limit of detection was 10 CFU in this direct plating method.

Table 3. Distribution of cattle based on the level of E. coli **O157 counts.**

		Number of cattle (%)		
Type of shedder	CFU/swab	Total	Bull	Steer
Non-shedder	0–10^1	54 (59.34)	7 (63.64)	47 (58.75)
Low-shedder	10^1–10^2	4 (4.40)	0 (0)	4 (5.00)
	10^2–10^3	12 (13.19)	1 (9.09)	11 (13.75)
Medium-shedder	10^3–10^4	5 (5.49)	0 (0)	5 (6.25)
	10^4–10^5	10 (10.99)	0 (0)	10 (12.50)
Super-shedder	10^5–10^6	2 (2.20)	1 (9.09)	1 (1.25)
	>10^6	4 (4.40)	2 (18.18)	2 (2.50)

Presence of Super-shedders in Pens Increases the Total Number of Bacteria in Herds

It has been shown that the presence of super-shedders increases the prevalence of E. coli O157 in hides and carcass by enhancing transmission of this pathogen [6], [7], [13]. However, it is not well known if the super-shedders increase the total number of this pathogen in herds. We determined if there was a correlation between the presence of super-shedders and the level of E. coli O157 in different pens. Pen 13, 14, and 17 had 3, 2, and 1 super-shedder in the group of cattle, respectively (Fig. 3A), resulting in a higher number of E. coli O157 in the pens compared to the pens without super-shedders (pen 15 and 16). To understand the role of super-shedders in the transmission of E. coli O157, the total number of E. coli O157 bacteria excreted from super-shedders was calculated to determine what percentage of this pathogen was shed from super-shedders. E. coli O157 excreted from super-shedder accounted for more than 95% of total bacteria in the herds (Fig. 3B). These data confirm that super-shedders serve as a high risk factor for the transmission of this pathogen to other animals (Fig. 2 or Table 3).

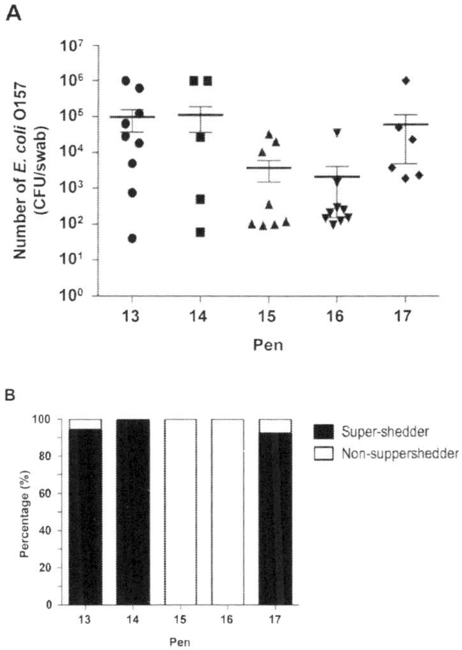

Figure 3. Super-shedders increase the total number of E. coli **O157 in pens with high density.** (A) The total number of E. coli O157 was calculated from individual animals in the five pens. Super-shedders increased the total number of E. coli O157 in the pens (pen 13, 14, and 17), and the average number of E. coli O157 was 100 fold less when a super-shedder was not identified in the pens (pen 15 and 16). The results are represented as mean ± SEM. (B) The percentage of E. coli O157 shed from super-shedders in the pens. Black bars represent the percentage of E. coli O157 isolated from super-shedders and white bars represent the percentage of E. coli O157 from non super-shedders.

In addition, it suggests that removing super-shedders in cattle herds can be an effective method to reduce the number of this pathogen in herds, thus identification of super-shedders is a critical control point to reduce potential outbreaks caused by this pathogen.

Bulls are More Susceptible to E. coli O157 than Steers

To determine whether castration may affect the prevalence of E. coli O157, the number of this pathogen was counted in bulls and steers at the RAJ. The number of super-shedders was significantly different (P=0.022) between bulls (27.27%, n=11) and steers (3.75%, n=80), indicating bulls are more susceptible to become super-shedders. However, it was remarkable that the prevalence of E. coli O157 in both cattle groups was similar at 36.36% and 41.25% respectively (Table 3). Taken together, even though steers and bulls are exposed to this pathogen in the same environment, probably by transmission from super-shedders, bulls are more likely to develop into super-shedders. This data suggest that steers may need to be separated from bulls to reduce total number of this pathogen in herds.

Brahman Cattle are More Resistant to E. coli O157

Results from previous inoculation studies of E. coli O157 with steers showed that certain cattle were resistant to this pathogen, despite repeated inoculation [32], suggesting animal factors were critical for colonization at the RAJ. We hypothesized that genetic factors play a significant role in the prevalence of E. coli O157 in animals. To assess the role of the genetic factors, we examined six different breed groups to identify if there is a correlation between breed groups and resistance to E. coli O157. The breed group 6 excreted the lowest number of E. coli O157 among the groups. As shown in Fig. 4A, the level of E. coli O157 was the lowest in breed group 6 compared to other groups, indicating that Brahman calves were more resistant to E. coli O157 than others. When calves were progeny of dams or sires from group 6, the calves had the lowest number of E. coli O157 compared to other groups (Fig. 4B and 4C). Notably, we could not observe a linear relationship between genetic composition and E. coli O157 resistance in calves, suggesting the resistance against pathogens may not be additive, but existing only in 100% Brahman cattle.

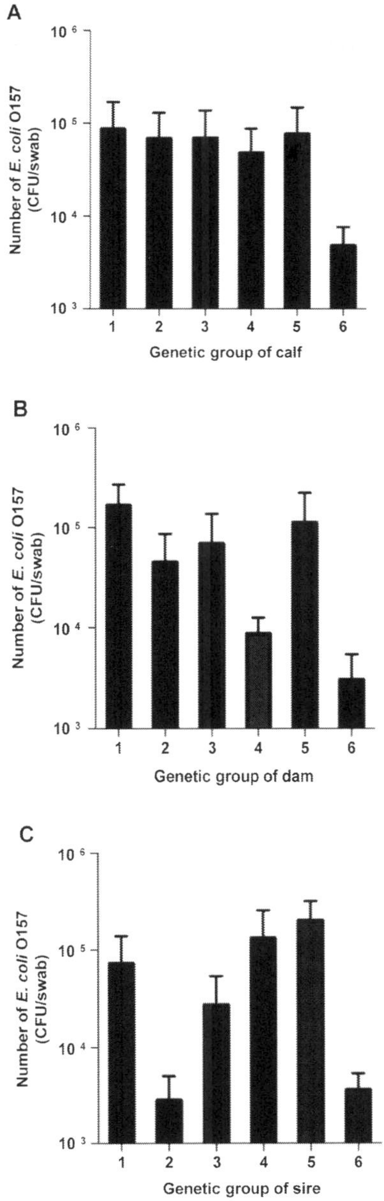

Figure 4. Brahmans are more resistant to E. coli O157 than other animals containing different genetic background. All calves (A), dams (B), and sires (C) were classified into the same breed groups. Overall, resistance to O157 was shown in breed group 6.

One Dominant E. coli O157 Strain was Prevalent Among Cattle

Although we observed that animal factors play key roles in determining the levels of E. coli O157 in cattle, we could not remove the possibility that a different level of this pathogen among cattle may have been mediated by different E. coli O157 strains. Bacterial factors are critical for prevalence and persistence of this pathogen in hosts; therefore, the different levels of E. coli O157 among cattle could have resulted by difference of bacterial strains rather than animal factors. Thus, we conducted molecular subtyping to eliminate the possibility that super-shedders carried a well-adapted E. coli O157 strain to cattle, while low shedders carried not-well-fitted strains. Strains isolated from low, medium, and super shedders of each pen were analyzed by multiplex PCR to compare strains. Multiplex PCR amplified the rfbE gene, which is specific for the O157 serotype, and three virulence genes, stx1, stx2, and hlyA. Unlike the EDL933 strain, all of the E. coli O157 isolates contained only stx2 and hlyA genes without stx1 (Fig. 5A). Furthermore, PFGE analysis with XbaI digestion identified only one type of PFGE pattern in the pens (Fig. 5B). Strains isolated from different cattle shedding low, medium, or super number of E. coli O157 displayed 100% similarity, indicating they are the same strains. Only one homogenous E. coli O157 strain was dominant among the cattle, indicating one specific strain containing only the stx2 gene was originated from one common source. These results support that bacterial factors were not major factors making super-shedding cattle in this experiment.

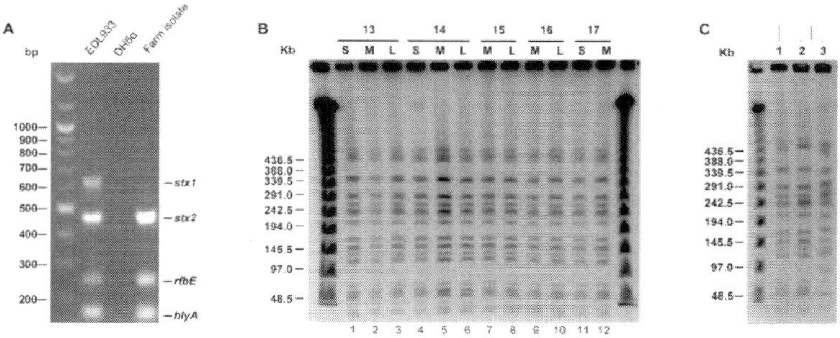

Figure 5. All of the E. coli O157 isolates were identical, indicating the strain originated from one common source. (A) Multiplex PCR was conducted using primers amplifying stx1, stx2, rbfE, and hlyA. Strains used were E. coli O157:H7 EDL933 (lane 1), DH5α (lane 2), and KCJ1266 (lane 3). (B) XbaI-digested

PFGE analysis. Strains used were isolated from the pens indicated on top of the line from animals shed low, medium, and super number of E. coli O157. Strains used; lane 1: KCJ1266, lane 2: KCJ1225, lane 3: KCJ1165, lane 4: KCJ1152, lane 5: KCJ1154, lane 6: KCJ1168, lane 7: KCJ1137, lane 8: KCJ1142, lane 9: KCJ1138, lane 10: KCJ1144, lane 11: KCJ1131, lane 12: KCJ1120 (C) The E. coli O157 strain isolated from animals raised at a commercial feedlot facility was probably transmitted from super-shedders. XbaI-digested PFGE analysis was conducted using the strains isolated from a super-shedder at NFREC and two animals at a commercial feedlot facility. Strains used; lane 1: KCJ1266, lane 2: KCJ1430, lane 3: KCJ1432.

Farm-to-farm Transmission of E. coli O157 via Animals

After the pen study at the NFREC, 80 steers were transported to a commercial feedlot X, and we traced the dominant E. coli O157 strain, which was first identified at the NFREC, to study if farm-to-farm transmission is mediated by cattle. After six months in feedlot X, we collected rectal swab samples (n=44) from cattle used in the pen study and additional rectal swab samples (n=19) were collected from random cattle raised in the feedlot. Swab samples were directly, without enrichment, plated on CT-SMAC plate and E. coli O157 strains were isolated. Sorbitol negative colonies were further identified by multiplex PCR and PFGE was conducted to evaluate similarity among strains. Unexpectedly, we identified only 2 steers that carried E. coli O157 strains, which were positive for rbfE, stx1, stx2, and hlyA. This is an unexpected result because normally the prevalence of E. coli O157 is high at feedlots compared to cattle on pasture. It is not known at this time point why the prevalence of E. coli O157 in feedlot X was very low compared to previous studies. However, we isolated the same E. coli O157 strain, which was dominant at NFREC, from the E. coli O157 positive cattle in feedlot X. As shown in Fig. 5C, one strain had 96.6% similarity, suggesting this strain probably originated from cattle transported from NFREC. It is not known at this time whether the strain originated from super-shedders or not, although it is likely that the strain originated from the super-shedders because they accounted for 90% of total bacteria at NFREC. It is noteworthy that we also isolated one strain showing 70.2% similarity compared to the dominant NFREC strain that may have originated from other sources, probably from cattle in feedlot X because it was not existing at NFREC. Therefore, these data strongly support our hypothesis that cattle transmit pathogens between farms and can be a critical control point to intervene E. coli O157 at the pre-harvest level.

DISCUSSION

We report here the role of animal and environmental factors in the prevalence of E. coli O157. In addition to the bacterial factors, animal genetic and physiological factors contribute to the prevalence of this pathogen in cattle. Brahman breed among the Angus-Brahman multibreed excreted the lowest level of E. coli O157, suggesting this breed is less prone to colonization of this pathogen. Bulls were more susceptible for colonization of this pathogen when compared with steers. These data indicate animal factors significantly contribute to the generation of super-shedders. The presence of super-shedders increased the number of bacteria in the pens and cattle. Super-shedders found in 6.6% of total cattle were responsible for more than 90% of the total number of pathogen in herd, increasing the chance of animal-to-animal and farm-to-farm transmission by either direct or indirect contact.

PFGE results from the commercial feedlot X isolates were in agreement with the data acquired from the pen isolates at the NFREC (Fig. 5C). The E. coli O157 strains isolated from the commercial feedlot X had high similarity, 96.6%, compared to the strains isolated from the NFREC dominant strain, indicating the commercial feedlot isolates were clonal variants of NFREC strain. Interestingly, we also isolated E. coli O157 showing lower similarity (70.2%) compared to the NFREC strains, indicating cattle likely acquired this strain from other cattle at the commercial feedlot X. Although it is plausible that this strain originated from NFREC, we believe that this strain was introduced from the commercial feedlot X. Taken together our data demonstrate the farm-to-farm transmission of E. coli O157 via cattle.

In addition to our data showing the transmission of E. coli O157 strains between farms, we observed a profound effect on animal-to-animal transmission of E. coli O157 strains in the pens. E. coli O157 is present in cattle at the prevalence ranging from 0% to 61% [3]; however, the prevalence of this pathogen at NEFREC farm was 40.66% (37 of 91 cattle). Furthermore, when we investigated the prevalence of E. coli O157 from grazing cattle at the same farm (i.e., NFREC), prevalence of this pathogen was 1.1% (1 out of 91 animals; Oh and Jeong's unpublished data). Although the high prevalence of E. coli O157 in the pens could be mediated by other risk factors such as transmission vectors (i.e., wild animals, insects, and soil) and diet, it is unlikely because the two groups of cattle were raised in the same environmental conditions, except cattle density (pen vs. pasture). Previous studies [10], [32] showed that contaminated water could be a major source for E. coli O157 contamination in the confined environment;

however, water was negative for E. coli O157 at NFREC (data not shown), eliminating the possibility that the cattle in the pans were contaminated via water. Thus, this unusual high prevalence of E. coli O157 was probably caused by the high density of cattle in the pens that may increase animal-to-animal transmission. Taken together cattle were probably the major source of E. coli O157 contamination via animal-to-animal transmission, and we suggest that maintaining cattle at the high density in the pens likely in part increased the prevalence of E. coli O157.

Based on an extensive E. coli O157 enumeration analysis, we have identified heterogeneous levels of this pathogen among cattle. Six cattle carried high level of E. coli O157 (>10^5 CFU/swab) that corresponds to the top 6.6% and 59.34% of cattle did not carry this pathogen even though they were housed in the same pens. Consistent with these data, previous research has shown that some cattle harbor and shed E. coli O157 at higher concentration than others [2], [6], [12]. In this study, we observed the role of super-shedders at high cattle density. Super-shedders were present in 3 pens, and the total number of bacteria in those pens was about 55 fold higher than pens without super-shedders. Super-shedders were responsible for more than 90% of the total number of pathogen in the pens. These data suggest that a super-shedder could increase not only the mean level of O157 among cattle but also the risk of E. coli O157 transmission to other cattle. Thus, identification and segregation of super-shedders from uninfected cattle may be a practical strategy to reduce the E. coli O157 prevalence in cattle and human disease.

Previous research has shown that bacterial factors may determine super-shedders by showing that diverse E. coli O157 strains were observed, but a few of them with particular phage types such as PT 21/28 are more likely to associate with super-shedding cattle via alteration in gene expression of E. coli O157 [9]. Unlike the previous findings, our data indicated that only one type of E. coli O157 strains was predominant among cattle in 37 cattle without bacterial strain preference (Fig. 5B), demonstrating that animal factors likely determine the colonization or shedding of this pathogen, but not bacterial factors.

We obtained multiple pieces of data indicating that the levels of E. coli O157 in cattle were modulated by multiple factors. These data include genetic factors (Fig. 4), physiological characteristics (i.e., castration, Table 3), and cattle density. Among the six breed groups examined, calves in the breed group 6 excreted the lowest number of E. coli O157. In addition, super-shedders were not identified in the group 6. As the breed group 6 was composed of Brahman and high percent Brahman cattle, it is reasonable to conclude that Brahman carry genetic factors that confer resistance against E. coli O157. Our findings are supported by a previous study where Riley et al. [35] found that 17.3% of Angus beef cattle (n=52) were positive for E. coli O157 while 10.1% of Brahman (n=109) beef cattle were positive with this pathogen by using an overnight enrichment method. The finding suggested that breed difference might affect the prevalence of E. coli O157 in the herd. Further studies regarding identification of genetic loci that confer resistance to E. coli O157 shedding and colonization will give us insights to develop intervention technologies to reduce E. coli O157 at the pre-harvest levels.

In addition to the genetic factors, castration decreased the susceptibility of male calves against this pathogen (Table 3). This was supported by data showing that 27.27% of bulls were super-shedders, whereas only 3.75% of steers were super-shedders. However, prevalence and colonization of E. coli O157 is probably influenced by many factors including rumen microflora [36], age [2], immune stress, the amount of feed intake, and unknown animal factors; thus, genetic variation and castration are likely associated with the prevalence of E. coli O157 directly or indirectly. Further studies, including identification of genetic loci, will be necessary to identify animal factors that directly govern the prevalence of pathogens. Our principal finding from this study is that animal factors, including genetic factors and castration, influence the prevalence of E. coli O157 in cattle. Super-shedders play a critical role in animal-to-animal and farm-to-farm transmission of E. coli O157. Identification of animal factors and controlling super-shedders will undoubtedly contribute to the development of intervention technologies to reduce outbreaks caused by this pathogen.

ACKNOWLEDGMENTS

We are grateful to Mara Brueck, Matthew Taylor, and Drs. Je Chul Lee, Won Sik Yeo, Man Hwan Oh, and Sung Cheon Hong for the technical assistance.

FUNDING STATEMENT

This work was supported by the Institute of Food and Agricultural Sciences, and Emerging Pathogens Institute at University of Florida, grants to KCJ. The funders had no role in study design, data collection and analysis, decision to publish, or preparation of the manuscript.

REFERENCES

1. Besser RE, Lett SM, Weber JT, Doyle MP, Barrett TJ, et al. (1993) An outbreak of diarrhea and hemolytic uremic syndrome from Escherichia coli O157:H7 in fresh-pressed apple cider. Jama 269: 2217–2220.
2. Cray WC Jr, Moon HW (1995) Experimental infection of calves and adult cattle with Escherichia coli O157:H7. Applied and environmental microbiology 61: 1586–1590.
3. Elder RO, Keen JE, Siragusa GR, Barkocy-Gallagher GA, Koohmaraie M, et al. (2000) Correlation of enterohemorrhagic Escherichia coli O157 prevalence in feces, hides, and carcasses of beef cattle during processing. Proc Natl Acad Sci U S A 97: 2999–3003.
4. Shere JA, Bartlett KJ, Kaspar CW (1998) Longitudinal study of Escherichia coli O157:H7 dissemination on four dairy farms in Wisconsin. Appl Environ Microbiol 64: 1390–1399.
5. Nataro JP, Kaper JB (1998) Diarrheagenic Escherichia coli . Clinical microbiology reviews 11: 142–201.
6. Arthur TM, Brichta-Harhay DM, Bosilevac JM, Kalchayanand N, Shackelford SD, et al. (2010) Super shedding of Escherichia coli O157:H7 by cattle and the impact on beef carcass contamination. Meat science 86: 32–37.
7. Arthur TM, Nou X, Kalchayanand N, Bosilevac JM, Wheeler T, et al. (2011) Survival of Escherichia coli O157:H7 on cattle hides. Applied and environmental microbiology 77: 3002–3008.
8. Callaway TR, Carr MA, Edrington TS, Anderson RC, Nisbet DJ (2009) Diet, Escherichia coli O157:H7, and cattle: a review after 10 years. Curr Issues Mol Biol 11: 67–79.
9. Chase-Topping ME, McKendrick IJ, Pearce MC, MacDonald P, Matthews L, et al. (2007) Risk factors for the presence of high-level shedders of Escherichia coli O157 on Scottish farms. J Clin Microbiol 45: 1594–1603.
10. Jeong KC, Kang MY, Kang J, Baumler DJ, Kaspar CW (2011) Reduction of Escherichia coli O157:H7 shedding in cattle by addition of chitosan microparticles to feed. Applied and environmental microbiology 77: 2611–2616.
11. Rangel JM, Sparling PH, Crowe C, Griffin PM, Swerdlow DL (2005) Epidemiology of Escherichia coli O157:H7 outbreaks, United States, 1982–2002. Emerging infectious diseases 11: 603–609.

12. Matthews L, Low JC, Gally DL, Pearce MC, Mellor DJ, et al. (2006) Heterogeneous shedding of Escherichia coli O157 in cattle and its implications for control. Proceedings of the National Academy of Sciences of the United States of America 103: 547–552.
13. Arthur TM, Keen JE, Bosilevac JM, Brichta-Harhay DM, Kalchayanand N, et al. (2009) Longitudinal study of Escherichia coli O157:H7 in a beef cattle feedlot and role of high-level shedders in hide contamination. Applied and environmental microbiology 75: 6515–6523.
14. Lingwood CA (1996) Role of verotoxin receptors in pathogenesis. Trends in microbiology 4: 147–153.
15. Lim JY, Li J, Sheng H, Besser TE, Potter K, et al. (2007) Escherichia coli O157:H7 colonization at the rectoanal junction of long-duration culture-positive cattle. Appl Environ Microbiol 73: 1380–1382.
16. Davis MA, Rice DH, Sheng H, Hancock DD, Besser TE, et al. (2006) Comparison of cultures from rectoanal-junction mucosal swabs and feces for detection of Escherichia coli O157 in dairy heifers. Appl Environ Microbiol 72: 3766–3770.
17. Naylor SW, Low JC, Besser TE, Mahajan A, Gunn GJ, et al. (2003) Lymphoid follicle-dense mucosa at the terminal rectum is the principal site of colonization of enterohemorrhagic Escherichia coli O157:H7 in the bovine host. Infection and immunity 71: 1505–1512.
18. Jeong KC, Kang MY, Heimke C, Shere JA, Erol I, et al. (2007) Isolation of Escherichia coli O157:H7 from the gall bladder of inoculated and naturally-infected cattle. Vet Microbiol 119: 339–345.
19. Keen JE, Laegreid WW, Chitko-McKown CG, Durso LM, Bono JL (2010) Distribution of Shiga-toxigenic Escherichia coli O157 in the gastrointestinal tract of naturally O157-shedding cattle at necropsy. Applied and environmental microbiology 76: 5278–5281.
20. Dziva F, van Diemen PM, Stevens MP, Smith AJ, Wallis TS (2004) Identification of Escherichia coli O157 : H7 genes influencing colonization of the bovine gastrointestinal tract using signature-tagged mutagenesis. Microbiology 150: 3631–3645.
21. Abe A, Heczko U, Hegele RG, Brett Finlay B (1998) Two enteropathogenic Escherichia coli type III secreted proteins, EspA and EspB, are virulence factors. The Journal of experimental medicine 188: 1907–1916.

22. Bretschneider G, Berberov EM, Moxley RA (2007) Reduced intestinal colonization of adult beef cattle by Escherichia coli O157:H7 tir deletion and nalidixic-acid-resistant mutants lacking flagellar expression. Veterinary microbiology 125: 381–386.
23. Dean-Nystrom EA, Bosworth BT, Moon HW, O'Brien AD (1998) Escherichia coli O157:H7 requires intimin for enteropathogenicity in calves. Infection and immunity 66: 4560–4563.
24. Sharma VK, Sacco RE, Kunkle RA, Bearson SM, Palmquist DE (2012) Correlating levels of type III secretion and secreted proteins with fecal shedding of Escherichia coli O157:H7 in cattle. Infection and immunity 80: 1333–1342.
25. Sheng H, Lim JY, Knecht HJ, Li J, Hovde CJ (2006) Role of Escherichia coli O157:H7 virulence factors in colonization at the bovine terminal rectal mucosa. Infection and immunity 74: 4685–4693.
26. Barkocy-Gallagher GA, Arthur TM, Rivera-Betancourt M, Nou X, Shackelford SD, et al. (2003) Seasonal prevalence of Shiga toxin-producing Escherichia coli, including O157:H7 and non-O157 serotypes, and Salmonella in commercial beef processing plants. J Food Prot 66: 1978–1986.
27. Laegreid WW, Elder RO, Keen JE (1999) Prevalence of Escherichia coli O157:H7 in range beef calves at weaning. Epidemiol Infect 123: 291–298.
28. Heuvelink AE, van den Biggelaar FL, de Boer E, Herbes RG, Melchers WJ, et al. (1998) Isolation and characterization of verocytotoxin-producing Escherichia coli O157 strains from Dutch cattle and sheep. J Clin Microbiol 36: 878–882.
29. Ferens WA, Hovde CJ (2011) Escherichia coli O157:H7: animal reservoir and sources of human infection. Foodborne pathogens and disease 8: 465–487.
30. Franz E, Semenov AV, van Bruggen AH (2008) Modelling the contamination of lettuce with Escherichia coli O157:H7 from manure-amended soil and the effect of intervention strategies. Journal of applied microbiology 105: 1569–1584.
31. Franz E, van Bruggen AH (2008) Ecology of E. coli O157:H7 and Salmonella enterica in the primary vegetable production chain. Critical reviews in microbiology 34: 143–161.

32. Shere JA, Kaspar CW, Bartlett KJ, Linden SE, Norell B, et al. (2002) Shedding of Escherichia coli O157:H7 in dairy cattle housed in a confined environment following waterborne inoculation. Appl Environ Microbiol 68: 1947–1954.
33. Elzo MA, West RL, Johnson DD, Wakeman DL (1998) Genetic variation and prediction of additive and nonadditive genetic effects for six carcass traits in an Angus-Brahman multibreed herd. J Anim Sci 76: 1810–1823.
34. Zadik PM, Chapman PA, Siddons CA (1993) Use of tellurite for the selection of verocytotoxigenic Escherichia coli O157. J Med Microbiol 39: 155–158.
35. Riley DG, Gray JT, Loneragan GH, Barling KS, Chase CC Jr (2003) Escherichia coli O157:H7 prevalence in fecal samples of cattle from a southeastern beef cow-calf herd. Journal of food protection 66: 1778–1782.
36. Zhao T, Doyle MP, Harmon BG, Brown CA, Mueller PO, et al. (1998) Reduction of carriage of enterohemorrhagic Escherichia coli O157:H7 in cattle by inoculation with probiotic bacteria. Journal of clinical microbiology 36: 641–647.
37. Perna NT, Plunkett G, 3rd, Burland V, Mau B, Glasner JD, et al (2001) Genome sequence of enterohaemorrhagic Escherichia coli O157:H7. Nature 409: 529–533.
38. Valadez AM, Debroy C, Dudley E, Cutter CN (2011) Multiplex PCR detection of Shiga toxin-producing Escherichia coli strains belonging to serogroups O157, O103, O91, O113, O145, O111, and O26 experimentally inoculated in beef carcass swabs, beef trim, and ground beef. Journal of food protection 74: 228–239.
39. Bai J, Shi X, Nagaraja TG (2010) A multiplex PCR procedure for the detection of six major virulence genes in Escherichia coli O157:H7. Journal of microbiological methods 82: 85–89.

CHAPTER 6

Assessment of Heat Stress in 7-week Old Dairy Calves with Non-invasive Physiological Parameters in different Thermal Environments

Levente Kova´cs[1,2], Fruzsina Luca Ke´ze´r [1,2], Ferenc Ruff[3], Viktor Jurkovich[4], Otto´ Szenci[1]

[1]MTA–SZIE Large Animal Clinical Research Group, U¨ llő-Do´ra major, Hungary,
[2]Institute of Animal Husbandry, Faculty of Agricultural and Environmental Science, Szent Istva´n University, Go¨do¨llő, Hungary,
[3]Department of Methodology, Hungarian Central Statistical Office, Budapest, Hungary,
[4]Department of Animal Hygiene, Herd Health and Veterinary Ethology, University of Veterinary Medicine, Budapest, Hungary

ABSTRACT

We estimated thermal stress in 7-week old Holstein bull calves during a warm episode in summer to study acute physiological responses of calves to

Citation: Kovács L, Kézér FL, Ruff F, Jurkovich V, Szenci O (2018) Assessment of heat stress in 7-week old dairy calves with non-invasive physiological parameters in different thermal environments. PLoS ONE 13(7): e0200622. https://doi.org/10.1371/journal.pone.0200622

heat stress. Data were collected over a 5-day period: day 1 (control), day 2 (heat stress), and a 3-day post-stress period in shaded ($n = 8$) and unshaded ($n = 8$) thermal environments. On the control day, both groups were shaded. Thermal environment was characterized by relative humidity, ambient temperature, and the temperature–humidity index (THI). Physiological variables included respiratory rate, rectal temperature, ear skin temperature and heart rate. Correlations between animal-based and meteorological indices were calculated, and ambient temperature correlated slightly better with physiological measures than THI. Rectal temperature was the only animal-based parameter that showed stronger correlations with the thermal indices when calculated for the shaded than for the unshaded environment [$r = 0.42$ vs. $r = 0.47$, $P = 0.032$ (ambient temperature), $r = -0.39$ vs. $r = -0.45$, $P = 0.012$ $P = 0.015$ (relative humidity), $r = 0.41$ vs. $r = 0.46$, $P = 0.022$ (THI)]. No differences were found between groups during the control day for any of the physiological parameters. During days 2 and 3, average and maximal values of respiratory and heart rates were higher in unshaded calves than in shaded ones. Maximal respiratory rates were in average by 25.9, 17.8 and 10.1 breaths/min lower in shaded calves than in unshaded calves for days 2, 3 and 4, respectively ($P < 0.001$, $P < 0.001$ and $P = 0.024$). Maximal heart rate was 127.4 ± 8.5 vs. 99.2 ± 6.3 beats/min on the heat stress day ($P < 0.001$), and 121.0 ± 6.9 vs. 103.4 ± 7.7 beats/min on day 3 ($P = 0.006$) in unshaded and shaded calves, respectively. Maximal body temperatures were higher measured either in the rectum or on the ear skin in unshaded calves than in shaded ones (with 0.5 and 1.6°C, $P = 0.040$ and $P = 0.018$, respectively), but only on the heat stress day. Based on our results, shading of young calves may be adequate for alleviating acute heat stress in continental regions. Ambient temperature is appropriate to estimate acute heat stress in dairy calves.

INTRODUCTION

A persistent rise in global mean temperature has severe impacts on natural and societal systems [1]. It has been shown that the increased tendency of heat-related mortality causes a considerable economic loss for the dairy industry in regions where high environmental temperatures represent a concern [2, 3]. Although animal-friendly housing technologies have been introduced to alleviate weather-related heat exposure in dairy calves [4], young animals are commonly kept in individual polyethylene hutches with a limited outdoor space. Although these pens allow calves freedom to select a comfortable resting place other than in the hutch, animals are exposed to direct sunlight and are at risk during warm episodes of summer.

Based on ambient temperature and relative humidity, thermos-hygrometric indices are used for the assessment of the thermal environment in cattle. The temperature-humidity index (THI) represents the combined effects of air temperature and relative humidity to estimate the level of thermal stress in adult cows [5].

In dairy [4, 6] and beef calves [7], the effects of thermal stress can be assessed using rectal temperature and respiratory rate. It is not known if there are physiological measures other than rectal temperature or respiratory rate indicative for acute heat stress. In this study, we monitored a typical warm episode that occurs more and more frequently in Central Europe during summer to assess acute physiological responses to heat stress of dairy calves in shaded and unshaded environments. We measured rectal temperature, respiratory rate, heart rate and ear skin temperature to assess the associations between physiological indices, ambient temperature, relative humidity and THI. We hypothesized that respiratory rate, ear skin temperature and heart rate will show higher correlations with meteorological parameters than rectal temperature, and correlations between animal-based and meteorological measures will differ based on the thermal environment (shaded vs. unshaded). We also hypothesized lower acute stress responses of calves kept in shaded hutches compared to non-shaded ones.

MATERIALS AND METHODS

All sampling procedures and experimental manipulations with animals were specifically approved by the Pest County Government Office, Hungary, Department of Animal Health (Permit Number: PE/EA/1973-6/2016).

Animals and experimental design

The experiment was carried out at a large-scale dairy farm in Hungary (N47°18’191” E18°48’336”), which has a herd of 1000 lactating Holstein cows. The farm was visited between August 15 and 20 in 2016. As a field study, our experiment was timed based on the expected meteorological patterns of the regional weather forecast. The 6-day study started with a 24-h habituation to the study environment when preparation of the shading structure over the hutches of both shaded (n = 8) and unshaded calves (n = 8) were done. On day 1 (control, maximum of 28.3°C for the hutch environment) all calves were shaded for 24-h from 0:00 o’clock until 24:00

o'clock and shade was removed from the unshaded group at 24:00 o'clock. Day 2 was considered as the heat stress day (maximum of 37.7°C for the shaded hutch environment), and days 3–5 covered a post-stress period (day 3: maximum of 30.3°C, day 4: maximum of 26.5°C, and day 5 maximum of 24.3°C for the shaded hutch environment).

On the study farm, Holstein calves were housed individually in 1.65 × 1.20 m plastic calf hutches (Calf-Tel ECO, Hampel Animal Care, WI, USA) with a 1.60-m^2 exercise pen that were arranged in rows based on their sex (heifer row and bull row) and birth date. To eliminate the possible effect of individual-related (i.e. sex, age and body weight) and environmental factors (i.e. row orientation) on our results, bull calves with similar age and body weight (means ± SD; age = 46.7 ± 2.4 days, body weight = 74.3 ± 2.6 kg) one week before weaning from the same row were selected for the study and assigned to the shaded or unshaded groups. The shading structure measured 32.5 × 3.4 m and covered the experimental hutches and pens as well. After removing the shading structure above unshaded calves, their hutches were exposed to direct sunlight 5 m from the shaded environment. A green raschel net (Nortene Texanet, Celloplast S.A.S, Ballée, France), located 1.9 m above the ground was used as shading material that provided 80% shading, according to the manufacturer. Calves received one feeding of 3.8 liters of milk replacer in the morning (5:00 o'clock) and had ad libitum access to chopped alfalfa hay and the starter grain diet (Purina calf starter, Cargill, USA), which met the requirements for preweaned Holstein calves [8]. Water from a plastic bucket (7.6 liters), filled twice a day, was provided throughout the study. The diet did not change throughout the experiment.

Meteorological Data

Ambient temperature and relative humidity were assessed for each calf by using the VOLTCRAFT DL-181THP device (Conrad Electronic SE, Hirschau, Germany) placed in the back of the hutches and with the Testo 175 H1 (Testo Inc., Sparta, USA) device that were fitted onto the shading structure 1 m above the ground above the exercise pens. A 30-min recording frequency was chosen for both devices between day 1 0:00 o'clock (first recording) and day 5 24:00 o'clock (last recording). The position of the calves was retrospectively determined based on video recordings of two day/night outdoor network bullet cameras (Vivotek IP8331, VIVOTEK Inc., Taiwan) installed above the experimental area to allow the selection of the appropriate temperature and humidity values (recorded in the hutch or in the exercise pen) for the calculation of meteorological metrics. The thermal environment of the calves was characterized by temperature, relative humidity and THI. The THI was calculated by the formula described by Bianca et al. [9]:

$$THI = (0.35 \times Tdb + 0.65 \times Twb) \times 1.8 + 32$$

where T_{db} = dry bulb temperature and T_{wb} = wet bulb temperature.

Physiological Data

Respiratory rate (breaths/min) was recorded by counting the movements of the abdominal muscles in the flanks during respiration [4], while calves were in a lying posture between day 1 0:00 o'clock (first observation) and day 5 24:00 o'clock (last observation) with a 4-h sampling frequency. Immediately after respiratory rate observation, rectal temperature was measured with a 10-sec digital thermometer (Digi-Vet SC 12; Jørgen Kruuse A/S, Langeskov, Denmark). Ear skin temperature was measured parallel with rectal temperature using the Testo 830 T2 infrared thermometer (Testo Inc., Sparta, USA).

Heart rate was recorded continuously between day 1 0:00 o'clock and day 5 24:00 o'clock using a Polar electrode belt with two integrated electrodes, a compatible Polar H7 HR sensor and a Polar V800 HR receiver (POLAR, Kempele, Finland). Devices were fitted to animals on day 0 and were removed from the calves after day 5 then heart rate data were transmitted to the Polar FlowSych program. The analysis of heart rate was performed with the Kubios HRV standard software (version 2.2, Biomedical Signal Analysis Group, Department of Applied Physics, University of Kuopio, Finland) using equal lengths of 5-min samples recorded for undisturbed lying posture of the animals. Posture was recorded using the HOBO Pendant G data logger (Onset Computer Corporation, Bourne, MA), which was validated for dairy calves [10].

Statistical Evaluation

All statistical analyses were performed in the R–3.0.2. statistical environment and language [11]. Data were tested for constant variance (Levene's test) and the Shapiro–Wilk test was used for testing the equality of error variances both for shaded and unshaded groups.

For the evaluation of the associations between animal-based heat stress indicators and thermal parameters, data recorded from both shaded and unshaded calves were used. Thermal indices were calculated from meteorological data recorded by devices fitted in the hutch or in the pen based on the location of the calves. Pearson correlations were calculated based on individual observations between day 1 0:00 o'clock and day 5 24:00 o'clock [for heart rate with 30-min sampling frequency (n = 3840); for respiratory rate, rectal and ear skin temperature with 4-h sampling frequency (n = 480)]. We tested whether a difference between correlation coefficients exists according to the environment (shaded or unshaded). For this purpose, the Fisher-type *z*-transformation based *z*-test was used at the significance level of 0.05.

To evaluate the effect of shading on the thermal environment of the animals, maximal and average values of selected thermal indices were calculated for all experimental days and compared between unshaded and shaded groups in the hutch environment, using all recorded meteorological data with 30-min sampling frequency. Statistical comparisons between environments were made with the *t*-test at the significance level of 0.05.

The effect of shading was tested by comparing physiological responses of shaded and unshaded groups by using the individuals' maximum and average values of respiratory rate, heart rate, ear skin and rectal temperatures calculated for each experimental day. All recorded animal-based data were used in this analysis, irrespective of the location of the calves (measured either in the hutch or the pen). Statistical comparisons between groups were made with the *t*-test. Significance was set at the level of 0.05.

RESULTS AND DISCUSSION

Relationships between thermal and physiological measures

Generally, animal-based heat stress indicators showed strong positive correlation with ambient temperature and THI (**Figs 1** and **2**). The THI formula used in our study was found to be appropriate for continental weather characteristics (low daytime relative humidity) in dairy cows [12,13], and in this formula relative humidity (wet bulb temperature) is weighted to a lower degree. This could be the explanation of the similar (or slightly higher in cases of ear skin temperature and heart rate) correlations of ambient temperature with physiological indices as THI in our study. Based on these findings, the use of simple temperature can be suggested for the estimation of acute heat stress experienced by dairy calves.

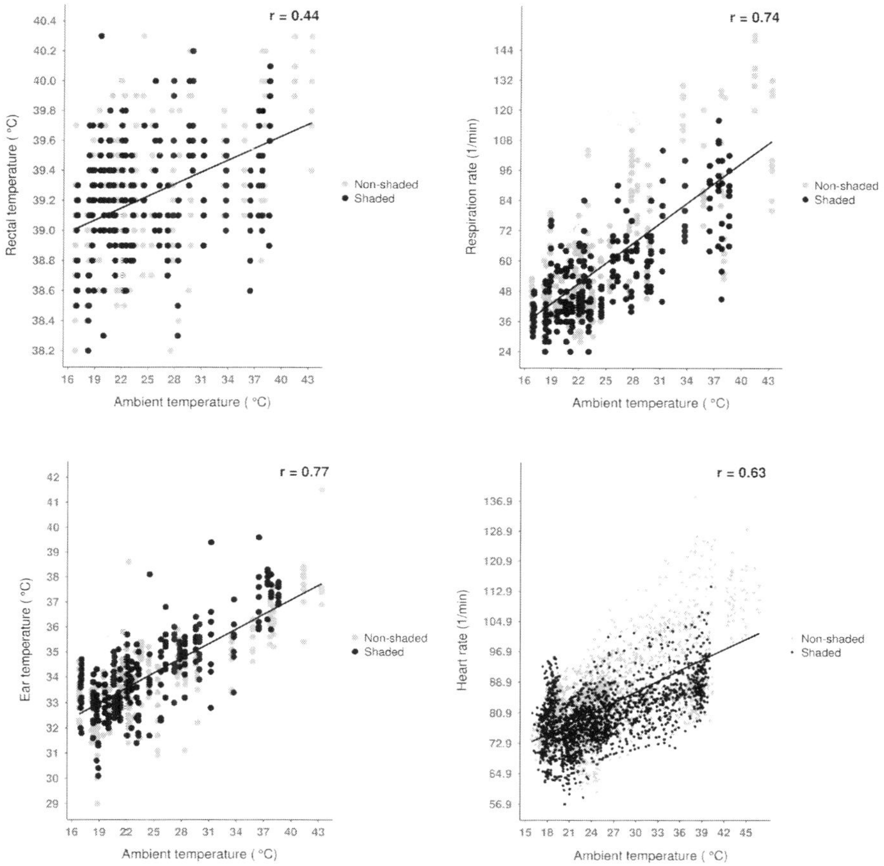

Figure 1. Relationship between rectal temperature, respiratory rate, ear temperature, heart rate and ambient temperature in dairy calves.

Points represent individual observations (n = 480 for rectal temperature, ear temperature and respiratory rate; n = 3840 for heart rate) between day 1 0:00 o'clock and day 5 24:00 0'clock; lines represent simple linear regression equations, and 'r' represents the correlation coefficient.

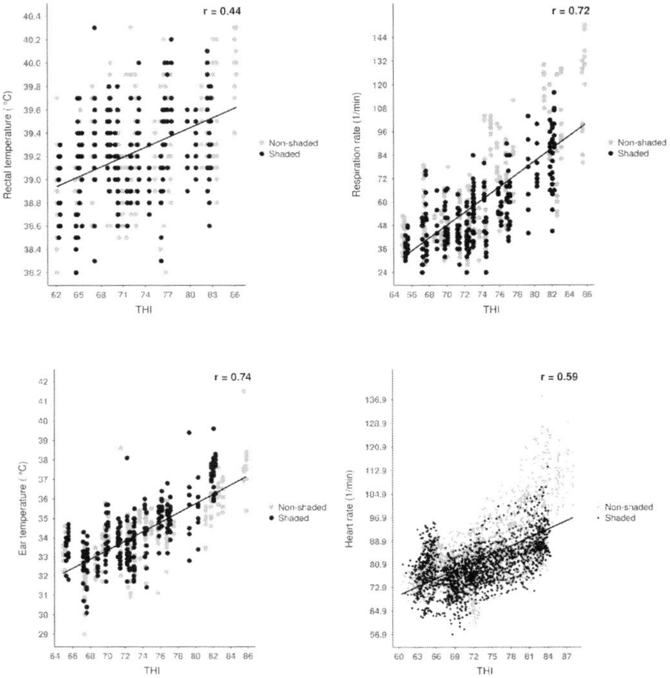

Figure 2. Relationship between rectal temperature, respiratory rate, ear temperature, heart rate and relative humidity in dairy calves.

Points represent individual observations (n = 480 for rectal temperature, ear temperature and respiratory rate; n = 3840 for heart rate) between day 1 0:00 o'clock and day 5 24:00 0'clock; lines represent simple linear regression equations, and 'r' represents the correlation coefficient.

Animal-based measures showed a negative correlation with relative humidity (**Fig 3**). During the experimental period, relative humidity and ambient temperature showed a converse pattern with increasing relative humidity and decreasing ambient temperature during the nighttime and increasing ambient temperature during the daytime (see **Fig 4A** and **4B** later). Thus, the negative relationship between relative humidity and animal-based indices assuredly mirrored the diurnal changes of ambient temperature as well not the direct effect of relative humidity itself. Dikmen and Hansen [5] found also negative correlation (r = –0.38) between rectal temperature and relative humidity in dairy cows. Effects of relative humidity on physiological measures should be estimated using climate-controlled chambers, where constant temperature and fluctuating humidity values could be set up.

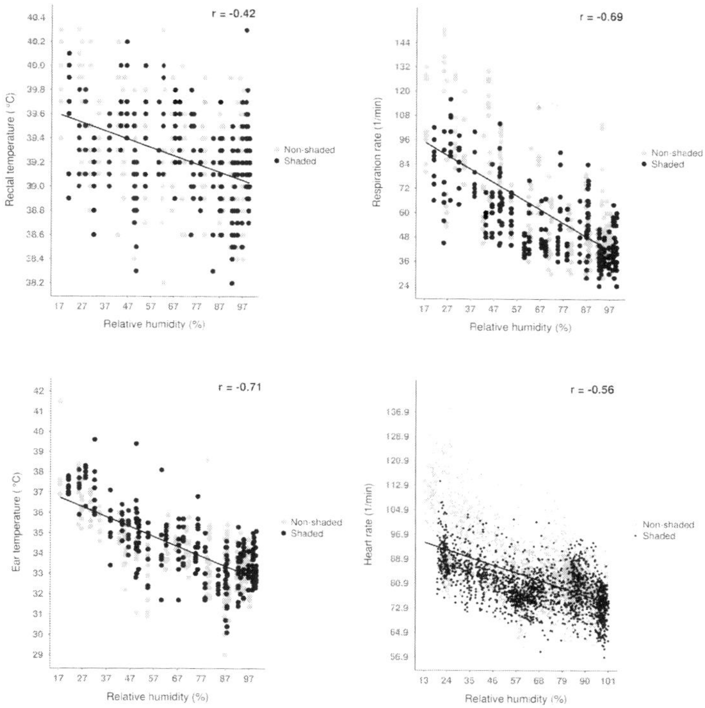

Figure 3. Relationship between rectal temperature, respiratory rate, ear temperature, heart rate and THI in dairy calves.

Points represent individual observations (n = 480 for rectal temperature, ear temperature and respiratory rate; n = 3840 for heart rate) between day 1 0:00 o'clock and day 5 24:00 0'clock; lines represent simple linear regression equations, and 'r' represents the correlation coefficient.

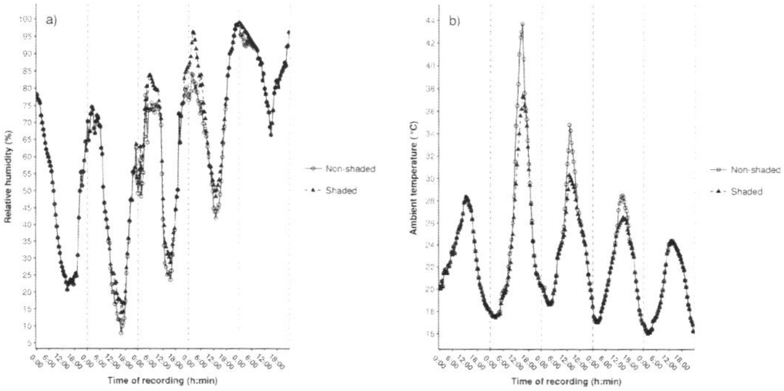

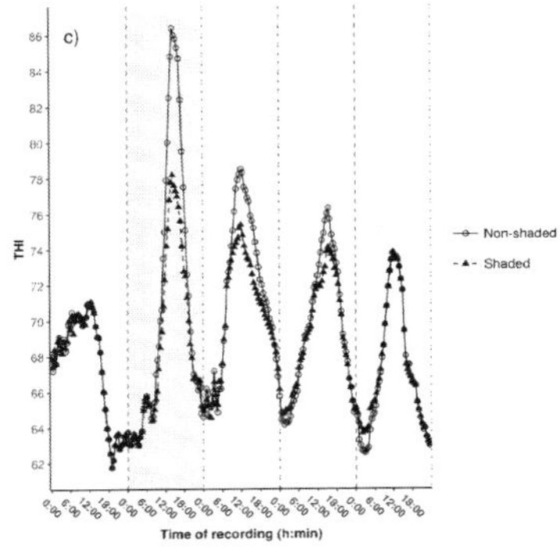

Figure 4. Changes in relative humidity (a), ambient temperature (b) and THI (c) in the shaded and unshaded hutch environment during the 5-day study period. Data are presented with a 30-min recording interval. The grey area between the first and the second dashed vertical lines represents day 2 (heat stress day).

The only correlation study that investigated temperature indices with respect to hyperthermia used only rectal temperature as an animal-based variable in adult cows [5]. A small number of studies on calves that measured rectal temperatures or respiratory rates used THI as environmental variable. It is also important to note, that data for the current study were collected in a continental region, in which low daytime humidity is a prevailing characteristic, while the referred experiments were done in tropical [14,15] or humid subtropical environments [4], therefore, comparison of the present results with earlier data is difficult.

It was hypothesized that rectal temperature would less correlate with meteorological parameters than other physiological measures. Our hypotheses were supported, since when data of both shaded and unshaded environments were assessed, rectal temperature showed medium strong, while other physiological parameters showed strong correlations with meteorological measures (**Figs 1–3**). This suggests that respiratory rate, heart rate or ear skin temperature could be more appropriate to estimate acute heat stress than rectal temperature in dairy calves. Our second hypothesis in that thermal environment would affect correlations between physiological and meteorological measures was partly supported. Interestingly, rectal temperature was the only animal-based parameter that showed stronger correlations with the thermal indices when calculated for the unshaded than for the shaded environment [r = 0.42 vs. r = 0.47, $P = 0.032$ (ambient temperature), r = –0.39 vs. r = –0.45, $P = 0.012$ $P = 0.015$ (relative humidity), r = 0.41 vs. r = 0.46, $P = 0.022$ (THI)]. Any other correlations showed no significant differences between shaded and unshaded environments ($P > 0.05$ for all cases). An earlier study reported an r = 0.52 correlation between THI and rectal temperature in dairy cows both recorded between 15:00 and 17:00 o'clock, under high ambient temperatures [5]. In the light of the authors' results, our findings point out that the associations between rectal temperatures and environmental heat stress indices could be stronger with higher heat load, which should be considered in future studies on calf thermal comfort.

Effect of shading on the thermal environment

Changes in relative humidity, ambient temperature and THI in the shaded and unshaded hutch environments over the study period are presented in **Fig 4A–4C**.

There were no statistical differences either for maximal or average values of relative humidity between shaded and unshaded environments during the experimental period. On day 1 (control) no differences were found in maximal and daily average values either for ambient temperature ($P = 0.875$ and $P = 0.920$) or THI ($P = 0.930$ and $P = 0.945$). Measurements between day 2 0:00 o'clock and day 4 24:00 o'clock proved the positive effect of shading on the thermal environment. On day 2, maximal temperature and THI were observed at 16:00 o'clock in the unshaded hutch environment that were higher than those recorded for the shaded one [43.7 ± 0.1 vs. 37.7 ± 0.1°C; $P = 0.005$ (**Fig 4B**) and 86.4 ± 0.1 vs. 78.2 ± 0.1; $P < 0.001$ (**Fig 4C**), respectively]. Reduced heat load in the shaded hutch environment compared to the unshaded one was reflected by daily averages of temperature as well on the heat stress day (28.5 ± 0.1 vs. 24.2 ± 0.1°C; $P < 0.001$) and THI (78.1 ± 0.1 vs. 71.3 ± 0.1; $P = 0.011$). Earlier studies have shown that shading reduces hutch temperature by around 1 to 2°C [6, 13]; however, in these experiments temperature was measured only at 15:00 o'clock. Maximums and daily averages of ambient temperature were lower in the shaded environment compared to the unshaded one on days 3 ($P < 0.001$ and $P = 0.005$) and 4 ($P = 0.008$ and $P = 0.012$), respectively. Maximal and average THI showed similar differences between the shaded and unshaded environments for the first two days of the post-stress period (day 3: $P = 0.003$ and $P = 0.014$; day 4: $P = 0.015$ and $P = 0.026$). On day 5 no differences were found in maximal and daily average values either for ambient temperature ($P = 0.395$ and $P = 0.420$) or THI ($P = 0.526$ and $P = 0.545$).

Effect of shading on physiological parameters

Recent studies on heat stress of dairy calves tested different housing methods [4, 16] or compared the stress-reducing efficiency of artificial vs. natural shade [14]; all using only rectal temperature and respiratory rate as indicators of heat stress. Here, the analysis was focused on ear temperature and heart rate as well that could be measured non-invasively and inexpensively both having the potential to assess heat stress in non-producing young calves. We were particularly interested in acute heat stress responses of calves during a typical warm episode in summer. Physiological reactions recorded over a longer period might influenced reactions by adaptation, development of calves during this early age of life, changes in feed (i.e. removal of milk during weaning) or meteorological effects such as a prolonged decline in temperature or cold and hot fronts. Maximal and daily average values of physiological measures recorded for the 5-day experiment are presented in **Figs 5** and **6**. The analysis of animal-based parameters proved that supplemental shade should be provided for young calves to reduce acute, severe heat stress. Physiological parameters reflected high stress level of calves without shade on both days 2 and 3, and differences between groups were more pronounced in respiratory rate and heart rate compared to rectal or skin temperatures.

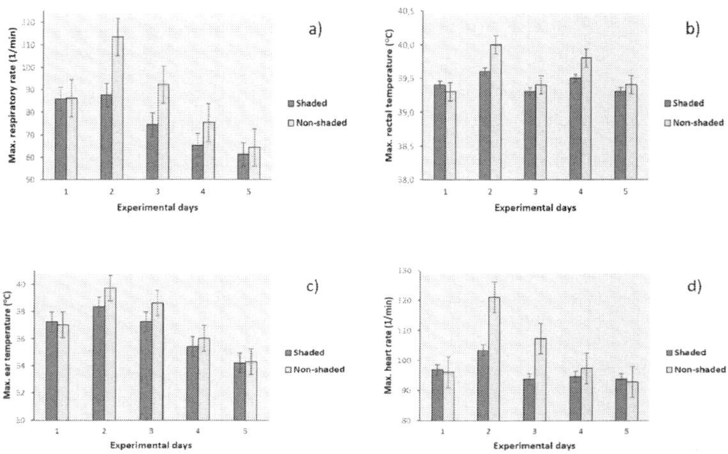

Figure 5. Maximal values of respiratory rate (a), rectal temperature (b), ear temperature (c) and heart rate (d) of shaded and unshaded calves for the five experimental days. Results are presented as means (±SD) based on individual data recorded with 30-min (for heart rate) and 4-h sampling frequency (for respiratory rate, rectal temperature and ear skin temperature). On day 1, calves for both groups were shaded. Day 2 = heat stress day.

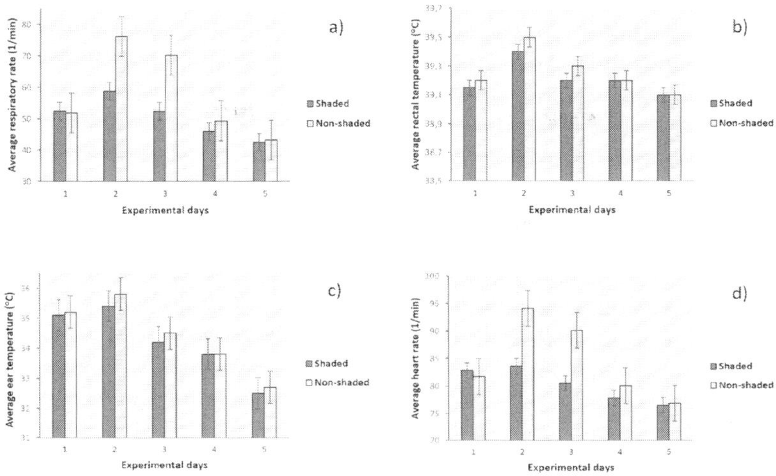

Figure 6. Daily averages of respiratory rate (a), rectal temperature (b), ear temperature (c) and heart rate (d) of shaded and unshaded calves for the five experimental days. Results are presented as means (±SD) based on individual data recorded with 30-min (for heart rate) and 4-h sampling frequency (for respiratory rate, rectal temperature and ear skin temperature). On day 1, calves for both groups were shaded. Day 2 = heat stress day.

Our results are in accordance to previous findings on immediate physiological responses to short-term, severe heat load. In ruminants, acute heat stress has proven to induce rapid physiological responses within 24 to 48 h [17–19]. In the present study, during the heat stress day, maximal temperature and THI exceeded 43°C and 86 units in the unshaded hutch environment, respectively. An earlier study reported marked attempts to acclimate to acute heat stress at 37.7°C in 4-week old bull calves (which was the maximal temperature in the shaded hutch environment in our study) based on systemic indicators of stress, serum thyroxine and cortisol concentrations [20].

Lower respiratory rates in calves in shaded hutches are consistent with earlier observations that have shown in average 10.4 breaths/min lower respiratory rates (47.3 vs. 57.7 breaths/min) for shaded calves than for calves in hutches without shade when measured at 15:00 o'clock [14]. In the present study, differences were greater; by 25.9 and 17.8 breaths/min higher maximal respiratory rates in average on days 2 and 3, respectively (**Fig 5A**), in unshaded calves compared to shaded ones ($P < 0.001$ for both days). Daily averages of respiratory rates differed also between groups either on day 2

($P = 0.008$) and day 3 ($P = 0.010$) (**Fig 6A**). Perhaps, calves of the present study were less adapted to extreme temperatures or to other features of their microenvironment. Another explanation for the substantial differences from previous results could be that the range of mean air temperature for individual sample days was 16.5 to 38.2°C in the cited paper, whereas animals were exposed to extremely high ambient temperatures during days 2 and 3 that ranged from 17.1 to 43.7°C for the unshaded hutch, and from 17.8 to 46.5°C for the unshaded pen environment in our study, respectively.

Although average rectal and ear temperatures did not reflect the positive effect of shading on day 2 ($P = 0.865$ and $P = 0.760$; **Fig 6B** and **6C**), significant, but not severe group differences were found for maximal body temperatures measured either in the rectum (0.5°C, $P = 0.046$) or on the ear skin (1.6°C, $P = 0.035$), respectively (**Fig 5B** and **5C**). Based on these findings sufficient daytime thermoregulatory ability of the experimental animals could be assumed (increased volume of water, and decreased feed intake) on the heat stress day. It was a limitation of our study that feed, and water intake were not measured. Previously, calf ear temperature reduced by about 0.3°C was found by positioning a shade cloth about 1.2 meters above the plastic calf hutches [14].

Following the heat stress day, differences in respiratory and heart rates were also meaningful with higher maximal and daily average values in unshaded animals for both parameters. Unshaded calves exhibited significantly higher maximal respiratory rates even on day 4 than calves with shading (**Fig 5A**, $P = 0.025$), suggesting their prolonged acclimatization to the challenging environment; however, based on this difference (10.1 breaths/min in average) a seriously impaired well-being of unshaded animals could not be assumed. Higher maximal and average heart rates of unshaded calves during day 2 ($P < 0.001$ and $P = 0.010$) and day 3 ($P = 0.005$ and $P = 0.012$) are possibly due to the increased physical activity associated with the high respiratory effort (**Fig 5D** and **Fig 6D**).

Although unshaded calves were exposed to significant daytime heat load during the afternoon hours on day 3 (maximal THI above 78, **Fig 4C**), even the maximal values of all animal-based parameters were in the physiological range. The available literature on critical cut-off points of different THIs of cattle focused only on adult cattle. Mader et al. [21], for example, determined 74 as a critical value of THI used in the present study in feedlot cattle based on panting scores, while others found THI 72 as critical in dairy cows [22]. In the light of previous findings, our results suggest that young calves would be expected to suffer less from heat stress than lactating cows that were also suffering from an additional metabolic heat output associated with milk production. Moreover, calves generate far less metabolic heat than cows and have a greater body surface area relative to internal body mass, which enables them to maintain thermal homeostasis more successfully than cows. However, cow studies usually assess production-related parameters of heat stress that cannot be measured in calves, which makes the comparison of cow and calf heat tolerance quite difficult.

It should be also noted, that despite extremely high daytime heat load observed for day 2 in this study, the drops of 18.2 and 16.4°C from daytime to nighttime temperatures on days 2 and 3, respectively, allowed calves to decrease core body temperature and thus the resumption of feed intake and normal physiological behavior. Although feeding and lying behavior were not investigated in this study, unshaded calves were usually in standing posture in the pen area at the feeding rack during the nighttime samplings, while shaded calves were mostly lying down in the hutches. It seems necessary to estimate thermal conditions which mean extreme stress for calves by determining upper critical values of meteorological indices. In these experiments, besides physiological measures, feed and water intake should be also recorded.

Our results emphasize that well-being of dairy calves kept in hutches exposed to direct sunlight can be impaired; nevertheless, the study might be limited due to the small number of calves used. It should be noted that the experimental environment and the arrangement of the calf hutches prevented the involvement of larger number of animals since bull calves were housed in three rows and each row had different orientation. To exclude orientation-related effects on thermal and physiological parameters calves from one row had to be selected. Age- and body weight-related differences between the study animals were also considered, thus calves from neighboring hutches were used (calves were placed into hutches based on the time of birth) that also limited the study population. Furthermore, we used only pre-weaned calves (up to 50 days of age) because weaning might have caused additional stress to the animals.

CONCLUSIONS

Thermal variables correlated well with animal-based heat stress indicators and, in case of rectal temperature all correlations were stronger in the unshaded than in the shaded thermal environment. Ambient temperature correlated better with physiological measures than THI therefore should be used as measure of acute heat stress in continental regions. Shading of preweaned dairy calves may be adequate to alleviate heat stress in continental regions. Respiratory rate and heart rate support that the well-being of unshaded dairy calves can be impaired even during decreasing thermal load following a short-term heat stress event. However, to provide practical recommendations on shading of dairy calves, further studies might be necessary using larger number of animals.

ACKNOWLEDGMENTS

Levente Kovács was supported by the following grants: 1) János Bolyai Research Scholarship of the Hungarian Academy of Sciences, Hungary [BO/00040/16/4], 2) OTKA Postdoctoral Scholarship of the National Research, Development and Innovation Office, Hungary [NKFIH-6493-1/2016], 3) Research Center of Excellence project of the National Research, Development and Innovation Office, Hungary [1476-4/2016/FEKUT], 4) New National Excellence Program of the Ministry of Human Capacities [ÚNKP-17-4-I/SZIE-7].

Viktor Jurkovich was supported by the grant of the Hungarian Ministry of Human Resources [11475-4/2016/FEKUT] and by the János Bolyai Research Scholarship of the Hungarian Academy of Sciences, Hungary.

FUNDING STATEMENT

Levente Kovács was supported by the following grants: 1) János Bolyai Research Scholarship of the Hungarian Academy of Sciences, Hungary [BO/00040/16/4], 2) OTKA Postdoctoral Scholarship of the National Research, Development and Innovation Office, Hungary [NKFIH-6493-1/2016], 3) Research Center of Excellence project of the National Research, Development and Innovation Office, Hungary [1476-4/2016/FEKUT], 4) New National Excellence Program of the Ministry of Human Capacities [ÚNKP-17-4-I/SZIE-7]. Viktor Jurkovich was supported by the grant of the Hungarian Ministry of Human Resources [11475-4/2016/FEKUT] and by the János Bolyai Research Scholarship of the Hungarian Academy of Sciences, Hungary.

REFERENCES

1. Edenhofer O, Pichs-Madruga R, Sokona Y. Climate Change 2014: Mitigation of Climate Change. Working Group III Contribution to the Fifth Assessment Report of the Intergovernmental Panel on Climate Change Cambridge Univ. Press, Cambridge, UK and New York, USA; 2014. pp. 1086.
2. Martin SW, Schwabe CW, Franti CE. Dairy calf mortality rate: the association of daily meteorological factors and calf mortality. Can J Comp Med. 1975;39: 377–388.
3. St-Pierre NR, Cobanov B, Schnitkey G. Economic losses from heat stress by US livestock industries. J Dairy Sci. 2003;86: E52–E77, Supplement.
4. Peña G, Risco C, Kunihiro E, Thatcher M-J, Pinedo PJ. Effect of housing type on health and performance of preweaned dairy calves during summer in Florida. J Dairy Sci. 2016;99: 1655–1662. 10.3168/jds.2015-10164
5. Dikmen S, Hansen PJ. Is the temperature-humidity index the best indicator of heat stress in lactating dairy cows in a subtropical environment? J Dairy Sci. 2009;92: 109–116. 10.3168/jds.2008-1370
6. Coleman DA, Moss BR, McCaskey TA. Supplemental shade for dairy calves reared in commercial calf hutches in a southern climate. J Dairy Sci. 1996;79: 2038–2043. 10.3168/jds.S0022-0302(96)76577-3
7. Theurer ME, Anderson DE, White BJ, Miesner MD, Larson RL. Effects of weather variables on thermoregulation of calves during periods of extreme heat. Am J Vet Res. 2014;75: 296–300. 10.2460/ajvr.75.3.296
8. NRC. Nutrient Requirements of Dairy Cattle. 7th rev. ed. Natl. Acad. Sci., 2001; Washington, DC.
9. Bianca W. Relative importance of dry and wet bulb temperature in causing heat stress in cattle. Nature. 1962;195: 252–252.
10. Bonk S, Burfeind O, Suthar VS, Heuwieser W. Technical note: Evaluation of data loggers for measuring lying behavior in dairy calves. J Dairy Sci. 2015;96: 3265–3271.
11. R Core Team. R: A language and environment for statistical computing R Foundation for Statistical Computing, Vienna, Austria; 2013. Available: http://www.R-project.org/ [accessed 12 February 2018].

12. Bohmanova J, Misztal I, Cole JB. Temperature-humidity index as indicator of milk production losses due to heat stress. J. Dairy Sci. 2007;90: 1947–1956. 10.3168/jds.2006-513
13. Reiczigel J, Solymosi N, Könyves L, Maróti-Agóts A, Kern A, Bartyik J. Examination of heat stress caused milk production loss by the use of temperature-humidity indices (In Hungarian). Hung. J. Vet. Sci. 2009;131: 137–144.
14. Lima PO, Souza JBF Jr, Lima RN, Oliveira FCS, Domingos HGT, Miranda MV. Effect of time of day and type of shading on the physiological responses of crossbred calves in tropical environment. J Anim Behav Biometeorol. 2013;1: 7–12.
15. Spain JN, Spiers DE. Effects of supplemental shade on thermoregulatory response of calves to heat challenge in a hutch environment. J Dairy Sci. 1996;79: 639–646. 10.3168/jds.S0022-0302(96)76409-3
16. Lammers BP, van Koot JW, Henrichs AJ, Graves RE. The effect of plywood and polyethylene calf hutches on heat stress. Appl Eng Agric. 1996;12: 741–745.
17. Maltz E, Silanikove N, Shalit U, Berman A. Diurnal fluctuations in plasma ions and water intake of dairy cows as affected by lactation in warm weather. J Dairy Sci. 1994; 77: 2630–2639. 10.3168/jds.S0022-0302(94)77204-0
18. Maltz E, Silanikove N. Kidney function and nitrogen balance of high yielding dairy cows at the onset of lactation. J Dairy Sci. 1996;79: 1621–1626. 10.3168/jds.S0022-0302(96)76525-6
19. Silanikove N. Effects of heat stress on the welfare of extensively managed domestic ruminants. Livest Prod Sci. 2000;67: 1–18.
20. Neuwirth JG, Norton JK, Rawlings CA, Thompson FN Ware GO. Physiologic responses of dairy calves to environmental heat stress Int J Biometeor. 1979;23: 243–254.
21. Mader TL, Davis MS, Brown-Brandl T. Environmental factors influencing heat stress in feedlot cattle. J Anim Sci. 2006; 84: 712–719.
22. Armstrong DV. Heat stress interaction with shade and cooling, J Dairy Sci. 1994;77: 2044–2050. 10.3168/jds.S0022-0302(94)77149-6

SECTION 4: CHICKENS

CHAPTER 7

Behavioural and Physiological Effects of Finely Balanced Decision-Making in Chickens

Anna C. Davies[1], Christine J. Nicol[1], Mia E. Persson[2], Andrew N. Radford[3]

[1]School of Clinical Veterinary Science, University of Bristol, Bristol, United Kingdom,

[2]Avian Behavioural Genomics and Physiology Group, Linko¨ping University, Linko¨ping, Sweden,

[3]School of Biological Sciences, University of Bristol, Bristol, United Kingdom

ABSTRACT

In humans, more difficult decisions result in behavioural and physiological changes suggestive of increased arousal, but little is known about the effect

Citation: Davies AC, Nicol CJ, Persson ME, Radford AN (2014) Behavioural and Physiological Effects of Finely Balanced Decision-Making in Chickens. PLoS ONE 9(10): e108809. https://doi.org/10.1371/journal.pone.0108809

of decision difficulty in other species. A difficult decision can have a number of characteristics; we aimed to monitor how finely balanced decisions, compared to unbalanced ones, affected the behaviour and physiology of chickens. An unbalanced decision was one in which the two options were of unequal net value (1 (Q1) vs. 6 (Q6) pieces of sweetcorn with no cost associated with either option); a finely balanced decision was one in which the options were of equal net value (i.e. hens were "indifferent" to both options). To identify hens' indifference, a titration procedure was used in which a cost (electromagnetic weight on an access door) was applied to the Q6 option, to find the individual point at which hens chose this option approximately equally to Q1 via a non-weighted door. We then compared behavioural and physiological indicators of arousal (head movements, latency to choose, heart-rate variability and surface body temperature) when chickens made decisions that were unbalanced or finely balanced. Significant physiological (heartrate variability) and behavioural (latency to pen) differences were found between the finely balanced and balanced conditions, but these were likely to be artefacts of the greater time and effort required to push through the weighted doors. No other behavioural and physiological measures were significantly different between the decision categories. We suggest that more information is needed on when best to monitor likely changes in arousal during decision-making and that future studies should consider decisions defined as difficult in other ways.

INTRODUCTION

Humans and non-human animals are faced with decisions in all aspects of their lives [1], [2]. The process of decision-making in humans is associated with changes in behaviour and physiology, including galvanic skin responses and heart-rate (HR) [3], indicative of increased arousal. Moreover, more difficult decisions lead to increases in such response parameters e.g. [4]–[7]. Although behavioural and physiological processes have been studied during decision-making in chickens [8], little is known about whether decisions of greater difficulty result in increased arousal in non-human species. Procedures involving decision-making are used widely in animal research, such as when testing preferences for environmental or social resources [9]–[11]. For example, choice tests have been used to determine chickens' preferred dustbathing substrate [12] and lighting [13]. Since the results of these tests can have practical, political and welfare implications, any procedural influences must be considered as part of an overall interpretation [14], [15].

A complicating factor is that decisions may be "difficult" in more than one way. A difficult decision might, for instance, involve finely balanced options (e.g. two options with the same net value: [16]), have the risk of a critical outcome (e.g. a risk of predation: [17]), have an ambiguous outcome (e.g. insufficient information available: [18]), vary in more than one dimension (e.g. cost, motivation and resource type: [19]) or require processing of a large amount of information (e.g. numerous options available: [20]). When options are finely balanced (e.g. the net benefit of accessing a large quantity of food with an associated cost is similar to accessing a small quantity of food with no associated cost), individuals are likely to become indifferent to the outcome, choosing both options at approximately equal frequency. When individuals are close to indifference, decisions may not be solved by simple prioritisation (i.e. choosing the alternative with the highest probability-weighted utility) e.g. [21], [22]. Significantly more cognitive effort is therefore invested in finely balanced problems [23] and such decision-making can result in behavioural indicators of frustration and anxiety [24]. In the present study we focussed on comparing the physiology and behaviour of chickens making decisions in which options were either finely balanced or unbalanced.

An unbalanced decision was one in which two options were of unequal net value (i.e. a small quantity of food vs. a large quantity of food). The first part of our study used a titration methodology, in which weighted push-doors were used, to obtain individually-determined points of indifference (i.e. a finely balanced decision). The second part of our study compared the behaviour (head movements and latency to choose) and physiology (surface body temperature and heart-rate variability (HRV)) of individual chickens when making finely balanced and unbalanced decisions. Head movement increases and surface body temperature decreases have been measured in chickens in response to both aversive stimuli (head movements: [25]; surface body temperature: [26]–[28]) and to a signalled palatable reward (head movements: [8], [29]; surface body temperature: [30]). Decreases in HRV, which provides more information on the source of cardiac stimulation (i.e. parasympathetic or sympathetic) than HR [31], occur in response to acutely stressful situations in birds [32]–[34].

We predicted that around the time of finely balanced decision-making, chickens would be more stressed, hence their HRV and surface body temperature would decrease (from levels at the start of the test) compared with an unbalanced alternative. We also predicted that hens would show increased arousal during decision-making (make more head movements) and have increased latencies to choose, due to increased cognitive demand when decisions were finely balanced.

METHODS

Ethics Statement

All work was conducted under UK Home Office licence (30/2779) and had University ethical approval. We also conducted the study in compliance with ASAB ethical guidelines. The hens were rehomed to small responsible free-range holdings after the study.

Animals, Housing and Husbandry

Sixteen Columbian Black Tail laying-hens were obtained at approximately 16–18 weeks old from a commercial pullet rearer in Devon. They were transported in poultry crates in a well-ventilated van for approximately 2 h prior to delivery. On arrival all birds were weighed, health-checked, treated with preventative red mite powder (diatomaceous earth: Oak Tree Poultry) and individually tagged for identification using numbered leg rings. They were housed in groups of four, in four out of eight available pens (0.96×1.2 m, 2 m high) in the same room (home room). During weekly cleaning, each group of four birds was switched to the opposite pen within the same room to avoid a housing side-bias. All birds were checked at least once per day and were weighed weekly.

Hens in all pens were fed *ad libitum* feed (Farmgate Layers Mash, BOCM Pauls, Ipswich, Suffolk, UK) via two external feed troughs (total length: 0.77 m), reached by an opening in the pen 0.15 m from floor level. Water was provided via a hanging drinker which was placed in the back corner of each pen (0.2 m high). A nest box (0.39×0.38×0.47 m) and a round perch (0.25 m high, stretching the width of the pen) were also provided. Wood shavings were used as bedding at a depth of 5–10 cm. The room temperature was kept at 19–22°C and the lighting schedule was 12 L: 12 D (light period 7 am–7 pm).

Experimental room and Procedure

The experimental room contained two pens (one on each side of the room) of the same size as the home room, but which contained no feeders, nest boxes or perches. The experimental pens could be joined by a Perspex tunnel

(1.79×0.24 m, 0.47 m high), which formed part of a T-maze apparatus. The experimental room was separated from the home room by solid wooden doors and a corridor, providing an area where hens could be tested away from the noise of conspecifics. Within the experimental room, a CCTV camera was attached to the ceiling above the test apparatus, which was connected to a computer on one side of the room. Another computer for electrocardiogram (ECG) monitoring was set-up on the other side of the room.

Aim 1 – Establishing Finely Balanced and Unbalanced Decisions

The first aim of this study was to devise a protocol for identifying finely balanced and unbalanced decisions in individual hens. This was achieved using the T-maze test apparatus described above, with incorporated push-doors at each end of the Perspex tunnel. Each push-door could be attached to an electromagnet to alter the force required to push through the door (figure 1). The first step to achieving this aim was to habituate and train the hens. This was followed by a titration procedure.

Figure 1. T-maze test apparatus consisting of a Perspex tunnel with push-doors and an attached wooden start-box. The tunnel connects the two pens in the experimental room. **A** indicates the rear of the start-box which was made of black plastic. **B** indicates the wooden side-doors which were removed to reveal wire mesh, through which the feed bowls could be viewed. **C** indicates the tunnel-door, which was raised using a pulley mechanism to allow access to the tunnel. **D** marks the push-doors which were located at either end of the tunnel. **E** marks the pen-door which was placed in the pen entrance once the hen entered the pen, to prevent her from re-entering the tunnel. **F** marks the electromagnet which could be altered to change the force required to open the push-door.

After a five day settling period, habituation to the ECG monitor, training to establish associations between feed bowls and reward quantity, and habituation to the T-maze test procedure (including push-doors) began. Habituation and training criteria needed to be satisfied at each stage before individuals could progress. Habituation and training took 4–5 weeks, depending on individual progression.

ECG recording habituation

We made sure birds were familiar with wearing a harness and an ECG monitor for later phases of the study. ECG was recorded as in [8] using non-invasive remote telemetric units [35] and ECG cables contained within a harness. Hens were gradually habituated to wearing the harness, initially using a simplified harness (with no monitor) for a short period of time. As habituation criteria were met (that hens were able to walk and behave normally in their home environment without moving backwards or stopping excessively), the amount of time wearing the harness was gradually increased in 15 min increments, to a maximum time of 6 h. The ECG cables were then added to the harness and finally the monitor (weighing approx. 100 g). Harness habituation was conducted in both the home and experimental rooms, including within the T-maze.

Training associations between feed bowl characteristics and food reward

The hens were individually trained to discriminate between two stimuli (feed bowls), each associated with a different quantity of food (either 1 (Q1) or 6 (Q6) pieces of sweetcorn). Sweetcorn was used instead of a more motivating stimulus like mealworms [36] because we previously found that HR increased considerably in anticipation of a mealworm reward [8] and we wanted to focus on arousal caused by decision-making. During preliminary studies, hens were motivated to eat the sweetcorn reward. We chose to manipulate food quantity rather than quality as we wanted to avoid individual differences in preference for different food types.

The two feed bowls used for each hen were of different size (90 or 140 mm diameter) and colour (green or blue) to aid discrimination of the reward quantity (Q1 and Q6). There were four possible combinations of how size and colour were allocated to reward quantity (figure 2), and bowl combinations were systematically allocated so that one hen from each pen was trained to each. Once hens approached both bowls without stopping or hesitating, discrimination training continued in the T-maze along with habituation to other aspects of the apparatus.

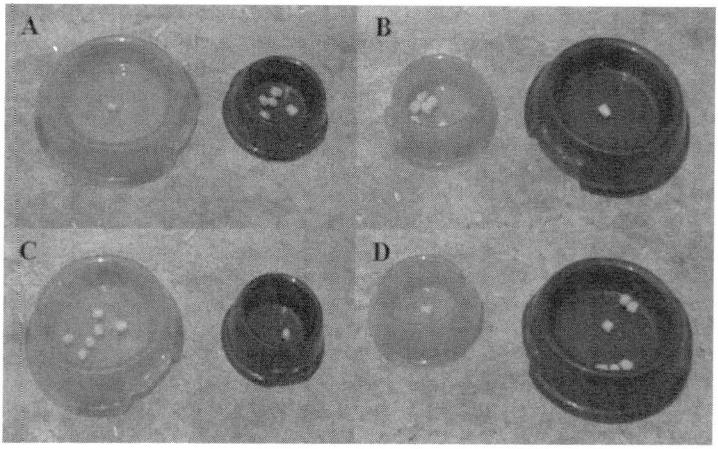

Figure 2. Different coloured and sized feed bowls were used to aid discrimination learning.

The feed bowls used were counter-balanced for colour and size, with four chickens being trained on each of the four possible combinations (**A–D**).

T-Maze Habituation

Initial habituation to the T-maze apparatus was conducted in home groups by connecting opposite pens in the home room with a tunnel. Three 3-h group sessions were sufficient to ensure all hens in each group had walked through the tunnel without showing fearful behaviour (stopping or hesitating whilst walking). Habituation then continued individually in the experimental room until hens became accustomed to leaving the start-box and walking through the tunnel with the push-doors (figure 1) fixed open (approximately seven sessions). A tunnel-door suspended by a pulley mechanism was then added to the start-box, to prevent hens from entering the tunnel for a short period at the start of each test (to allow for an initial assessment of behaviour and physiology during the testing phase). To begin with, hens were confined within the start-box for 5 s, but this was gradually increased (when they showed no escape attempts or excessive vocalisations) to 30 s (this took approximately five sessions).

Once accustomed to an initial confinement period, habituation to side-door removal began. The removable wooden side-doors of the start-box were kept in place at the start of each test to prevent hens from viewing the conditioned stimuli (feed bowls placed at both pen entrances). After a 10 s period within the start-box, side-doors were sequentially removed (to ensure that hens were looking at the food bowls on either side of the T-maze) and replaced, then both were removed simultaneously (viewing period). The tunnel-door was then raised and birds were allowed 10 s to leave the start-box (this took approximately six sessions).

When hens had successfully reached criteria in all other aspects of the T-maze procedure, the push-doors (with no weight applied) were introduced. It took approximately eight sessions to train all hens (using sweetcorn) to push through the doors to access the feed bowls. During habituation, a mixture of unidirectional and free-choice trials was conducted and both side-door removal order and the location of each reward quantity (Q1 or Q6) relative to the start-box were systematically varied to ensure equally balanced training. By the end of the training period, hens were well habituated to handling and to all other aspects of the T-maze (including the start-box), having completed approximately 50 individual training trials. If side biases became evident during habituation, additional training was conducted.

Prior to discrimination training in the T-maze, hens were food deprived for an individually-determined period of time to standardise hunger motivation. We selected the minimum food deprivation period (tested during the training phase) that ensured each hen would perform multiple (up to 8) consecutive tests. The required period of food deprivation varied between hens, ranging from 80 to 390 min. As we were using a within-subjects design, the inter-hen range of food deprivation times should not influence treatment-based findings.

When the doors were unweighted and hens consistently chose the Q6 option in the T-maze (at least 90% of the time over consecutive 12 sessions), they were deemed able to discriminate between the two stimuli and this was defined as an unbalanced decision. The mean percentage of Q6 choices made across the 12 consecutive sessions was 97%.

Titration Phase

The aim of the titration phase was to detect the point at which the two options (Q1 and Q6) had the same net value (i.e. a finely balanced decision). This was done by gradually increasing the force required for hens to push through the Q6 push-door (Q1 door force was always zero), to find the point at which Q6 and Q1 were selected with approximately equal frequency (25–75% of choices) across 12 consecutive sessions. One hen's balance point fell outside the acceptable threshold and the hen was excluded from further testing. A finely balanced decision was defined for the remaining 15 hens – the mean percentage of Q6 choices made across the 12 consecutive sessions was 49% – with the electromagnetic force that had to be applied to the Q6 door varying greatly between individuals (figure 3). The door force at the point of indifference can be seen in figure 4.

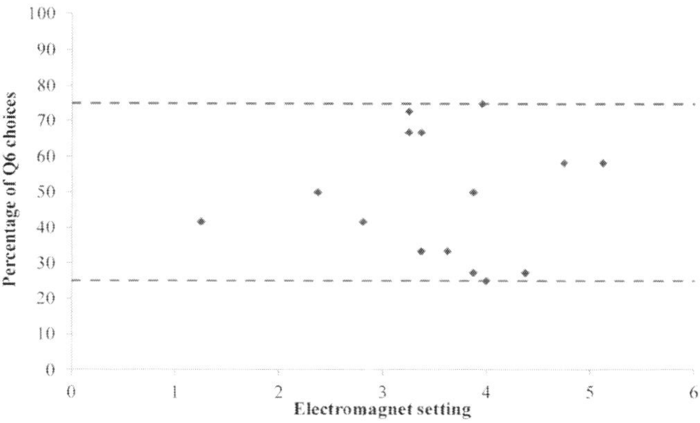

Figure 3. The electromagnet setting at which individual hens chose Q6 within our defined threshold of 25–75% (dashed line) of choices across 12 consecutive trials, measured at the end of the titration phase.

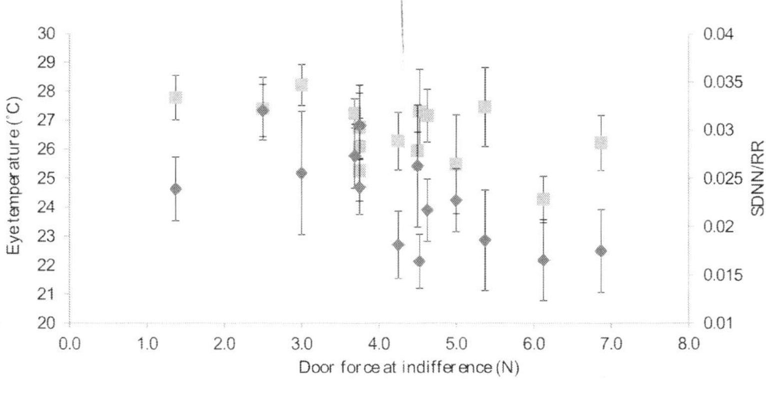

Figure 4. The association between the door force (in Newtons) at indifference and mean ±1 SE SDNN/RR (dark grey) and eye temperature (light grey) during the viewing period when making finely balanced decisions.

Aim 2 – Behavioural and Physiological Responses to Balanced and Unbalanced Decisions

Once the unbalanced and finely balanced decisions had been individually defined, each hen underwent 10 consecutive days of testing during which aspects of their behaviour and physiology were monitored. On each day (which involved either an unbalanced or a finely balanced decision), hens were food deprived as for training, before being given two unidirectional (forced) trials (one to either option) followed by one free-choice test (when the door weights remained as in the forced trials). The forced trials allowed hens to 'assess' whether the free-choice test would be unbalanced or finely balanced. The test protocol is outlined in figure 5. Prior to starting each free-choice test, the ECG monitor and a stopwatch were activated simultaneously. The test commenced when hens were placed in the start-box and confined for 10 s with the wooden side-doors in place, so that initial (start-box) physiological measures could be taken without conditioned stimuli (feed bowls) being visible. The side-doors were sequentially removed from the start-box for 5 s and replaced, and then both side-doors were simultaneously removed for a 10 s viewing period. The tunnel-door was then raised, allowing access to the push-doors ('the push-door period'). Once hens had accessed their chosen feed bowl via the push-door, the pen-door was closed and they were confined within the pen for 90 s (to allow consumption of their reward: 'the pen period'). If a hen failed to enter the tunnel within the first 60 s of each test, she was gently encouraged into the tunnel and the tunnel-door was replaced. A maximum time of 300 s was given for hens to enter a pen, after which time the test was stopped and the hen was removed from the T-maze.

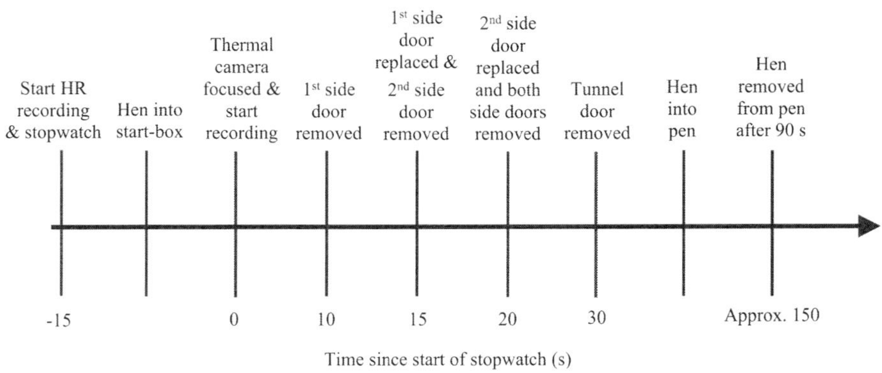

Figure 5. The test procedure during the testing phase.

A stopwatch was used to monitor the precise time at which each event occurred so that accurate start-box and viewing period HRV and temperatures could be extracted from the data files.

Only one free-choice test was given each day to ensure that hunger motivation was kept as constant as possible throughout the experiment. Tests were carried out on each individual at the same time of day across the testing phase. Five tests were conducted per hen for both finely balanced and unbalanced decisions during the testing period. The order in which each hen experienced each decision category was systematically alternated, with no more than two consecutive tests from either decision category. The side of the T-maze where Q1 and Q6 were presented, the order of unidirectional trials and the order of side-door removal were also systematically alternated using the same criteria. Side biases were investigated and checked as the experiment progressed. The unidirectional trials given ensured that hens visited both sides of the T-Maze.

For each free-choice test, the following measures were taken during the push-door period: the latency to the push-door, the number of switches between push-doors and the number of attempts made to enter the pen. The side of the T-maze which hens first tried to enter was recorded as their 'first push' and their 'ultimate choice' was also noted. The latency to the push-door and latency to the pen were both recorded as outlined in figure 5. A CCTV camera was fixed above the start-box and video was continuously recorded using WebCCTV software. The videos were analysed using The Observer Software version 10.0 to measure the number of head movements during the 10 s viewing period.

Throughout each free-choice test, the ECG was recorded continuously onto a micro-SD flash card which was inserted into the monitor. The monitor communicated with a base unit (attached to a computer via USB connection) and was controlled using RVC Telemetry Software version 1.5. HRV data were extracted using Spike 2 Software (version 6) from three 10 s periods: start-box, viewing and pen. As we were primarily interested in whether more difficult decisions induced arousal, we extracted two measures of HRV which are influenced by both branches of the autonomic nervous system [37] – mean interbeat interval (RR) and the quotient of the standard deviation of interbeat intervals and RR (i.e. the coefficient of variance (SDNN/RR)). Measures of HRV were taken at the start of each test (start-box period) to check for individual differences and to account for the influence of handling on arousal.

Surface body temperature was also recorded during the start-box and viewing periods using a thermal video camera (FLIR SC305), and was extracted using FLIR ResearchIR Software version 1.2 SP2. The rear surface of the wooden start-box was made from black plastic (through which we were able to detect heat radiation from the chicken), allowing us to record surface body temperature without hens being able to view the experimenter. The emissivity of the black plastic was calculated to be 0.292. A clear image of the side of the head was taken from each video to obtain the eye and maximum head temperature (hottest part of exposed skin). Video was recorded at 3 frames/second so it was usually possible to obtain clear images even if hens were moving.

Statistical Analysis

The data were analysed using *IBM SPSS Statistics 19*. Fifteen hens were given 10 free-choice tests each. HRV data were not collected from one individual as she did not behave normally when wearing the monitor during training. This hen did not wear the monitor during testing, hence we did not collect HRV data from her reducing the HRV sample to 14 individuals. Temperature, head movement and latency data were collected for all 15 individuals. Prior to conducting analyses we investigated the possibility of interactions by visual, inspection of plotted results for all measured variables and found no evidence of such interactions. For each set of data, the assumptions of parametric testing were checked and data were transformed if possible, then analysed using paired-samples t-tests. Where transformations were not possible or unsatisfactory, Wilcoxon tests were used. For analyses, a mean of each measure (% Q6 choices, RR intervals, SDNN/RR, head movements, maximum head and eye temperature, latency to first push and pen, number of switches made, number of attempts to enter the pen) was taken from the five repeats to each hen, to look for differences between finely balanced and unbalanced decisions during the relevant periods (start-box, viewing, push-door, pen). Unless otherwise stated, means ± SE are presented (in figures or text) for parametric data and the median, interquartile range (IQR) and range are presented for non-parametric data as appropriate. A measure of effect size is given alongside significant results. As means were used for the analyses, the range of individual hen standard deviations is given where appropriate.

RESULTS

Aim 1 – Establishing Finely-Balanced and Unbalanced Decisions

To validate the titration procedure, the percentage of Q6 choices made during the testing phase was calculated. Overall, significantly more Q6 choices were made in the unbalanced condition (median: 100; IQR: 60–100; range: 60–100%) than in the finely balanced condition (median: 60; IQR: 40–80; range: 0–100%; Wilcoxon signed rank test: z=2.83, n=15, p<0.01, r=0.52).

Aim 2 – Behavioural and Physiological Responses to Balanced and Unbalanced Decisions

Start-box period

There were no significant differences between treatments in RR intervals (range of individual standard deviation: finely balanced =3.5–22.4 ms, unbalanced =4.3–10.7 ms; paired samples t-test: t_{13} =0.43, p=0.68, figure 6*a*) or in SDNN/RR (range of individual standard deviation: finely balanced =0.004–0.03, unbalanced =0.004–0.03; paired samples t-test: t_{13} =0.72, p=0.49, figure 6*b*). There were also no significant differences in start-box maximum head temperature (range of individual standard deviation: finely balanced =0.9–5.2°C, unbalanced =1.2–4.1°C; paired samples t-test: t_{14} =0.91, p=0.38, figure 7*a*) or eye temperature (range of individual standard deviation: finely balanced =1.3–3.8°C, unbalanced =1.4–4.2°C; paired samples t-test: t_{14} =0.71, p=0.49, figure 7*b*) between finely balanced and unbalanced decisions.

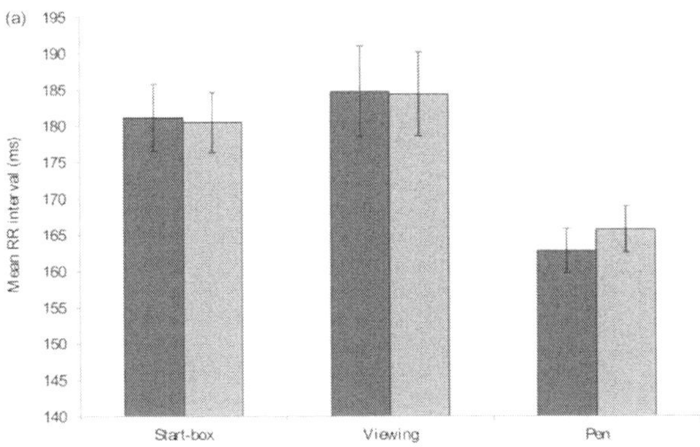

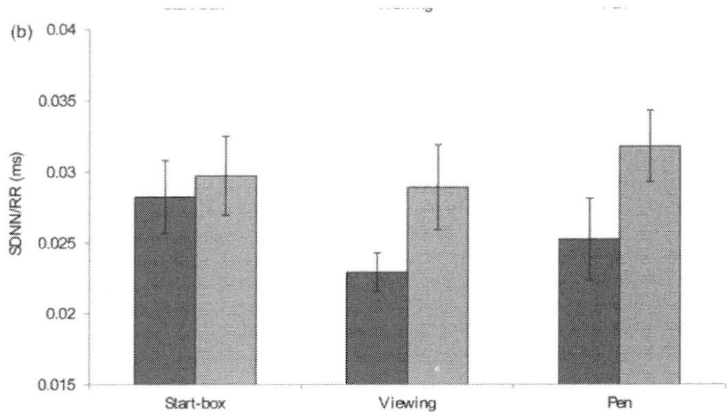

Figure 6. Mean ±1 SE (a) RR interval, (b) coefficient of variance (SDNN/RR), during each period within the test when making finely balanced (dark grey) and unbalanced (light grey) decisions.

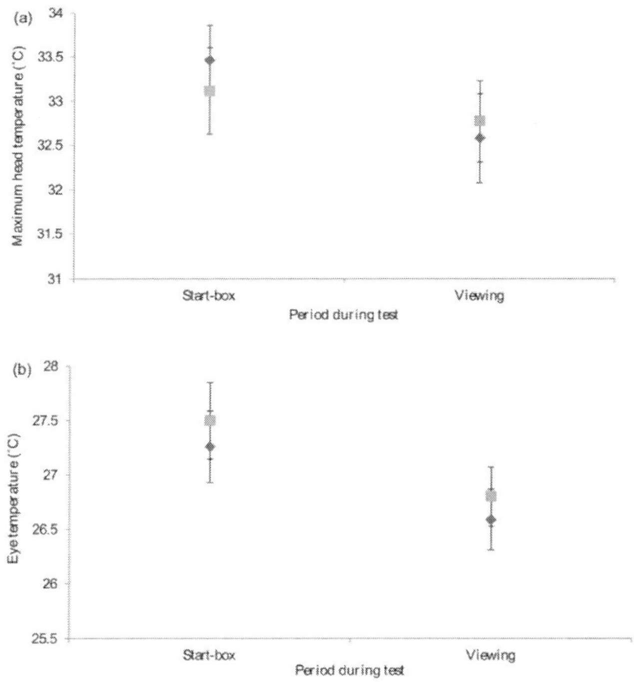

Figure 7. Mean ±1 SE (a) maximum head temperature, (b) eye temperature, during the start-box and viewing periods when making finely balanced (dark grey) and unbalanced (light grey) decisions.

Viewing period

During the viewing period, there was no significant difference in RR between finely balanced and unbalanced decisions (range of individual standard deviation: finely balanced =2.4–15.0 ms, unbalanced =2.7–13.8 ms; paired samples t-test: t_{13} =0.28, p=0.79, figure 6*a*). However, there was a strong trend for a lower SDNN/RR when making a finely balanced decision compared to an unbalanced one (range of individual standard deviation: finely balanced =0.006–0.01, unbalanced =0.002–0.05; t_{13}=2.10, p=0.055, figure 6*b*). There was a significant negative correlation between the door weight at indifference and SDNN/RR during the viewing period (Pearson correlation coefficient =−0.68, n=14, p=0.008, figure 4). There were no significant differences between treatments in the number of head movements (range of individual standard deviation: finely balanced =0.7–3.8, unbalanced =0.5–2.9; t_{14} =0.26, p=0.80) or in the maximum head (range of individual standard deviation: finely balanced =1.3–4.2°C, unbalanced =1.2–3.7°C; t_{14} =0.65, p=0.52, figure 7*a*) and eye temperatures (range of individual standard deviation: finely balanced =1.1–3.8°C, unbalanced =1.5–3.7°C; t_{14} =0.81, p=0.43, figure 7*b*). There was also a significant negative correlation between the door weight at indifference and the eye temperature during the viewing period (Pearson correlation coefficient =0.55, n=15, p=0.033, figure 4).

Push-door period

Latency to the first push on a door was not significantly influenced by experimental condition (range of individual standard deviation: finely balanced =1.0–36.5 s, unbalanced =0.4–50.1 s; paired-samples t-test: t_{14} =0.08, p=0.94). Hens did make significantly more switches between push-doors (finely balanced: median: 0.2; range: 0–0.6; unbalanced: median: 0; range: 0–0; Wilcoxon signed rank test: z=2.59, n=15, p=0.01, r=0.47) and significantly more attempts to enter the pen (finely balanced: median: 1.6; range: 1–6.2; unbalanced: median: 1; range: 1–1.4; z=3.06, n=15, p<0.01, r=0.56, figure 8*a*) in the finely balanced treatment compared to when the decision was unbalanced.

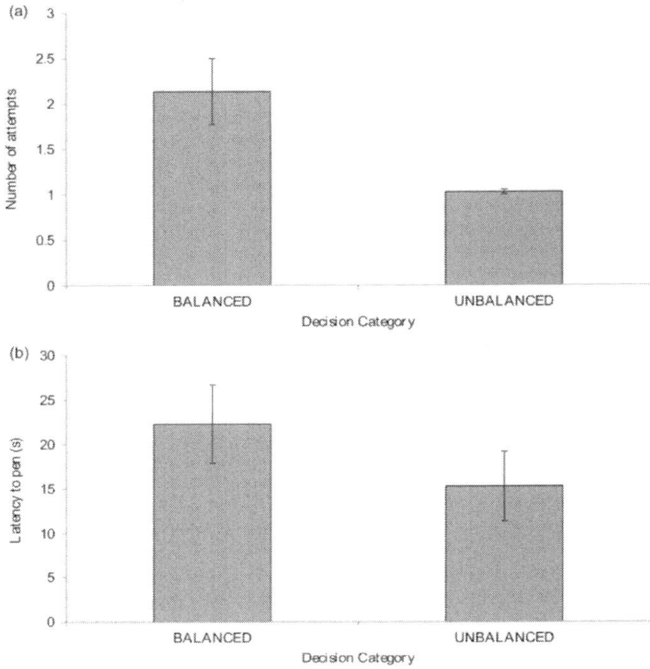

Figure 8. Mean ±1 SE (a) number of attempts made, (b) latency to pen.

Pen period

Hens took significantly longer to reach the pen when making finely balanced decisions compared to unbalanced ones (range of individual standard deviation: finely balanced 1.6–56.5 s, unbalanced =0.5–68.5 s; paired samples t-test: t_{14} =2.73, p=0.02, eta squared =0.35, figure 8*b*). Once in the pen, the RR interval was significantly shorter (range of individual standard deviation: finely balanced =3.0–20.8 ms, unbalanced =1.9–9.9 ms; t_{13} =3.22, p=0.007, eta squared =0.46, figure 6*a*) and the SDNN/RR was significantly lower (range of individual standard deviation: finely balanced =0.001–0.05, unbalanced =0.005–0.03; t_{13} =2.46, p=0.029, eta squared =0.34, figure 6*b*) when hens had made finely balanced decisions compared to unbalanced ones.

DISCUSSION

Our titration methodology, based on the idea that two options are substitutable when they are of equal net value [16], was successful in determining the cost (the force needed to open the access door) required for hens to choose a larger food reward at an equivalent frequency to a smaller food reward. The electromagnetic force required to generate finely balanced choices varied between individuals, highlighting the importance of conducting this protocol on an individual hen basis. Finely balanced decisions between two substitutable options are likely to be inherently more difficult than unbalanced decisions [16], [23], [24], so our method allowed subsequent testing of behavioural and physiological correlates of this aspect of decision difficulty in hens. The titration method we have developed could also prove useful in examining the effect of this aspect of decision difficulty in other species.

Three significant physiological and behavioural differences were found between the finely balanced and unbalanced conditions, but all of these occurred outside the 'viewing period' (when we expected hens to make their decision and that arousal caused by decision-making would be detectable) after hens had left the start-box. Measures were taken during the push-door and pen periods to monitor the effect of the decision consequence on arousal. Under finely balanced conditions, hens took longer to reach the pen and there were differences in the two measures of HRV: mean RR interval was significantly shorter and the coefficient of variance (SDNN/RR) was significantly lower when hens had made finely balanced decisions. These differences are likely to be artefacts of the additional time and effort caused by hens having to push against the weighted door, of them switching more often between push-doors, and of additional attempts to enter the pen in the finely balanced condition. We had predicted that decreases in measures of HRV in the pen and longer latencies to pen in the finely balanced condition would be accompanied by a decrease in HRV during the viewing period and an increased latency to the first push, if affected by the difficulty of the decision. Although the coefficient of variance was influenced by experimental condition, the difference was not quite statistically significant, and there were no treatment-related differences in latency to the first push and the RR interval, suggesting that the observed differences in pen measures were unlikely to be caused by the difficulty of decision-making. We did find however, that the door weight at indifference was negatively correlated with some measures of arousal during the viewing period (eye temperature and the coefficient of variance), suggesting that hens pushing heavier doors may have anticipated the extra weight.

We found no other significant influences of experimental condition on the eye and maximum head temperature change or on the number of head movements made during decision-making. This could be interpreted in several ways. First, although the perceived difficulty of a decision affects human behaviour and physiology [4]–[7], it might not result in the same responses around the time of decision-making in other animals or more specifically in birds. However, previous research has shown that some non-human animals, such as dolphins, gorillas and honey bees, behave similarly to humans during other types of difficult tasks [18], [38], [39]. The key difference may therefore relate to the manner in which a decision is defined as difficult. In the aforementioned studies, the type of difficult decisions examined tends to involve a degree of uncertainty, for example as a result of ambiguous cues. And in the studies of human arousal during decision-making, difficult decisions often involved a degree of risk [5], [6], [7]. These are factors that we are currently examining in separate work, but here we explored behavioural and physiological effects of an aspect of decision difficulty that has not previously been examined in other species (i.e. a decision between two options of equal net value).

One reason why the birds were not aroused when making finely balanced decisions may have been because they adopted a simple non-arousing strategy to deal with the situation. When humans must make decisions but are indifferent to the outcomes, they can adopt strategies such as choosing randomly, consistently choosing one option (status quo maintenance) or alternating between options e.g. [40]–[42]. Although potentially costly in terms of time [23], the use of non-evaluative decision strategies may be less arousing than evaluative decision-making in humans, though to our knowledge this has not been tested. Possibly our finely balanced decisions were difficult, but not stressful enough to result in physiological arousal.

One additional possibility is that the overall method and the behavioural and physiological measures we took were not sensitive enough to detect changes during the relevant period. However, we have previously found strongly significant changes in these measures in our related studies with chickens [8] and others have found differences in these measures in anticipation of and in response to positive and negative stimuli [25], [27], [29], [30]. Often during human decision-making experiments, antecedents and consequences of decision-making are monitored, rather than directly measuring changes during the process, due to the difficulty of identifying the exact point at which decisions are made. For the purpose of our current experiment, we considered that the birds' decision period most likely coincided with the 10-s viewing period, when the hens could view options but not yet access them. It is possible that hens did not make their choice during our defined decision period, or that changes in arousal occurred later than we had predicted. The coefficient of variance (HRV) showed a strong trend (p=0.055) towards a significant difference between our experimental conditions which suggests that some decreases in HRV occurred during the viewing period. It is possible that a time-lag between decision-making and the onset of physiological arousal exists.

Although the titration methodology we developed was successful in identifying finely balanced and unbalanced decisions in individual chickens, to progress work in this area more information is needed on when best to monitor changes in arousal during decision-making. Additionally, future work could aim to identify whether such finely balanced decisions result in behavioural and physiological arousal in other species, and to explore alternative definitions of difficult decisions. As experimental protocols involving decision-making are used widely in animal welfare research [9]–[11], it is essential that all procedural influences are taken into consideration for an accurate interpretation of results.

ACKNOWLEDGMENTS

We would like to thank Professor Mike Mendl for helpful discussions and comments on the manuscript and the Animal Services Unit at the University of Bristol for caring for the hens.

Funding Statement

This study was funded by the BBSRC (Grant reference: BB/F016662/1). The funders had no role in study design, data collection and analysis, decision to publish, or preparation of the manuscript.

REFERENCES

1. Saaty (1988) What is the analytic hierarchy process? In: Mitra G, Greenberg HJ, Lootsma FA, Rijckaert MJ, Zimmerman HJ, editors. Mathematical models for decision support. NATO ASI Series Vol 48. Springer Berlin Heidelberg. pp.109–121.
2. McFarland DJ (1977) Decision making in animals. Nature 269: 15–21.
3. Crone EA, Somsen RJM, Van Beek B, Van Der Molen MW (2004) Heart rate and skin conductance analysis of antecendents and consequences of decision-making. Psychophysiology 41: 531–540.
4. Gerard HB (1967) Choice difficulty, dissonance and the decision sequence. J Pers 35: 91–108.
5. Mann L, Janis IL, Chaplin R (1969) Effects of anticipation on predecisional processes. J Pers Soc Psychol 11: 10–16.
6. Critchley HD, Mathias J, Dolan RJ (2001) Neural activity in the human brain relating to uncertainty and arousal during anticipation. Neuron 29: 537–545.
7. Van Harreveld F, Rutjens BT, Rotteveel M, Nordgren LF, van der Pligt J (2009) Ambivalence and decisional conflict as a cause of psychological discomfort: feeling tense before jumping off the fence. J Exp Soc Psychol 45: 167–173.
8. Davies AC, Radford AN, Nicol CJ (2014) Behavioural and physiological expression of arousal during decision-making in laying hens. Physiol Behav 123: 93–99.
9. Kirkden RD, Pajor EA (2006) Using preference, motivation and aversion tests to ask scientific questions about animals' feelings. Appl Anim Behav Sci 100: 29–47.
10. Jensen MB, Pedersen LJ (2008) Using motivation tests to assess ethological needs and preferences. Appl Anim Behav Sci 113: 340–356.
11. Fraser D, Nicol CJ (2011) Preference and motivation research. In: Appleby MC, Mench JA, Olsson IAS, Hughes BO, editors. Animal Welfare 2nd Edition. CABI. pp.183–199.
12. Shields SJ, Garner JP, Mench JA (2004) Dustbathing by broiler chickens: a comparison of preference for four different substrates. Appl Anim Behav Sci 87: 69–82.

13. Widowski TM, Keeling LJ, Duncan IJH (1992) The preferences of hens for compact fluorescent over incandescent lighting. Can J Anim Sci 72: 203–211.
14. Bateson M (2004) Mechanisms of decision-making and the interpretation of choice tests. Anim Welf 13: 115–120.
15. Browne WJ, Caplen G, Statham P, Nicol CJ (2011) Mild environmental aversion is detected by a discrete-choice preference testing method but not by a free-access method. Appl Anim Behav Sci 134: 152–163.
16. Sørensen DB, Ladewig J, Ersbøl AK, Matthews L (2004) Using the cross point of demand functions to assess animal priorities. Anim Behav 68: 949–955.
17. Lima SL, Dill LM (1990) Behavioural decisions made under the risk of predation: a review and prospectus. Can J Zool 68: 619–640.
18. Perry CJ, Barron AB (2013) Honey bees selectively avoid difficult choices. Proc Natl Acad Sci USA 110: 19155–19159.
19. Cooper JJ, Mason GJ (2000) Increasing costs of access to resources cause re-scheduling of behaviour in American mink (Mustela vison): implications for the assessment of behavioural priorities. Appl Anim Behav Sci 66: 135–151.
20. Churchland AK, Kiani R, Shadlen MN (2008) Decision-making with multiple alternatives. Nat Neurosci 11: 693–702.
21. Manski CF (1975) Maximum score estimation of the stochastic utility model of choice. J Econom 3: 205–228.
22. Varian HR (1990) Goodness-of-fit in optimizing models. J Econom 46: 125–140.
23. Moffatt PG (2005) Stochastic choice and the allocation of cognitive effort. Exp Econ 8: 369–388.
24. Ventricelli M, Focaroli V, De Petrillo F, Macchitella L, Paglieri F, Addessi E (2013) How capuchin monkeys (Cebus apella) behaviourally cope with increasing delay in a self-control task. Behav Process 100: 146–152.
25. Zimmerman PH, Buijs SAF, Bolhuis JE, Keeling LJ (2011) Behaviour of domestic fowl in anticipation of positive and negative stimuli. Anim Behav 81: 569–577.
26. Cabanac M, Aizawa S (2000) Fever and tachycardia in a bird (Gallus domesticus) after simple handling. Physiol Behav 69: 541–545.

27. Edgar JL, Lowe JC, Paul ES, Nicol CJ (2011) Avian maternal response to chick distress. Proc R Soc B 278: 3129–3134.
28. Edgar JL, Nicol CJ, Pugh CA, Paul ES (2013) Surface temperature changes in repsonse to handling in domestic chickens. Physiol Behav 119: 195–200.
29. Moe RO, Nordgreen J, Janczak AM, Spruijt BM, Zanella AJ, et al. (2009) Trace classical conditioning as an approach to the study of reward-related behaviour in laying hens: A methodological study. Appl Anim Behav Sci 121: 171–178.
30. Moe RO, Stubsjøen SM, Bohlin J, Flø A, Bakken M (2012) Peripheral temperature drop in response to anticipation and consumption of a signaled palatable reward in laying hens (Gallus domesticus) . Physiol Behav 106: 527–533.
31. Von Borell E, Langbein J, Després G, Hansen S, Leterrier C, et al. (2007) Heart rate variability as a measure of autonomic regulation of cardiac activity for assessing stress and welfare in farm animals – A review. Physiol Behav 92: 293–316.
32. Kjaer JB, Jørgensen H (2011) Herat-rate variability in domestic chicken lines genetically selected on feather pecking behaviour. Genes Brain Behav 10: 747–755.
33. Cyr NE, Dickens MJ, Romero ML (2009) Heart-rate and heart-rate variabilirty responses to acute and chronic stress in a wild-caught passerine bird. Physiol Biochem Zool 82: 332–344.
34. Korte SM, Ruesnik W, Blokhuis HJ (1999) Heart-rate variability during manual restraint in chicks from high- and low-feather pecking lines of laying hens. Physiol Behav 65: 649–652.
35. Lowe JC, Abeyesinghe SM, Demmers TGM, Wathes CM, McKeegan DEF (2007) A novel telemetric logging system for recording physiological signals in unrestrained animals. Comput Electron Agr 57: 74–79.
36. Bruce EH, Prescott NB, Wathes CM (2003) Preferred food rewards for laying hens in behavioural experiments. Br Poult Sci 44: 345–349.
37. Task Force of the European Society of Cardiology, North American Society of Pacing and Electrophysiology (1996) Heart rate variability: standards of measurement, physiological interpretation and clinical use. Circulation 93: 1043–1065.

38. Smith DJ, Schull J, Strote J, McGee K, Egnor R, et al. (1995) The uncertain response in the bottlenosed dolphin (Tursiops truncatus). J Exp Psychol 124: 391–408.
39. Suda-King C, Bania AE, Stromberg EE, Subiaul F (2013) Gorillas' use of the escape response in object choice memory tests. Animal Cognition 16: 65–84.
40. Mandler M (2005) Incomplete preferences and rational intransitivity of choice. Games Econ Behav 50: 255–277.
41. Eliaz K, Ok EA (2006) Indifference or indecisiveness? Choice theoretic foundations of incomplete preferences. Games Econ Behav 56: 61–86.
42. Danan E (2010) Randomization vs. Selection: How to choose in the absence of preference? Manage Sci 56: 503–518.

SECTION 5: HORSES

CHAPTER 8

Effects of Pre-Conditioning on Behavior and Physiology of Horses During a Standardised Learning Task

Kate Fenner[1], Holly Webb[2], Melissa J. Starling[2], Rafael Freire[1], Petra Buckley[1], Paul D. McGreevy[2]

[1]School of Animal and Veterinary Sciences, Charles Sturt University, Wagga Wagga, NSW, Australia,

[2]Faculty of Veterinary Science, University of Sydney, NSW, Australia

ABSTRACT

Rein tension is used to apply pressure to control both ridden and unridden horses. The pressure is delivered by equipment such as the bit, which may restrict voluntary movement and cause changes in behavior and physiology. Managing the effects of such pressure on arousal level and behavioral

Citation: Fenner K, Webb H, Starling MJ, Freire R, Buckley P, McGreevy PD (2017) Effects of pre-conditioning on behavior and physiology of horses during a standardised learning task. PLoS ONE 12(3): e0174313. https://doi.org/10.1371/journal.pone.0174313

indicators will optimise horse learning outcomes. This study examined the effect of training horses to turn away from bit pressure on cardiac outcomes and behavior (including responsiveness) over the course of eight trials in a standardised learning task. The experimental procedure consisted of a resting phase, treatment/control phase, standardised learning trials requiring the horses ($n = 68$) to step backwards in response to bit pressure and a recovery phase. As expected, heart rate increased ($P = 0.028$) when the handler applied rein tension during the treatment phase. The amount of rein tension required to elicit a response during treatment was higher on the left than the right rein ($P = 0.009$). Total rein tension required for trials reduced ($P < 0.001$) as they progressed, as did time taken ($P < 0.001$) and steps taken ($P < 0.001$). The incidence of head tossing decreased ($P = 0.015$) with the progression of the trials and was higher ($P = 0.018$) for the control horses than the treated horses. These results suggest that preparing the horses for the lesson and slightly raising their arousal levels, improved learning outcomes.

INTRODUCTION

Training can be stressful for horses but little is known about how best to prepare them for effective learning. Some riders currently prepare their horses to be ridden by chasing them around the round pen [1] and lunging [2]. Unfortunately, these exercises are designed to tire, rather than to mentally engage, the horse and can possibly therefore compromise learning from the outset of the lesson. Human studies reveal that cognitive impairment results from both anxiety-induced increases in arousal level and exercise-induced fatigue [3]. While exercise-induced fatigue has been reported to produce both positive and detrimental cognitive effects [4], it is known that chasing the horse should always be avoided [5]. Inducing fear prior to learning is not commensurate with good training but is encouraged by some of the world's most popular trainers [2].

Learning theory describes how animals absorb, process and retain information through stimulus-response-reinforcement chains [6, 7], together with the emotional and environmental factors that influence this [8]. Many horse trainers and equestrian coaches are unaware of how horses learn or of the mechanisms that underpin positive and negative reinforcement and punishment [9]. This knowledge gap compromises coaches' ability to study the emergent equitation science literature and to convey information about learning theory and training to students and horse owners [9].

Riders and handlers need to understand the relationship between distress and cognition, to recognise events that horses may find stressful and to identify behavioral responses that suggest that a horse is overly emotional. These clues tell trainers when to reduce stressors in training. Emotional reactivity refers to physiological, chemical and behavioral changes resulting from so-called emotional stimulation [10]. Such stimulation can arise from a variety of novel or otherwise potentially frightening stimuli in the horse's environment. We know that emotionally reactive horses can be difficult to handle [11] and horse-human relationships improve when the horses' emotional reactivity is minimized [12]. Repeated exposure to increases in emotional level impairs cognitive function [13]. However, any such impairment is attenuated when the stressor is predictable [14]. The opposite of a highly emotionally reactive horse may be one that is either inherently relaxed, has been systemically desensitized to pressure, or may even be in a state of learned helplessness as a result [15]. Learned helplessness refers to a state where the horse has repeatedly been unable to escape aversive stimuli, such as bit or leg pressure, and eventually simply stops trialling any behaviors in an attempt to do so [15].

Any apparatus that applies pressure or constrains a horse's movement has the potential to compromise welfare [16]. In equitation, increases in rein tension primarily provide directional and deceleration cues from the handler or rider to the horse and, secondly, modify the head and neck position [17]. Reins tension converts to pressure on various parts of the horses' mouth and face and the timely release of this pressure communicates to the horse that it has made the correct response. This process is known as negative reinforcement, which is the removal of aversive stimuli to reward a desired response [18]. So, it is recommended that rein tension be applied minimally and consistently and as appropriate to the task. For effective and humane negative reinforcement, pressure-release relies on good timing and consistency [19]. Inconsistency and delays in timing may inadvertently punish the horse, untrain established responses and impede the acquisition of desirable ones [20].

For most horses, accommodating the bit is uneventful and they become habituated to having the bit in their mouth [21]. While habituation to the presence of the bit is desirable, habituation to additional bit pressure from rein use compromises effective training because it means that horses no longer respond to light pressure. This can lead to the need for excessive rein use and severe bits [19] that can threaten horse welfare. Mean rein tension ranges have been reported between 3 to 20 N by Heleski, McGreevy, Kaiser,

Lavagnino, Tans, Bello and Clayton [17], but higher ranges of 40 to 75 N have also been reported [22, 23]. To avoid habituation to bit pressure and assure animal welfare, rein tensions used in training and competition should be kept to a minimum.

Previous rein tension studies have described the role rein cues play in training horses to slow down [24] and avoid bit pressure [18] but none have reported the physiological and behavioral responses that accompany increased rein tension during a standardised learning task. Physiological parameters, such as heart rate, in combination with behavioral scoring may help to reveal the complexity of equine responses to bit pressure. Such responses may have important implications for welfare during training and competition [25]. Equally, some physiological and behavioral responses are of importance to competitors since they may compromise performances.

The aim of this study was to determine the effect of a pre-conditioning exercise, namely giving to bit pressure, on arousal level (measured by heart rate and behavior) and learning outcomes (measured by rein tension, time, step number and behavior) of horses. The pre-conditioning treatment was designed to engage the horse with the lesson. The purpose of the treatment was to achieve a moderately elevated, stable heart rate that was maintained throughout the pre-conditioning and subsequent trials, followed by a return to resting levels at the conclusion of the trial phase. It was hypothesized that the pre-conditioning would prepare the horses to respond appropriately to bit pressure and improve their performance in the trials that immediately followed.

MATERIALS AND METHODS

The protocol and conduct of this study were approved by the Charles Sturt University Animal Care and Ethics Committee, New South Wales, Australia (ACEC protocol number 14/030).

Selection of Horses

Eligible horses were selected from a population of riding horses. Most of the selected horses had been previously been taught to back-up from pressure on the nose with a head collar, two reins together from the ground or saddle, voice cues as well as other cues, such as tapping on the forelegs or chest. However all of them were naive to backing from rein tension from a single rein in the absence of all other signals. Horses were randomly assigned to either the treatment or control group, using an Excel generated randomization spreadsheet. They were recreational, pleasure riding or companion horses, and were not engaged in high-level competition. Horses that were active in adult riding club, pony club, trail riding, and riding school activities and that were habituated to wearing a girth and a bridle with a simple snaffle bit were eligible for inclusion.

Sixty-eight horses of various breeds were selected for the study. They comprised 35 geldings, 4 stallions and 29 mares. Horses had mixed training histories, ranging from a basic start to many years of consistent training (mean duration of training 3.1 ± 2.3 years). They were between 94.5 cm (9.3 hh) and 174.8 cm (17.2 hh) in height with a mean height of 149.3 ± 15.24 cm (14.3 ± 1.2 hh), and were aged from 2 to 25 years (mean age 9.8 ± 6.1 years). Sixty-two of the horses were housed in paddocks full-time, while six occasionally spent either the day or the night in a stable.

Data Collection

Data collection took place in an arena or property that was familiar to each horse. Fenced arenas were used where available and a safe, flat, fenced area, familiar to the horses, was chosen if no arena was available.

Immediately prior to data collection, each horse was fitted with a Centaur Trainology Rein Tension Device (Centaur Consulting, Vijverhof 27, 3734 DB Den Dolder, The Netherlands) attached to a full-cheek snaffle bit, held in place by an open bridle with no throatlatch, brow band or noseband. A webbing head collar was fitted underneath the bridle and leather reins were connected to the bit via the rein tension device. A surcingle was fastened on each horse and a Polar heart rate monitor (RS800CX) was attached with electrode pads placed under the surcingle with one behind the left elbow and the other at the withers. Video recordings were taken for each horse for the entire duration of the procedure. A Sony HDR-PJ790 camera (Sony Corporation of America, Sony Drive, Park Ridge, NJ 07656, USA), placed 10 m from the experimental area, was panned horizontally as the horse moved between experimental phases (see Fig 1). Baseline and recovery phases took place in the same area, approximately 10m from the test corridor. Each of the four experimental phases was recorded, resulting in a total of approximately 30 minutes of footage per horse.

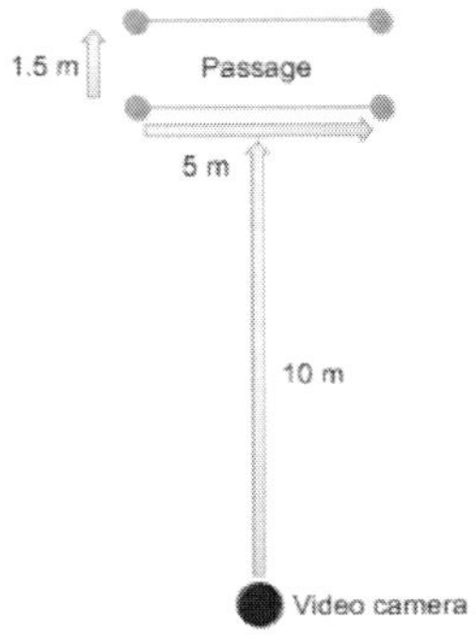

Figure 1. Experimental area.

The passage was 5 m x 1.5 m. The camera was 10 m from the passage and turned as the horse moved between phases.

Behavior observations

A behavioral observation sheet was developed for this study (see Table 1). Behavior was counted and coded using the Observer XT (Noldus Information Technoloty bv, Wageningen, The Netherlands, v11.5, 2013). A possible limitation of this study is that the person scoring the observer was not blinded to the treatments. The videos were divided into resting, treatment/control, trials and recovery phases and all of the trial phases were scored initially, followed by the recovery, resting and finally the treatment phases. In addition, a second person also scored a random sample of 20 videos that matched the initial findings. It was concluded that this resulted in a very low risk of bias. An improvement in future studies could be to have the all scoring completed by a person blinded to treatment.

Table 1. Behavioral observation sheet developed for this study.

Behavior	Definition
Head toss	Quick upward vertical movement of the head.
Handler contact	Touching the handler with the muzzle.
Paw	Hoof being lifted off the ground and extended forward before the toe is quickly dragged backward against the ground.
Yawn	Deep inhalation with the mouth wide open and jaws either opened or moved from side to side.
Chew on bit	Licking of the lips and opening the mouth so that a gap is apparent between the mandible and maxilla.

Heart rate

Horse heart rate was recorded at one-second intervals for the entire experimental period, starting with the resting phase and finishing when the recovery phase was completed. The data were stored in the Polar unit and later downloaded to the Polar Equine Pro Trainer 5 software for analysis.

Rein tension

Rein tension data, expressed in Newtons (N), were sampled at a rate of 100 Hz with a resolution of 21.6 +/- 1.8 Newton/Bits. This resulted in a large number of readings per horse from which one reading per second was taken for the purposes of this analysis. This was done by extracting each of the one hundredth readings, giving one reading per second. The rein tension monitors were calibrated at the start of each testing day. The data were live-streamed to a Windows based Hewlett Packard computer (HP Inc., 1501 Page Mill Road, Palo Alto, CA 94304, USA) within the testing area.

Experimental procedure

Resting phase

The horse was led from the stable block to the experimental area, using a webbing head collar and lead rope. The horse was bridled with rein tension monitors linking the reins to the bit and the reins were placed over the horse's neck. The video recorder and heart rate monitor were activated and the resting phase began. The handler loosely held the horse by a lead rope for 5 minutes.

Treatment and control phase

The 31 horses allocated to the treatment group were pre-conditioned with a negatively reinforced exercise known as 'give-to-the-bit' [26], a simple pressure-release activity that was repeated for 8 minutes. Pilot trials revealed that 8 minutes of treatment was optimal for horses to learn the lesson without becoming fatigued or restless. This involved the handler standing on one side of the horse, 50cm from the neck, facing the horse and positioned midway between the nose and shoulder, and applying tension to one rein, then releasing the tension when the horse turned its head to the side, laterally towards the handler, i.e., away from the pressure of the bit. Lateral movement, of more than 10cm, constituted a response and rein tension was released. If the horse did not move, lateral tension was gradually increased to the point that movement was induced. The horse was cued to give to pressure three times on one side and then the handler moved to the other side of the horse to repeat the exercise. Release of pressure was the only reinforcement used, the horses were not verbally praised or touched by the handler. The horse remained standing still during the treatment, with only lateral movement of the head and neck. Applying tension to one rein can signal the horse to move laterally. Horses that did move in the initial stages of treatment, stepped forwards. The average duration of each pre-conditioning treatment was 30 seconds per side.

Control horses (n = 37) were held for 8 minutes with the handler moving to the left and right side every minute, mimicking the movement of the handler in the treatment.

Trial phase

Immediately upon completion of the treatment and control phase, each horse was given to a different, trial handler (blinded to treatment allocation) and then led through the test corridor (5 m x 1.5 m) until the front feet were over the demarcated start line. The test corridor was 5 metres in length, marked with bollards at either end and designated by a fence and poles on the ground (See Fig 1) to guide movement in a straight line. The fence was used to prevent horses moving their hindquarters to the right when only one rein, in the first trial the left rein, was first picked up. The handler applied tension on the left rein in a caudal direction, cueing the horse to step backwards. When the horse stepped back, with any foot, all pressure was released. If the horse did not step backwards, rein tension was steadily increased until a step was taken. This procedure was repeated until the horse stepped out of the five-metre test corridor with the front feet.

The horse was then led, using only the lead rope attached to the head collar, around to the other end of the test corridor, up through the corridor and the handler halted the horse with the front feet beyond the start line, outside the corridor. The horse was led forwards and halted using the lead rope attached to the head collar and rein tension was not applied until the horse was halted and ready to begin the trial. The handler then moved to the right side of the horse and applied slight caudal tension on the right rein only, increasing tension steadily, if necessary, until the horse took a step backwards. As soon as the horse stepped back, the rein tension was released and the process repeated until the horse had exited the corridor with the front feet. Trials alternated between the left and right rein for a total of eight trials, beginning on the left, with four trials performed on each rein.

Recovery phase

The horse was held via a loose lead rope connected to the head collar for 5 minutes while recovery heart rate, heart rate variability and behavior were monitored.

Statistical analysis

A split plot experimental design was used in which horses from both treatment and control groups received eight trials. Heart rate, rein tension, time and steps were compared using a mixed model General Mixed Model with trial (8 trials in total), treatment (treated or control) and side (left or right) as factors and horse identity as a random effect (REML command, GenstatTM, 17th edition, VSN International Ltd, Waterhouse Street, Hemel Hempstead, UK). Prior to undertaking the REML analysis, the distribution of the data was visually inspected and tested for normality using a Shapiro-Wilk test. As all tests yielded a $P>0.1$, it was deemed that the data met assumptions for parametric analysis. Only the treatment.trial interaction was included in the final models. A visual inspection of the Polar heart rate data revealed some Type 1 errors [single transient spikes or troughs with great deviation from the surrounding data [27]] and corrections were made using the 'low' filter within the Polar software. Data were missing for the recovery phase of two horses.

The behavioral observation analysis used a generalised linear mixed model with binomial distribution for the number of events. A generalised model was used because behavioural elements were either seen/ not seen at each sampling point. Horse was used as a random effect and the probability of the behavioral event, as a proportion of the total number of observations, was reported.

RESULTS

Behavioral observation

As the trials progressed, horses tossed their heads significantly less frequently ($P = 0.015$). For both treatment and control groups, head tossing was significantly more frequent during the trial phase of the experiment than in any other phase ($P < 0.001$) and control horses tossed their heads significantly more ($P = 0.018$) than treated horses. Among the treated horses, 71 per cent tossed their heads compared with 89 per cent of control horses. The right hand trials elicited more head tossing than the left hand trials for both the control and treated horses (see Fig 2). Horses did not toss their heads when led forward using the head collar and lead rope. All head tossing observed occurred when tension was being applied to the reins.

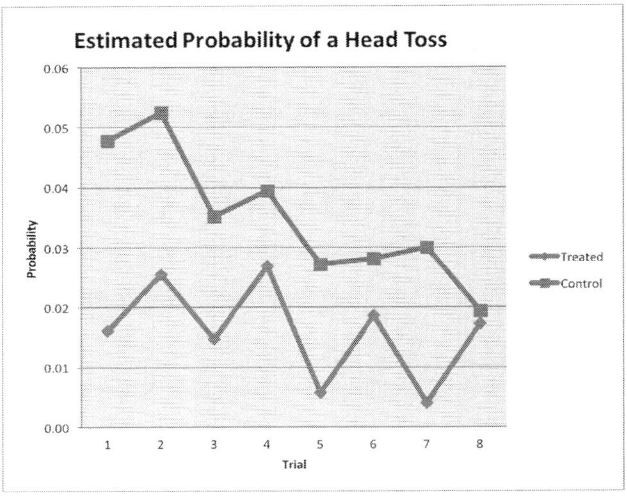

Figure 2. Estimated probability of a horse tossing the head during the trial phase of the trials.

Trials 1, 3, 5 and 7 –left rein, trials 2, 4, 6 and 8 –right rein. Treated horses n = 23, control horses n = 33.

The number of times the horses contacted the handler varied significantly (GLMM, $F_{3,\ 198.3} = 42.56$, $P < 0.001$) over the different phases of the experiment, with the horses contacting the handler more frequently in the resting and recovery phases than in the treatment phase. The frequency of this behavioral response was not significantly different between the treated and control allocated horses (GLMM, $F_{1,\ 59.4} = 0.91$, $P = 0.343$).

Only fourteen horses, across both the treatment and control groups displayed pawing behavior, making analysis of this trait meaningless. The same reservations held for yawning, where only 58 events occurred. Analysis of ‘chewing on the bit’ failed to converge, as only eighteen horses displayed the behavior.

Heart Rate Analysis

Treated horses had significantly higher heart rates during the treatment phase than the control horses (GLM, $F_{1, 66} = 5.01$, $P = 0.028$). While a significant difference was found in heart rate across the trials (1–8) of individual horses (REML, $F_{6, 462} = 4.07$, $P < 0.001$), this difference was not significant between the left (Trials 1, 3, 5 and 7) and the right sides (Trials 2, 4, 6 and 8) (REML, $F_{1, 462} = 0.15$, $P = 0.281$) nor between the treated and control horses (RML, $F_{7, 462} = 1.30$, $P = 0.246$) (see Fig 3). During the recovery phase, both the treated and control group horses' heart rates returned to their pre-trial resting rates (see Fig 3).

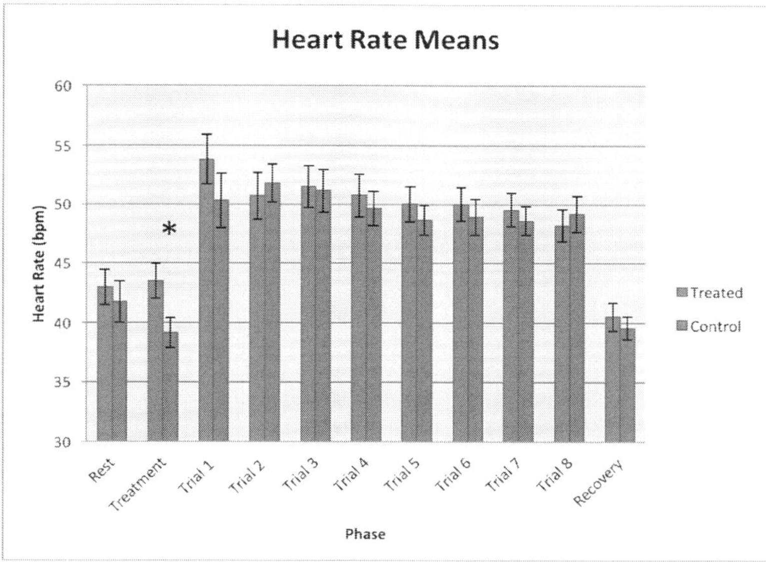

Figure 3. Heart rate means (bpm) for each of the experimental phases.

Treated horses had significantly higher heart rates during the Treatment phase ($P = 0.028$).

Rein tension analysis

Both total, the sum of each rein tension measurement taken per second during the trial, and mean rein tension per trial were analysed because the latency to complete each trial varied greatly. For comparative purposes in the current study, a total figure was considered more useful than the mean as it more faithfully represented the tension applied by the handler.

A significant difference between the mean rein tension required on the left and right reins was found during the treatment phase (t-test: t_{30} = 2.775, P = 0.009). Significantly more tension was required to elicit responses on the left rein (see Fig 4).

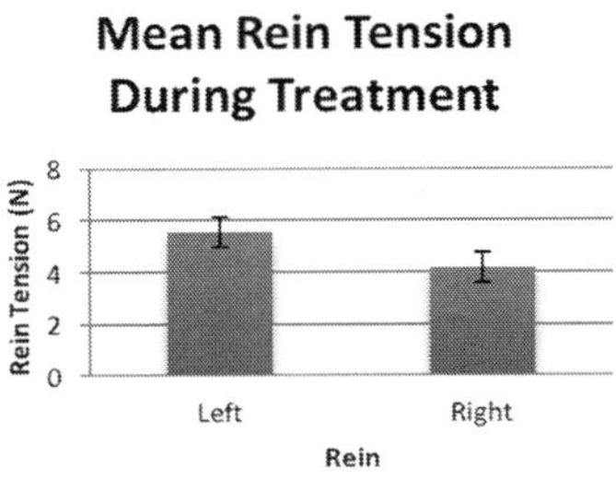

Figure 4. The significantly different (t-test, P = 0.009) mean rein tensions for left and right reins required to elicit a give-to-the-bit response during treatment (N = 31).

No significant difference in mean rein tension was found over the trials for each individual horse (REML, $F_{6, 462}$ = 0.92, P = 0.481) (see Fig 5).

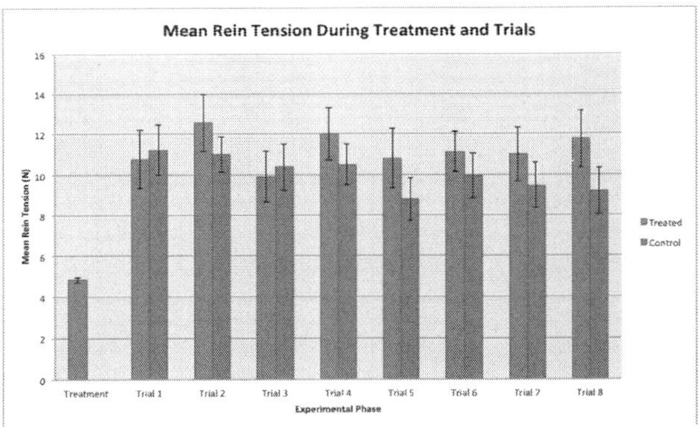

Figure 5. Mean rein tension (Newtons) during the treatment (n = 31) and trial (n = 68) phases of the experiment.

No significant differences were found between the trials (REML, P = 0.481) or between the treated and control groups (REML, P = 0.349).

A significant difference was found in the total rein tension exerted by the handler across the trials (REML, $F_{6, 462}$ = 6.10, P < 0.001). As the trials progressed, horses required less rein tension to back the 5 metres through the test corridor (see Fig 6).

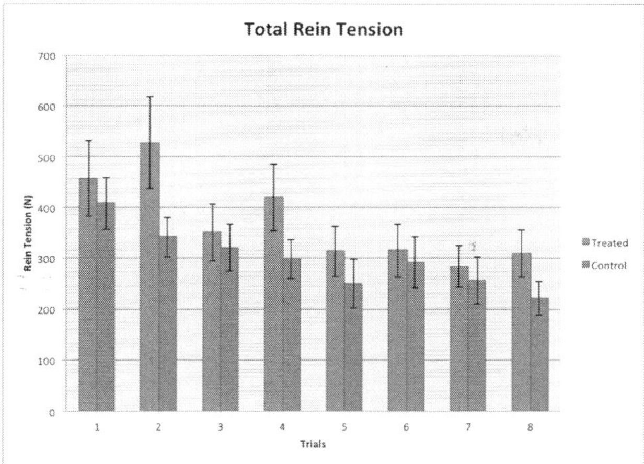

Figure 6. Total rein tension (in Newtons) required back-up the horse 5 metres (n = 68).

Significant differences found between the trials (REML, $P < 0.001$) but not between the treated and control groups (REML, $P = 0.146$) or between the left and right reins (REML, $P = 0.716$).

For both treated and control horses, there was a significant decrease in the number of rein tension events across all trials ($P < 0.001$), a slowly declining trend was detected from the following figures (Trial 1 = 43, Trial 2 = 31, Trial 3 = 30, Trial 4 = 29, Trial 5 = 26, Trial 6 = 24, Trial 7 = 25 and Trial 8 = 23 rein tension events) (see Fig 7). There was no significant difference in the mean number of rein tension events between the treatment and control groups ($P = 0.294$) during the trials.

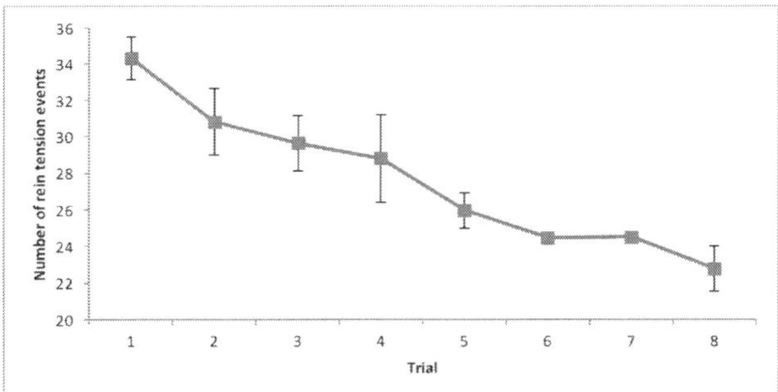

Figure 7. The decreasing ($P < 0.001$) mean number of rein tension events recorded for 68 horses during the standardised learning task (Trial 1–8) with standard error bars.

Time and steps

Time for each trial was measured from the first step taken by the horse at the start of the corridor until the horse finally exited the corridor. The number of steps the horse took with the front feet were counted while the horse reversed along the 5m corridor. Throughout the series of eight trials, all horses travelled through the test corridor faster in subsequent trials (REML, $F_{1, 41.7} = 41.67$, $P < 0.001$; Fig 8). However, no significant difference in time taken was found between the treated and control groups (REML, $F_{1, 41.7} = 0.77$, $P = 0.380$; Fig 8).

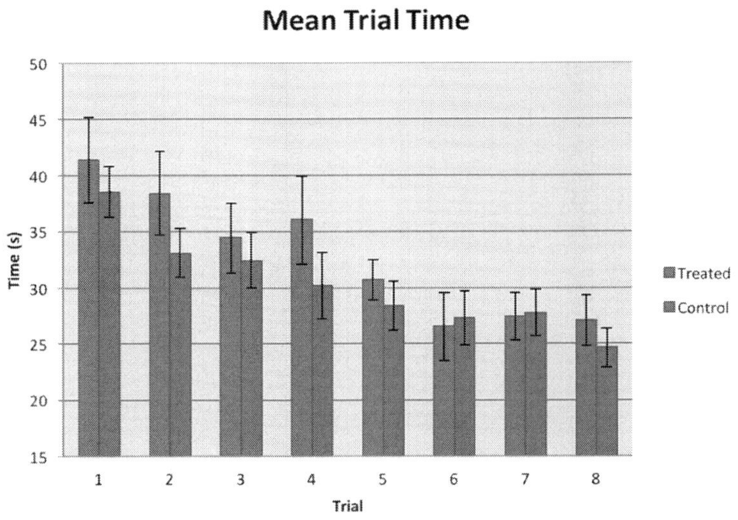

Figure 8. Mean trial time.

Significant differences were found between the trials (REML, $P < 0.001$) but not between the treated and control groups (REML, $P = 0.380$).

The number of steps taken by each horse decreased as the trials progressed (REML, $F_{6, 462} = 10.65$, $P < 0.001$; Fig 9), but there were no significant differences between the treatment and control allocated horses (REML, $F_{7, 462} = 6.35$, $P = 0.501$; Fig 9). There was a significant difference between the left and right sides ($F_{1, 462} = 0.27$, $P < 0.001$), with horses taking fewer steps when being handled on the right than the left side (see Fig 9).

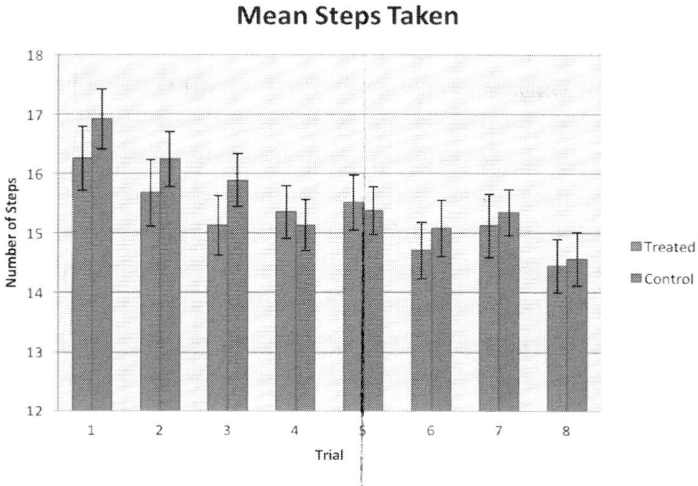

Figure 9. Mean number of steps taken per trial.

Significant differences were found between the trials (REML, $P < 0.001$) but not between the treated and control groups (REML, $P = 0.501$). Significant differences were also found between the left (Trials 1, 3, 5 and 7) and right (Trials 2, 4, 6 and8) sides (REML, $P < 0.001$).

DISCUSSION

To facilitate the production of safe riding horses, training that is consistent and adheres to learning theory is required [28]. Ideally, in any given lesson, the horse should learn the lesson in the fastest possible time with the least amount of associated stress, to produce a calm response. Developing easily accessible tools to assess the emotional state and arousal level [29] of the horse should optimize training, improve welfare and reduce wastage. An increase in heart rate can indicate increased arousal [30] as can behavioral parameters such as head tossing that, arguably, are easier to observe. This, together with appropriate reinforcement schedules (quantified by the use of rein tension meters to monitor the use of bit pressure and, more importantly, the release of such pressure) will help to define best practice in horse training.

In the current study, as the trials of the standardised learning task progressed, horses optimised their responses in several ways. They took fewer steps to complete the 5 m transit for each trial, indicating that they were increasing their stride length in each trial. They also completed the trials more quickly and with less rein pressure. Head tossing also reduced in a stepwise manner as the trials progressed, suggesting a reduction in stress. However, this was not reflected in the cardiac responses until the recovery phase. Head tossing occurred when rein tension was applied during the trials and the treated horses tossed their heads significantly less than the control horses. Pre-conditioning the horses to respond to rein tension resulted in fewer head tossing incidents during the trials phase of the lesson. Head tossing is seen as a conflict behavior that reduces perceived rideability and positive temperament traits in horses [31].

Effective use of negative reinforcement depends on timing the release of the applied pressure, and the difference between a good and an average trainer manifests chiefly in that ability. As McLean [32] argues, accurate timing of reinforcements should not only improve both training efficiency and performance by reducing the occurrence of conflict behaviors but also reduce horse wastage and improve welfare. The current findings, showing a significant reduction in the number of times horses tossed their heads as the trials progressed, indicate that horses learnt from the pre-conditioning exercise and that timing of the release was effective.

This study further supports the findings of others that bit pressure is aversive to horses. Christensen, Zharkikh, Antoine and Malmkvist [18] showed that horses voluntarily tolerated tensions of up to 11 N but did not habituate to greater tensions. In the current study, the required rein tensions steadily declined over the course of the eight trials suggests that, instead of habituating to bit pressure, the horses learned to avoid it. König Von Borstel and Glißman [33] and Heleski et al. [17] reported mean rein tension ranges between 9.1 ± 1.6 N and 21.7 ± 1.3 N and 3 to 20 N, respectively. These values correspond to the tensions recorded in the current study for the treatment and trials. The values in the present study were lower than those published by Clayton, Singleton, Lanovaz and Cloud [22] and Preuschoft, Witte, Recknagel, Bar, Lesch and Wuthrich [23] who reported ranges of 40 to 75 N. As the learning trials progressed, the mean required rein tension decreased. Similarly, the number of rein tension events decreased across trials, demonstrating that horses responded swiftly to the onset of pressure. This indicates that regardless of treatment, the horses did not habituate to pressure, but rather responded sooner to lighter pressure applied by the handler. This confirms that the handler's pressure-release was providing negative reinforcement for the required operant response.

A few studies have examined rein tension at different gaits in ridden horses [[34]; [35]; [36]; [37]], while another measured tensions applied to a mechanical horse [38]. Only one study has investigated tensions required for learning a new lesson [37]. This showed that higher rein tensions were required for horses being long-reined or driven from the ground (using long lines connected to the bit) through a set pattern, than those being ridden through the same pattern. However, the authors pointed out that the long-reined horses may not have been familiar with this exercise and proposed that this may have accounted for the higher tensions required. While it is probable that long-reining itself does require more tension than riding, given the weight and length of the rein, resulting in a delayed and partial release of tension at best, it would be illuminating to repeat that study over an extend period to see if required rein tension reduced as the lesson was learned.

In a recent study of negative reinforcement by Ahrendt, Labouriau, Malmkvist, Nicol and Christensen [39] horses showed a significant decrease in the pressure required to complete each trial; in this case, the operant task was to step laterally from pressure applied to the hindquarter. Interestingly, these researchers used horses that were completely naïve to the exercise but still found significant laterality in the results, with more pressure being required on the right side of the horse. It was thought that this could be due to the position of the experimenter, who used the right hand to apply the pressure to both sides of the horse. The horses in the current study required more pressure on the left rein than the right during treatment which may be the result of habituation to bit pressure prior to the study or an inherent bias.

For the current study, it was decided not to randomize the order of the left and right trials, although it may useful to do so in future work. While the horses were all well-handled, horses are conventionally handled more on the left from the ground than the right [40]. The decision not to randomize the trial order was taken after carrying out a pilot trial with a small number of horses in which rein tension and heart rate parameters were monitored. This showed that horses that began trials on the right side required considerably higher rein tensions and had higher heart rates than their counter-parts beginning on the left rein. Ahrendt et al. [39] did randomize sides in their study but nevertheless found more pressure was required on the right side of the horse to obtain the desired response and concluded that learning was not transferred from one side to the other.

Interestingly, during treatment in the current study, horses required significantly less tension on the right to respond appropriately in the simple pressure-release task of give-to-the-bit. Our findings suggest that this may be the result of the horses trialling the newly learned give-to-the-bit behavior more on the right than the left. This could result in horses learning the initial give-to-the-bit lesson, on the right, with less pressure being required initially but then more pressure being requiring when being cued to move backwards from a stimulus with caudal, rather than lateral, rein tension [26]. As most experienced horses generally have had more handling on the left [40], the current cohort may reflect an acquired decrease in sensitivity to the left rein, as found during the treatment phase of the experiment.

The trialling of the newly learned behavior trained here may not occur to the same extent in a less formal training situation. Under non-experimental conditions, where the variables would not have to be isolated and measured, the handler could simplify the lesson by adding another form of reinforcement. This would possibly be positive reinforcement for each spontaneous step back or another form of negative reinforcement such as placing a hand on the horse's chest, tapping lightly on the cannon bone with a dressage whip, moving in front of the horse or verbally cueing the horse to move back. Future studies could use an alternative follow-up lesson, such as the horse learning to load on to a trailer [41], to laterally yield the hindquarters using an algometer [39] or to walk over a bridge or tarpaulin. By not relying on bit pressure, these exercises would be significantly different from the treatment task. It would also be interesting to introduce other possible means of engaging the horse with alternative pressure-release exercises, such as head lowering from a halter [42] or a hindquarter yielding exercise.

Rietmann et al. [14] used a ground-work exercise (involving backing-up) that was similar to the current study, but considerably more challenging as the horses had to back-up for a full three minutes and various forms of reinforcement were used. Unfortunately, mixing positive, secondary positive and negative reinforcement makes the reinforcement schedule impossible to quantify and no attempt was made to do so. Rietmann et al. [14] reported that resting mean heart rates increased by 166 per cent with the initial backwards-walking exercise and that these rates came down to 85 per cent after some backing training was completed. However, such results are in stark contrast to the results of the current study, where mean heart rate rose by only 16 percent between the baseline collection period and the backing-up exercise. Rietmann et al. [14] suggest that the decrease they observed after training was the result of the horses now recognizing the predictability of the environment and learning that the stressor, the pressure from the handler, was predictable and controllable. Their horses walked backwards for a full three minutes, whereas those in the current study had to back-up for only 5 metres before walking forwards and around through the corridor again, to start the next trial. This more timely reward in the release of pressure may have improved learning outcomes, as the decrease in parameters including time, rein tension required, head tossing and steps taken indicate. The decrease in head tossing incidents suggests that horses are less stressed when given a pre-conditioning exercise in preparation for the lesson. Future research to accurately define the point at which the horse is optimally aroused for a given task should be considered a priority [43]. This optimal arousal state could be designated the engagement zone.

The collection of benchmark rein tension data is an important step towards improving horse welfare through trainer, coach and rider education. A quantitative definition of appropriate or ideal 'contact' and an understanding of how to correctly apply negative reinforcement so that horses become more responsive to pressure over time instead of desensitized to it will improve horse welfare. Whilst trainers, coaches and riders can all benefit from being made more aware of tension and also be incentivised to reduce overall rein tension over time [44], specifically targeting coaches and trainers has the potential to disseminate information widely and rapidly given their extension capacity.

CONCLUSIONS

A simple pre-conditioning pressure-release exercise was used to engage the horses in an operant locomotory task. The exercise significantly increased heart rate, indicative of a moderate increase in arousal. Both the treated and control horses had similarly raised heart rates during the trials and both returned to baseline rates immediately following the trials. Both groups learned to avoid bit pressure by stepping back with longer strides and moving more quickly across the course of the trials. The treated horses exhibited significantly less head tossing than their untreated counterparts during the trial phase of the experiment. This suggests that engaging the horse prior to training may lay the foundation for a better learning experience.

ACKNOWLEDGMENTS

Grateful thanks are extended to the following people: Ms Dione Sloane for horse handling, Dr Mick O'Neil for assistance with statistics and Dr Michelle Hyde for her assistance with the writing of the article. Andrew McLean is thanked for advice on the experimental design.

FUNDING STATEMENT

Author Kate Fenner prior to being affiliated with Kandoo Equine. Kate Fenner's affiliation during the time of the study was Charles Sturt University. Kandoo Equine it provides Kate Fenner a salary. The specific roles of these authors are articulated in the 'author contributions' section. Kandoo Equine played no part in the study. During the time of the study, the authors received no specific funding.

REFERENCES

1. Henshall C, McGreevy PD. The role of ethology in round pen horse training—A review. Applied Animal Behaviour Science. 2014;155:1–11.
2. Anderson C. Lessons Well Learned. Vermont, USA: Exisle Publishing Limited; 2009.
3. Nibbeling N, Oudejans R.R.D, Ubink E.M, Daanen H.A.M. The effects of anxiety and exercise-induced fatigue on shooting accuracy and cognitive performance in infantry soldiers. Ergonomics. 2014;57(9):1366–79. 10.1080/00140139.2014.924572
4. Lambourne K, Tomporowski P. The effect of exercise-induced arousal on cognitive task performance: A meta-regression analysis. Brain Research. 2010;1341:12–24. 10.1016/j.brainres.2010.03.091
5. Goodwin D, McGreevy PD, Waran N, McLean AN. How equitation science can elucidate and refine horsemanship techniques. The Veterinary Journal. 2009;181:5–11. 10.1016/j.tvjl.2009.03.023
6. Cooper JJ. Comparative learning theory and its application in the training of horses. Equine Clinical Behaviour. 1998;27:39–43.
7. Cooper JJ. Comparative learning theory and its application in the training of horses. Equine Clinical Behaviour. 1998;(27):39–43.
8. Starling MJ, Cody D, Branson N, McGreevy PD. Conceptualising the Impact of Arousal and Affective State on Training Outcomes of Operant Conditioning. Animals. 2013;3(2):300 10.3390/ani3020300 [PMC free article]
9. Warren-Smith AK, McGreevy PD. Equestrian coaches' understanding and application of learning theory in horse training.(Report).Anthrozoos. 2008a;21(2):153.
10. Hall C. Emotional behavior in the rat. I. Defecation and urination as measures of individual differences in emotionality. Journal of Comparative Psychology. 1935;18(3):385–403.
11. McGreevy PD, Henshall C, Starling MJ, McLean AN, Boakes RA. The importance of safety signals in animal handling and training. Journal of Veterinary Behavior: Clinical Applications and Research. 2014;6:382–7.
12. Fureix C, Pagès M, Bon R, Lassalle J- M, Kuntz P, Gonzalez G. A preliminary study of the effects of handling type on horses' emotional reactivity and the human–horse relationship. Behavioural Processes. 2009;82(2):202–10. 10.1016/j.beproc.2009.06.012

13. McEwen BS. Stress and hippocampal plasticity. Annu Rev Neurosci. 1999;22:105–22. 10.1146/annurev.neuro.22.1.105
14. Rietmann TR, Stuart AEA, Bernasconi P, Stauffacher M, Auer JA, Weishaupt MA. Assessment of mental stress in warmblood horses: heart rate variability in comparison to heart rate and selected behavioural parameters. Applied Animal Behaviour Science. 2004;88(1):121–36.
15. Hall C, Goodwin D, Heleski C, Randle H, Waran N. Is There Evidence of Learned Helplessness in Horses? Journal of Applied Animal Welfare Science. 2008;11(3):249–66. 10.1080/10888700802101130
16. McGreevy P, Warren-Smith A, Guisard Y. The effect of double bridles and jaw-clamping crank nosebands on temperature of eyes and facial skin of horses. Journal of Veterinary Behavior: Clinical Applications and Research. 2012;7(3):142–8.
17. Heleski CR, McGreevy PD, Kaiser LJ, Lavagnino M, Tans E, Bello N, et al. Effects on behaviour and rein tension on horses ridden with or without martingales and rein inserts. The Veterinary Journal. 2009;181(1):56–62. 10.1016/j.tvjl.2009.03.011
18. Christensen JW, Zharkikh TL, Antoine A, Malmkvist J. Rein tension acceptance in young horses in a voluntary test situation. Equine veterinary journal. 2011;43(2):223 10.1111/j.2042-3306.2010.00151.x
19. McGreevy PD. The fine line between pressure and pain: Ask the horse. The Veterinary Journal. 2011;188(3):250–1. 10.1016/j.tvjl.2010.10.011
20. McGreevy PD, McLean AN. Punishment in horse-training and the concept of ethical equitation. Journal of Veterinary Behavior: Clinical Applications and Research. 2009;4(5):193–7.
21. Christensen JW, Rundgren M, Olsson K. Training methods for horses: habituation to a frightening stimulus. Equine Veterinary Journal. 2006;38(5):439–43.
22. Clayton HM, Singleton WH, Lanovaz JL, Cloud GL. Measurement of rein tension during horseback riding using strain gage transducers. Experimental Techniques. 2003. May-Jun May/Jun 2003:34–6.
23. Preuschoft H, Witte H, Recknagel S, Bar H, Lesch C, Wuthrich M. Effect of common head gear on horses. Dtsch Tierarztl Wochenschr. 1999;106(4):169–75. Epub 1999/06/04.

24. Warren-Smith AK, Curtis RA, Greetham L, McGreevy PD. Rein contact between horse and handler during specific equitation movements. Applied Animal Behaviour Science. 2007;108(1–2):157–69.
25. Valera M, Bartolomé E, Sánchez MJ, Molina A, Cook N, Schaefer A. Changes in Eye Temperature and Stress Assessment in Horses During Show Jumping Competitions. Journal of Equine Veterinary Science. 2012;32(12):827–30.
26. Fenner K. Bridling and Give to the Bit (Multimedia DVD). Private Lesson Series. Sydney, Australia2009.
27. Parker M, Goodwin D, Eager RA, Redhead ES, Marlin DJ. Comparison of Polar ® heart rate interval data with simultaneously recorded ECG signals in horses. Comp Exerc Physiol. 2009;6(4):137–42.
28. McGreevy PD, McLean AN. Roles of learning theory and ethology in equitation. Journal of Veterinary Behavior: Clinical Applications and Research. 2007;2(4):108–18.
29. McBride S, Mills DS. Psychological factors affecting equine performance. BMC Veterinary Research. 2012;8:180 10.1186/1746-6148-8-180 [PMC free article]
30. Hall C, Huws N, White C, Taylor E, Owen H, McGreevy PD. Assessment of ridden horse behaviour. Journal of Veterinary Behavior: Clinical Applications and Research. 2013;8(2):62–73.
31. König von Borstel U, Pasing S, Gauly M. Towards a more objective assessment of equine personality using behavioural and physiological observations from performance test training. Applied Animal Behaviour Science. 2011;135(4):277–85.
32. McLean AN. Short-term spatial memory in the domestic horse. Applied Animal Behaviour Science. 2004;85(1):93–105.
33. König Von Borstel U, Glißman C. Alternatives to conventional evaluation of rideability in horse performance tests: suitability of rein tension and behavioural parameters. PloS one. 2014;9(1):e87285 10.1371/journal.pone.0087285 [PMC free article]
34. Egenvall A, Eisersiö M, Roepstorff L. Pilot study of behavior responses in young riding horses using 2 methods of making transitions from trot to walk. Journal of Veterinary Behavior: Clinical Applications and Research. 2012;7:157–68.

35. Kuhnke S, Dumbell L, Gauly M, Johnson JL, König Von Borstel U. A comparison of rein tension of the rider's dominant and non-dominant hand and the influence of the horse's laterality. Comp Exerc Physiol. 2010;7:57–63.
36. Randle H, Edwards H, Button L. The effect of rider position on the stride and step length of the horse at canter. Journal of Veterinary Behavior: Clinical Applications and Research. 2010;5(4):219–20.
37. Warren-Smith AK, Curtis R, Greetham L, McGreevy PD. Rein contact between horse and handler during specific equitation movements / Amanda K. Warren-Smith, R. Curtis, L. Greetham, P.D. McGreevy. Applied Animal Behaviour Science. 2007;108(1–2):157–69.
38. Randle H, Button A, A. The effect of different rein types on the rein tension applied when taking up a 'medium contact'. Journal of Veterinary Behavior: Clinical Applications and Research. 2011;6:295–.
39. Ahrendt LP, Labouriau R, Malmkvist J, Nicol CJ, Christensen JW. Development of a standard test to assess negative reinforcement learning in horses. Applied Animal Behaviour Science. 2015;169:38–42.
40. McGreevy PD, Rogers LJ. Motor and sensory laterality in thoroughbred horses. Applied Animal Behaviour Science. 2005;92:337–52.
41. Shanahan S. Trailer Loading Stress in Horses: Behavioral and Physiological Effects of Nonaversive Training (TTEAM). Journal of Applied Animal Welfare Science. 2003;6(4):263–74. 10.1207/s15327604jaws0604_1
42. Warren-Smith A, Greetham L, McGreevy PD. Behavioral and physiological responses of horses (Equus caballus) undergoing head lowering. Journal of Veterinary Behavior: Clinical Applications and Research. 2007;2/3:59–67.
43. Pierard M, Hall C, König Von Borstel U, Averis A, Hawson LA, McLean AN, et al. Evolving protocols for research in equitation science. Journal of Veterinary Behavior: Clinical Applications and Research. 2015;10:255–66.
44. Hawson LA, Salvin HE, McLean AN, McGreevy PD. Riders' application of rein tension for walk-to-halt transitions on a model horse. Journal of Veterinary Behavior: Clinical Applications and Research. 2014;9(4):164–8.

SECTION 6: ANIMAL MODELS

CHAPTER 9

A Pig Model of the Preterm Neonate: Anthropometric and Physiological Characteristics

Yvonne A. Eiby[1], Layne L. Wright[1], Viskasari P. Kalanjati[1], Stephanie M. Miller[1], Stella T. Bjorkman[1], Helen L. Keates[2], Eugenie R. Lumbers[1,3], Paul B. Colditz[1], Barbara E. Lingwood[1]

[1]UQ Centre for Clinical Research, The University of Queensland, Brisbane, Queensland, Australia,

[2]School of Veterinary Science, The University of Queensland, Brisbane, Queensland, Australia,

[3]School of Biomedical Sciences and Pharmacy and the Hunter Medical Research Institute, The University of Newcastle, Newcastle, New South Wales, Australia

BACKGROUND

Large animal models are an essential tool in the development of rationally-based new clinical therapies for preterm infants. We provide a description

Citation: Eiby YA, Wright LL, Kalanjati VP, Miller SM, Bjorkman ST, Keates HL, et al. (2013) A Pig Model of the Preterm Neonate: Anthropometric and Physiological Characteristics. PLoS ONE 8(7): e68763. https://doi.org/10.1371/journal.pone.0068763

of the newborn pig as a model of the preterm neonate in terms of growth parameters, physiology and the requirement for intensive care over a range of gestational ages.

Methods

Twenty-nine litters of piglets (n=298) were delivered by caesarean section at six timepoints during gestation from 91d to 113d (term=115d). Two groups, at 91 and 97d gestation, also received maternal glucocorticoid treatment. At four of these timepoints, piglets (n=79) were ventilated, sedated and monitored using standard neonatal intensive care techniques for up to 8 h in various experimental protocols.

Results

Body weight increased from mean 697 g (SD 193) at 91d gestation to 1331 g (SD 368) at 113d gestation. Piglets delivered at 97d gestation were able to be resuscitated and kept alive for at least 8 h on respiratory support after surfactant administration. Maternal glucocorticoid treatment 48 h and 24 h hours prior to delivery reduced the requirement for ventilator support and improved cardiovascular stability.

Conclusion

The pig provides a relevant model for the study of human preterm physiology and for investigation of novel therapies to improve outcomes.

INTRODUCTION

The preterm human neonate is subject to a number of acute diseases that are associated with a high risk of adverse outcomes. In the developed world, the incidence of preterm birth has increased over the last two decades but advances in neonatal care have reduced mortality [1]. Treatment with maternal glucocorticoids to accelerate fetal lung maturation and increase lung compliance [2], is now standard procedure and has been an important contributor to reductions in mortality and morbidity [3]. However the incidence of disability in survivors of preterm birth has not decreased.

Common serious clinical problems that preterm infants still face are respiratory distress syndrome [3], brain injury [4]_ENREF_1, cardiovascular instability [5], suboptimal growth [6] and necrotising enterocolitis (NEC) [7]. Adequate treatments have not yet been identified and those used in everyday practice have not been characterised nor fully tested. The development of novel therapies is hampered by deficiencies in knowledge of the relevant physiology. Increasing the knowledge of preterm physiology, and the evaluation of novel and existing clinical interventions, initially requires appropriate animal models.

An ideal animal model of human preterm physiology should produce viable neonates with maturation characteristics similar to a very preterm human infant, and have a body size that allows comparative monitoring, blood sampling and clinical interventions. Large animals such as sheep, dogs, pigs and cattle have been used to model adult physiology due to their large size and accessibility. The chronically cannulated sheep has been extensively used to study fetal physiology, particularly fetal hemodynamics, renal and lung function, endocrine function, sex differences and the effects of maternal glucocorticoid exposure. However the youngest use of a lamb ex utero is 125 days (0.85 of gestation) and survival was <3 h [8]. Below this gestation lung immaturity results in mortality, whereas many other organ systems are relatively mature and more representative of late preterm physiology [9].

The baboon model of the preterm neonate has been used extensively to study the impact of respiratory therapies [10]–[12]. Animals delivered at 125d (term approx. 185d gestation; 68%) are similar to a 26wk infant in that they require surfactant, ventilatory and hemodynamic support [12]. This non-human primate model has advantages because of the baboon's similarity to the human. However the availability of these animals is very restricted and the cost is very high, limiting the feasibility of this model.

Research using the preterm pig has been more limited, with reported studies relating only to NEC in late preterm piglets. The youngest reported caesarean delivery of piglets is at 105d (0.91 gestation), by which time piglets are physiologically similar to a late preterm infant and do not require mechanical ventilation [13]. There is no information available regarding the viability and physiology of piglets at earlier gestations that may be equivalent to very preterm infants.

Aim

The aim of this study was to describe the pig model of the preterm neonate in terms of its growth, physiology and requirement for intensive care over a range of gestational ages. In addition, the effect of maternal glucocorticoid treatment on some of these factors is described.

METHODS

Ethics Statement

This study was carried out in strict accordance with the recommendations of the Australian Code of Practice for the Care and Use of Animals for Scientific Purposes (7th Edition). The protocol was approved by The University of Queensland Animal Ethics Committee (Approval numbers: UQCCR/998/08/NHMRC, UQCCR/999/08/NHMRC, PRC/684/07 and PRC/RBWH/687/07/NHMRC). All surgery was performed under anaesthesia and all efforts were made to minimize suffering.

Animals

Artificially inseminated pregnant sows (Large white x Landrace) were sourced from a commercial piggery owned by The University of Queensland at Gatton, Australia. Piglets were delivered by Caesarean section at six gestational ages: 91d, 94d, 97d, 100d, 104d and 113d (term 115d) (Table 1). At each gestational age one to seven litters (11–73 piglets) were studied. Body weights and sex only were recorded at 3–24 h post-birth (postnatal age 1 day; PNA 1d) from 45 complete litters of pigs born by spontaneous vaginal delivery at term. Two additional groups of preterm piglets (at 91d and 97d) were exposed to maternally administered glucocorticoids (betamethasone, 0.19 mg/kg body wt, i.m.; Celestone Chronodose; Schering-Plough, USA) at 48 h and 24 h prior to delivery. This dose/kg is equivalent to that given to women presenting with threatened preterm labour.

Table 1. Gestational ages, sex ratios, body and organ weights of preterm piglets delivered by caesarean section.

Gestational age (days) Term=115d	% gestation	No. Litters	No. Female and male piglets (F:M)	Body (g)	Heart (g)	Brain (g)	Liver (g)	Kidney (g)	Brain:liver
91	79	4	20:20	697±193 (40)	5.2±1.3 (34)	18.6±1.7 (29)	16.4±4.5 (29)	2.9±0.8 (29)	1.20±0.31 (27)
94	82	1	5:6	902±71 (11)	5.1±0.5 (5)	–	–	4.0±1.1 (5)	–
97	84	5	17:23	881±213 (41)	7.0±1.6 (33)	21.9±2.2 (17)	19.0±4.3 (17)	4.3±1.3 (30)	1.20±0.29 (17)
100	87	2	13:14	841±171 (27)	5.6±2.2 (18)	24.8±2.9 (21)	20.4±5.3 (18)	3.4±1.0 (18)	1.28±0.30 (18)
104	90	2	12:12	1095±225 (24)	7.6±2.3 (15)	26.0±1.8 (15)	23.8±5.9 (15)	4.5±1.4 (15)	1.15±0.26 (15)
113	98	7	40:33	1331±368 (73)	11.6±3.3 (53)	30.9±2.9 (25)	46.8±12.0 (20)	5.1±1.6 (49)	0.72±0.21 (19)
PNA 1d	100	45	-	1512±329 (460)	–	–	–	–	–
91+ GC	79	4	12:24	715±149 (36)*	5.2±1.3 (31)	18.6±1.7 (20)	19.8±7.8 (27)	3.3±0.8 (27)	1.25±0.35 (20)
97+ GC	84	4	20:27	1021±180 (46)	7.1±1. 3 (45)	23.5±2.0 (21)	35.9±6.2 (21)#	4.5±1.0 (36)	0.67±0.11 (20)

Weights are means ± SD (n). PNA (postnatal age) indicates piglets born at term by spontaneous vaginal delivery and weighed 3–24 h post-birth. GC indicates animals treated with maternal glucocorticoids prior to birth.

* $p=0.010$, interaction effect between sex and maternal glucocorticoid treatment.

\# $p<0.001$, effect of maternal glucocorticoid treatment.

Caesarean Section

Pregnant sows (280–350 kg) were premedicated i.m. with 400 mg azaperone (Stresnil, i.m.; Janssen, Australia) and 1000 mg ketamine (approximately 3 mg/kg; Ketamil, i.m.; Troy Laboratories, Australia). An ear vein was catheterised for administration of anaesthesia which was induced with 200 mg of alfaxalone (approximately 0.6 mg/kg; Alfaxan-CD RTU, i.v.; Jurox, Australia), followed by intubation with a 14–16 mm endotracheal tube, with additional alfaxalone as required to allow intubation. The total administered dose of alfaxalone was 300–700 mg. Sows breathed spontaneously and were maintained with 2% isoflurane (Attane Isoflurane USP; Minrad, USA) in O_2. Inhalational anaesthetic was used to facilitate rapid recovery of the piglets from the effects of maternal anaesthesia. Additionally, the i.v. drugs used to premedicate and induce anaesthesia will have minimal effects on piglet physiology as the dose is low and there is a substantial period of time (at least 2 hours) between dosing and commencement of piglet experimentation. Throughout surgery, saline (2–3L of 0.15 M NaCl) was administered via an ear vein and the following variables were monitored: arterial blood pressure by Doppler (Parks Medical Electronics Inc; Aloha, OR, USA), O_2 saturation via pulse oximetry (Masimo; Masimo Corporation, Irvine, CA, USA), end tidal isoflurane and end tidal CO_2 concentrations (Capnomac Anaesthesia Monitor; Datex-Ohmeda Inc, Madison, WI, USA). There was no significant hypotension or hypoxemia.

Caesarean delivery was performed via a ventral midline incision. Following incision into the linea alba the uterine horn was partially exposed. Complete exposure of the uterine horns was avoided so that uterine blood flow was not compromised by stretching or occlusion of the uterine arteries. Piglets were individually removed from the uterus at varying time intervals (up to 20 min) according to experimental requirements. Medication could be delivered to the fetus (including anaesthesia if required) by injecting into the umbilical vein prior to cord clamping. After all piglets were delivered, the sow was euthanased by i.v. injection of sodium pentobarbital (60 ml Lethabarb; Virbac, Australia).

Piglet Resuscitation and Intensive Care

Where experimental protocols required piglet resuscitation, this was conducted in line with the current American Academy of Pediatrics (AAP) recommendations for human infants [14] but also took into account the requirements of the animal. Piglets were intubated and ventilated using a conventional neonatal ventilator and standard neonatal intensive care techniques. All ventilated piglets were sedated using a loading dose of fentanyl (5 μg/kg) and midazolam (0.2 mg/kg) followed by a maintenance infusion of fentanyl (2 μg/kg/h) and midazolam (6 μg/kg/h for <113d animals and 10 μg/kg/h for 113d animals). This protocol is similar to the current sedation protocol used in preterm infants [15]–[17]. Drug and replacement fluids (10% glucose at a rate of 3 mL/kg/h) were delivered via a 3.5FG dual lumen neonatal umbilical vein catheter (Argyle, Sherwood Medical, MO, USA).

Physiological Monitoring

A 3.5FG neonatal umbilical artery catheter (Argyle, Sherwood Medical, MO, USA) was inserted into the umbilical artery for continuous invasive measurement of arterial blood pressure (Transpac® disposable pressure transducer; Hospira, USA). Blood was sampled intermittently for immediate blood gas analysis (ABL815 Blood Gas Analyser; Radiometer®, Denmark) and other analyses as required by experimental protocols. ECG was monitored using alligator clips attached to the skin of the thorax. Rectal temperature was maintained at 39.0°C±1°C (normal body temperature for pigs) using an overhead radiant heater (IW920 mobile infant warmer; Fisher & Paykel, New Zealand), with the addition of an under-bed heater mat for animals delivered before 100d gestation. Pulse oximetry was used to measure oxygen saturation (Radical; Masimo Corporation, Irvine, CA, USA). Blood pressure and oximetry data were continuously captured from the clinical monitors using a 16 channel Powerlab and Lab Chart v7 software (AD Instruments, Australia). ECG and temperature data were fed directly to the Powerlab.

Additional physiological measures and surgical procedures were possible in preterm piglets. The measurements described below were part of a specific experimental protocol and results of this study will be reported in a subsequent publication. Continuous single channel EEG was recorded using conventional EEG needle electrodes placed subcutaneously in the C3–C4 position (BRM2; BrainZ Instruments, New Zealand). Peripheral blood flow (skin blood flow) was monitored with a Laser Doppler flowmetry standard surface probe (AD Instruments, Australia) attached to the hind limb with an adhesive ring. Recording of ventricular pressure and injection of coloured microspheres to measure regional blood flow [18], [19] was achieved using a dual lumen catheter inserted into the left ventricle via carotid artery cut down. The position of the ventricular catheter was confirmed using the pressure waveform.

Data Analyses

The effects of gestational age on body weight, relative organ weights (arcsine transformed) and brain:liver weight were detected using mixed model univariate ANOVA (where age was a fixed factor and litter was a nested random factor). At each age, the effect of sex on body weight and relative organ weights (arcsine transformed) was analysed using mixed model univariate ANOVA (sex was a fixed factor and litter was a nested random factor). The effect of maternal glucocorticoid treatment on body weight and relative organ weights (arcsine transformed) were analysed at 91d and 97d separately using univariate ANOVA (treatment and sex were fixed factors and litter was a nested random factor). The effect of gestational age, maternal glucocorticoid treatment and sex on hemodynamic and arterial blood gas parameters were analysed using univariate ANOVA (group and sex were fixed factors and litter was a nested random factor). For all analyses, significant differences are reported only where they existed independently of litter effects. Where an effect of group was detected Fisher's least significant difference (LSD) tests were used to determine which groups were different. The effect of litter size on body weight was analysed separately for near term (113d and PNA 1d) and preterm (91d, 94d, 97d, 100d and 104d) pigs using a Spearman's Rank correlation of body weight z scores and litter size. Reported values are means and SD unless stated otherwise. ANOVAs and correlations were conducted using SPSS Statistics version 20 (IBM, Chicago, IL, USA).

RESULTS

Litters and Sex Ratios

A total of 29 litters of Large white × Landrace piglets (n=298) were delivered by Caesarean section at six time points: 91d, 94d, 97d, 100d, 104d and 113d (term=115d). Details of number of litters and piglets are included in Table 1. A further 45 litters spontaneously delivered at term were used to determine term weights. Litter size ranged from 5–17 animals (10.3±2.7, n=74 litters). Litters were 53% male with the most extreme sex ratio 1:9 (female:male) (Table 1). Data reported are for non-glucocorticoid treated animals unless stated otherwise.

Growth

Body weight doubled between 91d and 113d (Table 1). Variability in body weight was high at all gestations as indicated by the high standard deviations (Table 1) and extensive scatter (Fig. 1), and this was the result of variation within each litter rather than large differences between litters. As a result there were often one or two animals in each litter with a birth weight close to the 10th centile for the whole group and approximately half of litters also had one piglet with a body weight more than 40% below the mean for that litter. In contrast, there was less variation in brain weight with proportionally lower standard deviations (Table 1 and Fig. 1).

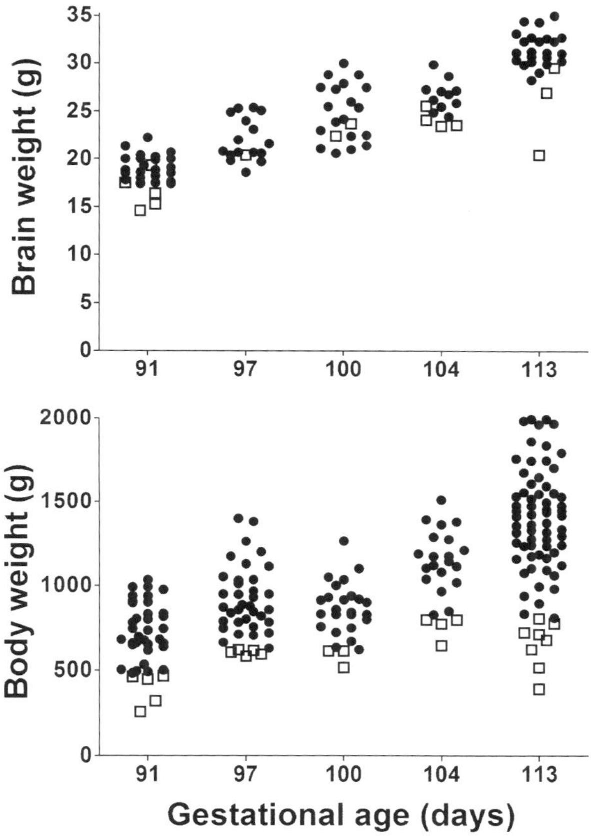

Figure 1. Brain and body weights of piglets over a range of gestational ages.

Closed circles represent normal body weight animals and open squares represent animals with body weight <10th centile. Not all piglets had brain weight recorded. Note the spread of weights within a given gestational age is greater for the body than for the brain implying that brain growth is relatively preserved even when body weight is low.

Males had significantly greater body weight compared with females at 91d (766±187 g vs. 627±177 g, p=0.038) and 113d (1450±289 g vs. 1233 g±39 g, p=0.019) only. Body weight was inversely correlated with litter size from 113d onwards (r=−0.236, p<0.01, n=533) but not at earlier gestations (r=−0.023, p=0.734, n=225). The weight of the heart, brain and kidney relative to total weight did not increase from 91d to 113d (Fig. 2A, B, D). In contrast, the relative weight of the liver was constant from 91–104d, but had increased significantly between 104d and 113d (p=0.004) (Fig. 2C).

The brain:liver weight ratio was significantly higher ($p<0.001$) in animals born at less than 113d of gestation (1.13±0.35) compared with animals born at 113d (0.71±0.21). There was no significant effect of sex on relative organ weights at any gestational age.

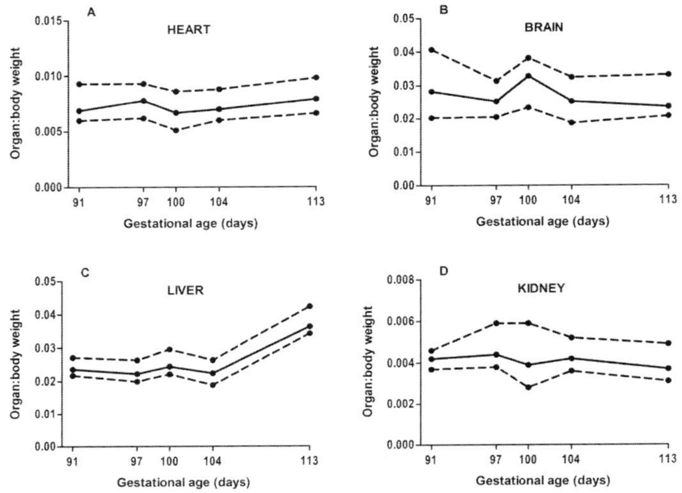

Figure 2. Relative organ growth from 91d to 113d of gestation (term=115d) for A) heart, B) brain, C) liver, D) kidney.

The solid line is the 50th centile and the dashed lines are the 10th and 90th centiles. Note the weight of all organs increased proportionally with body weight, except for the liver which increased significantly relative to body weight between 104d and 113d ($p=0.004$). Data from 94d was not included as only one litter was studied and few organ weights were recorded.

Physiology and Maturation

There was substantial variability, both within and between litters, in survival rates and requirements for additional ventilatory and cardiovascular support.

113d gestation

These near term piglets appeared similar to a term animal with open eyes, good muscle tone and well developed skin and hair. Animals had adequate thermoregulation requiring minimal intervention to keep them warm. Respiration was spontaneous upon delivery and animals were hemodynamically stable. During experimental manipulation requiring general anaesthesia, intubation with a 3.0 cuffed endotracheal tube and ventilation at 15/5 cmH_2O (peak inspiratory pressure/positive end expiratory pressure) at 30 breaths/minute in 21% O_2 was generally adequate to maintain arterial pCO_2 at 35–45 mmHg and oxygen saturation at 95% or above. Survival of these ventilated near term animals was 93% (28/30).

104d gestation

At 104d, piglets had open eyes, good muscle tone and well developed skin and hair. Thermoregulation was poor, requiring external heat sources to maintain body temperature at 38–39°C. All animals breathed spontaneously.

100d gestation

The eyelids of piglets at 100d gestation were partially fused, muscle tone was reduced, the skin was thin and some hair was present. Thermoregulation was poor and external heat sources were required. Most piglets autoresuscitated and breathed spontaneously and did not require mechanical ventilation. However in one litter, several animals exhibited intermittent apnea and required stimulation.

97d gestation

At 97d piglets had fused eyelids, poor muscle tone and thin skin with minimal hair. Thermoregulation was very inadequate and it was difficult to maintain normothermia with overhead radiant heaters and heated mats. Hypothermia was reduced by delivery into a warm towel, drying the animal and then inserting it into a plastic bag that was open at either end. This reduced evaporative heat loss, allowed subsequent access to the head and umbilical cord, and is a practice used in the care of the very preterm human infant. All piglets required ventilation to survive. Prior to intubation, active suctioning of the oropharynx was required to remove lung liquid. Immediate intubation was then required with a size 2.0 mm uncuffed paediatric endotracheal tube (ETT) (Argyle, Tyco Healthcare Group, Mansfield, MA, USA), and administration of surfactant (4 ml/kg, Survanta, Abbott Laboratories, Columbus, Ohio USA) via the ETT was necessary to improve lung compliance and avoid death due to respiratory insufficiency or pneumothorax.

Table 2 describes the physiology of piglets at 97d gestation compared to near term piglets (113d). Hemodynamic parameters, mean arterial pressure (MAP) and heart rate, are the average for a ten minute period occurring 30–40 min after resuscitation. MAP was lower and less stable compared with near term animals (mean 30 mmHg v. 36.5 mmHg; Table 2) (Fig. 3). Heart rate was similar in the two age groups. Arterial blood gas parameters from the first blood gas sample, typically taken 30–60 min post-birth, reflect the response to the initial ventilatory conditions rather than on-going maintenance. Ventilatory requirements were significantly higher for 97d animals than near term animals with peak inspiratory pressures of 25–30 cmH_2O, respiratory rate of 60bpm and FiO_2 up to 100%. However ventilatory requirements in these untreated preterm piglets were extremely variable and because some animals had poor saturation and very low pO_2, the initial FiO_2 was set quite high. In addition, during resuscitation some piglets exhibited symptoms consistent with pulmonary hypertension and so we decided to begin resuscitation with a high FiO_2 in attempt to avoid this. As a result a high mean pO_2 was recorded at the first blood gas with a very high SD (Table 2), but this was reduced to more normal levels as quickly as the piglets' physiology allowed and was similar to term levels during experimentation (Table 2). Blood pH and arterial base excess (ABE) were also highly variable but on average piglets were more acidotic and had higher pCO_2 compared to near term animals. At this gestational age, permissive hypercapnia (pCO_2 of 45–55 mmHg) was used, but in the most immature piglets initial and sometimes ongoing pCO_2 was above 70 mmHg. Pneumothorax, detected by clinical signs including chest asymmetry and reduced air entry, only occurred in one piglet and was secondary to severe respiratory failure requiring extremely high peak inspiratory pressures. The significant base excess appears normal for neonatal piglets and we have observed this in the many spontaneously delivered term piglets that we have studied over many years.

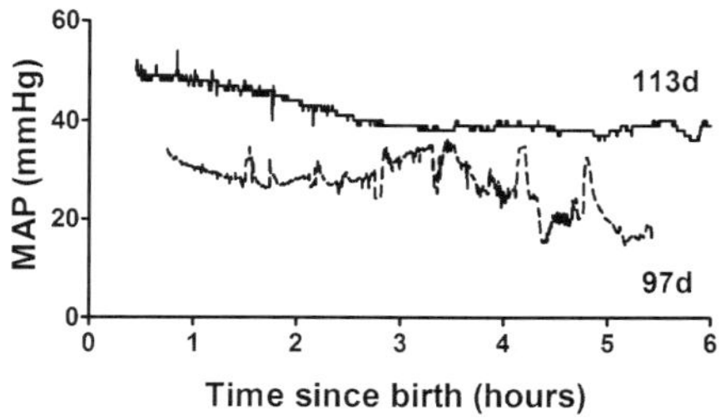

Figure 3. Mean arterial blood pressure (MAP) for a preterm (97d) and a term (113d) piglet (term=115d) for six hours following delivery by Caesarean section.

Note the preterm piglet has a lower and less stable mean arterial pressure (MAP).

Table 2. Physiology of preterm and near term piglets.

Physiological Variable		Preterm (97 d)	Preterm+GC (97 d)	Near term (113 d)	P value
Mean arterial pressure (mmHg)		30.0±0.9 (15)*	29.2±0.8 (15)*	36.5±1.4 (14)	<0.001
Heart rate (bpm)		178±11 (15)	177±27 (15)	173±7 (16)	ns
pCO_2 (mmHg)	initial	38.8±19.8 (17)	28.5±5.2 (16)	31.4±12.2 (16)	ns
	maintenance	36.4±9.0 (17)	41.7±7.0 (16)	35.8±2.8 (15)	ns
pO_2 (mmHg)	initial	195±167 (17)	251±139 (16)*	121±51 (16)	0.025
	maintenance	147±71 (16)	112±36 (16)	111±32 (15)	ns
pH	initial	7.49±0.20 (17)	7.62±0.07 (16)* #	7.57±0.13 (16)	0.037
	maintenance	7.50±0.07 (17)*	7.46±0.07 (16)*	7.60±0.08 (16)	<0.001
ABE (mmol/L)	initial	3.7±3.9 (17)	7.1±2.4 (16)#	5.2±2.7 (16)	0.010
	maintenance	4.2±2.6 (16)	4.7±2.2 (16)	5.8±2.0 (15)	ns
FiO_2 (%)	initial	96±9 (17)*	75±27 (16)* #	35±19 (16)	<0.001
	maintenance	70±26 (17)*	41±14 (16)#	32±20 (15)	<0.001
Hb (mmol/L)		81±23 (17)	89±9 (16)	74±19 (16)	ns

Physiological parameters of piglets delivered by caesarean section at 97 d gestation (untreated), 97 d (treated with maternal glucocorticoids=GC) and 113 d gestation (term is 115 d). Mean arterial pressure and heart rate have been averaged over a 10 min period (approximately 30 min post-birth). Initial arterial blood gas parameters are from the first blood gas taken 30–60 min post-birth and so reflect the response to the initial ventilatory settings. Maintenance values were obtained prior to experimentation and reflect ongoing maintenance. GC indicates animals treated with maternal glucocorticoids prior to birth. pCO_2 is carbon dioxide partial pressure, pO_2 is oxygen partial pressure, ABE is arterial base excess, FiO_2 is fraction of inspired oxygen, Hb is haemoglobin. All values are mean ± SD (n).

*indicates significantly different to near term piglets ($p<0.05$).

#indicates a significant difference between untreated preterm piglets and preterm piglets treated with maternal glucocorticoids ($p<0.05$).

Difficulties were encountered with the measurement of pulse oximetry due to poor perfusion of the extremities. Survival was 83%, and 19 of the 23 resuscitated animals (not treated with maternal glucocorticoids) survived at least the 8 h required by the experimental protocol. There was no difference in survival between males and females.

94 d and 91 d gestation

Piglets had very thin skin, poor muscle tone and no hair, and could not thermoregulate. Ventilation was extremely difficult even with the use of surfactant and high inspiratory pressures. Piglets were hemodynamically unstable and hypoxaemia was common. Pneumothorax occurred in 2 of 7 animals and specifically only in the context of severe respiratory failure which was being treated with extremely high peak inspiratory pressures. At 94 d, resuscitation was attempted on six piglets from one litter but all of these piglets survived <40 min primarily due to cardiovascular collapse and/or severe respiratory failure. At 91 d, resuscitation was attempted on only one piglet and it survived for <3 h due to cardiovascular collapse. At post-mortem, care was required when working with internal organs due to their delicate structure.

Maternal Glucocorticoid Treatment

Delivery at 91 d gestation following a clinically relevant dose of maternal glucocorticoids, resulted in females with an increased body weight (799 g±107) compared with untreated females (627 g±177), and males (673 g±151) with a decreased body weight compared with untreated males (766 g±187) ($p=0.010$). At 97 d there was no significant effect of treatment on body weight ($p=0.301$). The relative weights of organs were not significantly affected by treatment at either 91 d or 97 d.

Ventilation of glucocorticoid exposed piglets was commenced at the same settings as for the untreated preterm piglets but FiO_2 was rapidly reduced based on oxygen saturation. At the first blood gas, glucocorticoid treated animals already had a lower oxygen requirement (lower FiO_2, Table 2). Nevertheless pO_2 at the first blood gas was higher than in the untreated preterm piglets, pCO_2 was lower and pH higher (Table 2) suggesting that further reductions were possible and inspiratory pressure could also be reduced. As in untreated preterm piglets ventilator settings were adjusted to obtain normal physiology as rapidly as possible (Table 2). Maintenance of glucocorticoid treated piglets was easier with a reduced requirement for higher peak inspiratory pressures (PIP) and FiO_2, reduced time on 100% oxygen, and more stable cardiovascular parameters. In the glucocorticoid treated group 16 of 19 survived (84%) and survival was similar in males and females.

DISCUSSION

An ideal animal model of preterm physiology should produce viable neonates with maturation characteristics similar to preterm infants. We have demonstrated that with the application of standard neonatal intensive care techniques, preterm piglets at 97 d gestation can survive for at least 8 h.

From 97 d gestation, piglets' lungs are sufficiently mature to allow successful resuscitation even without maternal glucocorticoid treatment. Piglets require similar levels of respiratory support to a human preterm infant and similar benefits of glucocorticoid exposure were seen [3]. Piglets delivered at 97 d have survived in our hands for the duration of experimental protocols lasting up to 8 h. To date we have not attempted to support longer survival but this is likely to be feasible as there is no major deterioration by 8 h of age.

The lower MAP and instability of blood pressure observed in the human preterm infant are also present in the preterm piglet [20], [21]. While there is a clear need to study the cardiovascular transition to ex utero life in the preterm baby there are few suitable models. The rat is born very immature and so makes this transition at an earlier developmental stage. The sheep heart is structurally more mature than the human infant as indicated by the proportion of binucleated myocytes being 70% [9] compared with about 10% in the human [22]. The preterm piglet may be the ideal model for the study of the preterm cardiovascular transitional period due to its similar physiology and also structural maturation (2–5% binucleated myocytes at 0.8 of gestation, increasing to approximately 10% at birth) (Kim MY, unpublished data).

We have not comprehensively evaluated brain development as the piglet has been previously validated as a superior model of normal human brain development. Specifically the timing of the perinatal brain growth spurt [23], patterns of cellular development and myelination [23], presence of gyri and sulci [24] are similar in piglets and human infants. In addition, understanding the pathophysiology of neonatal brain injury requires a model, such as the piglet, where the proportion of grey matter to white matter [24] and electroencephalography features [25] are more similar to human infants than that in lower order animals.

The larger size of piglets provides an advantage in physiological studies, enabling extensive monitoring and tissue and/or fluid sampling. Even at 91 d gestation, very preterm piglets are large enough to allow extensive continuous physiological monitoring including arterial blood pressure using umbilical arterial catheters, heart rate, pulse oximetry, respiration, ECG, ventricular pressure, EEG, peripheral blood flow and rectal temperature. Additionally, repeated blood and urine sampling is possible. This is not possible in small animal models.

Other advantages of the pig model are a consequence of the relatively large litter size. Firstly, large litters provide an opportunity for parallel experiments on interrelated questions to be answered in piglets from the same litters. Littermates can also be used as internal controls or to address sex differences. Secondly, pigs spontaneously produce growth restricted offspring. While many models of IUGR require surgical interventions to produce growth restricted fetuses [26], the presence of spontaneous growth restriction in piglets may provide advantages.

Comparison to the Human Neonate

Varying developmental rates across species means it is not always appropriate to calculate equivalent gestational stages based simply on mathematics [27]. In addition, in rats and sheep some organs develop more rapidly than in the human infant while others develop more slowly [9], [28], [29]. In contrast, we and others have shown that the relative maturity of the lungs, cardiovascular system and brain [22]–[25], [29] of the piglet is comparable to that of the human infant. This is a major advantage of the piglet model.

A piglet born at 101–104 d of gestation requires minimal intervention beyond tactile stimulation and maintenance of body temperature, and is similar to a preterm infant >32 weeks gestation [30]. At 97 d of gestation, without maternal glucocorticoid treatment, piglets are physiologically similar to a very preterm baby born at 28–30 wk gestation, requiring surfactant for adequate ventilation and with some cardiovascular instability, but both have high survival rates [30]. At 91–94 d of gestation, ventilating and stabilising piglets without glucocorticoid exposure was extremely difficult and survival rates were low. These animals are the physiological similar to a preterm infant born at 23–25 wk of gestation [30].

Maternal Glucocorticoid Treatment

Maternal glucocorticoid treatment plays a critical role in the management of preterm neonates so it is imperative that animal models demonstrate an appropriate response to treatment. Glucocorticoid treatment improved respiratory function with reduced requirement for higher PIP and FiO_2, lower pCO_2, and less acidosis at the first blood gas. Cardiovascular instability was also reduced. These positive effects of maternal glucocorticoid exposure on cardiopulmonary function in the preterm pig are similar to those observed in the human preterm neonate [31].

While no resuscitation was performed on 91–94 d piglets following maternal glucocorticoid treatment, improvements seen at 97 d suggest that survival rates at 91–94 d would be significantly increased by glucocorticoid exposure.

There is no clear consensus on the effects of clinically relevant doses of maternal glucocorticoid treatment on body and organ growth. There are no reported effects on fetal growth following a single course of treatment in humans [32], [33] however these studies combined the sexes possibly obscuring any opposing effects in male and females. Sheep studies using clinically relevant doses have shown conflicting results. One study found no effect on birth weight in a combined sex analysis [34], another observed reduced birth weight in both males and females with greater effects in females [35]. Another study showed a dose effect on growth in males only, with no effect on birth weight in females [36]. In the current study, exposure to clinically relevant doses of maternal glucocorticoids resulted in opposing sex specific effects on body weight. The weight of females increased and that of males decreased, but only for animals born at 91 d gestation. This surprising result was observed within every litter and remained statistically significant after the exclusion of outliers. This effect was not present when treatment was delayed by 6 d, suggesting that the sensitivity of growth to glucocorticoid administration may vary across gestation.

Growth Restriction

The current study did not aim to identify pathogenetic issues associated with intrauterine growth restriction as this has already been extensively discussed [37]–[40]. However a number of observations were made that may assist others in the study of this important area. In term or near-term piglets we observed a significant correlation between piglet body weight and litter size, as previously reported by Bauer [41]. However this relationship was absent at earlier gestations suggesting nutrient supply to a large litter may not become a growth limiting factor for the whole litter until late in gestation. Despite this observation half of preterm litters did include a piglet with a body weight more than 40% below the mean for that litter. This may be the result of piglets developing in a section of the uterus where a borderzone of vascular supply exists leading to altered placental development [41].

Despite considerable variation in body weight within litters, variability in brain weight was low suggesting sparing of the brain in growth restricted piglets as previously reported [41]. At term the normal brain:liver weight ratio of piglets that have a body weight above the 10th centile, is less than 1 [41] and a brain:liver weight of greater than 1 has been used to identify growth restriction. However, due to the rapid growth of the liver late in gestation, the normal brain:liver weight of piglets born at 104 d or earlier is greater than 1. Therefore, if this ratio is used to identify preterm piglets that are growth restricted a higher cut-off value will be required, perhaps 1.5.

Suggested Improvements in the Model

There was extensive variability in how quickly piglets oxygenated so there is a need for rigorous early monitoring and adjustment of ventilation. In addition, the difficulties in ventilation and incidence of pneumothorax found in the untreated preterm piglets may be reduced by increasing the positive end expiratory pressure (PEEP) to 6–7 cmH_2O. The glucocorticoid treated preterm piglets are less variable with reduced requirements for high FiO_2 and high inspiratory pressure. To avoid hyperventilation in this group, initial ventilatory settings should be further adapted by reducing the PIP to 18–20 cmH_2O, reducing the fraction of inspired oxygen (FiO_2) to 60–70% and increasing the positive end expiratory pressure (PEEP) to 6–7 cmH_2O.

Limitations and Cautions

The major limitations of the preterm pig model are associated with the large size of the sow (350 kg) and requirement for equipment suitable for handling such weights. Intubation of sows is technically difficult. A large number of expert staff, including an experienced neonatologist, is required to deal with resuscitation and initiation of intensive care for the large numbers of piglets at preterm Caesarean section. A further limitation of the model is the high PIP initially required to adequately ventilate preterm piglets. The sedation used may alter piglet physiology. However maternal intravenous sedation is administered at very low doses and only very early in the procedure so that maternal levels should be extremely low by the time the piglets are delivered. Sedation administered to the piglets is designed to reflect NICU practices so that the physiology observed will more closely align with that of a preterm baby. Despite these limitations the preterm piglet provides an animal model that more closely resembles the human preterm neonate than other models.

Conclusions

The pig provides a clinically relevant model of preterm neonatal physiology where the maturation of multiple organ systems is similar to the human early preterm infant. It can be used to test potential new treatments for major clinical problems such as respiratory distress syndrome, brain injury, cardiovascular compromise and NEC. Importantly this animal model provides the opportunity to explore major gaps in our understanding of the physiology of the preterm neonate. This knowledge is essential to the development of effective new therapies.

ACKNOWLEDGMENTS

We wish to thank Dr Ranald Cameron for performing the Caesarean surgeries, Tom Connolly for assistance with the handling of sows, and Susan Sullivan and Zoe Ireland for sharing data on some organ weights.

REFERENCES

1. Thompson JMD, Irgens LM, Rasmussen S, Daltveit AK (2006) Secular trends in socio-economic status and the implications for preterm birth. Paediatr Perinat Epidemiol 20: 182–187.
2. Jobe AH (2010) Lung maturation: The survival miracle of very low birth weight infants. Pediatr Neonatol 51: 7–13.
3. Jobe AH (2006) Prenatal corticosteroids : A neonatologist's perspective. Neoreviews 7: e259.
4. Aylward GP (2005) Neurodevelopmental outcomes of infants born prematurely. J Dev Behav Pediatr 26: 427–440.
5. Hasenfuss G, Teerlink JR (2011) Cardiac inotropes: current agents and future directions. Eur Heart J 32: 1838–U1835.
6. Johnson MJ, Wootton SA, Leaf AA, Jackson AA (2012) Preterm birth and body composition at term equivalent age: A systematic review and meta-analysis. Pediatrics 130: E640–E649.
7. Wu SF, Caplan M, Lin HC (2012) Necrotizing enterocolitis: Old problem with new hope. Pediatr Neonatol 53: 158–163.
8. Probyn ME, Hooper SB, Dargaville PA, McCallion N, Crossley K, et al. (2004) Positive end expiratory pressure during resuscitation of premature lambs rapidly improves blood gases without adversely affecting arterial pressure. Pediatr Res 56: 198–204.
9. Burrell JH, Boyn AM, Kumarasamy V, Hsieh A, Head SI, et al. (2003) Growth and maturation of cardiac myocytes in fetal sheep in the second half of gestation. Anat Rec Part A 274A: 952–961.
10. Coalson JJ, Kuehl TJ, Escobedo MB, Hilliard JL, Smith F, et al. (1982) A baboon model of bronchopulmonary dysplasia 2.Pathologic features. Exp Mol Pathol 37: 335–350.
11. Coalson JJ, Winter VT, Siler-Khodr T, Yoder BA (1999) Neonatal chronic lung disease in extremely immature baboons. Am J Respir Crit Care Med 160: 1333–1346.
12. Thomson MA, Yoder BA, Winter VT, Martin H, Catland D, et al. (2004) Treatment of immature baboons for 28 days with early nasal continuous positive airway pressure. Am J Respir Crit Care Med 169: 1054–1062.

13. Sangild PT, Siggers RH, Schmidt M, Elnif J, Bjornvad CR, et al. (2006) Diet- and colonization-dependent intestinal dysfunction predisposes to necrotizing enterocolitis in preterm pigs. Gastroenterology 130: 1776–1792.
14. American Academy of Pediatrics and American Heart Association (2011) Textbook of Neonatal Resuscitation. Elk Grove Village, IL. 300 p.
15. Davies M, Cartwight D, Inglis G (2008) Pocket Notes on Neonatology. Australia: Elsevier.
16. Royal Prince Alfred Hospital Newborn Care Drug Database website. Available: http://www.sswahs.nsw.gov.au/RPA/neonatal/default.htm. Accessed 2013 March 1.
17. Swart EL, Slort PR, Plotz FB (2012) Growing up with Midazolam in the Neonatal and Pediatric Intensive Care. Curr Drug Metab 13: 760–766.
18. Walter B, Bauer R, Gaser E, Zwiener U (1997) Validation of the multiple colored microsphere technique for regional blood flow measurements in newborn piglets. Basic Res Cardiol 92: 191–200.
19. Stevenson KM, Gibson KJ, Lumbers ER (1996) Effects of losartan on the cardiovascular system, renal haemodynamics and function and lung liquid flow in fetal sheep. Clin Exp Pharmacol Physiol 23: 125–133.
20. Logan JW, O'Shea TM, Allred EN, Laughon MM, Bose CL, et al. (2011) Early postnatal hypotension and developmental delay at 24 months of age among extremely low gestational age newborns. Arch Dis Child Fetal Neonatal Ed 96: F321–328.
21. Laughon M, Bose C, Allred E, O'Shea TM, Van Marter LJ, et al. (2007) Factors associated with treatment for hypotension in extremely low gestational age newborns during the first postnatal week. Pediatrics 119: 273–280.
22. Schmid G, Pfitzer P (1985) Mitoses and binucleated cells in perinatal human hearts. Virchows Arch B-Cell Molec Pathol 48: 59–67.
23. Pond WG, Boleman SL, Fiorotto ML, Ho H, Knabe DA, et al. (2000) Perinatal ontogeny of brain growth in the domestic pig. Proc Soc Exp Biol Med 223: 102–108.
24. Williams SM, Sullivan RKP, Scott HL, Finkelstein DI, Colditz PB, et al. (2005) Glial glutamate transporter expression patterns in brains from multiple mammalian species. Glia 49: 520–541.

25. Bjorkman ST, Miller SM, Rose SE, Burke C, Colditz PB (2010) Seizures are associated with brain injury severity in a neonatal model of hypoxia-ischemia. Neuroscience 166: 157–167.
26. Morrison JL (2008) Sheep models of intrauterine growth restriction: fetal adaptations and consequences. Clin Exp Pharmacol P 35: 730–743.
27. Sengupta P (2011) A scientific review of age determination for a laboratory rat: How old is it in comparsion with human age? Biomed Internatl 2: 81–89.
28. De Matteo R, Blasch N, Stokes V, Davis P, Harding R (2010) Induced Preterm Birth in Sheep: A Suitable Model for Studying the Developmental Effects of Moderately Preterm Birth. Reprod Sci 17: 724–733.
29. Dobbing J, Sands J (1979) Comparative aspects of the brain growth spurt. Early Hum Dev 3: 79–83.
30. Colditz PB (1997) The newborn. In: Beischer NA, Mackay EV, Colditz PB, editors. Obstetrics and the newborn. 3rd ed. London, UK: W.B. Saunders. 581–704.
31. Liggins GC, Howie RN (1972) Controlled trial of antepartum glucocorticoid treatment for prevention of respiratory distress syndrome in premature infants. Pediatrics 50: 515.
32. French NP, Hagan R, Evans SF, Godfrey M, Newnham JP (1999) Repeated antenatal corticosteroids: Size at birth and subsequent development. Am J Obstet Gynecol 180: 114–121.
33. Newnham JP, Moss TJ (2001) Antenatal glucocorticoids and growth: single versus multiple doses in animal and human studies. Semin Neonatol 6: 285–292.
34. Quaedackers JS, Roelfsema V, Fraser M, Gunn AJ, Bennet L (2005) Cardiovascular and endocrine effects of a single course of maternal dexamethasone treatment in preterm fetal sheep. Bjog-an International Journal of Obstetrics and Gynaecology 112: 182–191.
35. Miller SL, Sutherland AE, Supramaniam VG, Walker DW, Jenkin G, et al. (2012) Antenatal glucocorticoids reduce growth in appropriately grown and growth-restricted ovine fetuses in a sex-specific manner. Reprod Fert and Develop 24: 753–758.

36. Miller DR, Jackson RB, Blache D, Roche JR (2009) Metabolic maturity at birth and neonate lamb survival and growth: The effects of maternal low-dose dexamethasone treatment. J Anim Sci 87: 3167–3178.
37. Alvarenga ALN, Chiarini-Garcia H, Cardeal PC, Moreira LP, Foxcroft GR, et al. (2013) Intra-uterine growth retardation affects birthweight and postnatal development in pigs, impairing muscle accretion, duodenal mucosa morphology and carcass traits. Reprod Fert and Develop 25: 387–395.
38. Bauer R, Walter B, Ihring W, Kluge H, Lampe V, et al. (2000) Altered renal function in growth-restricted newborn piglets. Pediatr Nephrol 14: 735–739.
39. Bauer R, Walter B, Kluge H, Zahlten W, Zwiener U (1993) Catecholamine response and cardiovascular reaction to moderate hypoxia in intrauterine growth-retarded (IUGR) newborn piglets; Cosmi EV, DiRenzo GC, editors. 423–426 p.
40. Estrada J, Sommer J, Collins B, Mir B, Martin A, et al. (2007) Swine generated by somatic cell nuclear transfer have increased incidence of intrauterine growth restriction (IUGR). Cloning Stem Cells 9: 229–236.
41. Bauer R, Walter B, Hoppe A, Gaser E, Lampe V, et al. (1998) Body weight distribution and organ size in newborn swine (sus scrofa domestica) - A study describing an animal model for asymmetrical intrauterine growth retardation. Exp Toxicol Pathol 50: 59–65.

CHAPTER 10

New Perspectives in Behavioural Development: Adaptive Shaping of Behaviour Over a Lifetime?

Fritz Trillmich[1], Anja Günther[1], Caroline Müller[2], Klaus Reinhold[3], Norbert Sachser[4]

[1]Animal Behaviour, Bielefeld University, Morgenbreede 45, 33615 Bielefeld, Germany.
[2]Chemical Ecology, Bielefeld University, Universitätsstraße 25, 33615 Bielefeld, Germany.
[3]Evolutionary Biology, Bielefeld University, Morgenbreede 45, 33615 Bielefeld, Germany.
[4]Behavioural Biology, University of Münster, Badestraße 13, 48149 Münster, Germany

SUMMARY

The last years have seen an explosion of publications on consistent, i.e. repeatable, behavioural differences among individuals and their potential

Citation: von Engelhardt, N., Kowalski, G.J. & Guenther, A. The maternal social environment shapes offspring growth, physiology, and behavioural phenotype in guinea pigs. Front Zool12, S13 (2015). https://doi.org/10.1186/1742-9994-12-S1-S13

adaptive significance and evolutionary origin. Stable, repeatable traits imply reduced plasticity and this is particularly interesting in the case of behaviour, which is the most flexible trait of an animal in its interaction with the biotic and abiotic environment. While the importance of consistent individual differences has been widely acknowledged, the lifetime development of these differences and their potentially adaptive or maladaptive consequences for individuals only recently came into the focus of research [1,2]. As a result, widespread evidence for the existence of changes in behaviour and behavioural plasticity over the lifetime of individuals begins to accumulate [3]. This has raised novel questions concerning the overwhelming importance of ontogenetic and trans-generational influences on the development and fitness of behavioural phenotypes and the mechanisms influencing changes during an individual's lifetime.

Much theoretical work has been published on the importance of and limits to individual plasticity, mostly in the framework of the reaction norm. Less empirical work has actually studied, how the reaction norm comes about in an individual's lifetime and how or whether it may change over the lifetime [4]. This raises important questions about the adaptivity of the phenotypic plasticity of behaviour and the question when within a lifetime such plasticity might be particularly relevant. Do particular windows of plasticity exist, and if so why are they where they are within the lifetime of an organism? The potential lifetime plasticity also raises the question about the evolution of the reaction norm and, more recently, about the evolution of its plasticity, questions that so far have received but little attention. In these fields of investigation theory has moved far ahead of empirical studies. Often, if not always, individuals face uncertain environmental conditions throughout their life, forcing them to adapt plastically in order to maximise fitness. Under such uncertain conditions we expect natural selection to favour plasticity. However, plastic responses to the environment can only be adaptive, if informative environmental cues exist that allow prediction of future conditions with reasonably high probability [5]. Depending on the cues available and the ontogenetic stage of an individual we expect differential effects of such predictive cues: Early in life highly informative cues, whether derived from maternal influences or perceived directly, may potentially act in an organisational manner, whereas less informative cues or cues perceived at a later age may lead to more immediate plasticity (see [4] for a definition of different types of plasticity). This may lead to developmental influences on plasticity, i.e. early environments may particularly influence the reaction norm. This idea of plasticity of the reaction norm as a trait that develops and thereby influences the reaction norm expressed during the lifetime of an individual is rather novel [6]. However, it follows directly from Gene x Environment interactions (GxE) influencing the reaction norm. Along the developmental axis Gene x Age (GxA) and Individual x Age (IxA) interactions will then determine the expression of the developing reaction norm in interaction with the environments encountered by an individual (see [7]). As pointed out earlier [2] genes themselves provide an environment for other genes so that even more complex interactions are to be expected. As these interactions may change with the state and the age of an individual it becomes clear why ontogenetic processes are of prime relevance for understanding potential adaptive effects of plasticity.

Complex models of ontogenetic processes (mostly taken from quantitative genetics; like animal models) are needed to describe such interactions (see [7]) and thereby characterize the reaction norm's dependence on internal and external influences over the lifetime. We stress that these ontogenetic processes shape animal behaviour over the lifetime and - through trans-generational effects – even beyond. This view leads to several major questions which we have addressed in our 2014 workshop at the Center for Interdisciplinary Research (ZIF, Bielefeld, Germany) and treat to various depths here in our collection of papers:

1) why do we observe ontogenetic plasticity and how does it come about in the interaction of GxExA?
2) how do changes in the reaction norm and the observed plasticities influence the measurement of repeatabilities and correlations among different behaviours (and other traits) of the individual?
3) These processes involve substantial self-organisation (as observed even when iso-genetic individuals living in the same environment become differentiated; [8,9]).This self-organisation becomes perhaps particularly amenable to measurement during certain windows of time when individuals readjust their phenotypes. Can we predict when such time windows would be most adaptive with respect to critical life stages or conditions of the environment that provide cues with particularly high predictive power (like diurnal or annual changes of conditions)? Does such plasticity of the reaction norm shrink with age or disappear completely?
4) Lastly, can we understand the mechanisms behind the width and plasticity of reaction norms for different traits beyond the answers that theory and animal models can provide?

In the collection of papers in this special issue, Brommer & Class [7] explore the plasticity of behaviour in response to age. They make use of the theory of reaction norms to predict how age-related changes in behaviour may affect repeatability and hence heritability of behaviour. Furthermore, they elaborate how developmental plasticity can affect covariances with other behaviours. The importance of particular life stage for the uptake of environmental cues and adjustment in behaviours is further discussed by Fawcett & Frankenhuis [10]. They model how cue frequency, cue reliability and cue importance may differ during life stages based on the life history of the organism. Based on differences in environmental cues, the authors predict when selection processes should lead to the evolution of 'sensitive windows'- periods during development in which organisms show increased susceptibility to (certain) environmental cues. Del Giudice [11] further explores the idea of plasticity as a developing trait that may be modulated by the environment an organism experiences. He presents a model that explicitly includes humans showing how cue reliability can affect the evolution of plasticity. Thereby a call of Fawcett & Frankenhuis [10] is answered who suggested that such a model should be formulated. In a further conceptual paper, Han & Dingemanse [12] focus on one of the most important environmental cues, the macronutrients contained in the diet, and suggest how diet composition may exert pleiotropic effects on personality and behavioural syndromes both, short term and over evolutionary time. Employing the same statistical framework as [7] to explain how GxA interaction may affect covariances between behaviours and repeatability/ heritability, Han & Dingemanse show how GxE interactions may affect these properties. Groothuis & Taborski [13] then point out different mechanisms by which environmental influences may interact with sensitive windows to affect the organism's performance. They critically discuss frequently used experimental designs by behavioural ecologists and neurobiologists and highlight the importance of full, reciprocal designs for understanding the consequences of environmental conditions across development.

Behavioural changes are often believed to be mediated by hormonal changes. Hau & Goyman [14] review evidence for consistent behaviour-hormone relationships within and between individuals in birds. Focusing on two hormones that are well known to change predictably during the ontogeny of organisms (corticosterone and testosterone), they conclude that our understanding of hormone-behaviour relationships on an individual level is rather limited. They call for better characterisation of individual hormonal phenotypes, their responses to environmental change and covariation with behaviours by employing a reaction norm approach.

On the empirical side, Müller & Müller [15] and Würz & Krüger [16] explore how age related changes in behaviour affect repeatability of behaviour within and between distinct life stages in an insect and a bird species. As expected, both studies conclude that behaviours are more repeatable within life stages than between, but more surprisingly show that some behaviours only become repeatable when the individual reaches a certain life stage. While in the leaf beetle [15] correlations among behaviours are mostly stable even across life stages, this is not the case in zebra finches [16]. An empirical example for the importance of naturally occurring differences in early diet composition on the expression of behaviour later in life is given by Van Oers et al. [17]. Great tit chicks provided with fewer caterpillars during the first days after hatching exhibited a stronger stress response later in life. Krause & Naguib [18] demonstrate how early diet affects the development of sexual ornaments and behaviours later in life. Although birds can catch up in expression of sexual ornaments quickly, other traits, like body mass, exploration and activity are affected for much longer, revealing a possible trade-off in resource allocation to compensate for poor early conditions. Two more empirical studies investigate the development of personality traits in infants, a life history stage for which data on personality are extremely scarce, of three different mammalian species. Hudson et al. [19] investigate how locomotion behaviour and separation calls develop in cats and mound-building mice, showing that both behaviours are repeatable and may be used to enhance our understanding of personality during early ages in future studies. Von Engelhardt et al. [20] study how development from birth to independence is affected by maternal social environment during pregnancy. Comparable to the study by Hudson et al. [19], they show that locomotion behaviour and separation calls are not correlated but that separation calls one day after birth predict activity in an unknown environment later in life.

Trans-generational effects are the focus of the study of Kinnally & Capitanio [21]. Rhesus macaques reared in a nursery show stronger physiological stress responses and more emotional behaviour when their fathers were nursery reared compared to offspring of fathers that were reared by their mother. This paper demonstrates that the (adaptive?) shaping of behaviour does not end within the lifetime of an organism but extents even to future generations. Mechanisms that lead to differences in personality may also be involved in evolutionary processes (through processes like genetic assimilation). Kappeler & Fichtel [22] describe the ‘lemur-syndrome’, a set of behavioural, morphological and ecological traits unique to Madagascan lemurs. They hypothesise how trans-generational effects of chronic maternal stress due to adverse environmental conditions and developmental consequences may have led to evolutionary stable life history adjustments. In contrast, Oldfield et al. [23] suggest how a broadly influential underlying mechanism might have been modified during evolution permitting its re-use in different environmental circumstances across a wide array of vertebrate taxa.

Brust et al. [24] review literature on laboratory mice (accounting for more than 70% of animals used in biological studies with fundamental nature) development and point out that even for this extensively studied species knowledge is scarce about most life stages and in particular about the environmental cues potentially influencing development of the behavioural phenotype during ontogeny. In search for potential mechanisms, Hennessy et al. [25] review evidence of hormonal, immunological and behavioural plasticity in guinea pigs. They explore in which way social experiences shape the responsiveness of the hypothalamic-pituitary-adrenocortical axis and immune-related activity during different stages of life. Ultimately these processes lead to predictable behavioural changes suggesting lifelong adaptive shaping of the organism. Shaping of behaviours and hormones across multiple generations to adapt organisms to man-made living conditions is called domestication. Kaiser et al. [26] describe how domestication has reshaped physiological and behavioural profiles and their development during ontogeny in guinea pigs as compared to their wild congeners, the cavy. They show that domestication processes affected suites of traits thereby altering behavioural syndrome structure between the two forms and discuss the evolutionary significance of these differences.

While behavioural and evolutionary biologists have acknowledged variability within and between individuals as important, behavioural neuroscientists have aimed to reduce variability as much as possible for decades in so-called translational (applied and pre-clinical) research. Standardisation procedures encompass housing conditions, experimental procedures, species studied, sex, age and even genetic background. In their contribution, Macrì & Richter [27] warn against such practices, pointing out the potential pitfalls of reduced reproducibility of results. They also stress that such a narrow comparative view, neglecting the fact of behavioural plasticity will reduce the option to translate and apply such research to human clinical problems. In the last paper of this collection, Crews et al. [28] strongly remind us that old concepts, particularly on the norms of reaction and the concept of plasticity published decades ago should not be forgotten as otherwise we may reinvent the wheel or repeat old mistakes. By means of five examples they exemplify: The interaction of heredity and environment cannot be understood outside the context of development!

Overall, this impressive collection provides a broad view of the importance of the ontogeny of the reaction norm and of the potential for development of its plasticity during the lifetime, thereby stressing the pervasive importance of the study of ontogenetic processes.

REFERENCES

1. Stamps JA, Groothuis TGG. Developmental perspectives on personality: Implications for ecological and evolutionary studies of individual differences. Philosophical Transactions of the Royal Society Series B. 2010;**365**(1560):4029–4041. doi: 10.1098/rstb.2010.0218.
2. Groothuis TGG, Trillmich F. Unfolding personalities: The importance of studying ontogeny. Developmental Psychobiology. 2011;**53**(6):641–655. doi: 10.1002/dev.20574.
3. Sachser N, Kaiser S, Hennessy MB. Behavioural profiles are shaped by social experience: when, how and why. Philosophical Transactions of the Royal Society Series B. 2013;**368**(1618):20120344. doi: 10.1098/rstb.2012.0344. [PMC free article]
4. Stamps JA. Individual differences in behavioural plasticities. Biol Rev Camb Philos Soc. 2015. [PubMed]
5. Dall SR, McNamara JM, Leimar O. Genes as cues: phenotypic integration of genetic and epigenetic information from a Darwinian perspective. Trends in Ecology & Evolution. 2015;**30**(6):327–333. doi: 10.1016/j.tree.2015.04.002.
6. Pluess M, Belsky J. Prenatal programming of postnatal plasticity? Development and Psychopathology. 2011;**23**(1):29–38. doi: 10.1017/S0954579410000623.
7. Brommer JE, Class B. The importance of genotype-by-age interactions for the development of repeatable behavior and correlated behaviors over lifetime. Frontiers in Zoology. 2015;**12**(Suppl 1):S2.
8. Lewejohann L, Zipser B, Sachser N. "Personality" in laboratory mice used for biomedical research: A way of understanding variability? Developmental Psychobiology. 2011;**53**(6):624–630. doi: 10.1002/dev.20553.
9. Freund J, Brandmaier AM, Lewejohann L, Kirste I, Kritzler M, Krüger A. et al.Emergence of individuality in genetically identical mice. Science. 2013;**340**(6133):756–759. doi: 10.1126/science.1235294.
10. Fawcett RW, Frankenhuis WE. Adaptive explanations for sensitive windows in development. Frontiers in Zoology. 2015;**12**(Suppl 1):S3.
11. Del Giudice M. Plasticity as a developing trait: exploring the implications. Frontiers in Zoology. 2015;**12**(Suppl 1):S4.

12. Han CS, Dingemanse NJ. Effect of diet on the structure of animal personality. Frontiers in Zoology. 2015;**12**(Suppl 1):S5.
13. Groothuis TGG, Taborsky B. Introducing biological realism into the study of developmental plasticity in behaviour. Frontiers in Zoology. 2015;**12**(Suppl 1):S6.
14. Hau M, Goymann W. Endocrine mechanisms, behavioral phenotypes and plasticity: known relationships and open questions. Frontiers in Zoology. 2015;**12**(Suppl 1):S7.
15. Müller T, Müller C. Behavioural phenotypes over the lifetime of a holometabolous insect. Frontiers in Zoology. 2015;**12**(Suppl 1):S8.
16. Würz Y, Krüger O. Personality over ontogeny in zebra finches: long-term repeatable traits but unstable behavioural syndromes. Frontiers in Zoology. 2015;**12**(Suppl 1):S9.
17. van Oers K, Kohn GM, Hinde CA, Naguib M. Parental food provisioning is related to nestling stress response in wild great tit nestlings: implications for the development of personality. Frontiers in Zoology. 2015;**12**(Suppl 1):S10.
18. Krause TE, Naguib M. Zebra finch males compensate in plumage ornaments at sexual maturation for a bad start in life. Frontiers in Zoology. 2015;**12**(Suppl 1):S11.
19. Hudson R, Rangassamy M, Saldaña A, Bánszegi O, Rödel HG. Stable individual differences in separation calls during early development in cats and mice. Frontiers in Zoology. 2015;**12**(Suppl 1):S12.
20. von Engelhardt N, Kowalski GJ, Guenther A. The maternal social environment shapes offspring growth, physiology, and behavioural phenotype in guinea pigs. Frontiers in Zoology. 2015;**12**(Suppl 1):S13.
21. Kinnally EL, Capitanio JP. Paternal Early Experiences Influence Infant Development through Non-Social Mechanisms in Rhesus Macaques. Frontiers in Zoology. 2015;**12**(Suppl 1):S14.
22. Kappeler PM, Fichtel C. Eco-evo-devo of the lemur syndrome: Did adaptive behavioral plasticity get canalized in a large primate radiation? Frontiers in Zoology. 2015;**12**(Suppl 1):S15.
23. Oldfield RG, Harris RM, Hofmann HA. Integrating resource defence theory with a neural nonapeptide pathway to explain territory-based mating systems. Frontiers in Zoology. 2015;**12**(Suppl 1):S16.

24. Brust V, Schindler PM, Lewejohann L. Lifetime development of behavioural phenotype in the house mouse (Mus musculus) Frontiers in Zoology. 2015;**12**(Suppl 1):S17.
25. Hennessy MB, Kaiser S, Tiedtke T, Sachser N. Stability and change: Stress responses and the shaping of behavioral phenotypes over the life span. Frontiers in Zoology. 2015;**12**(Suppl 1):S18.
26. Kaiser S, Hennessy MB, Sachser N. Domestication affects the structure, development and stability of biobehavioural profiles. Frontiers in Zoology. 2015;**12**(Suppl 1):S19.
27. Macrì S, Richter H. The Snark was a Boojum – reloaded. Frontiers in Zoology. 2015;**12**(Suppl 1):S20.
28. Crews D, Weisberg SA, Sahotra S Sarkar. Hazards Inherent in Interdisciplinary Behavioral Research. Frontiers in Zoology. 2015;**12** (Suppl 1):S21.

CHAPTER 11

Effects of a Copper-Deficient Diet on the Biochemistry, Neural Morphology and Behavior of Aged Mice

Silvia Bolognin[1], Federica Pasqualetto[1], Carla Mucignat-Caretta[2], Janez Scancar[3], Radmila Milacic[3], Pamela Zambenedetti[4], Bruno Cozzi[5], and Paolo Zatta[1]

[1]CNR-Institute for Biomedical Technologies, Metalloproteins Unit, Department of Biology, University of Padova, Padova, Italy,
[2]Department of Human Anatomy and Physiology, University of Padova, Padova, Italy,
[3]Department of Environmental Sciences, Jozef Stefan Institute, Ljubljana, Slovenia,
[4]Division of Pathology, General Hospital, Dolo (VE), Italy, 5 Department of Experimental Veterinary Science, University of Padova, Legnaro (PD), Italy

ABSTRACT

Copper dyshomeostasis has been suggested as an aetiological risk factor for some neurodegenerative diseases, such as Alzheimer's disease. However,

Citation: Bolognin S, Pasqualetto F, Mucignat-Caretta C, Scancar J, Milacic R, Zambenedetti P, et al. (2012) Effects of a Copper-Deficient Diet on the Biochemistry, Neural Morphology and Behavior of Aged Mice. PLoS ONE 7(10): e47063. https://doi.org/10.1371/journal.pone.0047063

the precise mechanism at the base of this involvement is still obscure. In this work, we show the effects of a copper-deficient diet in aged CD1 mice and the influence of such a diet on: a) the concentration of various metal ions (aluminium, copper, iron, calcium, zinc) in the main organs and in different brain areas; b) the alteration of metallothioneins I-II and tyrosine hydroxylase immunopositivity in the brain; c) behavioural tests (open field, pole, predatory aggression, and habituation/dishabituation smell tests). Our data suggested that the copper-deficiency was able to produce a sort of "domino effect" which altered the concentration of the other tested metal ions in the main organs as well as in the brain, without, however, significantly affecting the animal behaviour.

INTRODUCTION

Solid widely recognized evidence suggests that altered metal ion homeostasis may be implicated in the aetiology of several neurodegenerative disorders, particularly Alzheimer's disease (AD) [1], [2], [3] While metal alteration has been clearly established as the primary cause of the disease for few and rare neurodegenerative disorders (e.g. Wilson's disease), its involvement is still a controversial issue for the remaining common disorders. The brain tightly regulates metals trafficking as an important part of its normal activity, and disruption of this delicate equilibrium may have detrimental effects for the whole brain functioning. According to some studies, an increase in brain copper (Cu) concentration could be a risk co-factor for the progression of AD [4]. Cu is an essential trace metal, acting as an important catalytic cofactor in several redox reactions involved in growth and development [5]. However, several studies suggested that it might interact with β-amyloid 1–42 ($A\beta_{1-42}$), and catalyse hydrogen peroxide (H_2O_2) generation, using a variety of biological reducing agents as substrates [6], [7]. In the absence of sufficient levels of detoxifying enzymes, including catalase, superoxide dismutase and glutathione peroxidase, H_2O_2 can further react with Cu to generate highly toxic hydroxyl radicals via the Fenton reaction, becoming a potential neuropathogenic factor. Interestingly, Aβ and amyloid precursor protein (APP) are both metalloproteins that strongly bind Cu.

According to several findings, Cu concentration was lower in the central nervous system (CNS) of AD patients in comparison to age match controls [8], [9] although it increased focally in the senile plaques [10]. AD, in addition, seems to be characterized by an excess of Cu in the brain extracellular space [11] and by an intracellular decrease with respect to the age-matched controls [8]. Nevertheless, pure analytical data are highly controversial, and evidence for a Cu-AD connection is mainly indirect. Cells over-expressing APP exposed to high Cu levels decreased the rate of Aβ secretion [12], an effect replicated in AD animal models by elevating Cu levels in the brain both by genetic [13] as well as by dietary supplementation [14]. Moreover, dietary Cu supplementation, in a transgenic mouse model for AD, increased bioavailable Cu levels in the brain and lowered Aβ levels, suggesting that impaired Cu homeostasis could be linked to the pathology [14]. In fact, increased intracellular Cu availability inhibited the accumulation of Aβ oligomers and protein tau phosphorylation [15]. Nevertheless, a prospective double-blind study in patients with mild AD failed to demonstrate that a 8 mg Cu daily supplementation produced an effect on tau and phospho-tau levels in cerebrospinal fluid (CSF) that represent relevant markers for AD [16]. Diversely, during the 12 months of treatment, $A\beta_{1-42}$ markedly decreased in the control group compared to the Cu supplemented group. However, none of the above studies have demonstrated the existence of a Cu deficiency condition in AD patients [17]. Anyway, other set of studies reported detrimental effects upon dietary Cu: the addition of Cu (0.12 ppm) to drinking water worsened Aβ accumulation and learning deficit in a cholesterol-fed rabbit model of AD, suggesting that Cu may negatively influence Aβ clearance from the brain [18]. Epidemiological studies hinted that high Cu dietary intake, associated with a high saturated and trans fat diet, accelerated cognitive decline [19]. In addition, alteration of APP processing, enhanced Aβ production and tau phosphorylation were detected in young triple transgenic (3xTg-AD) mice after chronic Cu exposure [20].

The metal dishomeostasis, often observed in association with AD, does not correspond in most cases to a dramatic and generalised metal imbalance in the brain, but rather to focal alteration of a certain metal in few specific and critical brain areas or even cellular compartments (for a review see [21]). Consequently, neither the total metal burden of the brain nor the relative concentration of the main biometals is greatly perturbed in several instances.

This complex scenario prompted us to investigate the effects of a chronic Cu deficient diet in aged mice and to assess the relationship with other metal ions in the brain and other selected organs. We chose this experimental model as we wanted to unravel whether a deficiency of an essential metal in aged mice was able to produce a detectable effect in terms of behavioural and biochemical alterations. It has been indeed demonstrated that the dietary intake of essential micronutrients is usually inadequate in elderly [22]. This gap is worsened by loss of appetite, lack of teeth, reduced mineral intestinal absorption and decreased requirement of energy. We also tested whether metal ions act independently or may be linked to promote a sort of domino effect in which changes in the concentration of one element may alter the others, with a consequent variation in oxidative stress markers and animal behaviour.

MATERIALS AND METHODS

Animal Care and Diet

Twenty-three male CD-1 mice (15 month old) were purchased from Harlan (Udine, Italy). They were housed one mouse per cage under standard laboratory conditions (23±1°C, 55±5% humidity, 12 h light/dark cycle) and had access to bi-distilled water and food ad libitum. They were allowed to acclimatize for 1 month before starting the treatment. Mice were randomly divided in two groups for the Cu-deficient (CuD) diet (n=12) and Cu-adequate diet (n=11) and treated for 12 weeks (Fig. 1A). Animals were sacrificed after the 12 weeks of treatment by decapitation. The diet with (6 mg of cupric carbonate per kilo) and without Cu was purchased from Mucedola (Settimo Milanese, Italy). Controls were run to check the Cu concentration of the diet upon arrival. All mice were treated in accordance with the policy of the Italian Ministry of Health and European Community laws on the use of animals in research (116-92, art. 5). The procedures used were approved by the Institute of Vallisneri, University of Padova.

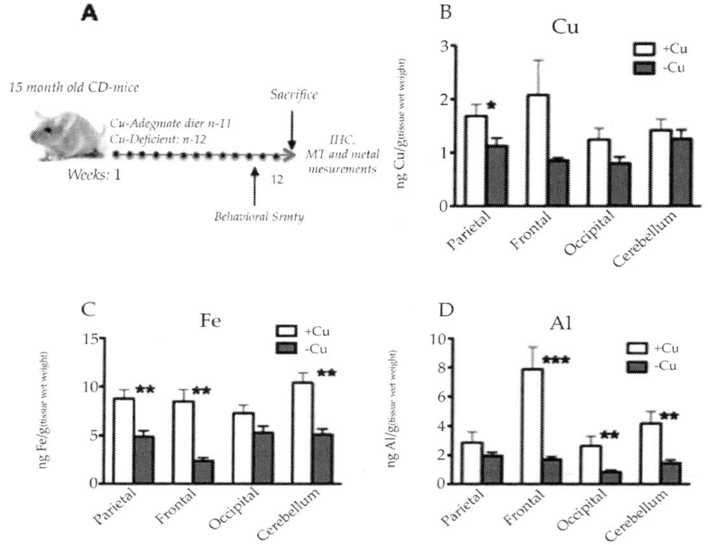

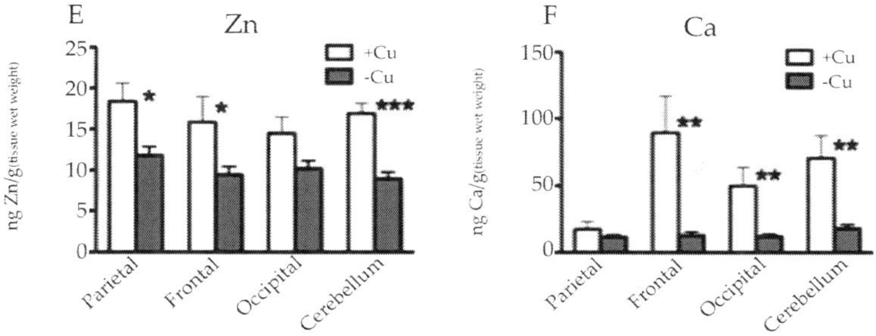

Figure 1. Metal content in the brain. (A) Schematic representation of the outline of the study. 15 month old CD1 mice were fed with a Cu-adequate (CuA) or Cu-deficient (CuD) diet for 12 weeks. Behavioural tests were performed the last 2 weeks of the treatment. Mice were then sacrificed and metal determination, IHC as well as biochemical investigations were performed. (B) Cu, (C) Fe, (D) Al, (E) Zn, and (F) Ca content (μg/g wet tissue) in the parietal, frontal, occipital cortex, and cerebellum. *$p<0.05$; **$p<0.01$; ***$p<0.001$.

Tissue Sampling

The brains were divided into two halves, one of which was frozen immediately after the decapitation for biochemical measurements and the other one used for immunohistochemistry (see below for details). Other organs were removed, weighted and immediately frozen until biochemical determination of metal ion and metallothionein (MT) concentrations.

Metal Analyses

Fresh tissues were mineralized in HNO_3 Suprapur (Merk, Milan, Italy) at 70°C for 24 h. After digestion, the solution was brought to the final volume using milliQ water and filtered with 0.2 μm pore size (Sigma, Milan, Italy).

The concentration of Cu, iron (Fe), calcium (Ca), and zinc (Zn) in tissues was determined by a Perkin-Elmer A100 flame atomic absorption spectrophotometer, using metal ion standard solutions for instrument calibration (Sigma, Milan, Italy).

The concentrations of Al were determined in helium mode by inductively coupled plasma mass spectrometry (ICP-MS) on an ICP-MS instrument (Agilent 7500ce, Tokyo, Japan), using the Babington nebuliser and a Scott-type spray chamber. A nickel sampler and skimmer with 1.0 and 0.4 mm cone orifices, respectively, were used. Treatment of data was performed with the Agilent ChemStation software. ICP-MS operating conditions for determination of Al are described elsewhere [23]. Before analysis samples were diluted with MilliQ water, so that the measured Al concentrations ranged between 1 and 100 μg L^{-1}.

Metallothionein Determination

Total MT concentration was determined following a silver-saturation assay, described in detail elsewhere [24], [25].

Immunohistochemistry (IHC)

The half-brains used for immunohistochemical analyses were fixed by immersion in buffered formalin, washed in phosphate saline buffer (PBS) 0.01 M pH 7.4, processed for paraffin embedding, sectioned sagittally at a thickness of 5 μm and mounted on slides with permanent positive charge (poly-L-lysine coated slides).

MT I-II IHC. Sections were incubated in 3% H_2O_2 in PBS for 10 min. Non-specific binding sites were blocked by a 30 min incubation with serum-free protein block (Dako Cytomation, Milan, Italy). Sections used for MT I-II staining were pre-treated with citrate buffer pH 6, then incubated overnight at 4°C with a monoclonal anti-MT I-II antibody raised in mouse (DakoCytomation, Milan, Italy) diluted 1:50. After washing, sections were reacted for 30 minutes with EnVision® kit, developed with 3,3′-diaminobenzidine (DakoCytomation, Milan, Italy) and counterstained with hematoxylin. The sections were then dehydrated, cover-slipped with balsam and observed. Glial Fibrillary Acidic Protein (GFAP) immunohistochemistry. Sections for GFAP staining were pre-treated with 0.1% trypsin in Tris/HCl pH 7.2 at 37°C for 10 min, then incubated 30 min at room temperature with an anti-GFAP polyclonal antibody raised in rabbit (DakoCytomation, Milan, Italy) with a 1:200 dilution. Further processing of the slides was the same described for MT I-II.

For the tyrosine hydroxylase (TH) immunohistochemistry sections were permeabilized with Triton X-100 2% in PBS 0.01 M, pH 7.2 for 15 min, blocked in 2% bovine serum albumin and Triton X-100 1% for one hour, incubated overnight in anti-TH (Santa Cruz Biotechnology N-19, Heidelberg, Germany) raised in rabbit diluted 1:100, followed by anti-rabbit Alexa 594 (Molecular Probes, Eugene, Oregon, USA) secondary antibody, 1:350, for 40 min at 37°C. The intensity of the staining was evaluated by grading – (faint labelling of cell bodies), +/− (detectable cell bodies and fibers), + (moderate labelling), ++ (intense labelling) and +++ (very intense) by two independent observers. MT I-II and GFAP sections were photographed with an Olympic BX51 photomicroscope. TH slides were photographed using the resident software of a Leica epifluorescence microscope. All images were acquired with the same parameters in TIFF format. The images were mounted unaltered, and lettered using Corel Draw 12.

The specificity of the antibodies has been verified by replacing the primary antibody with normal swine serum. Under these conditions there was no immunostaining.

Behavioural Tests

The behavioural tests were run in an isolated room during five days, with a four-day pause between days 2 and 3. Mice were weighted every day and tested neurologically with special emphasis on: general condition, deambulation, posture, righting from the side, placing reaction of hindlimbs, geotaxic reaction, avoiding of borders and equilibrium. Tests were chosen to explore different cognitive, sensory, motor and emotional domains; data were analyzed with mixed design ANOVA using Statistica software version 5 '97 edition (www.statsoft.com). The significant level was set at $p<0.05$.

Open-Field test

This test was performed on day 1 to evaluate locomotor and exploratory activities, and emotional reactivity to a novel environment [26]. Mice were individually moved to a plastic cage (55×33×20 cm), with opaque walls and black floor, and video-recorded for 3 min. The software SMART 2.5 (2 Biological Instrument, Varese, Italy) was used to quantify: a) the overall distance travelled (in cm), as an index of motor activity; b) the number of rearings with both forepaws on the lateral walls, as an index of exploratory activity; c) the time spent without moving, as an index of freezing; and d) the number of fecal boli and urine drops, as an index of autonomic activation.

Pole test

This test was performed to evaluate bradykinesia and motor performance [27], [28]. Mice were placed head upward on a vertical pole (1,5 cm diameter×50 cm height), the latency (in seconds) to climb down until the four paws were on the ground was recorded. The maximum time was 120 seconds. This test was repeated on days 1, 2 and 3, thus providing a measure for long-term contextual and procedural memory.

Predatory aggression test

The emotional reactivity to a different animal species was tested on day 1, by putting an earthworm (Lombricus terrestris) on the homecage floor. The latency (in seconds) to the first attack was recorded, then the test was stopped. The maximum time was 10 min. This test requires the sensory detection of the prey, the selection of the appropriate behavioral pattern of attack and its correct execution [26].

Habituation/Dishabituation smell test

From the second day onward, mice were tested for their ability to discriminate novel olfactory stimuli and their short-term memory [29]. Mice were moved to an open-field arena and video-recorded as described above. Each animal was left undisturbed for 5 min to habituate. A plastic well (20×7 mm) containing 10 μl of water was then fixed to the floor (10 cm from the short and 16 cm from the long wall) for 2 min, and then removed for 1 min. This procedure was repeated 3 times to provide habituation, seen as a lack of interest for the introduction of already known stimuli. Then a well containing 10 μl thymol (1:1000 in liquid paraffin) was introduced for three times. When a new odour is presented, the mouse should be aware of it and explore it for a longer time. After the third thymol presentation, a well containing camphor (1:1000 in liquid paraffin) was presented for three times. The area around the well (17×18 cm) was considered the target zone, in which the following measures were taken: latency to the first entry into the target zone (in seconds), distance travelled in cm, time spent in it (in seconds), resting time (in seconds), number of entries. Data were analyzed with a three-way mixed design ANOVA, for the factors Group (CuA vs CuD), odor (no odor, thymol, camphor), repetition (the first, second, third).

Statistical Analysis

The statistical analyses referred to the behavioural tests have been described in the former paragraph. Analyses on metal and MT concentration were performed using T-test, GraphPad software and they are presented as mean as + SEM. Results were labeled as follow: * $p<0.05$, ** $p<0.01$, and *** $p<0.001$.

RESULTS

Metal Concentration

Metal ion concentrations were measured in several organs to assess the effect of the CuD diet on the general metal balance. A widespread decrease of metal content was observed in each of the four brain regions tested and for each metal investigated (Figure 1B–F). In particular, CuD mice showed a significant reduction of Fe, Al, Zn, and Ca in the frontal areas of the brain, concurrent with a highly statistically significant decrease of Fe, Al, Zn and Ca in the cerebellum. Al was markedly decreased also in the occipital area (Figure 1D). Altogether, the frontal area and the cerebellum appeared to be the two regions most affected by the metal depletion.

Figure 2 shows Cu concentration in the other main organs: liver was the only one showing a significant Cu decrement (p=0.004). Fe was strongly curtailed in lungs (p=0.012), small (p=0.018) and large intestine (p=0.0012) and muscle (p=0.013) (Figure 3); Zn in kidneys (p=0.013) and liver (p=0.040) (Figure 4); and Ca in spleen (p=0.004) (Figure 5).

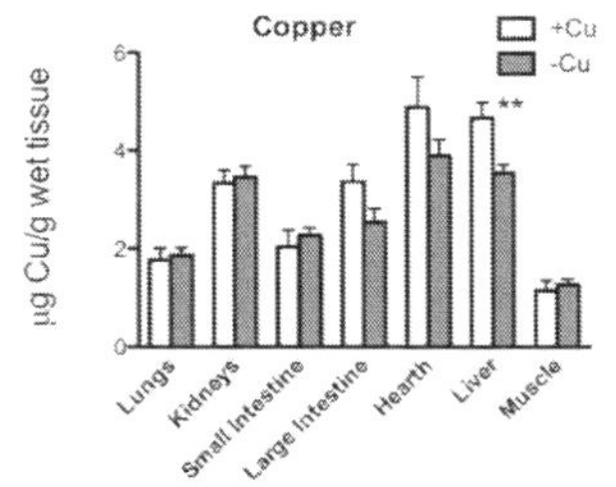

Figure 2. Cu content. Cu (µg/g wet tissue) in the lungs, kidneys, small intestine, large intestine, heart, liver, and muscle. ** $p<0.01$.

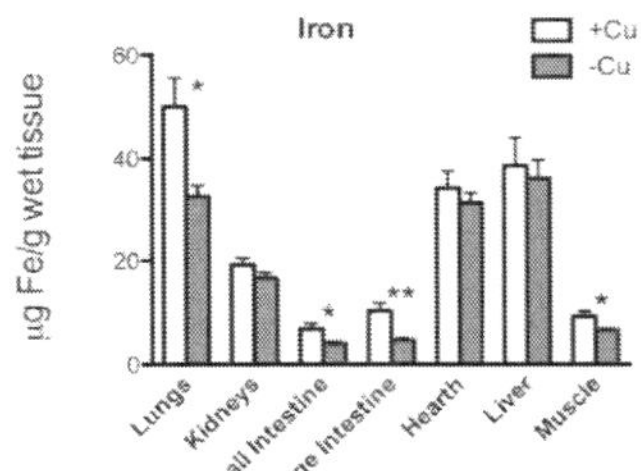

Figure 3. Fe content. Fe (µg/g wet tissue) in the lungs, kidneys, small intestine, large intestine, heart, liver, and muscle. * $p<0.05$; ** $p<0.01$.

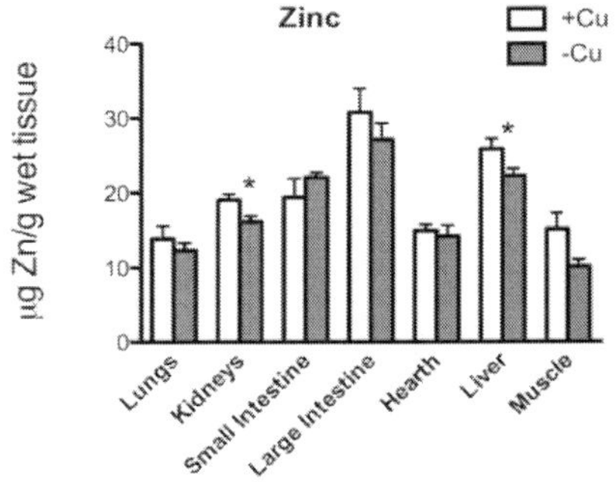

Figure 4. Zn content. Zn (µg/g wet tissue) in the lungs, kidneys, small intestine, large intestine, heart, liver, and muscle. * $p<0.05$.

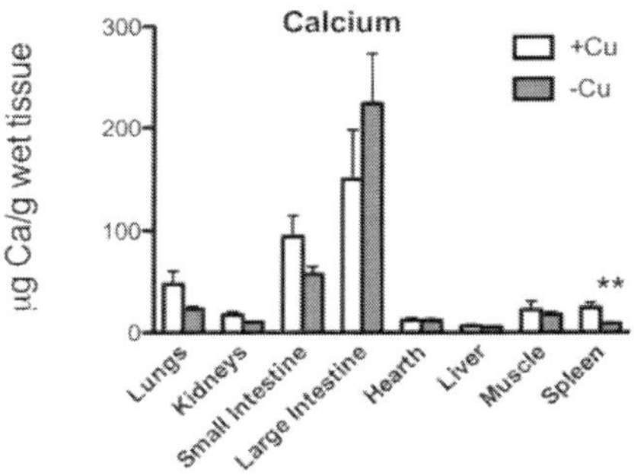

Figure 5. Ca content. Ca (µg/g wet tissue) in the lungs, kidneys, small intestine, large intestine, heart, liver, muscle, and spleen. ** $p<0.01$.

MT

The total amount of MT was significantly diminished only in the frontal ($p=0.0002$) and parietal ($p=0.020$) areas of the brain (Figure 6). Negligible was the difference in the MT concentration between CuA and CuD mice in the other organs (data not shown).

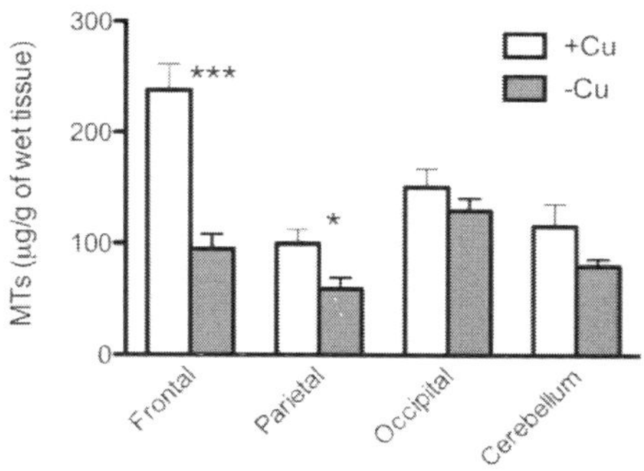

Figure 6. Metallothionein (MT) content in the brain. MT (µg/g wet tissue) in the parietal, frontal, occipital cortex, and cerebellum. *$p<0.05$; ***$p<0.001$.

Immunohistochemistry

Sections treated for immunohistochemistry were evaluated independently by three observers (see Table 1).

Table 1. Immunohistochemistry – Average of three independent evaluations.

Specimen	Tyrosine Hydroxylase	Metallothioneins I/II	GFAP
CTL 1	n.d.	+++	+++
CTL 2	n.d.	++	++
CTL 3	+/−	+++	++
CTL 4	n.d.	+	+++
CTL 5	++	++	++/+++
CTL 6	++	n.d.	++
CTL 7	++/+++	++	+++
CTL 8	n.d.	n.d.	n.d.
CTL 9	+	++	+++
CTL 10	++	+++	+++
CTL 11	+/++	++	+++

CTL 12	n.d.	n.d.	n.d.
EXP 1	++	+++	+++
EXP 2	++	+++	+++
EXP 3	+	++	+++
EXP 4	+++	++	+++
EXP 5	++/+++	++	++
EXP 6	+	+++	+++
EXP 7	+/−	++	++
EXP 8	++	++	++
EXP 9	n.d.	n.d.	n.d.
EXP 10	++	++	++
EXP 11	+/−	+	++
EXP 12	n.d.	n.d.	n.d.

CTL=control mice; EXP=copper deficient mice; n.d. not determined because of damaged specimen.

GFAP immunohistochemistry

GFAP-immunoreactive (-ir) elements were evident in both groups of mice without difference in distribution and density. Positive glial cells were mostly located in the white matter of the capsule system surrounding and crossing the basal ganglia complex, in the fimbria and hippocampus, in the pyramidal tract in the mesencephalon and in various fiber tracts of the medulla and in the white matter and cortex of the cerebellum. Positive elements were sometimes (but not always) found in the locus cœruleus of the CuD group but not in the controls (data not shown).

MT I-II immunohistochemistry

MT I-II-ir cells were abundant and similarly localized in the CuD- and in the control group (Figure 7 A–D). MTI-II-ir elements were often associated to fiber tracts, including the capsule system of the forebrain, pyramidal tract and other tracts of the brainstem, and cerebellar peduncles (Figure 7 A, B) and white matter. Positive elements were mostly evident in the caudate nucleus and putamen, hippocampus, fimbria, lateral nuclei of the septal area, central and ventral tegmental area of the mesencephalon, locus cœruleus (Figure 7 C D), motor nucleus of the facial nerve and other nuclei of the pons-medulla. Some (but not all) of the CuD animals showed a higher density of positive elements in the cochlear and vestibular nuclei. However, this latter result could be related to imperfect matching of sections, as differences of a few μm lateral to the sagittal plane may alter precise comparison of topography of smaller nuclear complexes.

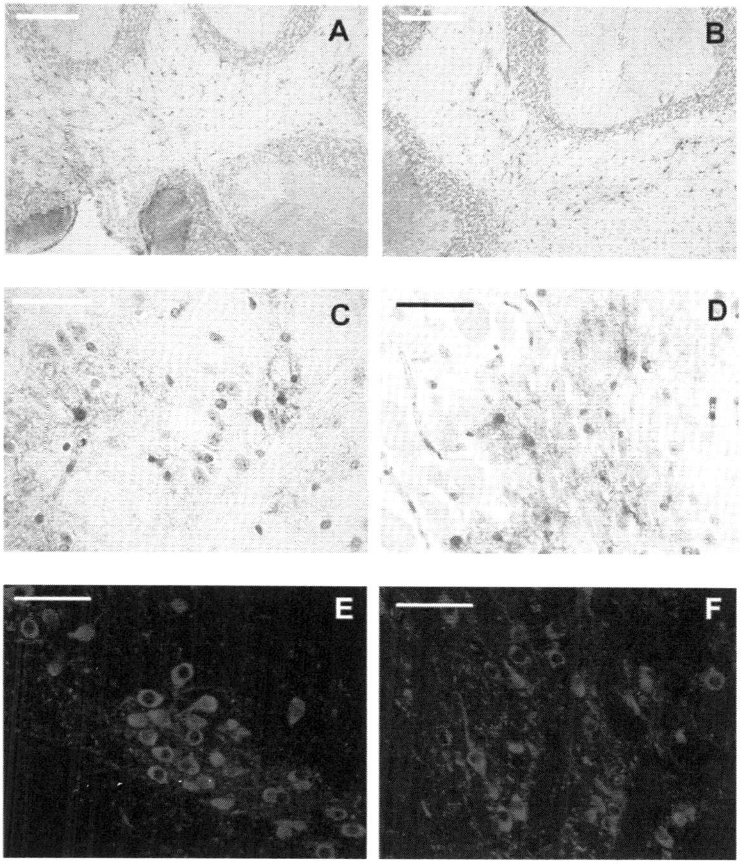

Figure 7. Representation of Immunohistochemical results. The images are relative to sagittal sections of the brain. Left column (A, C, E): CuD mice. Right column (B, D, F): control group. (A, B) MT I-II immunohistochemistry, cerebellum; scale bar=200 μm; C, ventral tegmental area of the mesencephalon, scale bar=50 μm; (D) locus cœruleus, scale bar=50 μm; (E, F) TH-immunohistochemistry, (E) substantia nigra, scale bar=50 μm.

TH immunohistochemistry

Representative images are shown in Figure 7 E, F. TH-ir neurons in the substantia nigra-ventral tegmental area appeared slightly more intense in the brains of mice receiving the CuD diet, showing more intense cell bodies and labelled fibers.

Behavioural Tests

In the open field test, no difference was detected between controls and CuD mice in the distance travelled, F[1], [21]=0.071, p=0.792, in the number of rearings, F[1], [21]=0.022, p=0.883, in the resting time, F[1], [21]=0.0237, p=0.879, in the number of urine drops, F[1], [21]=0.913, p=0.350 and in the number of fecal boli, F[1], [21]=1.434, 0=0.244.

The pole test (Figure 8) suggested that no significant difference was present between the groups that behaved in the same way, F[1], [21]=0.049, p=0.826 the first day, F[1], [21]=0.406, p=0.531 the second day, F[1], [21]=0.161, p=0.692 the third day.

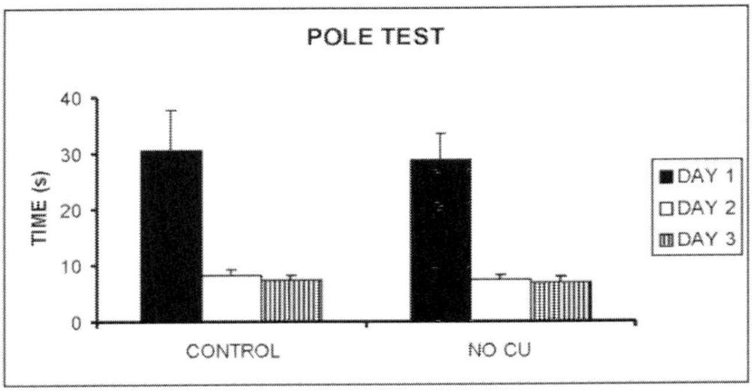

Figure 8. Pole test. Means+SEM are shown. Both groups can learn efficiently to climb down from the pole.

The predatory aggression test did not show any difference in the latency to attack the prey, F[1], [21]=1.347, p=0.258.

The habituation-dishabituation test showed no difference between the groups in each variable considered. Concerning the distance travelled in the target zone, the factors Odor and Repetition were significant, F[2], [42]=9.489, p<0.0005 and F[2], [42]=10.679, p<0.0005 respectively. These results are best explained by the interaction Odor×Repetition, F[4,84]=2.772, p<0.05: when water was presented, a similar distance was travelled in each of the three presentations. When camphor was presented, both the second and the third presentation induced mice to move for a shorter distance ($p<0.005$ compared to the first presentation). The latency to the first entry into the target zone was similar between the groups, since only the factor Repetition is different, F[2], [42]=7.564, $p<0.005$, with a shorter latency to enter the target zone at the first presentation, compared to the second ($p<0.005$) and third ($p<0.01$).

When considering the overall time spent in the target zone, that embrace the time spent moving and resting, only the factor Repetition was significant, F[2], [42]=10.033, $p<0.0005$, with the first presentation inducing a longer permanence in the proximity of the new stimulus ($p<0.002$ and $p<0.001$ compared to the second and third presentation, respectively). A similar result was obtained for the resting time in the target zone; the factor Repetition resulted significant, F[2], [40]=6.884, $p<0.005$, with the first presentation inducing a longer resting in the target zone, $p<0.005$ and $p<0.01$ compared to the second and third presentation.

When considering the number of entries in the target zone, the factor odor was significantly different, F[2], [42]=10.143, $p<0.0005$, with the first condition (no odor) inducing a higher number of entries ($p<0.005$ and $p<0.001$ compared to thymol and camphor, respectively). The factor Repetition was different, $F(2,42)=25.660$, $p<0.05$, with the first presentation inducing a higher number of entries, compared to the second and the third ($p<0.05$ for both).

DISCUSSION

Several studies have evaluated the effect of Cu deprivation and have concluded that this element is essential for brain growth and maturation [30], [31]. On the other hand, to the best of our knowledge, the investigations related to the effects of experimental diets on aged mice are scarce, notwithstanding its relevance for human nutrition strategies in old age. In addition, it is well established that a deficiency or excess in one metal can also influence the uptake of the others [32], [33], [34]. Our results confirmed this cross-interaction and, interestingly, demonstrated that also the concentration of Al was markedly altered in the brain of CuD animals compared to the controls.

Severe Cu deficiency can determine changes in Fe metabolism and can impair the correct absorption of dietary Fe [32]. Accordingly, we found decreased level of Fe in the brain (Figure 1C), particularly in the occipital area and in the cerebellum, as well as in the large and small intestine. It has been demonstrated that dietary Cu-deficiency induces anaemia that does not improve with Fe supplementation, however, it is fully reversed upon Cu supplementation [35].

In our experimental study we did observe neither liver nor intestinal Fe accumulation as described by other authors [32], [36]. The mechanism for Fe accumulation may be related to a decrease of Cu-enzymes such as ceruloplasmin and hephaestin that are required for Fe transport [37], [38]. Symptoms of anaemia and low plasma Fe level were recorded in weanling rats within days after treatment with a CuD diet [39]. One possible explanation for the discrepancy could be due to the different age of the animals. Pyatskowit and Prohaska [40] fed CuD diet to two cohorts of mice and rat pups: only the rat pups showed lower plasma and brain levels of Fe. Noteworthy, humans with Cu deficiency and anaemia may have normal plasma Fe concentrations [41]. In our series of aged mice, we observed no weight change between the two groups, and Cu was curtailed only in the frontal cortex.

Thus, Prohaska's laboratory has widely demonstrated the effect of Cu-deficiency in mice and rats when this nutritional restriction is imposed throughout gestation as well as lactation: as a result, brain Cu concentration resulted dramatically affected by the diet [42]. Notably, if the diet began shortly after birth the effect in terms of Cu-brain alteration was strongly limited [43]. This could explain the fact that, also in our experimental model, Cu seemed to be the metal less affected by the diet. On the other hand, Ca concentration was greatly affected by the diet (Figures 1F; 5). Studies on intracellular Ca homeostasis affected by Cu deficiency are limited. Ca mobilization from intracellular stores has been reported to be influenced by Cu deficiency as Cu deficient platelets showed a diminished rinse in cytosolic Ca upon agonist administration [44]. Beside, Cu deficiency impaired the architecture of mitochondria [45], the activity of the permeability transition pore [46], and altered ATP synthase [47], [48] suggesting that Cu deficiency may exert effects in the electron transport chain and oxidative phosphorylation system causing Ca mobilization.

Independently of metals being a primary cause or a consequence of the disease mechanism, a change in a single metal ion will upset the whole elemental homeostatic pool, resulting in a significant imbalance in the elemental levels in the body (serum, CSF and brain) as a sort of domino effect. The effect of an increase or decrease of a single metal is not restricted to the initiating metal alone, and such changes affect the total elemental and charge distribution pattern in the nervous system.

The role played by MT in the detoxification of heavy metals and the distribution of essential metals is clearly established but in recent years there has grown a wider appreciation of MT functions. The extensive presence of MT within the CNS and their altered expression as a common feature of several neurodegenerative diseases support the hypothesis of an important neurological function played by these proteins. These ubiquitous low molecular weight cysteine-rich proteins are present in mammals with four isoforms. Our MT I-II IHC data indicated that the metal-bearing system did not seem to be modified in the brain of the mice fed with the CuD diet (Figure 7 A–D). In fact, the presence and the distribution of MT I-II were similar in the experimental series and in the control group. On the other hand, the total content of MT in the brain was dramatically affected by the diet (Figure 6). We speculate that the decrease of MT content in the frontal area could be due to the correlated decrease of Cu and Zn in that area. Many studies have shown the critical role of MT in the regulation of zinc homeostasis [49], [50], [51]. Particularly, under oxidative stress conditions, Zn can be released from MT to exert protection. Thus, if an oxidative reaction triggers Zn release from MT, a dynamic change in Zn concentration can occur [52]. Concerning trivalent metals, alteration in their content is not likely solely to be due to alteration of these proteins and other transporter proteins must be involved. The dramatic decrease of total MT content, especially in the brain, could be detrimental in the light of their function as free radical scavenger. It has been demonstrated that animals as well as cell culture deficient in MT isoforms exhibited greater susceptibility to oxidative stress caused by different stimula [53], [54]. The mechanism by which MT exert a protective function is not clear. However, a reaction between the MT thiolate cluster and hydroxyl radicals has to be envisaged [55].

Furthermore, our findings indicated that the CuD diet did not influence the distribution of the classical astrocyte supporting elements, these latter identified by GFAP immunohistochemistry. Also the TH system showed only minor modifications in the two groups of mice (Figure 7 E, F).

Mice fed either diet were similar in their general appearance and did not show any overt sign of distress. The open field test showed that CuD mice had no motor impairment and did not differ from controls in their exploratory activity, since both groups recognized the open field as a new environment and behaved consequently.

The pole test evaluates motor performance, but its repetitions in different days allowed also the assessment of long-term memory. Again, both groups of mice behaved similarly: the first day both groups took a long time to reach the ground, while the subsequent days both groups were equally fast in reaching the ground. This means that mice had no motor deficit, but also that both groups could learn, retaining the information for one or more days (four days between the second and third repetition of the test).

The predatory aggression test confirmed that no deficit was present in the sensory (mainly olfactory) perception of the prey, in the emotional reactivity to the prey, or in the motor execution of the attack sequence. This test refers to a complex sensory-motor and cognitive-emotional domain mainly mediated by basal forebrain structures and olfactory projection areas [26].

The habituation-dishabituation test was designed to evaluate the discrimination of novel olfactory stimuli and their retention in short-term memory [29]. The distance travelled in the target zone showed a progressive shortening in both groups, most probably due to the mice habituating to the test procedure. The latency to the first entry into the target zone showed that both groups of mice entered equally fast in the area around the new stimulus, while both groups approached slower the stimulus, the second and third time it was presented. The overall time spent in the target zone confirmed that both groups of mice spent more time in the target zone the first time that a stimulus was presented. Therefore, both groups were able to discriminate the novel stimulus, which promptly elicited a fast exploratory behaviour, while the second and third presentation of the same stimulus induced a similarly slower response.

Although the array of alterations produced by the CuD diet in the aged mice appeared to be relevant, we did not observed any sign of impairments in their sensory, motor, emotional and cognitive behavioural patterns, compared to age-matched controls. The scenario is anyhow further compounded by the close interconnection among the different metals that can somehow create a sort of domino effect. In conclusion, the fundamental biochemical mechanisms linking brain biometal metabolism, environmental metal exposure, and neurodegenerative pathophysiology appear to be rather complicated and warrant further investigation.

FUNDING STATEMENT

This work was supported by PRIN 2007 grants from the Italian Ministry of Education, University and Research. The funders had no role in study design, data collection and analysis, decision to publish, or preparation of the manuscript.

REFERENCES

1. Gaeta A, Hider RC (2005) The crucial role of metal ions in neurodegeneration: the basis for a promising therapeutic strategy. Br J Pharmacol 146: 1041–1059.
2. White AR, Barnham KJ, Bush AI (2006) Metal homeostasis in Alzheimer's disease. Expert Rev Neurother 6: 711–722.
3. Drago D, Bolognin S, Zatta P (2008) Role of metal ions in the abeta oligomerization in Alzheimer's disease and in other neurological disorders. Curr Alzheimer Res 5: 500–507.
4. Bonda DJ, Lee HG, Blair JA, Zhu X, Perry G, et al. (2011) Role of metal dyshomeostasis in Alzheimer's disease. Metallomics 3: 267–270.
5. Zatta P, Frank A (2007) Copper deficiency and neurological disorders in man and animals. Brain Res Rev 54: 19–33.
6. Cuajungco MP, Faget KY, Huang X, Tanzi RE, Bush AI (2000) Metal chelation as a potential therapy for Alzheimer's disease. Ann N Y Acad Sci 920: 292–304.
7. Opazo C, Huang X, Cherny RA, Moir RD, Roher AE, et al. (2002) Metalloenzyme-like activity of Alzheimer's disease beta-amyloid. Cu-dependent catalytic conversion of dopamine, cholesterol, and biological reducing agents to neurotoxic H(2)O(2). J Biol Chem 277: 40302–40308.
8. Deibel MA, Ehmann WD, Markesbery WR (1996) Copper, iron, and zinc imbalances in severely degenerated brain regions in Alzheimer's disease: possible relation to oxidative stress. J Neurol Sci 143: 137–142.
9. Loeffler DA, DeMaggio AJ, Juneau PL, Brickman CM, Mashour GA, et al. (1994) Ceruloplasmin is increased in cerebrospinal fluid in Alzheimer's disease but not Parkinson's disease. Alzheimer Dis Assoc Disord 8: 190–197.
10. Lovell MA, Robertson JD, Teesdale WJ, Campbell JL, Markesbery WR (1998) Copper, iron and zinc in Alzheimer's disease senile plaques. J Neurol Sci 158: 47–52.
11. Crouch PJ, Barnham KJ, Bush AI, White AR (2006) Therapeutic treatments for Alzheimer's disease based on metal bioavailability. Drug News Perspect 19: 469–474.

12. Borchardt T, Camakaris J, Cappai R, Masters CL, Beyreuther K, et al. (1999) Copper inhibits beta-amyloid production and stimulates the non-amyloidogenic pathway of amyloid-precursor-protein secretion. Biochem J 344 Pt 2: 461–467.
13. Phinney AL, Drisaldi B, Schmidt SD, Lugowski S, Coronado V, et al. (2003) In vivo reduction of amyloid-beta by a mutant copper transporter. Proc Natl Acad Sci U S A 100: 14193–14198.
14. Bayer TA, Schafer S, Simons A, Kemmling A, Kamer T, et al. (2003) Dietary Cu stabilizes brain superoxide dismutase 1 activity and reduces amyloid Abeta production in APP23 transgenic mice. Proc Natl Acad Sci U S A 100: 14187–14192.
15. Crouch PJ, Hung LW, Adlard PA, Cortes M, Lal V, et al. (2009) Increasing Cu bioavailability inhibits Abeta oligomers and tau phosphorylation. Proc Natl Acad Sci U S A 106: 381–386.
16. Kessler H, Pajonk FG, Bach D, Schneider-Axmann T, Falkai P, et al. (2008) Effect of copper intake on CSF parameters in patients with mild Alzheimer's disease: a pilot phase 2 clinical trial. J Neural Transm 115: 1651–1659.
17. Quinn JF, Crane S, Harris C, Wadsworth TL (2009) Copper in Alzheimer's disease: too much or too little? Expert Rev Neurother 9: 631–637.
18. Sparks DL, Schreurs BG (2003) Trace amounts of copper in water induce beta-amyloid plaques and learning deficits in a rabbit model of Alzheimer's disease. Proc Natl Acad Sci U S A 100: 11065–11069.
19. Morris MC, Evans DA, Tangney CC, Bienias JL, Schneider JA, et al. (2006) Dietary copper and high saturated and trans fat intakes associated with cognitive decline. Arch Neurol 63: 1085–1088.
20. Kitazawa M, Cheng D, Laferla FM (2009) Chronic copper exposure exacerbates both amyloid and tau pathology and selectively dysregulates cdk5 in a mouse model of AD. J Neurochem 108: 1550–1560.
21. Bolognin S, Drago D, Messori L, Zatta P (2009) Chelation therapy for neurodegenerative diseases. Med Res Rev 29: 547–570.
22. Ames BN (2006) Low micronutrient intake may accelerate the degenerative diseases of aging through allocation of scarce micronutrients by triage. Proc Natl Acad Sci U S A 103: 17589–17594.
23. Murko S, Milacic R, Scancar J (2007) Speciation of Al in human serum by convective-interaction media fast-monolithic chromatography with

inductively coupled plasma mass spectrometric detection. J Inorg Biochem 101: 1234–1241.

24. Scheuhammer AM, Cherian MG (1986) Quantification of metallothioneins by a silver-saturation method. Toxicol Appl Pharmacol 82: 417–425.
25. Scheuhammer AM, Cherian MG (1991) Quantification of metallothionein by silver saturation. Methods Enzymol 205: 78–83.
26. Mucignat-Caretta C, Bondi M, Caretta A (2004) Animal models of depression: olfactory lesions affect amygdala, subventricular zone, and aggression. Neurobiol Dis 16: 386–395.
27. Sedelis M, Schwarting RK, Huston JP (2001) Behavioral phenotyping of the MPTP mouse model of Parkinson's disease. Behav Brain Res 125: 109–125.
28. Kurosaki R, Muramatsu Y, Kato H, Araki T (2004) Biochemical, behavioral and immunohistochemical alterations in MPTP-treated mouse model of Parkinson's disease. Pharmacol Biochem Behav 78: 143–153.
29. Pankevich DE, Baum MJ, Cherry JA (2004) Olfactory sex discrimination persists, whereas the preference for urinary odorants from estrous females disappears in male mice after vomeronasal organ removal. J Neurosci 24: 9451–9457.
30. Penland JG, Prohaska JR (2004) Abnormal motor function persists following recovery from perinatal copper deficiency in rats. J Nutr 134: 1984–1988.
31. Gybina AA, Tkac I, Prohaska JR (2009) Copper deficiency alters the neurochemical profile of developing rat brain. Nutr Neurosci 12: 114–122.
32. Reeves PG, DeMars LC (2004) Copper deficiency reduces iron absorption and biological half-life in male rats. J Nutr 134: 1953–1957.
33. Monnot AD, Zheng G, Zheng W (2012) Mechanism of copper transport at the blood-cerebrospinal fluid barrier: influence of iron deficiency in an in vitro model. Exp Biol Med (Maywood) 237: 327–333.
34. Yoshihara D, Fujiwara N, Kato S, Sakiyama H, Eguchi H, et al. (2012) Alterations in renal iron metabolism caused by a copper/zinc-superoxide dismutase deficiency. Free Radic Res 46: 750–757.
35. Roeser HP, Lee GR, Nacht S, Cartwright GE (1970) The role of ceruloplasmin in iron metabolism. J Clin Invest 49: 2408–2417.

36. Chen H, Huang G, Su T, Gao H, Attieh ZK, et al. (2006) Decreased hephaestin activity in the intestine of copper-deficient mice causes systemic iron deficiency. J Nutr 136: 1236–1241.
37. Vulpe CD, Kuo YM, Murphy TL, Cowley L, Askwith C, et al. (1999) Hephaestin, a ceruloplasmin homologue implicated in intestinal iron transport, is defective in the sla mouse. Nat Genet 21: 195–199.
38. Harris ZL, Klomp LW, Gitlin JD (1998) Aceruloplasminemia: an inherited neurodegenerative disease with impairment of iron homeostasis. Am J Clin Nutr 67: 972S–977S.
39. Reeves PG, Ralston NV, Idso JP, Lukaski HC (2004) Contrasting and cooperative effects of copper and iron deficiencies in male rats fed different concentrations of manganese and different sources of sulfur amino acids in an AIN-93G-based diet. J Nutr 134: 416–425.
40. Pyatskowit JW, Prohaska JR (2008) Copper deficient rats and mice both develop anemia but only rats have lower plasma and brain iron levels. Comp Biochem Physiol C Toxicol Pharmacol 147: 316–323.
41. Miyoshi I, Saito T, Iwahara Y (2004) Copper deficiency anaemia. Br J Haematol 125: 106.
42. Pyatskowit JW, Prohaska JR (2008) Iron injection restores brain iron and hemoglobin deficits in perinatal copper-deficient rats. J Nutr 138: 1880–1886.
43. Prohaska JR, Bailey WR (1993) Persistent regional changes in brain copper, cuproenzymes and catecholamines following perinatal copper deficiency in mice. J Nutr 123: 1226–1234.
44. Johnson WT, Dufault SN (1993) Intracellular calcium mobilization in rat platelets is adversely affected by copper deficiency. Biochim Biophys Acta 1175: 263–268.
45. Wildman RE, Hopkins R, Failla ML, Medeiros DM (1995) Marginal copper-restricted diets produce altered cardiac ultrastructure in the rat. Proc Soc Exp Biol Med 210: 43–49.
46. Johnson WT, Johnson LK (2009) Copper deficiency inhibits Ca2+-induced swelling in rat cardiac mitochondria. J Nutr Biochem 20: 248–253.
47. Chao JC, Medeiros DM, Davidson J, Shiry L (1994) Low levels of ATP synthase and cytochrome c oxidase subunit peptide from hearts of copper-deficient rats are not altered by the administration of dimethyl sulfoxide. J Nutr 124: 789–803.

48. Medeiros DM, Davidson J, Jenkins JE (1993) A unified perspective on copper deficiency and cardiomyopathy. Proc Soc Exp Biol Med 203: 262–273.
49. Kang YJ (2006) Metallothionein redox cycle and function. Exp Biol Med (Maywood) 231: 1459–1467.
50. Maret W (1994) Oxidative metal release from metallothionein via zinc-thiol/disulfide interchange. Proc Natl Acad Sci U S A 91: 237–241.
51. Maret W (1995) Metallothionein/disulfide interactions, oxidative stress, and the mobilization of cellular zinc. Neurochem Int 27: 111–117.
52. Kang YJ, Jiang Y, Saari JT (2007) Changes in copper and zinc status and response to dietary copper deficiency in metallothionein-overexpressing transgenic mouse heart. J Nutr Biochem 18: 714–718.
53. Yang HY, Wang YM, Peng SQ (2010) Metallothionein-I/II null cardiomyocytes are sensitive to Fusarium mycotoxin butenolide-induced cytotoxicity and oxidative DNA damage. Toxicon 55: 1291–1296.
54. Koumura A, Hamanaka J, Shimazawa M, Honda A, Tsuruma K, et al. (2009) Metallothionein-III knockout mice aggravates the neuronal damage after transient focal cerebral ischemia. Brain Res 1292: 148–154.
55. Maret W (2008) Metallothionein redox biology in the cytoprotective and cytotoxic functions of zinc. Exp Gerontol 43: 363–369.

CHAPTER 12

Development and Validation of a Physiology-based Model for the Prediction of PharmaCokinetics/ toxicokinetics in Rabbits

Panteleimon D. Mavroudis[1], Helen E. Hermes[1], Donato Teutonico[1], Thomas G. Preuss[2], and Sebastian Schneckener[1]

[1]Bayer AG, Engineering & Technology- Systems Pharmacology, Leverkusen, Germany, [2]Bayer Crop Science, Environmental Safety, Monheim, Germany

ABSTRACT

The environmental fates of pharmaceuticals and the effects of crop protection products on non-target species are subjects that are undergoing intense review. Since measuring the concentrations and effects of xenobiotics on all affected species under all conceivable scenarios is not feasible, standard laboratory animals such as rabbits are tested, and the observed adverse

Citation: Mavroudis PD, Hermes HE, Teutonico D, Preuss TG, Schneckener S (2018) Development and validation of a physiology-based model for the prediction of pharmacokinetics/toxicokinetics in rabbits. PLoS ONE 13(3): e0194294. https://doi.org/10.1371/journal.pone.0194294

effects are translated to focal species for environmental risk assessments. In that respect, mathematical modelling is becoming increasingly important for evaluating the consequences of pesticides in untested scenarios. In particular, physiologically based pharmacokinetic/toxicokinetic (PBPK/TK) modelling is a well-established methodology used to predict tissue concentrations based on the absorption, distribution, metabolism and excretion of drugs and toxicants. In the present work, a rabbit PBPK/TK model is developed and evaluated with data available from the literature. The model predictions include scenarios of both intravenous (i.v.) and oral (p.o.) administration of small and large compounds. The presented rabbit PBPK/TK model predicts the pharmacokinetics (Cmax, AUC) of the tested compounds with an average 1.7-fold error. This result indicates a good predictive capacity of the model, which enables its use for risk assessment modelling and simulations.

INTRODUCTION

Environmental and human risk assessments are based on standardized biotests that are conducted on different species. To minimize the use of animals, the 3R principle (i.e., reduction, replacement, and refinement) requires that animal experiments are substituted with appropriate alternative test methods whenever possible [1–4]. This principle is already a current practice in REACH [5, 6], which is a binding European Union regulation regarding the production and use of chemical substances. Additionally, in the European pesticide regulation 1107/2009, it is clearly stated that the use of non-animal test methods and other risk assessment strategies should be promoted, and animal testing should be minimized as tests on vertebrates should be undertaken as a last resort. However, despite the regulations in place, the use of alternative methods in pesticide risk assessments is limited. Since pesticides are designed to act on biological systems, they frequently have specific modes of actions. For these specific modes of action, reliable *in silico* methods are much harder to establish than for chemicals for which the baseline toxicity is assumed. In this regard, mathematical modelling can be used for for translations between different exposure routes and species. This method reduces animal testing, which is otherwise conducted to understand the effect under multiple exposure or species situations.

Physiologically based pharmacokinetic (PBPK) modelling is a technique used to simulate the pharmacokinetics (PK) of a substance in different animals and tissues [7–10]. In the case of toxicant risk assessments, the term physiologically based toxicokinetic (PBTK) models is preferably used to emphasize the application of the models to compounds that cause toxic responses [11]. In PBPK/TK models, the tissues of organisms are simulated as a group of physiologically based compartments of different volumes, blood flows, and other tissue composition elements. These compartments provide a mechanistic framework for evaluating the absorption, distribution, metabolism, and excretion (ADME) of a drug or toxicant in the body. Compared to other type of models, PBPK/TK models separate the information on the physicochemistry of a compound from the characteristics of an organism [12]. Therefore, the physiological model parameters are independent of the compound-specific parameters. Consequently, the physiological parameters can be used across different compounds and the compound parameters can be used across species [13]. Active processes involve parts of the physiology and properties of the compound. Hence the process itself may be translated (for homologues pathways), but the actual kinetic parameterisation may be very different.

PBPK models are standard tools in pharmacology and are frequently used to support the prediction of clinical pharmacokinetics, drug-drug interactions, dose scaling for children, and formulation/adsorption development [14]. Regulatory agencies such as the European Medicines Agency (EMA) and the Food and Drug Administration (FDA) encourage the use of PBPK models, and these models are often part of the submission of new medical entities [15, 16]. Also, World Health Organization (WHO) and the United States Environmental Protection Agency (US-EPA) underlines the necessity to establish common principles for their application in chemical hazard and risk assessment [17]. Despite that the use of PBPK/TK models is encouraged in risk assessment by several recent reports and articles [18], their utilization is limited. This limited use is surprising given the potential for translating the information from standard laboratory species such as rats, rabbits, and mice to physiologically related small (wild) animals.

The aim of this work is to develop and validate a rabbit PBPK/TK model. The rabbit is a common laboratory animal as well as a relevant species for environmental risk assessment. Hence, more information on rabbit physiology is available than for other relevant wild species (e.g., vole or wood mouse). An extensive literature search was performed, and the gathered physiology information was integrated into a PBPK/TK model. Next, a set of compounds with increasing complexities regarding administration and clearance profiles was used for validation. The validation set included experiments with intravenous injection and oral administration, compounds with different excretion and metabolisation profiles (renal and hepatic clearance), and measurements of compound concentrations in different tissues. By considering a diverse set of compounds, different aspects of the model were validated, such as general physiology, renal versus hepatic (i.e., metabolic) clearance, and properties of the gastrointestinal tract (GIT). The model was validated by comparing the model predictions to measured data. The discrepancies between observations and the simulation were resolved by optimizing the parameters or retrieving independent information from the literature.

MATERIALS AND METHODS

Model structure and parameterization

For a comprehensive PBTK model as described here, a stepwise approach is typically used to inform the parameters. The first level retrieves the parameter values from the literature to describe the general properties of the species and is independent of the compound to be modelled. Second, the modelling of a specific compound starts by informing the model with the physicochemical properties of the molecule. This information is readily available or can be calculated independently [19]. Third, the active processes (e.g., transport and metabolism) are parameterized based on the information of *in vitro* or *in vivo* assays. Information from other species can be also used when there is no information for the species of interest. Lastly, if exposure data for a compound in the target species is available, this can be used to fine-tune the selected parameters. A schematic workflow of the model establishment is shown in Fig 1. In our model, physiological parameters of rabbit were introduced in the PK-Sim® framework in order to represent rabbit physiology (first level). If parameters of rabbit physiology were not available in the literature, the corresponding parameters of a pre-existing PBPK model were used. For this, the model of the taxonomically related mouse was used.

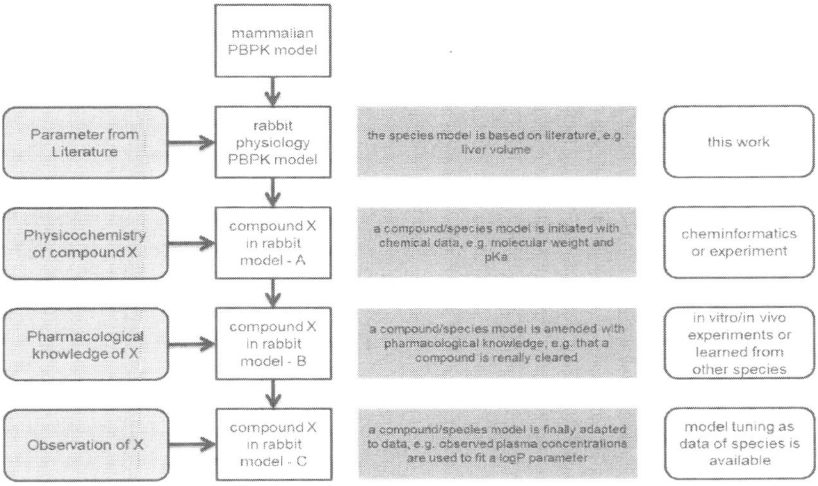

Figure 1. Workflow of the rabbit PBPK/TK model development.

Starting with a generic mammalian PBPK model on top, rabbit specific physiology parameters are added (first level). Next, physicochemistry from a compound (second level), followed by information regarding active processes (third level) are introduced from literature, in-house experiments or transferred from other species. Lastly, PK data are used to identify additional parameter values of the model (fourth level).

https://doi.org/10.1371/journal.pone.0194294.g001

The rabbit model is based on the mammalian physiological framework implemented in the PK-Sim® software platform [20]. This whole-body physiology-based pharmacokinetic framework provides a detailed mathematical representation of ADME processes in humans and several common laboratory animals. In essence, the mammalian body is divided into compartments representing the relevant organs or tissues as well as the arterial and venous blood pools that connect the different organs via blood flow (Fig 2A). Organs are further divided into sub- compartments that describe the vascular space (divided into plasma and red blood cells) as well as the avascular space (divided into interstitial and cellular space) (Fig 2B and 2C). The characteristics of the administered compound (small or large molecules such as proteins) determine its distribution by either diffusion through membranes (Fig 2B) or by pores and endosomal clearance (Fig 2C). A substantial amount of experimental information has been used to inform and calibrate the biological and physiological processes included in the model. The model has successfully been used to describe the uptake and distribution of a wide range of compounds in the species included to date [10, 21–26].

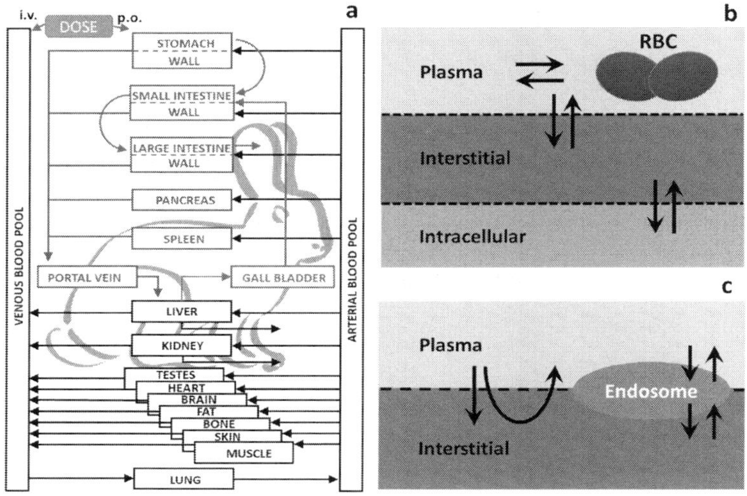

Figure 2. Rabbit PBPK model structure.

a: Generic model structure implemented in the PK-Sim® software suite. b: The sub-compartments of the tissues and the distribution scheme for the case of small molecules (RBC: red blood cells), c: The sub-compartments of the tissues and the distribution scheme for the case of large molecules.

https://doi.org/10.1371/journal.pone.0194294.g002

Based on the current animal models implemented in the PK-Sim® software suite, the model for a 2.5 kg rabbit was built by adjusting the values of organ volumes, haematocrit, specific blood flow rates, GIT compartment pH, GIT compartment dimensions, GIT compartment transit/emptying times, and the effective surface areas of the GIT-small intestine compartments according to rabbit physiology values found on literature. For most of the parameters, there was more than one value available from the literature. In most cases, the relative differences among the different sources were not large. The sources that provided information for as many tissues as possible were selected to achieve a consistent representation. Therefore, most organ volumes/specific blood flow rates were taken from [27], GIT-pH values were taken from [28], and GIT-proximal/distal radius and length values were taken from [29]. In the case of gastric emptying time, since the reported values in the literature had large deviations (i.e., from 20 min to 8 hrs), different values were chosen based on the experiment. Similarly, for the large/small intestinal transit rates, the full range of values was considered to set up the model. For the parameter values not found in the literature, the mouse PBPK model parameters were transferred and used, as it is the closest animal species to the rabbit among the species available in the physiology database (e.g., mouse retains a gallbladder unlike the rat).

Validation process

The rabbit model was informed, tested and verified using published data for a range of compounds in which the clearance pathways were known from either the literature or from in-house data. The choice of compounds took a hierarchical approach beginning with compounds cleared solely by either the kidney or the liver, which are the two main excretory organs, followed by the compounds cleared by both organs (i.v. administration). Subsequently, after assessing the physiology of the kidney and the liver, increasingly more complex clearance processes and administration protocols (p.o. administration) were simulated. For each simulation, the model prediction was first compared with the available data. In the case of deviations, either those parameters are known to vary among species or the parameters with no prior information were optimized. For all simulations, the partition coefficients and the cellular permeabilities were calculated based on the PK-Sim® Standard distribution model [22, 30]. In short, partition coefficients describe the partition of a compound between water and lipids or water and proteins. They are calculated using a mechanistic formula similar to that described in [31]. The product of organ surface area and permeability controls the rate of permeation across lipid cell membranes from plasma into organs. Calculation of the permeabilities was based on the semi-empirical formula first published in [32] that accounts for transcellular as well as paracellular passive transport for small molecules. For large molecules, the two pores transport model is used [33] without accounting for FcRn binding.

Sensitivity analysis

To evaluate the sensitivity of our model to parameter perturbations, a local sensitivity analysis was performed. At every step of the sensitivity analysis, one parameter *j* of the model was varied by 10% of its nominal value while the remaining model parameters were kept constant. Next, the sensitivity coefficients ($s_{i,j}$) for the AUC or C_{max} outputs were calculated as (Eq 1)

$$s_{i,j} = \frac{\partial f_i}{\partial x_j} \qquad (1)$$

where ∂f_i is difference in PK parameter (AUC or C_{max}) between perturbed and un-perturbed simulation and ∂x_j is the difference between varied and nominal parameter value. PK parameters of the sensitivity analysis were constrained to C_{max} and AUC as the most relevant PK parameters for evaluating drug toxicity/efficiency. The sensitivity analysis was restricted to parameters of the rabbit model while parameters specific for compounds were not considered.

Availability of rabbit PBPK/TK model

The rabbit PBPK/TK model presented in this work and the software PK-Sim® are available in the Open Systems Pharmacology Suite, which is now developed as an open source platform (https://github.com/Open-Systems-Pharmacology/Suite/releases/tag/v7.1.0). The data used to calibrate and validate the model were collected from the literature and referenced where used.

RESULTS

Model validation

First, a rabbit PBPK/TK model was developed, based on the pre-existing mammalian model structure and relative parameters retrieved from the literature or transferred from other mammalian PBPK/TK models. The rabbit PBPK/TK model structure with all considered organs and interactions is shown in Fig 2A. Second, to assess the reliability of the model, six different compounds (i.e., inulin, caffeine, ofloxacin, theophylline, paracetamol, and acyclovir) were modelled. Since inulin is solely filtered by the glomerulus of the kidney but is not secreted or reabsorbed by the tubulus or metabolized, it is well suited to validate the renal glomerular filtration rate (GFR) in the rabbit. To simulate renal excretion, the GFR was set to 100%, a value that is expected for renally cleared molecules. After determining the renal function of the rabbit, caffeine was tested since it is predominantly metabolized by the liver. The metabolic processes were always localized in the liver as hepatic clearance, and the respective reaction kinetics were optimized based on the available data. Next, more complex compounds such as ofloxacin and theophylline were used. To assess the rabbit GIT, oral administration was further added to test the model. The validation of oral administration was based on the compounds paracetamol, theophylline, and acyclovir. In the following sections, the detailed results for the compounds inulin and paracetamol are shown as they provide information for multiple tissues (inulin) and multiple oral formulations (paracetamol).The model results for the additional compounds are described in the supporting information and are only summarized here.

Inulin case study

Fig 3 compares the observed PK-data from 200 mg/kg inulin after i.v. administration in rabbits weighing 2.5 kg [34] to the PBPK/TK model simulations in different tissues. The dotted lines represent the rabbit model simulations with no parameter calibration. The rabbit PBPK model predicts the PK profiles very close to the experimental data for the muscle, heart, and bone tissues without any further parameter adjustments (Fig 3 dotted lines and Table 1). For the plasma, skin and lung, the model was adjusted to better agree with the observed data. Here, the specific glomerular filtration rate ($GFR_{specific}$) was increased from 0.6 to 0.8 [l/min/kg-kidney-weight] [35] to better explain the observed exposure in the plasma. Similarly, the hydraulic conductivity of the skin was increased in accordance with the inulin simulations performed in rats [21, 36]. Finally, to describe the PK profile in the lungs, the interstitial fraction of the lung was increased. These adaptions to the model are all in accordance with the literature. The simulation with the three adjusted parameters indicates a correct distribution of the compound among the different tissues of the rabbit (Fig 3-solid lines).

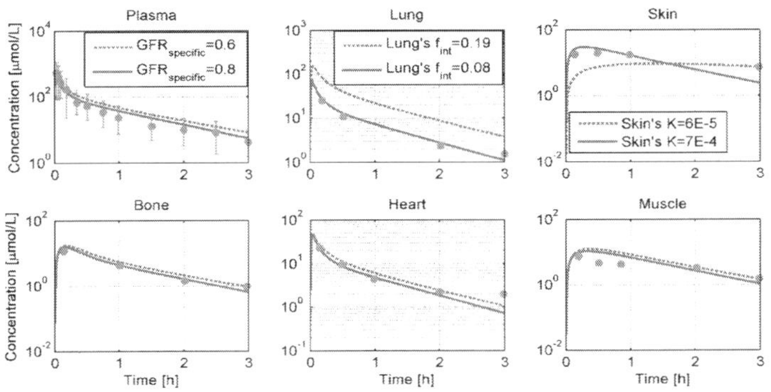

Figure 3. Simulations of the concentration profiles of inulin in venous blood plasma, heart, muscle, bone, lung, and skin, for 200 mg/kg intravenous administration in New Zealand white rabbits weighing 2.5 kg.

https://doi.org/10.1371/journal.pone.0194294.g003

The dotted lines represent the predicted profiles, while the solid lines show the simulated profiles after the $GFR_{specific}$, skin's hydraulic conductivity (*K*), and lung's fraction interstitial (f_{int}) were adjusted. The green dots are experimental observations from [34].

Table 1. Comparison of experimentally observed AUC, predicted AUC, and AUC after parameter optimization for available compartments, along with the relative errors, administration details and references.

Compounds	AUC observed	AUC Predicted	AUC Optimized	% AUC Error (Predicted)	% AUC Error (Optimized)	Tissue/Administration	Literature
Inulin	842.36	994.63	-	15.31	-	Venous Blood (40 mg/kg)/i.v.	[37]
	1345.61	1491.95	-	9.81	-	Venous Blood (60 mg/kg) i.v. i.v.	[37]
	6901.24	9383.03	7564.91	26.45	8.77	Plasma/i.v.	[34]
	1071.31	3643.40	1186.67	70.60	9.72	Lung/i.v.	[34]
	2438.59	1435.94	2242.76	69.83	8.73	Skin/i.v.	[34]
	660.24	941.45	760.29	29.87	13.16	Bone /i.v.	[34]
	873.24	1045.38	819.07	16.47	6.61	Heart/i.v.	[34]
	620.97	1090.23	884.54	43.04	29.80	Muscle/i.v.	[34]
Caffeine	5432.82	7747.60	5739.40	29.88	5.34	Venous Blood/i.v.	[39]
Theophylline	23871.43	63755.93	27257.72	62.56	12.42	Venous Blood/i.v.	[40]
	54651.84	57080.23	54084.85	4.25	1.05	Venous Blood/i.v.	[41]
Theophylline (PO)	217832.06	86553.63	192065.78	151.67	13.42	Venous Blood/Conventional Tablet	[41]
Paracetamol (PO)	2279.80	1836.71	2372.61	24.12	3.91	Venous Blood/Solution	[38]
	1712.76	1958.72	1581.04	12.56	8.33	Venous Blood/Rapid Tablet	[38]
	2225.68	401.05	1335.17	454.96	66.70	Venous Blood/Conventional Tablet	[38]
Acyclovir (PO)	6469.00	9089.22	7630.12	28.83	15.22	Venous Blood/Solution	[42]

https://doi.org/10.1371/journal.pone.0194294.t001

Inulin simulation efficiently described the data adopted from [37] and as such no further optimization was performed.

https://doi.org/10.1371/journal.pone.0194294.t001

The results from two additional inulin i.v. doses simulating the experiment of [37] are shown. These simulations indicate that the rabbit PBPK model explains the available observations without parameter calibration.

Paracetamol case study

For paracetamol, the work of [38] reports three oral formulations: 50 mg of paracetamol administered as an oral solution, a rapidly disintegrating tablet, and a conventional tablet. Fig 4 compares the observed plasma concentration data for the three different oral formulations with the simulation profiles. The dotted lines represent the model predictions based on the typical gastric emptying time (GET = 0.5 hr) of a rabbit and the dissolution shapes (DS). Regarding dissolution time (DT), the rapidly disintegrating tablet model predictions incorporated the values reported in the respective publication [38] (0.25 min). A typical DT (240 min.) was hypothesized for a conventional tablet since no other information was available. For the solution and the rapidly disintegrating tablet, the model predictions (dotted lines, Fig 4 and Table 1) are in close accordance with the observed data. Furthermore, we optimised our model by calibrating the GET, DT, and DS based on the available experimental evidence (solid line, Fig 4). The simulations after parameter calibration explain the data for a solution and a rapidly disintegrating tablet. For the case of a conventional tablet, the data point towards a slower dissolution.

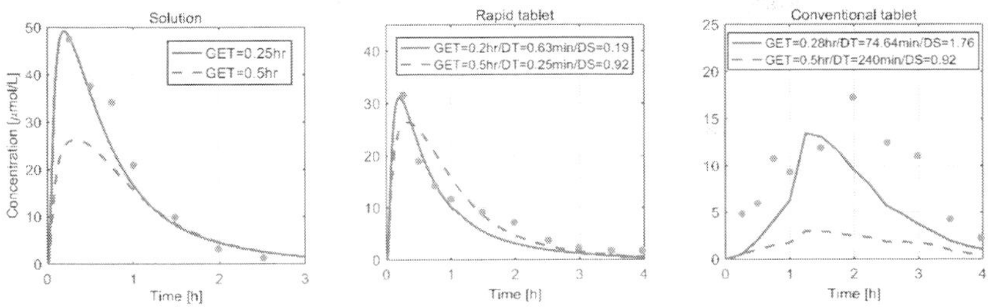

Figure 4. Plasma concentration profiles of three different formulations of 50 mg paracetamol in rabbit.

The green dots are data from [38], the dotted lines are the model predictions, and the solid lines represent simulations after parameter adaption. GET: Gastric emptying time (for solution/rapidly disintegrating tablet/ conventional tablet). DT: dissolution time (for rapid tablet/conventional tablet). DS: dissolution shape (for rapid tablet/conventional tablet).

https://doi.org/10.1371/journal.pone.0194294.g004

Predictive performance

Table 1 summarizes the model performance for a) a prediction based on the rabbit model solely informed by literature based values of rabbit physiology and compound related processes, as well as active transport parameters transferred from other species, and b) the simulations after the appropriate parameters were optimized given the observed data. The endpoint was the area under the curve (AUC) of the tissue concentration-time profiles. The table shows the observed, predicted and optimized AUC values along with the respective prediction errors. The mean fold error of the predicted AUC was 1.7, and with the exception of the tablet formulations, the AUC was always below 2. Similar results were obtained when the C_{max} (maximum concentration) predictions and observations were compared.

Sensitivity analysis

A sensitivity analysis was performed to further investigate the relationship between the parameter values and the model output for different compounds and administration strategies (i.v. and p.o. administration). Fig 5 illustrates the relative change in the AUC and C_{max} when the ten most sensitive parameters for each are changed. From green to yellow, the square of the sensitivity coefficient (Eq 1) increases denoting a higher sensitivity of the model output (AUC, C_{max}) to the respective parameter. The parameters listed are related to the distribution, clearance or uptake of the substance. For example, inulin is solely cleared by the GFR in the kidney and therefore, the kidney volume and the GFR (specific) are the most sensitive model parameters regarding inulin's AUC. Similarly, the AUC of caffeine, which is cleared by the liver, remains highly sensitive to the liver volume. On the other hand, C_{max} appears to be more sensitive to variation of parameters that are related to either the route of administration or the distribution of the drugs. For example C_{max} in paracetamol and acyclovir that are administered orally is more sensitive to gastric emptying time whereas in theophylline muscle volume plays a significant role, possibly as relevant volume of distribution. The sensitivity analysis was focused on rabbit-specific parameters and omitted compound specific parameters (e.g. V_{max}, K_m of active processes) as it is aimed to understand the impact of choosing alternative literature values of the rabbit physiology.

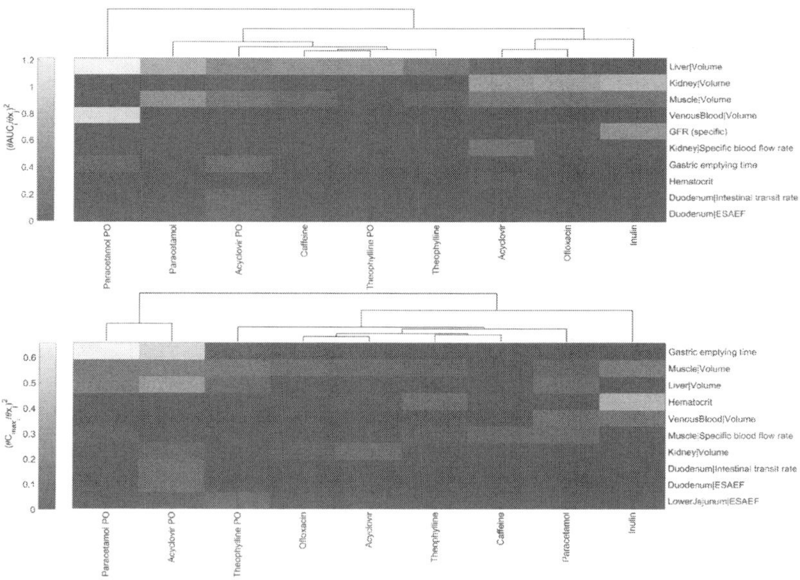

Figuge 5. Local sensitivity analysis showing the ten most sensitive parameters for i.v. and p.o. (PO) administration regarding the change in AUC and C_{max}.

ESAEF stands for Effective Surface Area Enhancement Factor. Upper panel shows sensitivity indices values for AUC variation, and the lower panel sensitivity indices values for C_{max} variation.

https://doi.org/10.1371/journal.pone.0194294.g005

DISCUSSION

PBPK/TK modelling is progressively gaining acceptance in risk assessment as it allows for the simulation of tissue-specific compound concentrations and as it is a suitable framework to translate *in vitro* experiments to *in vivo* exposures when considering ADME processes for relevant species [43–45]. In this work, a PBPK/TK model was developed for rabbits, and its predictive capacity was tested based on data from the current literature.

There are a number of reports, opinions and reviews that stress the significance of PBPK/TK modelling in risk assessment and propose the best modelling practices [46–53]. Although there may be deviations in model development strategies based on the questions of interest, the first critical step in PBPK modelling is the review of existing datasets for the relevant species. After informing the model parameter describing the rabbit physiology, the model was validated by testing its predictive capacity on PK data from selected compounds. The validation was conducted to gain more knowledge by either re-estimating the relevant parameters and observing how the predictions improved or by evaluating the sensitivities of the physiological parameters to certain administration scenarios. The rabbit physiology was validated by compounds of gradually increasing complexity regarding clearance, so that any model inconsistency between the simulation and observations could be traced back to the clearance mechanism. For the compounds tested in this work, there were no evidence for different metabolic/transport pathways between rabbit and other species. However, in a PBPK setting, inter-species differences regarding active processes (e.g. an extra transporter) can be considered by re-evaluating compound-related parameters.

The first compound tested was inulin, which is a "gold standard" for measuring GFR [54] and thereby renal function. In our PBPK/TK model, the filtrated amount depends on the fraction unbound (fu) of the compound, the weight of the kidney, and $GFR_{specific}$, which is the volume of the compound filtered per time per weight of the kidney (i.e., l/min/kg organ). These parameters in the rabbit PBPK/TK model were retrieved from the literature and included the deviation of the filtering capacity ($GFR_{specific}$) between individual animals [35]. Fig 3 shows the prediction of our model in dotted lines for six relevant rabbit tissues namely, plasma, lung, skin, bone, heart, and muscle. Interestingly, our model has an exceptionally high predictive capacity even before the experimental data are used to further optimize the model parameters. As Table 1 indicates, the % error of the PBPK/TK model predictions for this experiment is less than 70% (the mean value for all tissues is approximately 35%). To further inform the rabbit PBPK/TK model, additional data from the literature were considered, and the skin's hydraulic conductivity was recalibrated in accordance with the values observed in rats [36]. Finally, the interstitial fraction of the lungs (i.e., the interstitial volume of the lung over whole tissue volume) was re-estimated to better explain the data. In our panel of compounds, only concentration levels of inulin were observed in other tissues than blood plasma. This data was utilised to adjust parameters of the skin and the lung. As more data in those organs with other compounds become available, those parameters need to be re-evaluated.

In pharmacokinetics/toxicokinetics, there is no *a priori* threshold on an acceptable model error, but PBPK/TK models are generally accepted and considered useful when the prediction error on AUC or C_{max} is in the 2-fold or 3-fold range [55–57]. In the rabbit model presented in this study, a mean fold error of 1.7 was calculated for the AUC among all compounds tested and simulated (Table 1). This result illustrates the high predictive power of our PBPK/TK model in i.v. and p.o. administration scenarios. Specifically, the validation included the kidney function through inulin i.v. administration, liver function through caffeine i.v. administration, and simultaneous kidney and liver function through ofloxacin and theophylline i.v. administration. Additionally, the model was tested on an oral administration of paracetamol through solution, rapidly disintegrating tablet, and conventional tablet formulations. The model predictions for the three formulations are shown in dotted lines in Fig 4. The predictions for the solution and the rapidly disintegrating tablet are in close accordance with the observed data, whereas the 'conventional' tablet prediction suggests a significantly different dissolution time (DT). This result is reasonable considering the available experimental data. For the case of solutions, the model assumes an instantaneous availability of the drug in the stomach compartment of the GIT, whereas for both formulations of tablet, a Weibull function was used to empirically describe the tablet dissolution process. For the case of the rapidly disintegrating tablet, the information for the dissolution time found in the published literature (i.e., 0.25 min) [38] was included. However, the dissolution time for the 'conventional' tablet formulation of paracetamol was not reported. In practice, dissolution times are experimentally accessible *ex vivo* and could be used in modelling, as done for the rapidly disintegrating tablet. However, the parameters were fit to the data in this study.

As a final step in the PBPK/TK model development, a local sensitivity analysis was performed. A sensitivity analysis is described as a crucial step in the best practices for modelling in risk assessment [47] and is recently used increasingly in numerous mechanism-based modelling efforts in PK/PD [58, 59]. A sensitivity analysis not only evaluates the robustness of the PBPK/TK models but in a PBPK/TK setting where several literature values are available for certain physiologically-based parameters, it reveals whether the choice of alternative values would significantly impact the observed behaviour. This result is expected since most of the compounds tested were partially or totally cleared through metabolism in the liver. Therefore, a relative change in the liver mass or volume would highly affect the amount of drug cleared and as such the AUC. Similarly, for the compounds cleared renally, either through GFR (inulin) or tubular secretion and GFR (acyclovir), the parameters highly affecting the AUC are kidney volume and filtration capacity ($GFR_{specific}$). The parameters such as gastric emptying time, or GIT-length and the effective surface area are sensitive only in scenarios involving oral administration (i.e., paracetamol PO, acyclovir PO). In contrast, gastric emptying time appears to be the most sensitive parameter regarding C_{max} variation on compounds involving oral administration such as paracetamol and acyclovir. Furthermore, the muscle volume is the largest organ of the rabbit and the most important regarding distribution of drugs in the body hence it appears to be one of the most sensitive parameters. Overall, our analysis reveals sensitivities that are consistent with the biology and clearance pathways of each compound.

The workflow for using the rabbit PBTK model in risk assessment starts with identifying the physicochemical properties of a new active ingredient and its formulation. This step is followed by the use of *in vitro* assays to understand the active transport and metabolism. Ideally, before informing an initial PBTK model, TK data are generated in common species such as mice or rats in addition to the *in vitro* experiments. The translation of the toxicokinetics to the relevant species, such as rabbits and other wild animals, is then an exchange of the physiology-related parameters to the relevant physiology. The method can be used to predict the toxicokinetics of a new substance for different tissues and organs in rabbits. Based on our work, an average fold error of 1.7 can be expected for a compound of typical complexity.

In the present study, we demonstrated the development of a rabbit PBTK model based on data from the available literature. Specifically, a generic rabbit model based on well verified mammalian physiology was further informed by data from the available rabbit-related literature. The model yielded good TK predictions for different compounds with different exposure and clearance routes. By comparing predictions to measured toxicokinetic data, further species-specific parameters could be refined. Overall, this work presents a rabbit PBPK/TK model and its successful validation, and can additionally serve as a blueprint for the development of PBPK/TK models for additional species.

ACKNOWLEDGMENTS

The authors would like to acknowledge the valuable technical and scientific contributions of Virginie Ducrot, Christoph Niederalt, Juri Solodenko, Katrin Coböken, Tobias Kanacher, Jan Schlender and Lars Küpfer.

REFERENCES

1. Russell WMS, Burch RL. The principles of humane experimental technique. The principles of humane experimental technique. 1959.
2. Hartung T. Food for thought… on animal tests. *ALTEX*. 2008;25(1):3–16. .
3. Leist M, Hartung T, Nicotera P. The dawning of a new age of toxicology. *ALTEX*. 2008;25(2):103–14. .
4. Scholz S, Sela E, Blaha L, Braunbeck T, Galay-Burgos M, Garcia-Franco M, et al. A European perspective on alternatives to animal testing for environmental hazard identification and risk assessment. *Regul Toxicol Pharmacol*. 2013;67(3):506–30. doi: 10.1016/j.yrtph.2013.10.003 .
5. Lilienblum W, Dekant W, Foth H, Gebel T, Hengstler JG, Kahl R, et al. Alternative methods to safety studies in experimental animals: role in the risk assessment of chemicals under the new European Chemicals Legislation (REACH). *Arch Toxicol*. 2008;82(4):211–36. doi: 10.1007/s00204-008-0279-9 .
6. ECHA. European Chemicals Agency. The Use of Alternatives to Testing on Animals for the REACH Regulation 2011. Helsinki,Finland.: 2011.
7. Eissing T, Kuepfer L, Becker C, Block M, Coboeken K, Gaub T, et al. A computational systems biology software platform for multiscale modeling and simulation: integrating whole-body physiology, disease biology, and molecular reaction networks. *Frontiers in physiology*. 2011;2:4 doi: 10.3389/fphys.2011.00004 ; PubMed Central PMCID: PMC3070480.
8. Jones HM, Chen Y, Gibson C, Heimbach T, Parrott N, Peters SA, et al. Physiologically based pharmacokinetic modeling in drug discovery and development: a pharmaceutical industry perspective. *Clin Pharmacol Ther*. 2015;97(3):247–62. doi: 10.1002/cpt.37 .
9. Reddy M, Yang R, Andersen ME, Clewell HJ III III. *Physiologically based pharmacokinetic modeling: science and applications*: John Wiley & Sons; 2005.
10. Edginton AN, Theil FP, Schmitt W, Willmann S. Whole body physiologically-based pharmacokinetic models: their use in clinical drug development. *Expert opinion on drug metabolism & toxicology*. 2008;4(9):1143–52. doi: 10.1517/17425255.4.9.1143 .
11. Andersen ME. Toxicokinetic modeling and its applications in chemical risk assessment. *Toxicology letters*. 2003;138(1–2):9–27. .

12. Jusko WJ. Moving from basic toward systems pharmacodynamic models. *J Pharm Sci*. 2013;102(9):2930–40. doi: 10.1002/jps.23590 ; PubMed Central PMCID: PMC3743951.
13. Thiel C, Schneckener S, Krauss M, Ghallab A, Hofmann U, Kanacher T, et al. A systematic evaluation of the use of physiologically based pharmacokinetic modeling for cross-species extrapolation. *J Pharm Sci*. 2015;104(1):191–206. doi: 10.1002/jps.24214 .
14. Sager JE, Yu J, Ragueneau-Majlessi I, Isoherranen N. Physiologically Based Pharmacokinetic (PBPK) Modeling and Simulation Approaches: A Systematic Review of Published Models, Applications, and Model Verification. *Drug Metab Dispos*. 2015;43(11):1823–37. doi: 10.1124/dmd.115.065920 ; PubMed Central PMCID: PMC4613950.
15. Zhuang X, Lu C. PBPK modeling and simulation in drug research and development. *Acta Pharm Sin B*. 2016;6(5):430–40. doi: 10.1016/j.apsb.2016.04.004 ; PubMed Central PMCID: PMC5125732.
16. Zhao P, Zhang L, Grillo JA, Liu Q, Bullock JM, Moon YJ, et al. Applications of physiologically based pharmacokinetic (PBPK) modeling and simulation during regulatory review. *Clin Pharmacol Ther*. 2011;89(2):259–67. doi: 10.1038/clpt.2010.298 .
17. European Food Safety A. Modern methodologies and tools for human hazard assessment of chemicals. *EFSA Journal*. 2014;12(4):3638-n/a. doi: 10.2903/j.efsa.2014.3638
18. Meek ME, Barton HA, Bessems JG, Lipscomb JC, Krishnan K. Case study illustrating the WHO IPCS guidance on characterization and application of physiologically based pharmacokinetic models in risk assessment. *Regul Toxicol Pharmacol*. 2013;66(1):116–29. doi: 10.1016/j.yrtph.2013.03.005 .
19. Hillisch A, Heinrich N, Wild H. Computational Chemistry in the Pharmaceutical Industry: From Childhood to Adolescence. *ChemMedChem*. 2015;10(12):1958–62. doi: 10.1002/cmdc.201500346 .
20. Willmann S, Lippert J, Sevestre M, Solodenko J, Fois F, Schmitt W. PK-Sim®: a physiologically based pharmacokinetic 'whole-body' model. *BIOSILICO*. 2003;1(4):121–4. http://dx.doi.org/10.1016/S1478-5382(03)02342-4.
21. Willmann S, Schmitt W, Keldenich J, Dressman JB. A physiologic model for simulating gastrointestinal flow and drug absorption in rats. *Pharm Res*. 2003;20(11):1766–71. .

22. Willmann S, Schmitt W, Keldenich J, Lippert J, Dressman JB. A physiological model for the estimation of the fraction dose absorbed in humans. *J Med Chem*. 2004;47(16):4022–31. doi: 10.1021/jm030999b .
23. Willmann S, Lippert J, Schmitt W. From physicochemistry to absorption and distribution: predictive mechanistic modelling and computational tools. *Expert opinion on drug metabolism & toxicology*. 2005;1(1):159–68. doi: 10.1517/17425255.1.1.159 .
24. Willmann S, Hohn K, Edginton A, Sevestre M, Solodenko J, Weiss W, et al. Development of a physiology-based whole-body population model for assessing the influence of individual variability on the pharmacokinetics of drugs. *J Pharmacokinet Pharmacodyn*. 2007;34(3):401–31. Epub 2007/04/14. doi: 10.1007/s10928-007-9053-5 .
25. Thelen K, Coboeken K, Willmann S, Burghaus R, Dressman JB, Lippert J. Evolution of a detailed physiological model to simulate the gastrointestinal transit and absorption process in humans, part 1: oral solutions. *J Pharm Sci*. 2011;100(12):5324–45. Epub 2011/10/14. doi: 10.1002/jps.22726 .
26. Schmitt W, Willmann S. Physiology-based pharmacokinetic modeling: ready to be used. Drug discovery today *Technologies*. 2005;2(1):125–32. doi: 10.1016/j.ddtec.2005.01.001 .
27. Davies B, Morris T. Physiological parameters in laboratory animals and humans. *Pharm Res*. 1993;10(7):1093–5. .
28. Kararli TT. Comparison of the gastrointestinal anatomy, physiology, and biochemistry of humans and commonly used laboratory animals. *Biopharm Drug Dispos*. 1995;16(5):351–80. .
29. Lebas F, Coudert P, de Rochambeau H, Thébault R. The Rabbit: Husbandry, health and production *FAO Animal Production and Health Series No 21*. Rome: Food and agriculture organization of the United Nations; 1997.
30. Willmann S, Lippert J, Sevestre M, Solodenko J, Fois F, Schmitt W. PK-Sim®: a physiologically based pharmacokinetic 'whole-body'model. *Biosilico*. 2003;1(4):121–4.
31. Poulin P, Theil FP. A priori prediction of tissue:plasma partition coefficients of drugs to facilitate the use of physiologically-based pharmacokinetic models in drug discovery. *J Pharm Sci*. 2000;89(1):16–

35. doi: 10.1002/(SICI)1520-6017(200001)89:1<16::AID-JPS3>3.0.CO;2-E .

32. Prescott LF, Nimmo WS. *Novel drug delivery and its therapeutic application*: Wiley; 1989.
33. Ferl GZ, Wu AM, DiStefano JJ, 3rd. A predictive model of therapeutic monoclonal antibody dynamics and regulation by the neonatal Fc receptor (FcRn). *Ann Biomed Eng*. 2005;33(11):1640–52. doi: 10.1007/s10439-005-7410-3 .
34. Tsuji A, Nishide K, Minami H, Nakashima E, Terasaki T, Yamana T. Physiologically based pharmacokinetic model for cefazolin in rabbits and its preliminary extrapolation to man. *Drug metabolism and disposition: the biological fate of chemicals*. 1985;13(6):729–39. .
35. Michigoshi Y, Katayama R, Yamagishi N, Kato M, Saito J, Satoh H, et al. Estimation of glomerular filtration rate in rabbits by a single-sample method using iodixanol. *Lab Anim*. 2012;46(4):341–4. doi: 10.1258/la.2012.011065 .
36. Renkin EM, Gustafson-Sgro M, Sibley L. Coupling of albumin flux to volume flow in skin and muscles of anesthetized rats. *The American journal of physiology*. 1988;255(3 Pt 2):H458–66. doi: 10.1152/ajpheart.1988.255.3.H458 .
37. Michigoshi Y, Yamagishi N, Satoh H, Kato M, Furuhama K. Using a single blood sample and inulin to estimate glomerular filtration rate in rabbits. Journal of the American Association for Laboratory Animal Science: *JAALAS*. 2011;50(5):702–7. ; PubMed Central PMCID: PMC3189675.
38. Ishikawa T, Koizumi N, Mukai B, Utoguchi N, Fujii M, Matsumoto M, et al. Pharmacokinetics of acetaminophen from rapidly disintegrating compressed tablet prepared using microcrystalline cellulose (PH-M-06) and spherical sugar granules. *Chemical & pharmaceutical bulletin*. 2001;49(2):230–2. .
39. Beach CA, Mays DC, Sterman BM, Gerber N. Metabolism, distribution, seminal excretion and pharmacokinetics of caffeine in the rabbit. *The Journal of pharmacology and experimental therapeutics*. 1985;233(1):18–23. .
40. Celardo A, Traina GL, Jankowski A, Bonati M. Pharmacokinetics of theophylline and its metabolites in rabbits. *European journal of drug metabolism and pharmacokinetics*. 1985;10(4):279–88. .

41. El-Yazigi A, Sawchuk RJ. Theophylline absorption and disposition in rabbits: oral, intravenous, and concentration-dependent kinetic studies. *J Pharm Sci.* 1981;70(4):452–6. .
42. van Jaarsveld MF, Walubo A, du Plessis JB. Interaction between valproic acid and acyclovir after intravenous and oral administration in a rabbit model. *Basic & clinical pharmacology & toxicology.* 2007;101(6):434–40. doi: 10.1111/j.1742-7843.2007.00134.x .
43. Wiltse JA, Dellarco VL. U.S. Environmental Protection Agency's revised guidelines for carcinogen risk assessment: evaluating a postulated mode of carcinogenic action in guiding dose-response extrapolation. *Mutation research.* 2000;464(1):105–15. .
44. Chiu WA, Barton HA, DeWoskin RS, Schlosser P, Thompson CM, Sonawane B, et al. Evaluation of physiologically based pharmacokinetic models for use in risk assessment. *J Appl Toxicol.* 2007;27(3):218–37. Epub 2007/02/15. doi: 10.1002/jat.1225 .
45. Bessems JG, Loizou G, Krishnan K, Clewell HJ, 3rd, Bernasconi C, Bois F, et al. PBTK modelling platforms and parameter estimation tools to enable animal-free risk assessment: recommendations from a joint EPAA—EURL ECVAM ADME workshop. Regulatory toxicology and pharmacology: *RTP.* 2014;68(1):119–39. doi: 10.1016/j.yrtph.2013.11.008 .
46. Lipscomb JC, Haddad S, Poet T, Krishnan K. Physiologically-based pharmacokinetic (PBPK) models in toxicity testing and risk assessment. *Advances in experimental medicine and biology.* 2012;745:76–95. doi: 10.1007/978-1-4614-3055-1_6 .
47. Loizou G, Spendiff M, Barton HA, Bessems J, Bois FY, d'Yvoire MB, et al. Development of good modelling practice for physiologically based pharmacokinetic models for use in risk assessment: the first steps. Regulatory toxicology and pharmacology: *RTP.* 2008;50(3):400–11. doi: 10.1016/j.yrtph.2008.01.011 .
48. Louisse J, Bosgra S, Blaauboer BJ, Rietjens IM, Verwei M. Prediction of in vivo developmental toxicity of all-trans-retinoic acid based on in vitro toxicity data and in silico physiologically based kinetic modeling. *Archives of toxicology.* 2015;89(7):1135–48. doi: 10.1007/s00204-014-1289-4 .
49. Mumtaz M, Fisher J, Blount B, Ruiz P. Application of physiologically based pharmacokinetic models in chemical risk assessment. *Journal*

of toxicology. 2012;2012:904603 doi: 10.1155/2012/904603 ; PubMed Central PMCID: PMC3317240.

50. Hartmanshenn C, Scherholz M, Androulakis IP. Physiologically-based pharmacokinetic models: approaches for enabling personalized medicine. *J Pharmacokinet Pharmacodyn*. 2016;43(5):481–504. doi: 10.1007/s10928-016-9492-y .
51. Androulakis IP. Quantitative Systems Pharmacology: A Framework for Context. *Curr Pharmacol Rep*. 2016;2(3):152–60. doi: 10.1007/s40495-016-0058-x ; PubMed Central PMCID: PMCPMC4996481.
52. Block M. Physiologically based pharmacokinetic and pharmacodynamic modeling in cancer drug development: status, potential and gaps. *Expert Opin Drug Metab Toxicol*. 2015;11(5):743–56. doi: 10.1517/17425255.2015.1037276 .
53. Mager DE, Jusko WJ. Development of translational pharmacokinetic-pharmacodynamic models. *Clin Pharmacol Ther*. 2008;83(6):909–12. doi: 10.1038/clpt.2008.52 ; PubMed Central PMCID: PMCPMC2671003.
54. Jones GRD, Lim E-M. The National Kidney Foundation Guideline on Estimation of the Glomerular Filtration Rate. *The Clinical Biochemist Reviews*. 2003;24(3):95–8. PMC1853341.
55. Maharaj AR, Edginton AN. Physiologically based pharmacokinetic modeling and simulation in pediatric drug development. *CPT Pharmacometrics Syst Pharmacol*. 2014;3:e150 doi: 10.1038/psp.2014.45 ; PubMed Central PMCID: PMC4260000.
56. Poulin P, Jones RD, Jones HM, Gibson CR, Rowland M, Chien JY, et al. PHRMA CPCDC initiative on predictive models of human pharmacokinetics, part 5: prediction of plasma concentration-time profiles in human by using the physiologically-based pharmacokinetic modeling approach. *J Pharm Sci*. 2011;100(10):4127–57. doi: 10.1002/jps.22550 .
57. De Buck SS, Sinha VK, Fenu LA, Nijsen MJ, Mackie CE, Gilissen RA. Prediction of human pharmacokinetics using physiologically based modeling: a retrospective analysis of 26 clinically tested drugs. *Drug metabolism and disposition: the biological fate of chemicals*. 2007;35(10):1766–80. doi: 10.1124/dmd.107.015644 .
58. Hartmann S, Biliouris K, Lesko LJ, Nowak-Gottl U, Trame MN. Quantitative Systems Pharmacology Model to Predict the Effects

of Commonly Used Anticoagulants on the Human Coagulation Network. *CPT Pharmacometrics Syst Pharmacol*. 2016;5(10):554–64. doi:10.1002/psp4.12111;PubMedCentralPMCID:PMCPMC5080651.

59. Zhang L, Mager DE. Physiologically-based pharmacokinetic modeling of target-mediated drug disposition of bortezomib in mice. *J Pharmacokinet Pharmacodyn*. 2015;42(5):541–52. doi: 10.1007/s10928-015-9445-x ; PubMed Central PMCID: PMCPMC4620045.
60. Marangos MN, Zhu Z, Nicolau DP, Klepser ME, Nightingale CH. Disposition of ofloxacin in female New Zealand white rabbits. *Journal of veterinary pharmacology and therapeutics*. 1997;20(1):17–20. .
61. Karbownik A, Szalek E, Sobanska K, Polom W, Grabowski T, Biczysko-Murawa A, et al. The effect of sunitinib on the plasma exposure of intravenous paracetamol and its major metabolite: paracetamol glucuronide. *European journal of drug metabolism and pharmacokinetics*. 2015;40(2):163–70. doi: 10.1007/s13318-014-0191-z ; PubMed Central PMCID: PMC4426134.
62. Bienert A, Kaminska A, Olszewski J, Gracz J, Grabowski T, Wolc A, et al. Pharmacokinetics and ocular disposition of paracetamol and paracetamol glucuronide in rabbits with diabetes mellitus induced by alloxan. Pharmacological reports: *PR*. 2012;64(2):421–7.

SECTION 7: WILD ANIMALS

7.1 REPTILES

CHAPTER 13

Behaviour and Physiology: The Thermal Strategy of Leatherback Turtles

Brian L. Bostrom[1], T. Todd Jones[1,2], Mervin Hastings[3], and David R. Jones[1]

[1]Department of Zoology, University of British Columbia, Vancouver, British Columbia, Canada,

[2]Joint Institute for Marine and Atmospheric Research, University of Hawaii at Manoa, National Oceanic and Atmospheric Administration, Kewalo Research Facility, Honolulu, Hawaii, United States of America,

[3]Conservation and Fisheries Department, Ministry of Natural Resources and Labour, Government of the British Virgin Islands, Road Town, Tortola, British Virgin Islands

Citation: Bostrom BL, Jones TT, Hastings M, Jones DR (2010) Behaviour and Physiology: The Thermal Strategy of Leatherback Turtles. PLoS ONE 5(11): e13925. https://doi.org/10.1371/journal.pone.0013925

ABSTRACT

Background

Adult leatherback turtles (Dermochelys coriacea) exhibit thermal gradients between their bodies and the environment of ≥8°C in sub-polar waters and ≤4°C in the tropics. There has been no direct evidence for thermoregulation in leatherbacks although modelling and morphological studies have given an indication of how thermoregulation may be achieved.

Methodology/Principal Findings

We show for the first time that leatherbacks are indeed capable of thermoregulation from studies on juvenile leatherbacks of 16 and 37 kg. In cold water (< 25°C), flipper stroke frequency increased, heat loss through the plastron, carapace and flippers was minimized, and a positive thermal gradient of up to 2.3°C was maintained between body and environment. In warm water (25 – 31°C), turtles were inactive and heat loss through their plastron, carapace and flippers increased. The thermal gradient was minimized (0.5°C). Using a scaling model, we estimate that a 300 kg adult leatherback is able to maintain a maximum thermal gradient of 18.2°C in cold sub-polar waters.

Conclusions/Significance

In juvenile leatherbacks, heat gain is controlled behaviourally by increasing activity while heat flux is regulated physiologically, presumably by regulation of blood flow distribution. Hence, harnessing physiology and behaviour allows leatherbacks to keep warm while foraging in cold sub-polar waters and to prevent overheating in a tropical environment.

INTRODUCTION

Body temperature (T_B) has a pronounced effect on all metabolic processes and, for many ectotherms, maintaining T_B within a certain temperature range can be advantageous both behaviourally and metabolically. To regulate T_B ectotherms employ a thermal strategy that has both physiological and behavioural components which alter the rate of heat loss or gain [1]. However, for the majority of ectotherms metabolic heat is not integral to their thermal biology. Due to the lack of an internal heat source most fish, amphibians and reptiles maintain optimum physiological performance only across a narrow range of ambient temperatures [2]–[4]. Nevertheless some animals, such as the leatherback sea turtle, can maintain T_B near optimum over a large range of ambient temperatures which expands their thermal niche and greatly increases their global range [5].

Leatherback sea turtles are large oceanic pelagic reptiles that nest in the tropics where the water temperature (T_W) can be as high as 30°C [6] and spend extended periods of time foraging in cold northern waters that approach 0°C [5]. No other known reptile inhabits such a large ambient temperature range. Leatherbacks swimming in tropical waters have body temperatures (T_B) 1.2 – 4.3°C above ambient T_W [6]. In contrast, leatherbacks captured in foraging grounds off of Nova Scotia (Canada) typically maintained T_B of 24.3°C in surface water of 16.1°C for a thermal gradient (T_B–T_W) of at least 8.2°C [7]. Currently, the precise mechanisms involved in the leatherbacks' thermoregulatory ability are poorly understood.

Leatherbacks are thought to draw on a suite of physiological and behavioral adaptations to regulate their rate of heat loss and gain. Based on biophysical modeling it was concluded that large body size and use of peripheral tissues as insulation coupled with the ability to control heat flux via circulatory adjustments would allow leatherbacks to regulate T_B in both warm and cold waters [8]. Furthermore, leatherbacks are thought to possess counter-current heat exchangers in both the anterior and posterior flippers [9], have thick layers of adipose tissue surrounding cranial structures including the esophagus to prevent heat loss from the animal's head [10] and have peripheral layers of fat which include deposits of brown adipose tissue [11]. However, the T_B – T_W a leatherback maintains not only depends on the rate of heat loss, but also on the rate of heat production. Recently, the importance of behavioral adjustments (i.e. swimming activity) as a further thermoregulatory mechanism to maintain preferred T_B – T_W in different thermal environments was analyzed [12]. Taken together, results of previous studies have suggested integrated roles of large body size and physiological and behavioral adjustments in leatherback thermoregulation.

While different components of leatherback thermal biology have been measured and/or modelled in several studies, an holistic approach to quantifying the collective leatherback thermoregulatory response is lacking. In this study, we report the first empirical observations of the physiological and behavioral responses of leatherbacks to controlled variations in thermal environment. We apply our results to thermoregulation of leatherbacks, from juveniles to adults, in their natural environment, from tropical to polar seas.

RESULTS

1. **Thermal gradient:-** As tank water was cooled stepwise from the acclimation temperature of 25°C (for T_W change protocol see Figure 1) both study animals maintained a progressively larger thermal gradient between their body and the water (Table 1 and 2). At T_W 25°C the 37 kg turtle maintained a $T_B - T_W$ of 0.9°C which increased to 2.3°C in T_W of 16°C (Table 1). The 16 kg animal started off with a gradient of 1.5°C which increased to 2.0°C in the coldest water (Table 2). The 37 kg animal's thermal gradient decreased to 0.5°C and the 16 kg turtle's gradient was 0.8°C in water at 31°C. Overall a trend was found in both leatherbacks where cooler water led to the maintenance of larger thermal gradients, but $T_B - T_W$ gradients in both animals were greater when water temperature was reduced to a given T_W compared with being raised to the same T_W. Additionally, the larger turtle displayed a larger variation in thermal gradients (0.5 to 2.3°C) when compared with the smaller turtle (0.8 to 2.0°C) over the T_W range tested.

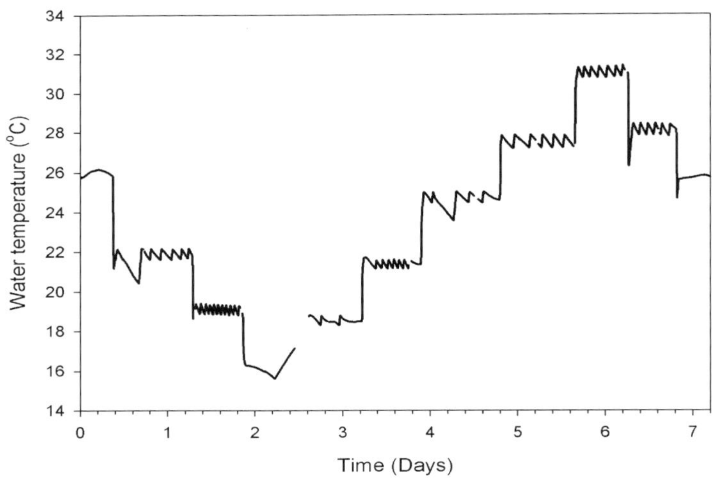

Figure 1. The leatherbacks were exposed to stepwise changes in water temperature.

The complete water temperature profile for the experiment performed on the 37 kg leatherback.

https://doi.org/10.1371/journal.pone.0013925.g001

Table 1. A leatherback's rate of heat production and heat loss depend on water temperature

Water temperature (°C)	Thermal Gradient T_B- T_W (°C)	Flipper stroke frequency (SPM)	Flipper Heat Loss Rate (W m²)	Plastron Heat Loss Rate (W m²)	Thermal Admittance (W m² K⁻¹)	Total Heat Loss (W kg⁻¹)	Flipper Heat Loss (% total heat loss)
25	0.9	3	6.5	12.8	14.4	0.24	16
22	1.4	21	5.8	16.0	11.4	0.29	12
19	1.6	25	11.0	19.5	12.0	0.38	18
16	2.3	29	4.6	26.2	11.5	0.44	6
19	1.0	19	2.8	10.9	10.4	0.19	9
22	0.7	5	5.4	10.7	14.4	0.20	16
25	1.1	3	5.7	10.9	10.4	0.21	17
28	0.7	3	16.5	13.8	20.6	0.32	32
31	0.5	3	14.6	13.5	29.5	0.30	29
28	0.5	8	14.3	14.0	26.9	0.31	28
25	0.9	6	2.4	9.4	10.4	0.16	9

Recorded and calculated values at each water temperature for the 37 kg leatherback.

https://doi.org/10.1371/journal.pone.0013925.t001

Table 2. Recorded and calculated values at each water temperature for the 16 kg leatherback

Water temperature (°C)	Thermal Gradient T_B- T_W (°C)	Flipper stroke frequency (SPM)	Flipper Heat Loss Rate (W m²)	Plastron Heat Loss Rate (W m²)	Thermal Admittance (W m² K⁻¹)	Total Heat Loss (W kg⁻¹)	Flipper Heat Loss (% total heat loss)
25	1.5	20	6.7	14.5	9.9	0.35	14
22	1.4	22	4.0	12.3	8.7	0.28	10
19	1.6	30	3.1	-	-	-	-
16	2.0	36	3.3	15.7	7.9	0.35	7
19	1.0	29	4.0	8.2	8.6	0.20	15
22	0.8	13	3.0	8.9	11.1	0.21	11
25	1.2	2	6.2	9.1	7.9	0.23	20
28	1.0	2	8.6	8.8	8.4	0.24	26
31	0.8	2	7.4	-	-	-	-
28	-	3	-	11.1	-	-	-
25	-	9	-	-	-	-	-

https://doi.org/10.1371/journal.pone.0013925.t002

2. **Swimming activity:-** In tank water 22°C and higher the 37 kg leatherback was nearly inactive swimming at a flipper stroke frequency ranging between 2 and 8 strokes per minute, SPM (Table 1). The 16 kg leatherback on the other hand became nearly

inactive in water that was any warmer 22°C (Table 2). Stroke frequency greatly increased in both leatherbacks after a drop in temperature from 22°C. After each reduction in T_W the animals stroke rate increased and the rate was maintained over the entire time the turtle was in that T_W. In the coldest water (16°C) the 37 and 16 kg leatherbacks maintained their highest average activity rates at 29 and 36 SPM, respectively.

3. **Heat loss:-** When T_W was 25°C or less the 37 kg leatherback lost <7 W m^{-2}, through the front flippers except when cooled to 19°C when heat flux from the front flippers, Q_F, rose to 11.0 W m^{-2} (Table 1). When warmed to 28°C heat flux increased substantially to 16 W m^2. A similarly high Q_F was maintained until the animal was re-cooled to 25°C. The 16 kg leatherback lost <4 W m^{-2} through its flippers when tank water was 22°C and below (Table 2). When at 25°C heat flux increased to 6.2 W m^{-2} and at 28°C increased a further 39% to 8.6 W m^{-2}. At 31°C Q_F of the smaller turtle dropped slightly to 7.4 W m^{-2}.

The 37 kg leatherback lost 12.8 W m^{-2} through the plastron at the acclimation temperature and this value steadily rose to 26.2 W m^{-2} in the coldest water (Table 1). Upon re-warming to 19°C heat flux through the plastron, Q_P, dropped to 10.9 W m^{-2}. This heat flux was maintained until 28°C when it increased slightly to 13.8 W m^{-2} remaining stable until returned to 25°C when plastron heat flux dropped to 9.4 W m^{-2}. The 16 kg leatherback followed a similar trend with heat flux highest in 16°C water (15.7 W m^{-2}) and a steady heat flux around 9 W m^{-2} from 19 through 28°C. The rate of heat flux from the carapace of the 16 kg leatherback was nearly the same as Q_P at each T_W tested.

When the 37 kg leatherback was in $T_W \leq 25$°C, thermal admittance (see Materials & Methods for explanation of thermal admittance) was between 10 and 14 W m^{-2}°C^{-1} (Table 1). When the tank water was warmed to 28°C, the thermal admittance nearly doubled to 20.6 W m^2 K^{-1} and in 31°C water reached 29.5 W m^2 K^{-1}, nearly triple the cold water value. On cooling the animal to its acclimation temperature of 25°C thermal admittance fell to 10.4 W m^2 K^{-1}. The 16 kg turtle had a constant value for thermal admittance around 9 W m^2 K^{-1} in all water temperatures tested. There was insufficient data, unfortunately, to calculate the thermal admittance in 31°C T_W for the smaller turtle.

4. **Surface area:-** The total surface area of the plastron and carapace, A_B, was 0.64 and 0.42 m^2 for the 42.0 and 22.3 kg leatherback carcasses, respectively. The area of the front and rear flippers, A_F, was 0.25 and 0.15 m^2 for the large and small turtle, respectively. Total body area (m^2), was fitted to a power function and found to scale with M_B (kg) as A_B=0.049 $M_B^{0.69}$ and the area of all four flippers scaled as A_F=0.014 $M_B^{0.77}$. Using these equations, at the time of the experiments, A_B for the smaller 16 kg turtle was 0.33 m^2 and the 37 kg turtle was 0.59 m^2. A_F was 0.12 m^2 for the small and 0.22 m^2 for the larger turtle.
5. **Total heat loss:-** At 25°C the 37 kg leatherback lost 0.24 W kg^{-1} from the body and flippers, and this total heat loss, q_T, rose steadily to 0.44 W kg^{-1} in the coldest water (Table 1). When warmed to 19°C, q_T dropped to 0.19 W kg^{-1} and this value varied little until 28°C when it rose to 0.32 W kg^{-1} (see Figure 1 for T_W profile). This q_T was held steady until the animal was again cooled to 25°C. The smaller turtle at 25°C had a q_T of 0.35 W kg^{-1} which fell to 0.28 W kg^{-1} in 22°C water but increased to 0.35 W kg^{-1} in 16°C water. Upon re-warming to 19°C, q_T dropped to 0.20 W kg^{-1} and then slowly increased to 0.24 W kg^{-1} at 28°C.
6. **Fraction of heat loss through the body and carapace:-** In the coldest water (16°C) both turtles lost about 7% of q_T through their flippers with the remaining 93% being lost through the body (ie. plastron and carapace, Table 1 and 2). The proportion of heat lost from the flippers increased as T_W rose in both turtles. In 28 and 31°C T_W around 30% of q_T was lost from the flippers compared with 70% from the plastron and carapace.

DISCUSSION

There has been considerable speculation that leatherbacks are endothermic and able to thermoregulate based upon the fact that their global range spans from cold northern foraging grounds to tropical nesting beaches. In the coldest water we tested (16°C) the 16 and 37 kg leatherback maintained a thermal gradient of 2.0 and 2.3°C, respectively, while in the warmest water (31°C) the thermal gradient was reduced to 0.5 and 0.8°C (Table 1 and 2). Therefore, we have shown for the first time, using juveniles in a controlled temperature environment, that leatherbacks possess the ability to hold and regulate their thermal gradient. Furthermore, since the heat energy to hold these thermal gradients is metabolically derived the animals are, by definition, endothermic.

1. **Physiological and behavioural responses to warm and cold water:-** In water colder than their acclimation temperature (T_W <25°C) the flipper stroke frequency of both leatherbacks increased as T_W decreased (Table 1 and 2). Since each flipper stroke causes water movement the activity rate is proportional to the power output by the turtle. Due to the inefficiency of metabolic processes, as power output increases heat production must increase as well. Therefore as T_W got colder endogenous heat production increased. In the coldest water flipper stroke frequency was highest and the largest thermal gradient was held. $T_B - T_W$ during cooling was different from that during re-warming and a similar hysteresis was seen in the relation between flipper stroke frequency and T_W. Further, $T_B - T_W$ varied in association with activity for $T_B - T_W$'s between 1 and 2.3°C. Since both leatherbacks maintained a stable thermal admittance during those trials the $T_B - T_W$'s were due largely to activity (Figure 2). The potential to use behavioral control of activity to regulate heat production and therefore T_B [12] has been confirmed in our study. Eckert (2002) found leatherbacks swim continuously which shows their potential to maintain high activity rates over very long periods of time [13]. Behavioral control of heat production has not been shown to be an integral thermoregulatory mechanism in any other reptile but it is probable that it may be used in other endothermic species such as lamnid sharks, tunas and billfish, especially when exposed to very cold waters. Behavioural control of heat production contrasts with other endotherms in which metabolic heat production is largely controlled autonomically.

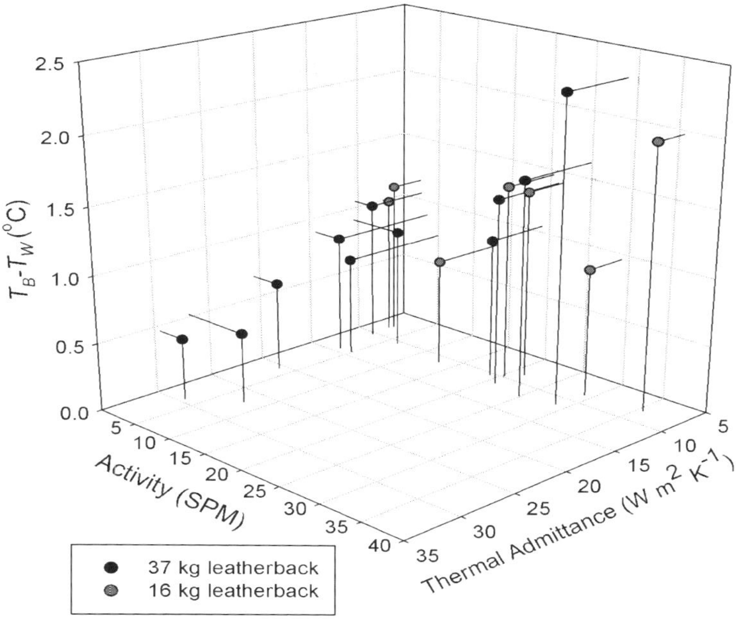

Figure 2. Leatherbacks thermoregulate by controlling both heat loss and heat gain.

A 3D image showing how activity (heat production) and thermal admittance (heat loss) affect the thermal gradient (T_B-T_W) held by juvenile leatherbacks. In cold water heat loss was minimized and activity was proportional to the thermal gradient held. In warm water activity was very low and thermal gradient was due to varying heat loss.

https://doi.org/10.1371/journal.pone.0013925.g002

Even though the surface area of the front and rear flippers combined represents 27% of the total surface area of each turtle we found that only 6–7% of total heat loss, q_T, came from the flippers in the coldest water (Table 1 and 2). Despite not having the ability to completely halt heat flux, the flippers are responsible for only a small fraction of total heat lost. In warm water heat flux increases substantially from the flippers to a rate nearly 3 times that in cold water and accounts for 30% of q_T. These observations suggest an effective control over flipper heat flux, either by utilization of counter-current heat exchangers or, at its simplest, a reduction in total blood flow to the flippers in cold water. A dense, intertwining network of arteries and veins in both the anterior and posterior flippers have been histologically described and provide anatomical evidence for heat exchangers [9]. Heat exchangers have evolved independently in lamnid sharks and scombrids, underscoring the primary importance of heat retention in maintenance of elevated body temperatures in each lineage [14], [15]. No evidence of heat exchangers in green and loggerhead sea turtles has been reported, but both species have been shown to greatly reduce blood flow to their flippers when exposed to T_W lower than their acclimation temperatures [16]. In addition, thermal effects on the whole circulation could have an important influence on heat flux. Nonetheless, despite the lack of clarity on the precise mechanism controlling heat loss from the flippers there is no doubt that the flippers are an important part of a leatherbacks thermal arsenal.

In contrast to the flippers, leatherbacks lost a substantial amount of heat energy through the plastron at all water temperatures (Table 1 and 2). Since the body was always warmer than tank water, heat was conducted from the core of the animal to the water. If heat passed to the water solely by conduction, heat flux should be directly proportional to the thermal gradient across the plastron (Eq. 3) with any deviation reflecting physiological changes such as a variation in blood flow. Therefore, the thermal admittance, the heat flux for a given thermal gradient, should be an accurate index of physiological blood flow changes a leatherback makes that affect Q_P. In $T_W \leq 25°C$ the 37 kg leatherback maintained a stable thermal admittance of around 12 W m^{-2} K^{-1} (Table 1) and at all temperatures tested the 16 kg leatherback also maintained a constant thermal admittance, although slightly lower at 9 W m^{-2} K^{-1} (Table 2). When in colder water, than the acclimation temperature, little heat energy was lost through their flippers so physiologically their thermal insulation was at a maximum due to little blood flow to the skin surface. In contrast, in water above the acclimation temperature, thermal admittance nearly tripled so more heat was lost convectively and blood flow to the skin must have increased. As well, $T_B - T_W$ correlated closely with thermal admittance when $T_B - T_W$ was below 1°C (Figure 2). Since activity was minimized the physiological changes affected the $T_B - T_W$ confirming the suggestion that leatherbacks could control T_B through blood flow [8].

Interestingly, the thermal admittance of the large leatherback was around 30% greater than that of the small turtle at 16°C. Therefore at a given gradient between T_B and T_W the larger turtle lost around 30% more heat per unit area from its plastron than the small turtle, regardless of the gradient between body and water. The smaller turtle had a thinner plastron and therefore we expected it to have a thinner layer of insulation and consequently a higher heat flux for a given $T_B - T_W$. Since thermal admittance is equal to thermal conductivity, *k*, divided by insulation thickness, *L*, (see Methods) either the smaller turtle has a less thermally conductive insulation layer, lower *k*, or the insulating layer is thicker than in the larger turtle. It is possible that the thermal gradient we measured between the body and tank water did not truly represent the thermal gradient directly across the plastron. T_B was measured as gastrointestinal tract temperature and although core body temperature is generally modelled as being homogenous due to blood flow [8], [12], this may not always be the case. Heat lost through the plastron could cause a temperature profile inside the turtle where the gastrointestinal tract of the animal is at T_B and peripheral tissues near the plastron are closer to T_W because of inadequate perfusion and the temperature approaches T_W closer to the plastron. In essence, since heat around the gastrointestinal tract is now passing through these cooler peripheral tissues largely by conduction, the tissues would have acted as insulation, increasing *L,* and leading to a lower thermal admittance in the smaller turtle. In cold water, the entire body cavity of the larger turtle is likely more homogenous in temperature than the smaller turtle because the plastron and carapace are thicker, will provide better insulation, and there is a larger volume of blood circulating endogenously produced heat.

In cold water, heat flux from the carapace was the same as from the plastron in the 16 kg turtle which suggests the heat flux may be homogenous over the body surface, $A_{B,}$ if the skin is not perfused. This would be expected if there is an even layer of insulation over A_B. Assuming the measured thermal gradient in the larger animal was a true representation of the thermal gradient directly across the plastron, the thermal conductivity, *k*, of leatherback shell can be estimated. The plastron of juvenile leatherbacks of similar mass to those used in these experiments is around 0.02 m thick (personal observations), so *k* is between 0.2–0.3 J s^{-1} K^{-1}, which is similar to the thermal conductivity recorded for whale blubber [17].

As the leatherbacks were in steady state at each T_W and held a stable T_B, the total rate of heat production must equal the total rate of heat loss. Therefore the sum of the heat lost from the plastron, carapace and flippers (ie. q_T) will allow prediction of total heat produced (Table 1 and 2). A caveat is that heat loss is modelled to occur evenly over the entire surface. Total heat production was greatest when the animals were most active and therefore in the coldest water. Despite activity falling as water was warmed from 19 to 25°C, heat loss was constant in both animals and this was probably due to the Q_{10} effect on basal metabolic processes as T_B was increasing. In 28 and 31°C T_W the 37 kg leatherbacks heat production rate further increased despite having a very low activity rate, again due to temperature effects on metabolism. In adult leatherbacks that maintain stable T_B's the heat production rate would be expected to more closely reflect rate of activity because basal metabolic rate should be constant.

2. **Effect of body mass on thermal gradients:-** At steady state $T_B - T_W$, the total rate of heat transfer, q_T, to the environment equals the rate of heat gained. In leatherbacks, heat is produced endogenously and therefore heat production must approximate the metabolic rate of the animal. Given that resting metabolic rate scales with body mass, M_B (kg), to the 0.83 power in leatherback sea turtles [18] and rate of heat transfer across an insulation layer is given by Eq. 3 then:

$$(T_B - T_W)kAL^{-1} = aM_B^{0.83} \quad (1)$$

where a is the proportionality coefficient. The body shape of juvenile leatherbacks is similar to those of sub-adults and adults so A scales with M_B to the 2/3 power and L will scale with M_B to the 1/3 power since M_B has dimensions of L^3. Therefore if k is constant then the thermal gradient scales with M_B to the 0.5 power as:

$$T_B - T_W = bM_B^{0.5} \quad \textbf{(2)}$$

where b is a coefficient that is proportional to the energy expenditure of the animal (ie. doubling b corresponds to twice the heat production).

The thermal gradients that adult leatherbacks hold can be predicted by scaling the thermal gradient held by juveniles. At 25°C the animals in this study maintained a low level of activity with $T_B - T_W$ around 1°C. We fitted these results to Eq. 2 to predict the $T_B - T_W$ that animals of different M_B could maintain with low endogenous heat production and found b=0.21. Figure 3 quantitatively shows the effect that varying heat production, changing b, has on the $T_B - T_W$ achieved by leatherbacks with minimized heat loss in cold water. The thermal gradient is predicted to be 1.7 and 2.5°C for juveniles of 16 and 37 kg with a heat production rate two times the resting level. This is close to the 2.0 and 2.3°C thermal gradient our animals held when in 16°C tank water when heat loss was double that at 25°C. An average thermal gradient of 8.2°C, measured for adults in 15°C water off of Nova Scotia [7], occurs if heat production is double that of a resting animal.

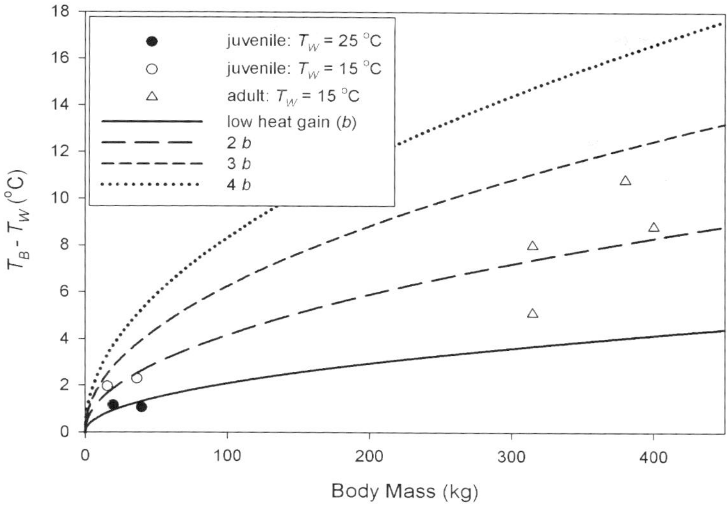

Figure 3. Predicting the thermal gradient adult leatherbacks could maintain.

The affect of mass and heat production on the thermal gradient held by leatherbacks were estimated using scaling equations.

https://doi.org/10.1371/journal.pone.0013925.g003

Metabolic rate of active leatherbacks on a beach was 1.51 W kg^{-1} [8], or five times resting metabolic rate [18]. If leatherbacks maintain such a high heat production rate, our model predicts that a 300 kg animal could hold a maximum $T_B - T_W$ gradient of 18.2°C. Therefore leatherbacks swimming in northern temperate waters must maintain substantially elevated metabolic rates to keep T_B stable. The predictions in Figure 3 assume a thickness of insulation which scales to the 1/3 power with M_B. Animals fatten while foraging so if the insulation layer increases, then for a given heat production rate, a larger thermal gradient will be maintained. James et al. (2005) found leatherbacks off Nova Scotia, Canada to have a 33% greater mass than nesting animals of the same carapace length [19]. If a substantial portion of this increased mass is sub-cutaneous fat insulation will be improved.

Due to scaling effects, the achievable thermal gradient is predicted to scale with M_B, at a given metabolic cost, to the 0.5 power (Eq.

2). Therefore if leatherbacks maintain their T_B above a minimum value only larger animals should be found foraging in colder waters. Eckert (2002) noted that animals <100 cm carapace length were not observed in T_W <26°C [13]. James et al. (2006) recorded a leatherback of 148 cm curved carapace length (around 500 kg) repeatedly diving into water of 0.4°C [5]. The model predicts that a 500 kg leatherback would have to have a metabolic rate twice that of our swimming juveniles (four times resting) to keep a body temperature of 20°C in these extremely cold waters.

In warm water our leatherbacks had a similar heat loss rate (Table 1 and 2) from the flippers as the plastron suggesting similar heat flux over their entire body surface when dumping heat in warm water. Presumably this was due to maximizing perfusion of the skin. When coupled with a minimization in flipper stroke frequency the 37 kg leatherback had a $T_B - T_W$ of only 0.5°C in the warmest water (31°C). The 16 kg leatherback although having a similar flipper stroke frequency as the 37 kg animal did not increase heat loss as water temperatures rose so its thermal gradient was greater ($T_B - T_W$=0.8°C). Since surface area scales with M_B to the 2/3 power and heat production scales to the 0.83, there is very little added heat production per surface area as an animal grows. Therefore, in the tropics, if an adult leatherback routed blood to its entire surface the potential for heat loss should be great enough that the animals will not be in danger of overheating, even when swimming. In the tropics, adult females had a T_B 1 – 4°C greater than ambient water [6] and T_B was correlated with $T_{W,}$ signifying high rates of heat loss in large leatherbacks.

3. **Other sea turtles:-** Sea turtles other than leatherbacks have never been shown to sustain T_B – T_Ws that are large enough to allow migration to cold temperate waters. Free swimming adult loggerhead turtles, for example, generally hold thermal gradients between stomach and water of only 1–2°C [20]. A loggerhead's T_B thus closely reflects $T_{W,}$ and in fact in the Western Atlantic, loggerheads rely on warm waters of the Gulf Stream to overwinter [21]. Kemp's ridleys and green turtles cease to feed and become semi-dormant in water of 15°C [22], whereas adult leatherbacks have been recorded actively feeding in waters as low as 0.4°C [5]. Hard shelled turtles are distributed throughout tropical and subtropical waters again suggesting an inability to maintain a

homeostatic T_B across a wide range of T_Ws. A lack of insulation and/or insufficient heat production must underlie the low $T_B - T_W$s held by these turtles.

Green turtles do seem to be capable of substantial heat production, at least in the short term, and can maintain pectoral muscles 8°C above T_W while actively swimming [23]. However, activity in green turtles is unpredictable and when a juvenile was exposed to water at 20°C it remain inactive for 30 minutes before starting swimming [24]. Green turtles of 7–11 kg decrease activity when T_W is below 20°C and are quiescent in water of 15°C [22]. Juvenile leatherbacks, on the other hand, increase activity as water temperature decreases to at least 16°C and at each T_W activity levels were constant. In a similar study to ours juvenile green turtles appear to use a thermal strategy much like leatherbacks and accomplish similar $T_B - T_W$s [24]. Juvenile greens between 2 and 60 kg held a $T_B - T_W$ of 2.2°C in 20°C water and a gradient of 1.7°C in 30°C. The greens had a higher thermal admittance in warm water and were more active in cooler water (20°C) but, as shown above, activity declines below 20°C. Despite superficial similarity between the green turtle data and that presented in this study, in reality, the green turtle results yield little insight into why leatherbacks are capable of traveling to sub-polar waters and greens are not because they did not expose their animals to low temperatures [24]. Due to constant swimming leatherbacks have a more reliable source of endogenous heat which is fuelled by their oceanic, pelagic lifestyle.

To maintain elevated thermal gradients retaining body heat is as important as producing it. Although the pectoral muscles of an actively swimming green sea turtle can be 8°C above T_W the rest of the body is only 1–2°C above T_W [23]. This suggests a lack of suitable insulation to maintain large $T_B - T_W$s. The measurement of similar internal and external carapace temperatures in a green sea turtle being exposed to intense solar radiation confirms that their carapace is a very poor insulator [24]. Leatherbacks are able to hold a larger gradient than other sea turtles by a combination of large size, a more efficient insulative layer and better control of heat production. Controlling heat loss and gain concurrently is a thermal strategy that allows leatherbacks to exploit the rich foraging grounds of sub-polar waters while avoiding overheating while actively swimming in tropical reproductive zones.

MATERIALS AND METHODS

1. Animals and Husbandry

Ethics Statement

These animals were held for research purposes and all animal care/research standards of the Canadian Council for Animal Care (CCAC) and the UBC Animal Care Committee (UBC Animal Care Protocol: A04-0323) were met.

Leatherback hatchlings were transported from nesting beaches on the British Virgin Islands to the Animal Care Centre, University of British Columbia, Vancouver, B.C., Canada, and raised for a two year period. The experiments reported here were performed on two animals, one weighed 16.1 kg at the start of the experiment and the other weighed 36.7 kg. Both animals were raised in T_W=25°C. Throughout the experiment the animals were held in a cylindrical holding tank 2 m in diameter and 1.5 m deep, filled with seawater supplied from the Vancouver Aquarium and Marine Sciences Centre (Vancouver, B.C., Canada). Leatherback turtles were obtained on Canada CITES Import permit CA05CWIM0039 and British Virgin Islands CITES Export certificate CFD062005.

Leatherbacks have an oceanic-pelagic lifestyle and do not recognize barriers. Consequently, the animals were tethered to the centre of their housing tank by a short length of monofilament fishing line attached to a custom made harness (Figure 4A). The animals could swim or dive without touching the walls or bottom of the tank. The animals were exposed to a 12 hour light/dark cycle. Water quality was maintained by a biological filter, UV sterilization and a protein skimmer. A reservoir tank of equal volume was plumbed into the holding tank. Water temperature of the reservoir was varied and mixing the water in the two tanks allowed T_W to be changed rapidly. T_W was maintained within ±0.25°C in the holding tank by a thermostat that controlled hot or cold water flow through a stainless steel heat exchanger. The animal was instrumented and put in the tank at the beginning of the experiment and disturbed only to repair instruments or for feeding which was attempted twice a day.

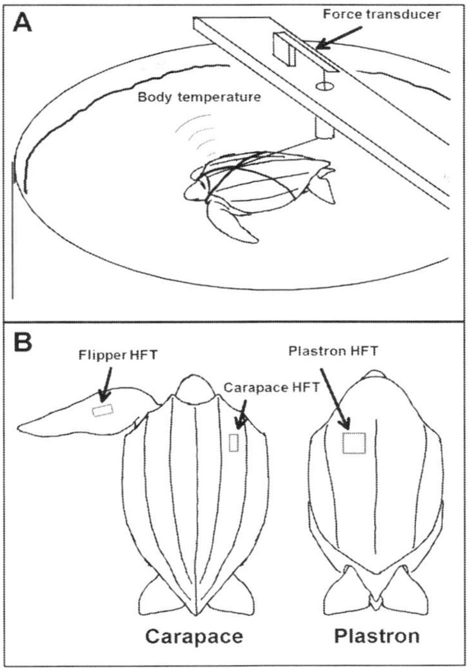

Figure 4. Experimental setup.

(A) An illustration of the turtles harnessed in their tanks. (B) The placement of the heat flux transducers (HFT) on the animals.

https://doi.org/10.1371/journal.pone.0013925.g004

2. Temperature regime

T_W was changed in a stepwise manner by 3°C increments or decrements starting from 25°C. Water was cooled to 16°C and then warmed from 16 to 31°C and cooled back to 25°C over several days (Figure 1). The lowest maintainable T_W was 16°C due to limitations in the experimental setup and equipment. T_W was changed by mixing the water in the holding tank with the reservoir tank and the change was completed within 20 minutes of commencing the mix. The water was maintained at each temperature for 11–22 hours for the 37 kg turtle and for 7–24 hours for the 16 kg turtle, long enough for T_B to become stable (i.e. T_B change <0.1°C hour^{-1}). In total, the experiment with the 16 kg leatherback took 6 days and the experiment with the 37 kg leatherback took 8 days.

3. Instrumentation

3.1. T_B and T_W recording

A thermocouple (90104, Mon-a-therm® General Purpose, Mallinckrodt Medical, St. Louis, MO, USA) was mounted 5 cm into the housing tank and connected to an electronic thermometer (Physitemp BAT – 12, Sensortek Inc., Clifton, NJ, USA). T_W was recorded every second. At the beginning of the experiment each animal was given a thermometer pill that was 2.5 cm in length and 1.0 cm in diameter (HT150036, CorTemp™ Equine EXSM Temperature Sensor, HQ Inc., Palmetto, FL, USA). The signal from the pill was picked up by a receiver (HT150001, CorTemp™ Data Recorder, HQ Inc.) housed in a waterproof container suspended 15 cm above the center of the tank. T_B was recorded every 10 seconds. The thermometer pill had a lifespan of around 3 days after which we briefly removed the turtle from the water and gave the animal a new pill. The animals passed the pills in 1–2 weeks. The electronic thermometer and thermometer pill were calibrated against a mercury thermometer (14-985B, FISHERbrand, Fisher Scientific Ltd., Nepean, ON, Canada).

3.2. Heat flux recording

Heat flux, Q (W m^2), was recorded with heat flux transducers (HFT's; Thermonetics Corp., La Jolla, CA, USA). HFT's are rectangular flat pads that produce a voltage directly proportional to the heat flux through the pad. We attached the transducers to the animals with a thin layer of cyanoacrylate glue. A 3.9×1.9 cm HFT was attached 1/3 of the way along the left front-flipper at the point of largest flipper width (see Figure 4B for placement of HFT›s). A 5.7×5.7 cm HFT was attached to the plastron 10 cm distally from the anterior edge and 2.5 cm to the right of each animal›s center line. A 3.9×1.9 cm HFT was attached only to the carapace of the 16 kg turtle. The HFT wires ran from the turtle, along the bottom of the tank and then out of the tank to a multiple channel signal conditioner (CyberAmp 320, Axon Instruments) and, after amplification, the signals were analog to digital, A – D, converted and recorded at a rate of 1 Hz. HFT's were attached so that a positive value of heat flux represented heat transferring from the turtle to the water.

3.3. Activity recording.

A wooden plank with a short section of 14 mm (internal diameter) PVC pipe inserted through it was placed across the tank. A length of monofilament fishing line attached to the animal's harness (Figure 1A) passed through the PVC pipe to a force transducer (FT03C, Grass Instrument Co., Quincy, MA, USA). The PVC pipe redirected the force exerted by the turtle vertically so that recorded force was unaffected by the direction in which the animal was swimming. The signal from the force transducer was A–D converted and recorded at a rate of 5 Hz. Flipper strokes per minute (SPM) were used as an index of activity.

4. Data recording and analysis

All analog signals were digitized with an analog to digital (A–D) converter (USB – 1208LS, Measurement Computing, Norton, MA, USA) and recorded on a notebook computer with TracerDAQ® (Measurement Computing). All feeding events and data collected when an animal had been disturbed due to re-instrumentation were discarded. Heat flux data was calibrated using calibration curves provided by the manufacturer. Acq*Knowledge*® software (version 3.7.5, BIOPAC Systems Inc., Santa Barbara, CA, USA) was used to detect peaks in the data from the force transducer. Each peak corresponded to a flipper stroke. This force data was then converted to strokes per minute (SPM) and averaged over the period the animal was held at a given temperature. All other data were taken when the animal was close to steady state which was considered to be when T_B was changing at a rate <0.1°C hour^{-1}. Therefore, in the 16 kg animal no data was included that was <6 hours after the water change, and in the 37 kg animal no data <7.5 hours was included. Water temperature in the holding tank fluctuated by ±0.25°C and all heat fluxes as well as the thermal gradient fluctuated in time with T_W oscillations (Figure 5). To get the steady state heat flux and thermal gradient maintained for a given T_W we averaged that small section of the data when T_W was ±0 0.075°C around the mean value for that trial. The actual period averaged was usually about 30 minutes.

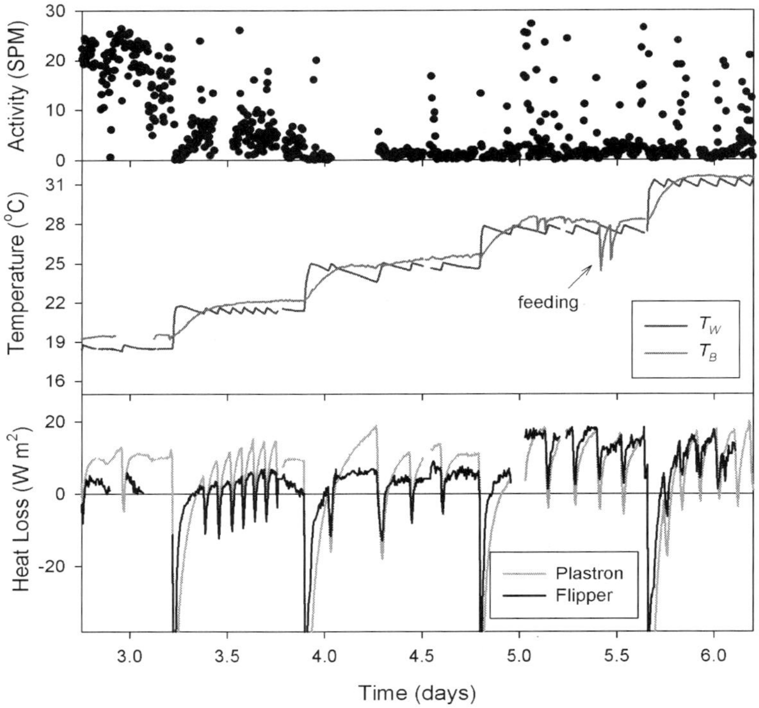

Figure 5. All data was simultaneously recorded.

Activity, water and body temperature and heat fluxes recorded simultaneously from the 37 kg leatherback during stepwise increases in water temperature.

https://doi.org/10.1371/journal.pone.0013925.g005

5. Calculations

5.1. Surface area

Surface area of both leatherbacks was measured post-mortem. When the large and small animals died of natural causes their masses were 42.0 and 22.3 kg, respectively. Half the carapace, half the plastron, the ventral side of the left front and rear flipper of each animal was covered in paper. The paper was then removed, laid flat and the area was determined by creating geometric shapes with an easily measured area. Total body area (A_B) was twice the measured area of plastron and carapace and total flipper area (A_F) was four times the measured area of both front and rear flippers. These areas were scaled allometrically with M_B in order to predict the surface area of the turtles during experimentation.

5.2. Total heat transfer rate.

The total rate of heat transfer, q_T (W), from the turtle to the tank was estimated as $A_B Q_P + A_F Q_F$ where Q_P and Q_F are heat flux (W m^2) from the plastron and flipper, respectively. The fraction of q_T that was lost through the flippers is $A_F Q_F / q_T$. The remaining heat loss was from the body.

5.3. Thermal admittance of the plastron.

The rate of heat transfer (W) across an insulation layer of thermal conductivity k (W m^{-1} K^{-1}) and thickness L (m) is given by:

$$q = Ak(T_B - T_W)/L \tag{3}$$

where $T_B - T_W$ is the thermal gradient from one side of the layer of insulation to the other, and A (m^2) is the area over which heat is lost. To give an index of the control of heat loss through the plastron exhibited by the turtles we divided measured Q_P from the HFT (ie., q/A) by the measured $T_B - T_W$. This gives a rate at which heat energy transfers across a given insulator of area 1 m^2 driven by a thermal gradient of 1°C and is referred to as the thermal admittance (W m^{-2}°C^{-1}). Thermal admittance is also equal to k/L.

ACKNOWLEDGMENTS

We thank A. Vanderhorst and S. Gopaul for turtle emergency care and B. Gillespie and V. Grant for mechanical assistance. We also thank Dr. Chris Harvey-Clark for veterinary services and Dr. Bryan Wallace for comments on an early draft of the manuscript.

Author Contributions

Conceived and designed the experiments: BLB TTJ DRJ. Performed the experiments: BLB TTJ MH. Analyzed the data: BLB. Contributed reagents/materials/analysis tools: DRJ. Wrote the paper: BLB.

REFERENCES

1. Hutchinson VH, Maness JD (1979) The role of behaviour in temperature acclimation and tolerance in ectotherms. Amer Zool 19: 367–384.
2. Angilletta MJ, Niewiarowski PH, Navas CA (2002) The evolution of thermal physiology in ectotherms. J Therm Biol 27: 249–268.
3. Brett JR (1971) Energetic responses of salmon to temperature. A study of some thermal relations in the physiology and freshwater ecology of sockeye salmon (Oncorhynchus nerkd). Amer Zool 11: 99–113.
4. Huey RB, Stevenson RD (1979) Integrating thermal physiology and ecology of ectotherms: A discussion of approaches. Amer Zool 19: 357–366.
5. James MC, Davenport J, Hays GC (2006) Expanded thermal niche for a diving vertebrate: a leatherback turtle diving into near-freezing water. J Exp Mar Biol Ecol 335: 221–226.
6. Southwood AL, Andrews RD, Paladino FV, Jones DR (2005) Effects of swimming and diving behavior on body temperatures of Pacific leatherbacks in tropical seas. Physiol Biochem Zool 78: 285–297.
7. James MC, Mrosovsky N (2004) Body temperatures of leatherback turtles (Dermochelys coriacea) in temperate waters off Nova Scotia, Canada. Can J Zool 82: 102–106.
8. Paladino FV, O'Connor MP, Spotila JR (1990) Metabolism of leatherback turtles, gigantothermy, and thermoregulation of dinosaurs. Nature 344: 858–860.
9. Greer AE, Lazelle JD, Wright RM (1973) Anatomical evidence for a countercurrent heat exchanger in the leatherback turtle (Dermochelys coriacea). Nature 244: 181.
10. Davenport J, Fraher J, Fitzgerald E, McLaughlin P, Doyle T, et al. (2009) Fat head: an analysis of head and neck insulation in the leatherback turtle (Dermochelys coriacea). J Exp Biol 212: 2753–2759.
11. Goff PG, Stenson GB (1988) Brown adipose tissue in leatherback sea turtles: A thermogenic organ in an endothermic reptile? Copeia 4: 1071–1075.
12. Bostrom BL, Jones DR (2007) Exercise warms adult leatherback turtles. Comp Physiol Biochem A147: 323–331.
13. Eckert SA (2002) Swim speed and movement patterns of gravid leatherback sea turtles (Dermochelys coriacea) at St. Croix, US Virgin Islands. J Exp Biol 205: 3689–3697.

14. Bernal D, Dickson KA, Shadwich RE, Graham JB (2001) Review: Analysis of the evolutionary convergence for high performance swimming in lamnid sharks and tunas. Comp Physiol Biochem A129: 695–726.
15. Block BA, Finnerty JR, Stewart AF, Kidd J (1993) Evolution of endothermy in fish: mapping physiological traits on a molecular phylogeny. Science 260, 210-214:
16. Hochscheid S, Bentivegna F, Speakman JR (2002) Regional blood flow in sea turtles: implications for heat exchange in an aquatic ectotherm. Physiol Biochem Zool 75: 66–76.
17. Kvadsheim PH, Folkow LP, Blix AS (1996) Thermal conductivity of minke whale blubber. J Therm Biol 21: 123–128.
18. Wallace BP, Jones TT (2008) What makes marine turtles go: A review of metabolic rates and their consequences. J Exp Mar Biol Ecol 356: 8–24.
19. James MC, Ottensmeyer A, Myers RA (2005) Identification of high-use habitat and threats to leatherback sea turtles in northern waters: new directions for conservation. Ecol Lett 8: 195–201.
20. Sato K, Sakamoto W, Matsuzawa Y, Tanaka H, Naito Y (1994) Correlation between stomach temperatures and ambient water temperatures in free-ranging loggerhead turtles, Caretta caretta. Mar Biol 118: 343–351.
21. Hawkes LA, Broderick AC, Coyne MS, Godfrey MS, Godley BJ (2007) Only some like it hot – quantifying the environmental niche of the loggerhead sea turtle. Divers Distrib 13: 447–457.
22. Moon D, MacKenzie DS, Owens DWM (1997) Simulated hibernation of sea turtles in the laboratory: I. Feeding, breathing frequency, blood pH, and blood gases. Comp Physiol Biochem A278: 372–380.
23. Standora EA, Spotila JR, Foley RE (1982) Regional endothermy in the sea turtle, Chelonia mydas. J Therm Biol 7: 159–165.
24. Heath ME, McGinnis SM (1980) Body temperature and heat transfer in the green sea turtle, Chelonia mydas. Copeia 4: 767–773.

CHAPTER 14

Blood Gases, Biochemistry, and Hematology of Galapagos Green Turtles (Chelonia Mydas)

Gregory A. Lewbart[1], Maximilian Hirschfeld[2], Judith Denkinger[2], Karla Vasco[2], Nataly Guevara[2], Juan García[3], Juanpablo Muñoz[2], and Kenneth J. Lohmann[4]

[1]North Carolina State University, College of Veterinary Medicine, Raleigh, North Carolina, United States of America,

[2]University San Francisco de Quito, Galapagos Science Center, Puerto Baquerizo Moreno, Galapagos, Ecuador,

[3]Galapagos National Park Service, Puerto Ayora, Galapagos, Ecuador,

[4]Department of Biology, University of North Carolina, Chapel Hill, North Carolina, United States of America

Citation: Lewbart GA, Hirschfeld M, Denkinger J, Vasco K, Guevara N, García J, et al. (2014) Blood Gases, Biochemistry, and Hematology of Galapagos Green Turtles (Chelonia Mydas). PLoS ONE 9(5): e96487. https://doi.org/10.1371/journal.pone.0096487

ABSTRACT

The green turtle, Chelonia mydas, is an endangered marine chelonian with a circum-global distribution. Reference blood parameter intervals have been published for some chelonian species, but baseline hematology, biochemical, and blood gas values are lacking from the Galapagos sea turtles. Analyses were done on blood samples drawn from 28 green turtles captured in two foraging locations on San Cristóbal Island (14 from each site). Of these turtles, 20 were immature and of unknown sex; the other eight were males (five mature, three immature). A portable blood analyzer (iSTAT) was used to obtain near immediate field results for pH, lactate, pO_2, pCO_2, HCO_3^-, Hct, Hb, Na, K, iCa, and Glu. Parameter values affected by temperature were corrected in two ways: (1) with standard formulas; and (2) with auto-corrections made by the iSTAT. The two methods yielded clinically equivalent results. Standard laboratory hematology techniques were employed for the red and white blood cell counts and the hematocrit determination, which was also compared to the hematocrit values generated by the iSTAT. Of all blood analytes, only lactate concentrations were positively correlated with body size. All other values showed no significant difference between the two sample locations nor were they correlated with body size or internal temperature. For hematocrit count, the iSTAT blood analyzer yielded results indistinguishable from those obtained with high-speed centrifugation. The values reported in this study provide baseline data that may be useful in comparisons among populations and in detecting changes in health status among Galapagos sea turtles. The findings might also be helpful in future efforts to demonstrate associations between specific biochemical parameters and disease.

INTRODUCTION

The green turtle (*Chelonia mydas*), also known as the black turtle in the Pacific Ocean, is a marine chelonian inhabiting oceans throughout the world [57]. The green turtle is currently listed as Endangered on the IUCN Red List, and no commercial use is permitted under CITES Appendix I. Major threats to green turtle populations include habitat destruction, pollution, disease, consumption of meat and eggs by local populations, fishing gear entanglement, and consumption of plastics and other anthropogenic materials [16], [37], [42], [43], [59]. Health assessments of green turtles may therefore have implications for wildlife biology and species conservation. Considerable research on natural history has been performed in this species and studies on the health parameters of green turtles, while still relatively limited, have increased dramatically in the last 5 years [1], [4], [5], [8], [17], [18], [22], [27], [30], [31], [39], [40], [45], [50]. A recent review summarizes the health of wild sea turtles and methods of assessment, including blood parameters [20].

Biochemical and hematology parameters, as measured in peripheral blood, are a useful diagnostic tool in animal health management [2], [3]. To determine the significance of alterations in biochemical and hematological values, it is essential to establish species-specific (or at least taxon-specific) normal values for parameters of interest. Reference intervals for certain chelonian species, including sea turtles [2], [5], [6], [8]–[10], [12], [15], [18], [19], [24], [33], [34], [39], [46], [60], have been widely investigated. Additionally, wild sea turtle blood gas values have been reported and analyzed [28], [55]. However, few studies have combined blood gas, biochemistry, and hematology parameters from the same geographic subpopulation of sea turtles. The present study evaluates selected blood gas, blood biochemical, and hematology parameters from 28 wild-caught green turtles (*Chelonia mydas*) in two coastal foraging areas adjacent to San Cristóbal Island, Galapagos, Ecuador. The current study appears to be the first health assessment study on Galapagos sea turtles and on Western Hemisphere green turtles south of the Equator.

MATERIALS AND METHODS

Ethics Statement

This study was performed as part of a population health assessment approved and supported by the Galapagos National Park Service (Permit # PC-35-12 to J. Denkinger) and approved by the Universidad San Francisco de Quito ethics and animal handling protocol. All handling and sampling procedures were consistent with standard vertebrate protocols and veterinary practices.

Turtle Capture and Sampling

Turtles were captured in shallow water within 100 m from the shore by swimmers using standard snorkel, mask, and fins. The animals were carried to shore and placed on the beach. Turtles were aligned with their heads facing the water so that the slight incline of the beach assisted efforts to obtain blood samples from the dorsal jugular vein. A few turtles were sampled while still partially in water when a receding tide exposed rocks, which prevented turtles from being carried safely onto the sandy beach.

Blood samples were obtained within approximately 5 minutes of capture. Fourteen turtles were captured, examined, and sampled from La Loberia Beach (0° 55′ 40″ S, 89° 36′ 43″ W) on 30 June, 2013 and 14 turtles were similarly captured and sampled on 1 July, 2013 at a second site at Punta Carola (0° 53′ 26″ S, 89° 34′ 46″ W). Both study sites are shallow bays of similar size, sheltered from wave exposure, and were identified as important foraging and resting areas where individual turtles show very high site fidelity [16]. To avoid capturing the same individual more than once, a line of white zinc oxide ointment was applied to each turtle's carapace after a blood sample had been obtained and before the animal was released. The line was clearly visible underwater and remained on for at least a day, but disappeared soon after (Hirschfeld, personal observation).

In addition to obtaining a blood sample from each turtle, photographs were taken of each side of the head, as well as the carapace. These images can be used to identify individuals in the future [35], [44], [47].

Blood Sample Collection and Handling

All turtles were manually restrained and blood samples of approximately 2.5 mLs were obtained from either the left or right dorsal jugular sinus using a heparinized 22 gauge needle attached to a 3.0 mL syringe. The blood was then immediately divided into sub-samples. Some subsamples were used for making blood films on clean glass microscope slides, some were stored on ice in sterile plastic vials for future analyses, and others were loaded into the CG-8+ and CG-4+ iSTAT cartridges within 10 minutes of sample collection.

Blood Gas and Biochemistry Parameters

The blood gas, electrolyte, and biochemistry results were obtained using an iSTAT Portable Clinical Analyzer (Heska Corporation, Fort Collins, Colorado, USA) with CG8+ and CG4+ cartridges. The iSTAT is a portable, handheld, battery-operated electronic device with the ability to measure a wide variety of blood gas, chemistry, and basic hematology parameters with only a few drops (0.095 mL) of whole, non-coagulated blood. The following parameters were measured and recorded: pH, lactate, pO_2, pCO_2, HCO_3^-, Hct, Hb, Na, K, iCa, and glucose. The iSTAT device analyzed the blood at 37°C then corrected pH, pO_2, and pCO_2 for body temperature once this information was entered. Because the validity of the iSTAT temperature corrections has been questioned by some authors [13], [28], we also manually calculated an independent set of corrections for pH, pO_2, pCO_2, iCa, and HCO_3^- based on the turtle cloacal temperature (Ti) at the time of sampling [5], [38], using the equations listed below. Both sets of values (i.e., those derived from auto-corrections from the iSTAT and those derived from independent calculations) are reported in Table 1. The methodology for obtaining the various measured and calculated values is summarized in Figure 1. For clarity, instant values obtained from the iSTAT are denoted by an 'I' subscript. Those that were auto-corrected for temperature by entering the turtle's body temperature into the iSTAT are denoted by 'A' subscript. Values that were manually corrected for temperature using the equations shown below are denoted by 'M' subscript.

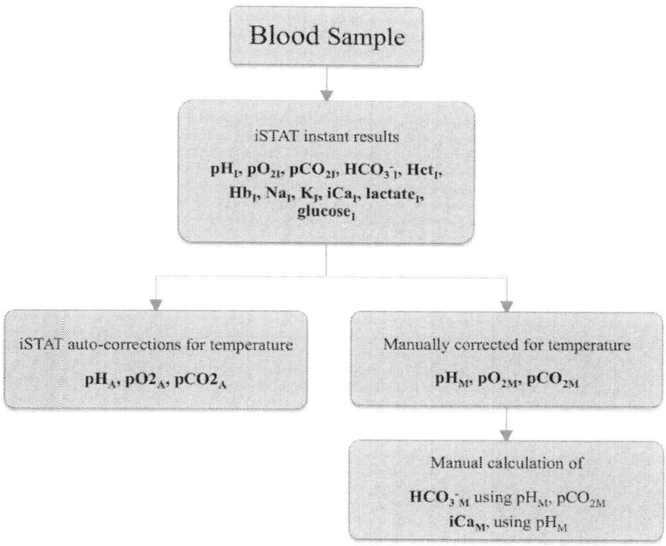

Figure 1. Flow chart of the iSTAT sample handling and calculations.

https://doi.org/10.1371/journal.pone.0096487.g001

Table 1. Mean, standard deviation, and range for blood gas and blood biochemical values for wild Galapagos green turtles

Analyte	n	Mean	SD	Min	Max
$HCO_3{}^-{}_I$ (mmol/L)	24	43.8	5.8	35.6	58.2
$HCO_3{}^-{}_M$ (mmol/L)	28	41.1	5.6	33.0	54.4
pH_A	28	7.477	0.085	7.273	7.626
pH_M	28	7.441	0.075	7.256	7.568
pCO_{2A} (mmHg)	24	46.5	7.1	32.4	65.4
pCO_{2M} (mmHg)	28	49.0	9.2	32.4	68.3
pO_{2A} (mmHg)	28	22	5	14	32
pO_{2M} (mmHg)	28	53	10	36	72
Na_I (mmol/L)	28	148	3	143	153
K_I (mmol/L)	28	3.4	0.5	2.7	4.3
iCa_I (mmol/L)	27	0.87	0.14	0.64	1.18
iCa_M (mmol/L)	27	0.79	0.12	0.57	1.06
Glu_I(Umg/dl)	28	60	9	46	82
Hct_I(L/L)	28	0.24	0.05	0.17	0.38
Hb_I (g/L)	28	80	16	58	129
Lac_I (mmol/L)	28	3.73	2.44	0.8	8.73

'I' subscript denotes values obtained through the instant iSTAT analysis, 'M' subscript indicates values manually corrected for temperature using standard equations (see Results and Figure 1), and 'A' subscript indicates values that were auto-corrected for temperature by the iSTAT after a turtle's cloacal temperature was entered into the iSTAT.

https://doi.org/10.1371/journal.pone.0096487.t001

$$pH_M = 0.014(\Delta T_{37\text{-}25}) + 0.005(\Delta T_{25\text{-}Ti}) + pH_I [\text{at temperatures} < 25^\circ C].$$

Anderson et al. (2011) [5] simplifies the above for Ti values below 25°C: $pH_M = 0.168 + 0.005(\Delta T_{25\text{-}Ti}) + pH_I$. (We note that a typographical error in [5] shows 0.0168 instead of 0.168.).

Because pH has an effect on iCa calculation, the formula below was used in the current study [5], [21], [32]:

$$iCa_M = iCa_I(1 + 0.53[pH_I - pH_M])$$

The following formulas [5] were used to correct pCO_2 and pO_2 for temperature: $pO_{2M} = pO_{2I}(10^{-0.0058\Delta T})$ and $pCO_{2M} = pCO_{2I}(10^{-0.019\Delta T})$

Forthecalculationofbicarbonatethefollowingadaptationofthe Henderson-Hasselbach Equation was used [36], [51]: $HCO_{3\text{-}M} = \alpha CO_2 \times pCO_{2M} \times 10^{(pHM - pKa)}$

The corresponding αCO_2 and pKa values were calculated using the formulas published by Stabenau and Heming (1993) [51].

Hematology

Heparinized whole blood was stored on ice immediately after collection and refrigerated overnight. Total erythrocyte and leukocyte counts (Table 2) were obtained within 24 hours using Natt Herick's stain and a Neubaeuer hemocytometer [11]. Hematocrit was determined using high-speed centrifugation of blood-filled microhematocrit tubes. Differential white blood cell counts were conducted by examining 100 white blood cells on a peripheral smear stained with Wright-Giemsa stain.

Table 2. Mean, standard deviation, and range for manually analyzed hematology values of wild Galapagos green turtles

Analyte	n	Mean	SD	Min	Max
Hct (L/L)	28	0.236	0.048	0.17	0.38
WBC ($x10^9$/L)	28	6.58	4.02	1.76	22.4
Heterophils (%)	28	16.4	6.6	8.0	35.0
Lymphocytes (%)	28	50.5	7.72	33.0	67.0
Monocytes (%)	28	12.0	5.73	6.0	32.0
Azurophils (%)	28	0.04	0.19	0	1.0
Eosinophils (%)	28	20.8	5.99	5.0	31.0
Basophils (%)	28	0.36	0.56	0	2.0
Heterophils ($x10^9$/L)	28	1.06	0.66	0.23	2.59
Lymphocytes ($x10^9$/L)	28	3.42	2.24	0.97	119
Monocytes ($x10^9$/L)	28	0.79	0.60	0.18	2.92
Azurophils ($x10^9$/L)	28	0.001	0.01	0	0.03
Eosinophils ($x10^9$/L)	28	1.41	1.07	0.37	5.16
Basophils ($x10^9$/L)	28	0.02	0.04	0.00	0.13

https://doi.org/10.1371/journal.pone.0096487.t002

Turtle Measurements and Body Temperature

A flexible measuring tape was used to determine curved carapace lengths (CCL). The sex of an immature sea turtle cannot be determined with confidence on the basis of an external examination [8]. Green turtles in the Galapagos have the slowest reported growth rate for this species [26]. Although the smallest nesting female recorded for the archipelago is 60.7 cm CCL, two peaks for nesting females exist at 80 and 95 cm CCL [59]. Furthermore, laparoscopic gonad assessment of green turtles has shown size at maturity is highly variable [20]. Therefore, we used a relatively large size of 80 cm CCL as a threshold to distinguish immature from mature individuals. Adults were sexed on the basis of the external sexual dimorphism of adult green turtles, while some smaller individuals could also be sexed as males by their evident secondary sexual characteristics [54], [57]. An EBRO® Compact J/K/T/E Thermocouple Thermometer was used to obtain all temperature readings (model EW-91219-40; Cole-Parmer, Vernon Hills, Illinois, USA 60061). Core body temperatures were recorded from the cloaca using the probe T PVC epoxy tip 24GA×3 ft in length.

Statistical Analysis

First, the iSTAT blood chemistry results of all overlapping values of the CG8+ and CG4+ cartridges (all analytes except lactate) were compared using paired t-tests. The iSTAT results for hematocrit were compared (paired t-test) to the results of manually determined hematocrit. Subsequently, we grouped values for each blood analyte by foraging site and compared them using Student's t-test. Group sizes for different age classes (mature/immature) as well as sex (male/female) were too small to perform statistical analysis. Linear regressions were used to examine the relationship between the turtles' body size or body temperature and the measured blood chemistry analytes. A standard alpha level of $p=0.05$ was used for all statistical tests using R statistical software, version 3.0.2 (R Development Core Team).

RESULTS

Turtle Demographics and Health Status

A total of 28 green turtles, 23 immature (CCL<80 cm) and 5 mature (CCL>80 cm) were sampled. Half of the animals (n=14) were captured at each study location. The mean CCL was 68.0 cm; values ranged from 41.8 cm CCL for the smallest immature to 84.6 cm CCL for the largest adult male. All of the mature sea turtles (n=5) were males and an additional three males of 79, 76.5 and 71.1 cm CCL were identified. Internal body temperature varied little among individuals, with a mean of 20.6°C and ranging between 19.4°C and 22.5°C. All 28 sea turtles appeared clinically healthy, had no barnacle or excessive algae growth on their carapaces and were bright, alert, and responsive. Many were observed feeding just prior to capture. One turtle had a large piece of monofilament fishing line on the right front flipper causing a strangulating lesion at the elbow that was deemed significant but had not compromised blood flow to the distal limb. The ligature was cut free of the turtle and discarded.

Blood Biochemical Analysis

Values obtained from the CG4+ and CG8+ cartridges were consistent and not statistically different; thus, all overlapping iSTAT values reported are from the first cartridge run for all samples (CG8+). Table 1 displays the biochemistry, blood gas, and hematology results. In a few samples the iSTAT blood analyzer was unable to calculate values for some of the parameters resulting in a slightly smaller n. The hematocrit figures generated by the iSTAT were not significantly different from the manually determined values. There were no statistical differences between the two sampling sites for any values of blood gas, electrolyte, biochemical, and hematology parameters. In addition, none of these values, with one exception, were correlated either with body size or body temperature.

The only blood analyte found to be positively correlated (r^2=0.172, $p<0.05$) with body length (CCL) was lactate (Figure 2). Male green turtles appeared to have a higher lactate concentration than the rest of the unsexed and immature animals, but low sample size and uncertainty of determining sex of smaller individuals prevented a comparative analysis.

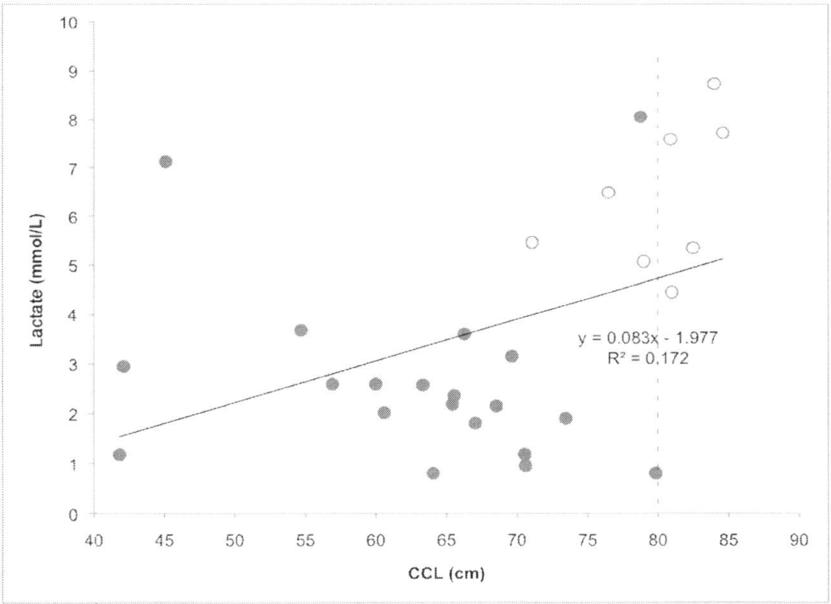

Figure 2. Scatter plot with linear regression line showing the correlation between sea turtle body size and blood lactate levels.

Closed circles indicate data for immature turtles of undetermined sex. Open circles indicate turtles that could be identified as male based on secondary sexual characteristics. The dotted line indicates the size threshold (80 cm CCL) used to classify individuals as immature or mature.

https://doi.org/10.1371/journal.pone.0096487.g002

DISCUSSION

Previous studies have reported baseline blood gas, biochemical, and hematology parameters for some chelonian species, including sea turtles. Clinicians desire species-specific baseline values for the parameters commonly measured by commercial blood gas and chemistry analyzers. When working with reptiles, species specificity of health data is especially important, due to the diverse environmental conditions that exist in different habitats. This study reports the first set of blood gas, biochemistry, and hematology values in Galapagos green turtles and, indeed, in any subequatorial Western Hemisphere green turtle population.

Field sampling of green turtles was performed after the nesting season, when a resident subpopulation appears to remain in the Galapagos [16], [26], [48]. In addition, the two foraging areas sampled are frequented by resident turtles, some of which have been observed repeatedly for up to 5 years [16]. Thus, the turtles sampled in our study were most likely resident turtles, although the possibility that some of the turtles were transient visitors rather than residents cannot be ruled out.

The small sample size (n=28) in the present study precluded the calculation of formal reference intervals, a process that would require a minimum of 120 individuals [23]. However, given the limited sample size available, and the lack of previously published data regarding biochemical, blood gas, and hematology parameters in resident Galapagos green turtles of the eastern tropical Pacific subspecies, these results provide a useful starting point for clinicians and researchers. All 28 turtles were judged to be clinically healthy and their blood parameters support this assessment.

In general, most of the blood parameters we recorded were similar to those reported previously for healthy green turtles [1], [17], [22], [30], [40]. For example, Harms et al. (2009) [30] reported iSTAT values obtained from five green turtles kept in captivity for several years. Relative to these, Galapagos green turtles had slightly lower pO_2, higher pCO_2, and slightly lower lactate. Similarly, Anderson et al. (2009) [5] reported median values for green turtles in North Carolina, U.S.A.; relative to the Galapagos turtles, most values were nearly equivalent, but significant differences existed in hematocrit and blood glucose. The hematology values we observed appear consistent with ranges reported for other clinically healthy green turtles and other chelonian species [1], [5], [8], [17], [25], [45], [56], [58].

In our study, parameter values affected by temperature were corrected based on published, standard formulas; these corrected values sometimes differed from auto-corrected iSTAT values (Table 1) but the differences did not appear to be clinically important. We conclude that iSTAT auto-corrected values are usually sufficient for clinical applications in the field, but also suggest that when accuracy is paramount, investigators studying animals with body temperatures below 37°C should be cautious about relying on auto-corrected iSTAT values for parameters that are influenced by temperature.

Galapagos green turtles in general had higher lactate values than other sea turtles [29], [32], [51], [52]. Only gillnet trapped green turtles [50], loggerhead turtles captured in shrimp trawlers, and cold-stunned Kemp's ridley turtles (sampled after 2 to 3 days of hospitalization) [28], [36], had higher lactate. Loggerheads caught using pound nets had low lactate levels when sampled initially, but lactate concentrations were significantly higher after 30 min [28]. Similarly, Berkson (1966) [7] found no increase in blood lactate of green turtles until 30 to 60 min after forced submergence. Since most voluntary dives are thought to be aerobic and no difference has been found in the dive time of green turtles differing in size, none of these factors readily explain the overall high blood lactate and its increase with body size [41], [49]; Figure 2.

Several factors can influence chelonian biochemistry values, including environmental conditions, age, and sex. Relative to males, females frequently possess higher albumin, calcium, cholesterol, phosphorus, and triglyceride values, which are generally attributed to vitellogenesis [9], [14], [46], [58]. We could not investigate differences between sexes with regard to calcium or other parameters in the Galapagos green turtles due to the difficulty of determining the sex of immature animals in the field [53].

Age also affects some biochemical parameters in sea turtles. Casale et al. (2009) [12] reported that juvenile loggerhead turtles had lower values of albumin, calcium, globulins, hematocrit, total protein, and triglycerides than did adult animals, a difference probably at least partly attributable to egg production in adult females. In green turtles, albumin, total protein, and triglyceride levels were found to increase with body size while calcium, glucose, potassium, and sodium did not [39].

A number of papers have examined differences between clinically healthy and ill or compromised green sea turtles. Aguirre et al. (2009) [1] determined, by examining hematology, biochemistry, and adrenocortical values, that green turtles afflicted with green turtle fibropapillomatosis (GTFP) were immunocompromised and stressed. Similarly, Anderson et al.

(2011) [5] reported that many important biochemical parameters (including calcium, glucose, total protein, and some electrolytes) were significantly lower, and uric acid and blood urea nitrogen higher, in cold-stunned green turtles versus healthy, wild individuals.

In a recent study with a very large sample size, Flint et al. (2010) [18] report blood reference intervals (RIs) for 290 Australian green turtles, 211 of which were deemed clinically healthy. All of the 25 turtles judged unhealthy possessed at least one value outside of the RIs. Aside from higher blood glucose levels in the Australian turtles, healthy green turtles in Australia and the Galapagos appear to have similar blood parameters.

In summary, data reported in this study represent an important step toward determining the normal range of values against which future blood gas and biochemistry results in green turtles can be compared. Such assessments are important for health monitoring and disease diagnostics. Because green turtles are an endangered species with importance in the wildlife biology research community and the aquarium/zoo industry, health assessments are important from the standpoint of sustainable conservation and management. These results add to a growing database of knowledge about health management in wild chelonian species. Future research should continue to establish reference values in this species and facilitate comparisons of blood values across age groups and disease states.

ACKNOWLEDGMENTS

We thank Diana Amoguimba, Eduardo Espinoza, Craig Harms, Tillie Laws, Carlos Mena, Philip Page, Kent Passingham, Carlos Valle-Castillo, Galo Quezada, Erich Stabenau, and Stephen Walsh for their support and assistance with this project.

Author Contributions

Conceived and designed the experiments: GAL MH JD KV JM KJL. Performed the experiments: GAL MH JD KV NG JG KJL. Analyzed the data: GAL MH JD KV NG JM KJL. Contributed reagents/materials/analysis tools: GAL MH JD KV NG JG KJL. Wrote the paper: GAL MH JD KV KJL.

REFERENCES

1. Aguirre AA, Balazs GH, Spraker TR, Gross TS (1995) Adrenal and hematological responses to stress in juvenile green turtles (Chelonia mydas) with and without fibropapillomas. Physiol Zool 68(5): 831–854.
2. Aguirre AA, Balazs GH (2000) Plasma biochemistry values of green turtles (Chelonia mydas) with and without fibropapillomas in the Hawaiian Islands. Comp Haematol Int 10: 132–137.
3. Aguirre AA, Lutz P (2004) Sea turtles as sentinels of marine ecosystem health: is fibropapillomatosis an indicator? Eco Health 1(3): 275–283.
4. Anderson NL, Wack RF, Hatcher R, Wack F (1997) Hematology and clinical chemistry reference ranges for clinically normal, captive New Guinea Snapping Turtles (Elseya novaeguineae) and the effects of temperature, sex, and sample type. J Zoo Wildl Med 28: 394–403.
5. Anderson ET, Harms CA, Stringer EM, Cluse WM (2011a) Evaluation of hematology and serum biochemistry of cold-stunned green sea turtles (Chelonia mydas) in North Carolina, USA. J Zoo Wild Med 42(2): 247–255.
6. Anderson ET, Minter LJ, EO Clarke III, Mroch RM, Beasley JF, et al. (2011b) The effects of feeding on hematological and plasma biochemical profiles in green (Chelonia mydas) and Kemp's Ridley (Lepidochelys kempii) sea turtle. Vet Med Internat doi: 10.4061/2011/890829.
7. Berkson H (1966) Physiological adjustments to prolonged diving in the Pacific green turtle (Chelonia mydas agassizii). Comp Biochem Phys 18: 101–119.
8. Bolton AB, Bjorndal KA (1992) Blood profiles for a wild population of green turtles (Chelonia mydas) in the Southern Bahamas: Size-specific and sex-specific relationships. J Wild Dis 28(3): 407–413.
9. Brenner D, Lewbart G, Stebbins M, Herman D (2002) Health survey of wild and captive Bog Turtles (Clemmys muhlenbergii) in North Carolina and Virginia. J Zoo Wild Med 33: 311–316.
10. Camacho M, Quintana MP, Luzardo OP, Estevez MD, Calabuig P, et al. (2013) Metabolic and respiratory status of stranded juvenile loggerhead sea turtles (Caretta caretta): 66 cases (2008–2009). J Am Vet Med Assoc 242: 396–401.
11. Campbell TW (1995) Avian Hematology and Cytology, 2nd Ed. Iowa State UniversityPress, Ames, IA. Pp. 7–11.

12. Casale AB, Camacho M, Lo´pez-Jurado LF, Juste C, Oro´ s J (2009) Comparative study of hematologic and plasma biochemical variables in Eastern Atlantic juvenile and adult nesting loggerhead sea turtles (Caretta caretta). Vet Clin Path 38: 213–218.
13. Chittick EJ, Stamper MA, Beasley JF, Lewbart GA, Horne WA (2002) Medetomidine, ketamine, and sevoflurane for anesthesia of injured loggerhead sea turtles: 13 cases (1996–2000). J Am Vet Med Assoc 221: 1019–1025.
14. Christopher MM, Berry KH, Wallis IR, Nagy KA, Henen BT, et al. (1999) Reference intervals and physiologic alterations in hematologic and biochemical values of free–ranging Desert Tortoises in the Mojave Desert. J Wild Dis 35: 212–238.
15. Deem SL, Norton TM, Mitchell M, Segars A, Alleman AR, et al. (2009) Comparison of blood values in foraging, nesting, and stranded loggerhead turtles (Caretta caretta) along the coast of Georgia, USA. J Wild Dis 45: 41–56.
16. Denkinger J, Parra M, Munoz JP, Carrasco C, Murillo JC, et al. (2013) Are boat strikes a threat to sea turtles in the Galapagos Marine Reserve? Ocean Coastal Mgt 80: 29–35.
17. Flint M, Morton JM, Limpus CJ, Patterson-Kane JC, Murray PJ, et al. (2010a) Development and application of biochemical and haematological reference intervals to identify unhealthy green sea turtles (Chelonia mydas). Vet J 185: 299–304.
18. Flint M, Patterson-Kane JC, Limpus CJ, Murray PJ, Mills PC (2010b) Health surveillance of stranded green turtles in Southern Queensland, Australia (2006– 2009): An epidemiological analysis of causes of disease and mortality. Eco Health 7: 135–145.
19. Flint M, Morton JM, Limpus CJ, Patterson-Kane JC, Murray PJ, et al. (2010c) Reference intervals for plasma biochemical and hematologic measures in loggerhead sea turtles (Caretta caretta) from Moreton Bay, Aust J Wild Dis 46: 731–741.
20. Flint M (2013) Free ranging sea turtle health. In: Wyneken J, Lohmann KJ, Musick JA (eds.) The Biology of Sea Turtles Vol. III. CRC Press, Inc., Boca Raton, FL. 379–397.
21. Fogh-Anderson N (1981) Ionized calcium analyzer with a built-in pH correction. Clin Chem 27: 1264–1267.

22. Fong CL, Chen HC, Cheng IJ (2010) Blood profiles from wild populations of green sea turtles in Taiwan. J Vet Med Anim Health 2(2): 8–10.
23. Geffre A, Friedricks K, Harr K, Concordet D, Trumel C, et al. (2009) Reference values: a review. Vet Clin Path 38: 288–298.
24. Gelli D, Ferrari V, Zanella A, Arena P, Pozzi L, et al. (2008) Establishing physiological blood parameters in the loggerhead sea turtle (Caretta caretta). Eur J Wild Res 55: 59–63.
25. Gibbons PM, Klaphake E, Carpenter JW (2013) Reptiles. In: Exotic Animal Formulary, Fourth Edition (Carpenter J, Marion C eds.), Elsevier Saunders, St. Louis, MO, pp. 84–170.
26. Green D (1993) Growth rates of wild immature green turtles in the Galapagos Islands, Ecuador. J Herp 27(3): 338–341.
27. Hamann M, Schauble CS, SimonT, Evans S (2006) Demographic and health parameters of green sea turtles Chelonia mydas foraging in the Gulf of Carpentaria, Australia. Endang Spec Res 2: 81–88.
28. Harms CA, Mallo KM, Ross PM, Segars A (2003) Venous blood gases and lactates of wild loggerhead sea turtles (Caretta caretta) following two capture techniques. J Wild Dis 39(2): 366–374.
29. Harms CA, Eckert SA, Kubis SA, Campbell M, Levenson DH, et al. (2007) Field anesthesia of leatherback sea turtles (Dermochelys coriacea). Vet Rec 161: 15–21.
30. Harms CA, Eckert SA, Jones TT, Dow Piniak WE, Mann DA (2009) A technique for underwater anesthesia compared with manual restraint of sea turtles undergoing auditory evoked potential measurements. J Herp Med Surg 19(1): 8–12.
31. Hasbu´n CR, Lawrence AJ, Naldo J, Samour JH, Al-Ghais SM (1998) Normal blood chemistry of free-living green sea turtles, Chelonia mydas, from the United Arab Emirates. Comp Haematol Int 8(3): 174–177.
32. Innis CJ, Tlusty M, Merigo C, Weber ES (2007b) Metabolic and respiratory status of cold-stunned Kemp's ridley sea turtles (Lepidochelys kempii). J Comp Physiol B 177: 623–630.
33. Innis CJ, Ravich JB, Tlusty MF, Hoge MS, Wunn DS, et al. (2009) Hematologic and plasma biochemical findings in cold-stunned Kemp's Ridley Turtles: 176 cases (2001–2005). J Am Vet Med Assoc 235: 426–32.

34. Innis CJ, Merigo C, Dodge K, Tlusty M, Dodge M, et al. (2010) Health evaluations of leatherback turtles (Dermochelys coriacea) in the Northwestern Atlantic during direct capture and fisheries gear disentanglement. Chelon
35. Conserv Biol 9(2): 205–222.
36. Jean C, Ciccione S, Talma E, Ballorain K, Bourjea J (2010) Photo-identification method for green and hawksbill turtles- First results from Reunion. Indian Ocean Turt News 11: 8–13.
37. Keller KA, Innis CJ, Tlusty MF, Kennedy AE, Bean SB, et al. (2012) Metabolic and respiratory derangements associated with death in cold-stunned Kemp's ridley turtles (Lepidochelys kempii): 32 cases (2005–2009). J Am Vet Med Assoc 240:317–323.
38. Koch V, Nichols WJ, Peckham H, De La Toba V (2006) Estimates of sea turtle mortality from poaching and bycatch in Bahia Magalena, Baja California Sur, Mexico. Biol Conserv 128: 327–334.
39. Kraus DR, Jackson DC (1980) Temperature effects on ventilation and acid-base balance of the green turtle. Am J Physiol 239: R254–R258.
40. Labrada-Martago´n V, Me´ndez-Rodriguez LC, Gardner SC, Lo´pez-Castro M, Zenteno-Savin T (2010a) Health indices of the green turtle (Chelonia mydas) along the Pacific coast of Baja California Sur, Mexico. I. Blood biochemistry values. Chel Conserv Biol 9(2): 162–172.
41. Labrada-Martago´n V, Me´ndez-Rodriguez LC, Gardner SC, Cruz-Escalona VH, Zenteno-Savin T (2010b) Health indices of the green turtle (Chelonia mydas) along the Pacific coast of Baja California Sur, Mexico. II. Body condition index. Chel Conserv Biol 9(2): 173–183.
42. Lutcavage ME, Lutz PL (1997) Diving Physiology. In: Lutz PL and Musick JA (eds.) The Biology of Sea Turtles. Mar Sci Ser CRC Press Boca Raton, FL.
43. Mancini A, Koch V, Seminoff JA, Madon B (2011) Small-scale gillnet fisheries provoke massive green turtle (Chelonia mydas) mortality: a case study from Baja California Sur, Mexico. ORYX 46 (1), 69e77.
44. Parra M, De^em SL, Espinoza E (2011) Green turtle (Chelonia mydas) mortality in the Galapagos Islands, Ecuador during the 2009–2010 nesting season. Mar Turtle News 130: 10–15.
45. Reisser J, Proietti M, Kinas P, Sazima I (2008) Photographic identification of sea turtles: method description and validation, with an estimation of tag loss. End Sp Res 5: 73–82.

46. Samour JH, Hewlett JC, Silvanose C, Hasbu´n CR, AlpGhais SM (1998) Normal haematology of free-living green sea turtles (Chelonia mydas) from the United Arab Emirates. Comp Haematol Int 8(2): 102–107.
47. Santoro M, Meneses A (2007) Haematology and plasma chemistry of breeding Olive Ridley Sea Turtles (Lepidochelys olivacea). Vet Rec 161: 818.
48. Schofield G, Katselidis KA, Dimopoulos P, Pants JD (2008) Investigating the viability of photo-identification as an objective tool to study endangered sea turtle populations. J Exp Mar Bio Ecol 360: 103–108.
49. Seminoff JA Jones TT, Marshall GJ (2006) Underwater behaviour of green turtles monitored with video-time-depth recorders: what's missing from dive profiles? Mar Ecol Prog Ser 322: 269–280.
50. Seminoff JA, Zarate P, Coyne M, Foley DG (2008) Post-nesting migrations of Galapagos green turtles Chelonia mydas in relation to oceanographic conditions: integrating satellite telemetry with remotely sensed ocean data. Endan Spec Res 4: 57–72.
51. Snoddy JE, Landon M, Slanvillain G, Southwood A (2009) Blood biochemistry of sea turtles captured in gillnets in the lower Cape Fear River, North Carolina, USA. J Wildlife Management 73(8): 1394–1401.
52. Stabenau EK, Heming TA (1993) Determination of the constants of the Henderson-Hasselbalch equation, (alpha) CO_2 and pKa, in sea turtle plasma. J Exp Bio 180(1): 311–314.
53. Stacy NI, Innis CJ, Hernandez JA (2013) Development and evaluation of three mortality prediction indices for cold-stunned Kemp's ridley sea turtles (Lepidochelys kempii). Conserv Physiol. doi: 10.1093/conphys/cot003.
54. Wabnitz C, Pauly D (2008) Length-weight relationships and additional growth parameters for sea turtles. In: Palomares, MLD, Pauly D (eds.) Fish Cent Res Rep 16(10): 92–101.
55. Wibbels T (1999) Diagnosing the sex of sea turtles in foraging habitats. In: Eckert, KL, Bjorndal KA, Abeu-Grobois FA, Donnelly M (eds.). Research and Management Techniques for the Conservation of Sea Turtles. IUCN/SSC Marine Turtle Specialist Group Publication No. 4 1–5.

56. Wolf KN, Harms CA, Beasley JF (2008) Evaluation of five clinical chemistry analyzers for use in health assessment in sea turtles. J Am Vet Med Assoc 233: 470–475.
57. Wood FE, Ebanks GK (1984) Blood cytology and hematology of the green sea\ turtle, Chelonia mydas. Herpetologica 40(3): 331–336.
58. Wyneken J, Lohmann KJ, Musick JA, eds. (2013) The Biology of Sea Turtles Vol. III. CRC Press, Inc., Boca Raton, FL. 379–397.
59. Yilmaz N, Tosunoglu M (2010) Hematology and some plasma biochemistry values of free–living freshwater turtles (Emys orbicularis and Mauremys rivulata) from Turkey. North-West J Zool 6: 109–117.
60. Zarate P (2002) Evaluacio´n de la actividad de anidacio´n de la Tortuga verde Chelonia mydas, en las islas Gala´pagos druante la temporada 2001. Fundacio´n Charles Darwin. Presentado al Parque Nacional Gala´pagos. Ecuador. 35 pp.
61. Zhang F, Gu H, Li P (2011) A review of chelonian hematology. Asian Herp Res 2(1): 12–20.

CHAPTER 15

Evaluating Group Housing Strategies for the Ex-Situ Conservation of Harlequin Frogs (Atelopus Spp.) Using Behavioral and Physiological Indicators

Shawna J. Cikanek[1], Simon Nockold[2], Janine L. Brown[3], James W. Carpenter[1], Angie Estrada[2], Jorge Guerrel[2], Katharine Hope[3], Roberto Ibáñez[2], Sarah B. Putman[3], and Brian Gratwicke[3]

[1]Department of Clinical Sciences, College of Veterinary Medicine, Kansas State University, Manhattan, Kansas, United States of America,

[2]Panama Amphibian Rescue and Conservation Project, Smithsonian Tropical Research Institute, Gamboa, Republic of Panama,

[3]Smithsonian Conservation Biology Institute, National Zoological Park, Front Royal, Virginia, United States of America

Citation: Cikanek SJ, Nockold S, Brown JL, Carpenter JW, Estrada A, Guerrel J, et al. (2014) Evaluating Group Housing Strategies for the Ex-Situ Conservation of Harlequin Frogs (Atelopus spp.) Using Behavioral and Physiological Indicators. PLoS ONE 9(2): e90218. https://doi.org/10.1371/journal.pone.0090218

ABSTRACT

We have established ex situ assurance colonies of two endangered Panamanian harlequin frogs, Atelopus certus and Atelopus glyphus, but observed that males fought with each other when housed as a group. Housing frogs individually eliminated this problem, but created space constraints. To evaluate the potential stress effects from aggressive interactions when grouping frogs, we housed male frogs in replicated groups of one, two, and eight. We measured aggressive behavioral interactions and fecal glucocorticoid metabolite (GC) concentrations as indicators of stress in each tank. In both small and large groups, frogs initially interacted aggressively, but aggressive interactions and fecal GCs declined significantly after the first 2 weeks of being housed together, reaching the lowest levels by week 4. We conclude that aggressive interactions in same-sex groups of captive Atelopus may initially cause stress, but the frogs become habituated within a few weeks and they can safely be housed in same-sex groups for longer periods of time.

INTRODUCTION

Amphibian biodiversity is being lost at an unprecedented rate [1] prompting the creation of *ex-situ* assurance colonies of endangered species as part of a global 'Amphibian Ark' effort coordinated through the IUCN [2]. *Atelopus* species are a high priority for rescue and assurance populations because of their susceptibility to the invasive fungal pathogen, *Batrachochytrium dendrobatidis* (*Bd*), which has devastated naïve upland amphibian communities throughout Panama [3], [4]. The Panama Amphibian Rescue and Conservation Project was created in response to these *Bd*-related declines and consists of two *ex-situ* facilities in Panama that house populations of amphibians; the El Valle Amphibian Conservation Center (EVACC) and the Smithsonian Tropical Research Institute›s Gamboa Amphibian Research Center (Gamboa ARC). Collectively, these facilities house five of the six Panamanian *Atelopus* species: *A. zeteki* Dunn, 1933; *A. varius* (Lichtenstein & Martens, 1856); *A. limosus* Ibáñez, Jaramillo & Solís, 1995; *A. certus* Barbour, 1923; and *A. glyphus* Dunn, 1931. The sixth known harlequin frog species, *A. chiriquiensis* Shreve, 1936, has not been observed since 1996 and may be extinct [5].

The Panama Amphibian Rescue Project aims to create assurance colonies of 20 endangered amphibian species and to grow the captive population of each species to a minimum effective population size of 500 individuals [6], [7]. However, space is a major limiting factor for *ex situ* programs, and housing frogs in groups would greatly expand our capacity to meet our population management goals. The AZA species survival plan for *A. zeteki* manages about 2,000 captive *Atelopus* and frogs are regularly housed in same-sex groups to limit the extended periods of amplexus observed in opposite sex groups (K. Murphy, pers comm). Many of these frogs have been raised in captivity in group-housing situations, and therefore acclimated to group conditions, but it was unclear how readily our wild-caught *Atelopus* would acclimate to group housing.

To evaluate the levels of stress associated with different housing scenarios for captive *Atelopus,* we developed behavioral and physiological indicators to quantify stress. Types of *Atelopus* calls include advertisement, release, territorial, and courtship with the most common being advertisement [8]. *Atelopus* males also use visual signals, such as semaphore foot-raising, to signal antagonistic behavior [9], and their territorial behavior has been well studied [10], [11], allowing us to compile an ethogram to observe and document aggressive interactions as a behavioral indicator of stress.

Glucocorticoids are a group of steroid hormones that can be used as indicators to evaluate stress, health status, and disease in many species including amphibians [12], [13]. Glucocorticoid release is the last step of a hormonal cascade that begins in the brain to help an animal react to a stressor [14], [15]. An animal's internal response to stress involves the activation of the hypothalamic-pituitary-adrenal axis (HPA), and the release of cortisol or corticosterone from the adrenal cortex [16]. These GCs can be measured in urine, feces, plasma, and blood [17]. Blood analyses are the most common, but not always the most practical because of the potential stress of sample collection that would be too invasive to employ on small-bodied, endangered *Atelopus,* e.g. [17], [18], [19]. Methods to evaluate corticosterone in amphibian urine have been developed, validated and tested on several species [20], [21], [22], but handling and manipulating individuals to collect the urine sample is still somewhat invasive, especially with smaller, more delicate frogs. By contrast, fecal pellets are readily collected from captive frogs without disturbance, so we adapted and validated existing fecal GC tests as a non-invasive approach to evaluate physiological stress responses in frogs. The goal of this study was to use behavioral and physiological indicators to determine if wild-caught *A. certus* and *A. glyphus* could be maintained in same-sex groups without compromising animal welfare.

METHODS

Husbandry

Facilities to house wild-caught amphibians from Central America were established in Gamboa, Panama. Collections of from the Darien region were made with permission from the Autoridad Nacional del Ambiente (ANAM) permits SE/A-130-10 and SE/A-42-11. The Animal Care and Use Committee of the Smithsonian National Zoological Park approved the project(#09–31). A total of 44 *A. certus* and 22 *A. glyphus* were housed individually in small plastic cages measuring 28×19×16.5 cm for at least 1 year before the start of this 5-week study. Cages were misted daily and enriched with native plant leaves (*Philodendron* spp.) and damp brown paper towels. Tanks were kept on metal racks with fluorescent overhead lighting for 12 hours per day and cleaned twice per week. At the start of the experiment, frogs were removed from the plastic cages and placed in large, numbered glass tanks (size 25×53×38 cm) with automated misting systems lightly spraying the tank interiors for 5 minutes every 2 hours. Cages had false bottoms installed (plastic egg crate covered in 0.5 mm screen mesh), keeping frogs (and fecal pellets) out of contact with any dirty water that may have pooled on the tank bottom. Ultraviolet-emitting lights supplemented the 12-hour overhead fluorescent lights for eight 45-minute intervals per day. Each tank was furnished with two potted plants (*Philodendron* spp.), rocks and a water basin. Frogs were randomly assigned to one of three treatments consisting of identical tanks housing one, two, or eight male *Atelopus* frogs, respectively. Each treatment was replicated 6 times in a completely randomized design. *A. glyphus* males (mean Snout Vent Length (SVL) =37 mm SD +/−1.8 mm, 3.76 g SD +/−0.53 g) were used in 2 full replicates, and *A. certus* males (mean SVL =32.3 mm SD +/−1.6 mm, 2.65 g SD +/−0.39 g) were used for the remaining 4 replicates. Black, opaque dividers were placed between tanks to limit visual cues from neighbors. Frogs were fed ad libitum with small crickets (*Acheta domesticus*) or fruit flies (*Drosophila melanogaster and D. hydei*) dusted with calcium or vitamin supplements 4 times per week. Frogs were weighed and measured (SVL) at the start and end of the experiment. We examined overall weight loss or gain as a measure of body condition, expressed as the relative change in mass as a percentage of the starting weight. We created cage cards using photographs of the unique pattern of dark spots clearly visible on their white ventral side to distinguish between individuals.

Behavior

A range of territorial and aggressive behaviors were recorded to assess the degree of conflict associated with each group size and defined using an ethogram (Figure 1). Aggressive interactions included fighting, mounting, release-call, stalking, and waving. A single observer (SN) tallied behavior in each tank for 5 minutes twice a day, in the morning between 0700–0830 hr and in the afternoon between 1400–1530 hr. The order of sampling was randomized to prevent any sequential bias due to time of day. All observations in a single week were summed and divided by the number of frogs in each tank to obtain a total number of aggressive interactions observed per frog per week.

Figure 1. Ethogram describing different types of aggressive interactions observed for *Atelopus*.

Fight: Combat involving mouth or front limbs, often flipping of opponent; Mount: >50% of initiators body covers the victim for >5 seconds; Release call: High pitched, weak, peep like call; maximum tally of one per individual; Physical contact: Any remaining forms of physical contact; Stalk: One individual actively follows/chases another for >5 seconds; Wave: Circular movements in front limbs.

https://doi.org/10.1371/journal.pone.0090218.g001

Fecal collection

Fecal material was removed manually and tanks were not changed for the duration of the experiment. Each week a single frog produces 5–9 fecal pellets with a mean weight of 0.038 g SD+/−0.026 g. Fecal pellets were stored at −20°C until extraction and analysis of GC metabolite concentrations. Collection began 1 week prior to moving frogs to the glass cages (week 0) to establish baseline GC concentrations. Every solid fecal sample was collected within 12 hours of being voided and all weekly samples from each enclosure were frozen together in 1.5 mL polypropylene tubes until processing. Samples from each cage were pooled by week to obtain a sufficient weight of fecal material for extraction.

Fecal glucocorticoid extraction and analysis

A fecal GC extraction method similar to Brown et al. [23], [24] was used for frog feces. Four solvent:water (v:v) ratios were tested to determine the best recovery: 90% ethanol:dH_2O, 80% ethanol:dH_2O, 90% methanol:dH_2O and 80% methanol:dH_2O. The subsequent extracts were serially diluted, analyzed on the cortisol EIA (described below), and compared to the standard curve for parallelism (90% ethanol: r^2 =0.988, $F(1,4)$ =316.64, $p<0.01$; 80% ethanol: r^2 =0.980, $F(1,4)$ =397.92, $p<0.01$, 90% methanol: r^2 =0.995, $F(1,5)$ =1093.27, $p<0.01$ and 80% methanol: r^2 =0.995, $F(1,4)$ =758.56, $p<0.01$).

For each method, the linear portion of the slope of the curve was similar to the standard curve (standards: −11.74; 90% ethanol: −11.84; 80% ethanol: −11.78; 90% methanol: −10.98 and 80% methanol: −12.16). Although all solvent ratios resulted in high steroid recovery, based on the maximum percent binding (%B) of a neat sample, 90% methanol:10% dH_2O was the optimal extraction method (90% ethanol, 41.36%B; 80% ethanol, 38.95%B; 90% methanol, 31.59%B; 80% methanol, 37.04%B). Thus, for this study, wet samples (0.05 g) were weighed and placed into 16×125 mm borosilicate tubes.

Five mL of 90% methanol:10% dH_2O was added to each sample, tubes were capped and then vortexed for 10 seconds. Samples were shaken on a large capacity mixer for 30 minutes (Glas-Col, Terre Haute, Indiana, speed 55, pulse rate 1/second) followed by centrifugation at 2500×g for 20 minutes. The supernatant was recovered, and 5 mL 90% methanol:10%

dH_2O was again added to each tube. The pellets were re-suspended and the samples were shaken on a large capacity mixer (30 seconds, speed 55, pulse rate 1/second) and centrifuged for 20 minutes. The supernatants were combined, evaporated to dryness under directed air, then reconstituted in 1 mL 100% methanol and placed in an ultrasonic cleaner water bath (Cole Parmer Instrument Company, Vernon Hills, Illinois) for 10 minutes and dried down. Fecal extracts were reconstituted in 1 mL preservative-free buffer (0.2 M NaH_2PO_4, 0.2 M Na_2HPO_4, 0.15 M NaCl; pH 7.0), sonicated for 15 minutes, transferred to polypropylene tubes and stored at −20°C until analysis. One hundred μL of ^{3}H-cortisol (c10,000 cpm/100 μL) was added to monitor extraction efficiency of each sample, which was determined to be 90±0.003% (mean ± SEM) based on recovery of radioactivity after extraction.

Two assays, a cortisol enzyme immunoassay (EIA; C.J. Munro, University of California, Davis, California) and a corticosterone radioimmunoassay (RIA; MP Biomedicals, Santa Ana, California), both of which have broad crossreactivity with fecal GC metabolites in numerous species [25], [26] were evaluated for use with Panamanian golden frog (*A. zeteki*) feces.

Both the corticosterone RIA and cortisol EIA demonstrated parallelism between serial fecal extract dilutions and the respective standard curve. Low matrix interference was indicated in the corticosterone RIA as a result of 88% recovery of known standard concentrations when diluted with equal parts fecal extract pool.

For the cortisol EIA, the average recovery was 91%. To compare longitudinal patterns, samples from 11 frogs were analyzed in both assays and the correlation between the two was calculated. The median correlation between the assays for individual fecal GC profiles was high at r=0.92 (range: 0.57–0.98).

Biological validity was shown by a frog that demonstrated a marked increase in GC concentrations within 4 days after ACTH injection (0.2 IU, IM) for both EIA (pre, 13.8 ng/g; post, 66.4 ng/g) and RIA (pre, 2.9 ng/g; post, 15.7 ng/g); the two profiles were correlated (r=0.93; p<0.001) (Figure 2). Thus, both assays were able to detect similar patterns of hormone excretion; however, the cortisol assay detected higher overall concentrations of metabolites (5-fold higher) and so was used in this study.

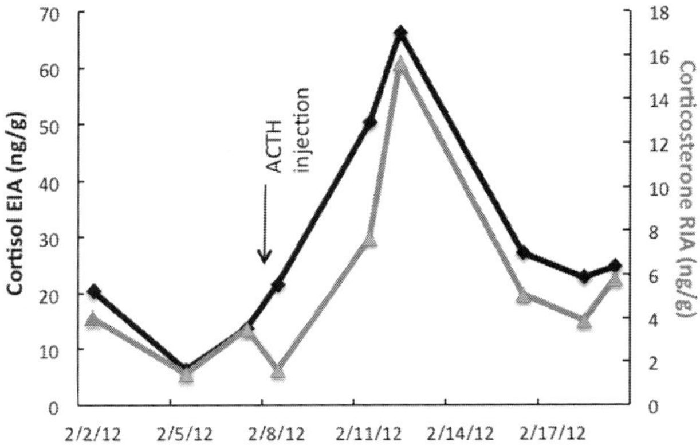

Figure 2. Cortisol EIA profile (black, diamond marks) and Corticosterone RIA profile (grey, triangle marks), in Atelopus feces following an ACTH challenge (0.2 IU, IM).

https://doi.org/10.1371/journal.pone.0090218.g002

The single-antibody cortisol EIA was based on the methodology of Munro and Lasley [27] and used a polyclonal antiserum (R4866) and horseradish peroxidase ligand (lot 051229, SCBI, Front Royal, Virginia). The cross-reactivities for R4866 are: cortisol 100.0%, prednisolone 9.9%, prednisone 6.3%, cortisone 5.0%; all other compounds cross-react with the antibody <1.0% [26]. The standard curve range for the assay was 0.78–20.00 ng/mL. Antiserum was diluted with coating buffer (0.015 M Na_2CO_3, 0.035 M $NaHCO_3$, pH 9.6) and adsorbed to NUNC Maxi-sorp flat-bottomed, 96-well microtiter plates overnight at 4°C. The plate was washed 5 times (0.05% Tween 20 in 0.15 M NaCl solution), then 50 μL of standards, internal controls and samples were loaded onto the plate in duplicate, followed by the addition of 50 μL diluted horseradish peroxidase solution. Assays were incubated at room temperature for 1 hour, washed 5 times and 100 μL of 2,2′-azinobis [3-ethylbenzothiazoline-6-sulfonic acid]-diammonium salt [ABTS] solution (0.04 M ABTS, 0.5 M H_2O_2 in 0.05 M citric acid buffer) added to every well. Absorbance was read on a spectrophotometer (MRX, Dynex Technologies, Chantilly, VA, with a 405 nm filter 405 and reference filter of 490 nm) until the optical density (OD) of the 0.00 ng/mL standard reached 1.0 (range: 0.9–1.1; desired OD reached within 20–30 min). Steroid concentrations were divided by the amount extracted and reported as ng/g feces. Samples weighing <0.01 g were excluded from the data set because low weight samples consistently exhibited higher values compared to heavier samples [28]. The inter-assay variation on two internal controls (high and low GC concentration) were 7.3 and 8.0% CV, respectively (n=16). Intra-assay variation between sample duplicates was <10% CV.

High performance liquid chromatography

High performance liquid chromatography (HPLC; Varian ProStar; Varian Analytical Instruments, Lexington, Massachusetts) was used to characterize the numbers and proportions of immunoactive hormone metabolites excreted in *Atelopus* feces. Three aliquots of pooled fecal samples were extracted as described above, omitting the ^{3}H tracer. The methanol extracts were pooled, dried down under directed air, resuspended in 500 µL PBS (0.03 M Na_2HPO_4, 0.02 M NaH_2PO_4, 0.15 M NaCl, 0.002 M NaN_3, pH: 5.0), filtered through a C18 Spice cartridge and evaporated to dryness. For chromatographic markers, approximately 14,000 cpm/mL of ^{3}H-cortisol and ^{3}H-corticosterone were each added to the extract. The extract was dried down then reconstituted in 300 µL methanol (HPLC Grade Methanol, Fisher Scientific, Pittsburgh, Pennsylvania) and sonicated for 15 minutes. Then, 50 µL of extract was loaded onto a reverse-phase C18 HPLC column (Agilent Technologies, Santa Clara, California) with a 20–80% linear gradient of HPLC Grade methanol:water over 80 minutes (1 mL/minute. flow rate, 1 mL fractions). A 50 µL aliquot of each fraction was analyzed for radioactivty using a multi-purpose β-radiation scintillation counter (LS 6500, Beckman Coulter, Brea, California). The remaining volume was dried down, reconstituted in 200 µL preservative-free phosphate buffer and analyzed in singlet in the cortisol EIA and corticosterone RIA.

Statistical analysis

Data were analyzed using SYSTAT 11 software. We performed a repeated measures ANOVA using GC values (ng/g) or feces extracted from pooled weekly fecal samples collected in each tank as the repeated measure, and group size (two frogs per tank vs. eight frogs per tank) was our experimental factor. For the group size of n=1, one week's worth of fecal pellets was sometimes not enough material to perform an extraction, resulting in too many missing values to be incorporated into the statistical analysis. For the behavioral analysis, we performed repeated measures ANOVA using total number of aggressive interactions observed per frog per week as the repeated measure and group size (two frogs per tank vs eight frogs per tank) as our experimental factor.

RESULTS

High performance liquid chromatography

Based on HPLC analysis, ^{3}H-cortisol eluted at fractions 39–41 and peaked at fraction 40, while peak ^{3}H-corticosterone eluted at fraction 45 (range: 44–46). Immunoactivity of fractions analyzed on the cortisol EIA indicated the presence of native cortisol with a peak at fraction 39 (15% of the total immunoactivity) and a smaller amount of immunoreactivity at fractions 44–45 (11%). Additional immunoreactivity was observed at fraction 13 (6%), and there were peaks of uncharacterized less polar metabolites at fractions 54, 59, 66, 75, and 79 (68% of total immunoactivity). Concentration of GC immunoactivity of HPLC-separated fractions in the corticosterone RIA was only about a tenth that of the cortisol EIA, with small peaks at 37-39 (3%) and 44–48 (11%), and also several less polar peaks of similar levels of immunoreactivity at fractions 50, 54, 59, 63, 65, 76, and 79 (86%).

Behavioral responses

The most common behavior observed was physical contact, which accounted for 28% of all aggressive interactions included in the ethogram, but for the purposes of this analysis, all aggressive interactions were pooled because no single behavioral response was recorded in high enough frequency to compare behavior types. When we placed groups of both two and eight *Atelopus* together, aggressive interactions were high at first, but declined significantly in subsequent weeks to a mean of almost zero aggressive interactions observed per frog per week (Figure 3B, Table 1B). There was no significant difference in relative number of aggressive interactions per frog, between groups of 8 and 2 frogs.

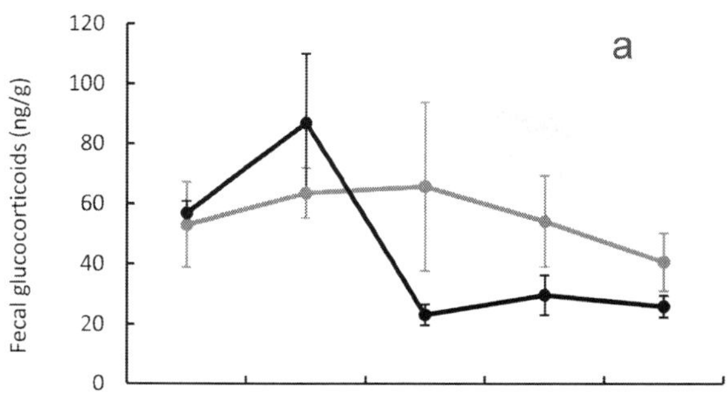

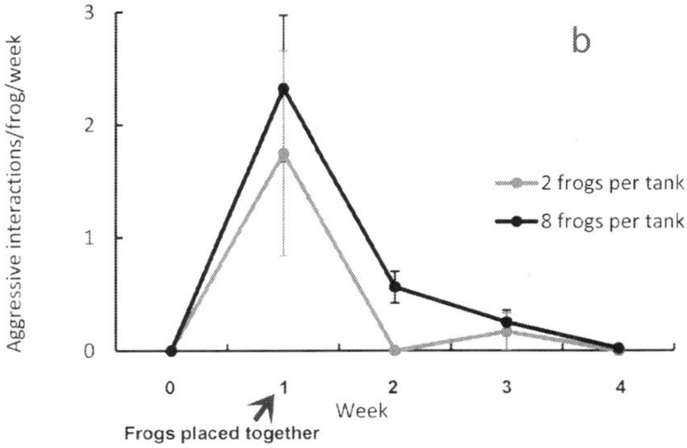

Figure 3. Fecal glucocorticoid concentrations immediately before and after male *Atelopus* **were grouped together at week 1** (ng cortisol/g ± SEM) changed significantly over time (p=0.04*), but there were no significant differences between groups sizes (A).

Frogs housed singly (mean =44.2 ng cortisol/g±7.4 SEM) could not be included in this analysis because of too many missing values. Aggressive interactions changed significantly over time (p<0.001***), but there were no significant differences between group sizes.

https://doi.org/10.1371/journal.pone.0090218.g003

Table 1. Repeated measures ANOVA testing the effects of time (weeks 0–4) and group size on fecal glucocorticoid levels in *Atelopus* housed together (groups of 8 vs 2)

a Fecal Glucocorticoids	F	df	P
Group size	1.217	1	0.296 NS
Week	2.768	4	0.04*
Week*Group size	1.728	4	0.163 NS
b Aggressive Interactions			
Group size	1.128	1	0.313 NS
Week	11.009	3	<0.001***
Week*Group size	0.278	3	0.841 NS

*(**a**). We omitted group size n = 1 from the analysis because there were too many missing values to run statistical comparisons. Repeated measures ANOVA testing the effects of time (weeks 1–4) and group size (8 vs 2) on aggressive interactions in Atelopus housed together (**b**).*

https://doi.org/10.1371/journal.pone.0090218.t001

Physiological responses

Similarly, there was a significant increase in fecal GC levels in groups of both two and eight frogs per tank when they were placed together for the first time. Fecal GC concentrations in groups of two and eight individuals rose initially and then declined to their lowest concentrations by week 4 (Figure 3A, Table 1A) when mean fecal glucocorticoid levels approximated the mean glucocorticoid levels of individually housed frogs 44.2±7.4 ng/g. Interestingly, the cortisol levels by the end of the experiment were both lower than the baseline (week =0) levels obtained when they were being housed individually in small plastic cages. Overall body condition measurements had a high degree of variability, but on average frogs lost 2.6% of their body mass over the 5 weeks SD +/−7.1%. There were no statistically significant differences in changes in body condition between frogs in the different group sizes (Kruskall-Wallace p 0.454).

DISCUSSION

Fecal glucocorticoid extraction methods

In comparative analyses, a cortisol EIA and corticosterone RIA were both capable of quantifying GC metabolites in *Atelopus* fecal extracts, and results were highly correlated. The cortisol antibody reacted with metabolites that co-eluted with both tritiated cortisol and corticosterone tracers, and also a number of less polar metabolites. The corticosterone RIA appeared to detect a small amount of native corticosterone, but also significant amounts of less polar metabolites not unlike that observed for the EIA, although at much lower levels. We decided to employ the cortisol EIA because it detected more immunoreactive mass, it is more portable for possible field studies, and does not rely on the use of radioactive tracers.

Behavioral and physiological responses to grouping

We demonstrated that wild-caught male *Atelopus* do interact aggressively when housed in both small and large groups, but over a relatively short period of time these frogs become accustomed to each other and reduce the frequency of their aggressive interactions. A similar pattern was observed in fecal GC concentrations, indicating that these aggressive interactions are likely associated with a physiological stress response. Interestingly, the mean baseline GC values pre-grouping were higher (60 ng/g) than might have been expected based on the mean GC values observed at the conclusion of the experiment (40 ng/g; Figure 3A). This may be connected to the fact that, prior to the group housing, the smaller holding cages were completely changed twice per week, involving frequent handling of the frog that may have slightly elevated GCs. Once the frogs were transferred to their larger glass cages with automated misting and draining systems we did not need to handle the frogs at all for the duration of the experiment. Many other studies have examined GC levels in amphibians; e.g., [17], [18], [20], [21], but this is the first time that fecal GCs have been used to evaluate stress in frogs. Our findings are similar to observations by others that group housing of cane toads led to increased urinary corticosterone concentrations that declined once they were moved to individual housing [21]. Measurements of GCs using other methods such urine sampling or buccal swabs can potentially give accurate, short-term measurements e.g. [20]. However, the frogs in this study were not used to handling, so invasive procedures like blood and buccal sample collection would have in themselves induced stress. These approaches would have permitted more fine-scale measurements of acute stress responses, but this was not our objective. Urine is an effective noninvasive approach, but housing conditions of the frogs in our study did not permit reliable sample collection. We also could not partition stress by individual frogs to see if some individuals were experiencing more or less stress than others. But the one major advantage to feces is that it represents a pooled sample over time, so acute fluctuations among and between individuals are dampened. Thus, we conclude that fecal steroid monitoring provides a minimally invasive option to researchers that can be applied, as we have demonstrated here, to evaluate and compare husbandry practices, and how they impact frog welfare over time, as efforts to build global amphibian arks for endangered amphibians continue to grow [28].

Our results are significant from a conservation perspective because they justify housing frogs in groups. Given that space is a major limiting factor in *ex-situ* conservation programs, it will greatly increase the number of frogs that can be held in captivity and managed for amphibian conservation and reintroduction efforts. We did not detect a statistically significant effect of group size; this may have been due to low statistical power from having just six replicates. Future studies that increase the number of replicates and treatment groups may shed light on optimal housing densities. It is important to note, however, that housing frogs in groups may have other consequences not addressed in this study. For example, group housing may lead to changes in body condition if smaller or non-dominant animals do not compete as well for food [29], but this was not observed in this experiment. Group housing may lead to increased buildup of gut parasite loads [29] or increased aggressive interactions during the breeding season [11]. Any of these could have an impact on the long-term health of an individual if not carefully managed and monitored by animal care staff, and should be considered carefully before making management changes.

AUTHOR CONTRIBUTIONS

Conceived and designed the experiments: SJC SN AE JG RI BG. Performed the experiments: SN SJC KH JG AE. Analyzed the data: SJC JLB JWC BG. Contributed reagents/materials/analysis tools: KH SBP JLB. Wrote the paper: SJC JLB BG JWC.

REFERENCES

1. Stuart SN, Chanson JS, Cox NA, Young BE, Rodrigues ASL, et al. (2004) Statusand Trends of Amphibian Declines and Extinctions Worldwide. Science 306: 1783–1786.
2. Zippel K, Johnson K, Gagliardo R, Gibson R, McFadden M, et al. (2011) The Amphibian Ark: a global community for ex situ conservation of amphibians. Herpetological Conservation and Biology 6: 340–352.
3. La Marca E, Lips KR, Lotters S, Puschendorf R, Iba´n˜ez R, et al. (2005) Catastrophic population declines and extinctions in neotropical Harlequin frogs (Bufonidae: Atelopus) Biotropica 37: 190–201.
4. Crawford AJ, Lips KR, Bermingham E (2010) Epidemic disease decimates amphibian abundance, species diversity, and evolutionary history in the highlands of central Panama. Proceedings of the National Academy of Sciences: doi: 10.1073/pnas.0914115107.
5. Lips K, Iba´n˜ez R, Bolan˜os F, Chaves G, Solı´s F, et al. (2008) Atelopus chiriquiensis. IUCN Red List of Threatened Species Version 20102 wwwiucnredlistorg Downloaded on 01 August 2010.
6. Schad K (2008) Amphibian Population Management Guidelines. Amphibian Ark Amphibian Population Management Workshop December 10–11, 2007 San Diego: Amphibian Ark, http://www.amphibianark.org/pdf/Aark material/ AArk Amphibian Population Management Guidelines.pdf. Accessed Feb 13, 2014.
7. Frankham R, Briscoe D, Ballou J (2002) Introduction to conservation genetics: Cambridge Univ Press.
8. Jaslow AP (1979) Vocalization and aggression in Atelopus chiriquiensis (Amphibia, Anura, Bufonidae). Journal of Herpetology 13: 141–145.
9. Lindquist ED, Hetherington TE (1996) Field Studies on Visual and Acoustic Signaling in the "Earless" Panamanian Golden Frog, Atelopus zeteki. Journal of Herpetology 30: 347.
10. Cocroft RB, McDiarmid RW, Jaslow AP, Ruiz-Carranza PM (1990) Vocalizations of eight species of Atelopus (Anura: Bufonidae) with comments on communication in the genus. Copeia: 631–643.
11. Crump ML (1986) Homing and Site Fidelity in a Neotropical Frog, Atelopus varius (Bufonidae). Copeia 1986: 438–444.
12. Goymann W, Wingfield JC (2004) Allostatic load, social status and stress hormones: the costs of social status matter. Animal Behaviour 67: 591–602.

13. Narayan EJ (2013) Non-invasive reproductive and stress endocrinology in amphibian conservation physiology. Conservation Physiology 1: 1–16.
14. Romero LM (2004) Physiological stress in ecology: lessons from biomedical research. Trends in Ecology & Evolution 19: 249–255.
15. Sapolsky RM, Romero LM, Munck AU (2000) How do glucocorticoids influence stress responses? Integrating permissive, suppressive, stimulatory, and preparative actions. Endocrine reviews 21: 55–89.
16. Cockrem JF, Barrett DP, Candy EJ, Potter MA (2009) Corticosterone responses in birds: Individual variation and repeatability in Adelie penguins (Pygoscelis adeliae) and other species, and the use of power analysis to determine sample sizes. General and comparative endocrinology 163: 158–168.
17. Sheriff MJ, Dantzer B, Delehanty B, Palme R, Boonstra R (2011) Measuring stress in wildlife: techniques for quantifying glucocorticoids. Oecologia 166: 869–887.
18. Belden L, Moore I, Mason R, Wingfield J, Blaustein A (2003) Survival, the hormonal stress response and UV avoidance in Cascades Frog tadpoles (Rana cascadae) exposed to UVB radiation. Functional Ecology 17: 409–416.
19. Gendron AD, Bishop CA, Fortin R, Hontela A (1997) In vivo testing of the functional integrity of the corticosterone-producing axis in mudpuppy (amphibia) exposed to chlorinated hydrocarbons in the wild. Environmental Toxicology and Chemistry 16: 1694–1706.
20. Narayan EJ, Molinia FC, Kindermann C, Cockrem JF, Hero J-M (2011) Urinary corticosterone responses to capture and toe-clipping in the cane toad (Rhinella marina) indicate that toe-clipping is a stressor for amphibians. General and comparative endocrinology 174: 238–245.
21. Narayan EJ, Cockrem JF, Hero J-M (2013) Sight of a Predator Induces a Corticosterone Stress Response and Generates Fear in an Amphibian. PLoS one 8: e73564.
22. Narayan EJ, Cockrem JF, Hero J-M (2011) Urinary corticosterone metabolite responses to capture and captivity in the cane toad Rhinella marina. General and comparative endocrinology 173: 371–377.
23. Graham LH, Brown JL (1996) Cortisol metabolism in the domestic cat and implications for non-invasive monitoring of adrenocortical function in endangered felids. Zoo Biology 15: 71–82.

24. Brown JL, Wasser SK, Wildt DE, Graham LH (1994) Comparative aspects of steroid hormone metabolism and ovarian activity in felids, measured noninvasively in feces. Biology of reproduction 51: 776–786.
25. Wasser SK, Hunt KE, Brown JL, Cooper K, Crockett CM, et al. (2000) A generalized fecal glucocorticoid assay for use in a diverse array of nondomestic mammalian and avian species. General and comparative endocrinology 120:
26. 260–275.
27. Young K, Walker S, Lanthier C, Waddell W, Monfort SL, et al. (2004) Noninvasive monitoring of adrenocortical activity in carnivores by fecal glucocorticoid analyses. General and comparative endocrinology 137: 148–165.
28. Munro C, Lasley B (1988) Non-radiometric assays: technology and application in polypeptide and steroid hormone detection. Non-radiometric methods for immunoassay of steroid hormones. New York: Allan Liss Corporation. pp. 289–329.
29. Millspaugh JJ, Washburn BE (2004) Use of fecal glucocorticoid metabolite measures in conservation biology research: considerations for application and interpretation. General and comparative endocrinology 138: 189–199.
30. Haislip NA, Hoverman J, Miller DL, Gray MJ (2012) Natural stressors and disease risk: does the threat of predation increase amphibian susceptibility to ranavirus? Canadian Journal of Zoology 90: 893–902.

SECTION 7: WILD ANIMALS

7.2. MAMMALS

CHAPTER 16

Fecal Near Infrared Spectroscopy to Discriminate Physiological Status in Giant Pandas

Erin E. Wiedower[1], Andrew J. Kouba[2], Carrie K. Vance[3], Rachel L. Hansen[2], Jerry W. Stuth[4], and Douglas R. Tolleson[5]

[1]Grazingland Animal Nutrition Lab, Blackland Research and Extension Center, Texas A&M University, Temple, Texas, United States of America,

[2]Department of Conservation and Research, Memphis Zoological Society, Memphis, Tennessee, United States of America,

[3]Department of Biochemistry and Molecular Biology, Mississippi State University, Mississippi, United States of America,

[4]Department of Rangeland Ecology and Management, Texas A&M University, College Station, Texas, United States of America,

[5]School of Natural Resources and the Environment, V Bar V Ranch, The University of Arizona, Camp Verde, Arizona, United States of America

Citation: Wiedower EE, Kouba AJ, Vance CK, Hansen RL, Stuth JW, Tolleson DR (2012) Fecal Near Infrared Spectroscopy to Discriminate Physiological Status in Giant Pandas. PLoS ONE 7(6): e38908. https://doi.org/10.1371/journal.pone.0038908

ABSTRACT

Giant panda (Ailuropoda melanoleuca) monitoring and research often require accurate estimates of population size and density. However, obtaining these estimates has been challenging. Innovative technologies, such as fecal near infrared reflectance spectroscopy (FNIRS), may be used to differentiate between sex, age class, and reproductive status as has been shown for several other species. The objective of this study was to determine if FNIRS could be similarly used for giant panda physiological discriminations. Based on samples from captive animals in four U.S. zoos, FNIRS calibrations correctly identified 78% of samples from adult males, 81% from adult females, 85% from adults, 89% from juveniles, 75% from pregnant and 70% from non-pregnant females. However, diet had an impact on the success of the calibrations. When diet was controlled for plant part such that "leaf only" feces were evaluated, FNIRS calibrations correctly identified 93% of samples from adult males and 95% from adult females. These data show that FNIRS has the potential to differentiate between the sex, age class, and reproductive status in the giant panda and may be applicable for surveying wild populations.

INTRODUCTION

Giant pandas (*Ailuropoda melanolueca*) are listed as critically endangered by the International Union for the Conservation of Nature (IUCN) [1] with an estimated 1,600 animals scattered across six mountain ranges and three provinces in central Asia [2]. There are also currently over 330 captive individuals kept in 50 breeding facilities and zoos worldwide [3]. Fragmented habitat and a growing number of anthropogenic influences threaten the remaining free-range populations and restrict interaction of wild pandas between protected areas.

Having a clear understanding of the panda population in the wild and demographics within individual nature reserves is critical for establishing management policies and appropriate levels of protection. However, there are numerous challenges to obtaining surveys for giant pandas in the wild. First, pandas occupy extremely mountainous areas with steep terrain such that conducting traditional habitat transects is very difficult. Second, the few pandas left are scattered across a large geographic area and are by nature extremely secretive [4]. Additionally, it is difficult to determine the number of unique individuals using non-invasive methods without also applying more costly molecular genetic analysis [5],[6]. Although observing pandas in the wild is difficult, locating panda sign (e.g. feces) is much easier owing to the 10 to18 kg of bamboo they eat each day [4] resulting in numerous fecal boli passed. Work by Schaller [4] and Pan et al. [7] suggest that the bite size of undigested bamboo leaves or culm in feces may be specific to individuals.

Over the last decade, significant progress has been made in developing genetic markers that have provided more current information on population estimates in specific nature reserves [8], [9],[10]. Genetic analysis has provided a wealth of information on gene flow and diversity within and between fragmented populations [8],[9] and can also be used to identify numbers and sex of individuals [5],[10]. While genetic analysis is a useful tool, the amount of effort and expense for processing samples is quite high. However, there may be other technologies that can help support wild panda population surveys. For example, near infrared reflectance spectroscopy (NIRS) could be applicable to sex and age determination, and perhaps even pregnancy status or individual identification.

Near infrared spectroscopy is a non-invasive analytical technique that has been widely used in agriculture to predict forage characteristics of domestic [11], as well as wild herbivores [12],[13],[14],[15]. Specifically, NIRS has been used to discriminate between and predict the nutrient composition of bamboo plant parts and species, with the goal of developing a method of studying giant panda foraging ecology [16]. Since Foley et al. [17] reviewed applications of NIRS in ecological research, his group has taken the technique a step further and directly predicted forage intake by marsupials from near infrared (NIR) spectra of *Eucalyptus* foliage [18],[19]. Moreover, analysis of stomach contents in dugongs (*Dugong dugon*; [20]) and esophageal extrusa in livestock [21] provided accurate determinations of major dietary components and a logical progression from the direct analysis of forage, to the indirect analysis of feces as a means of determining herbivore dietary characteristics [22],[23],[24],[25].

Fecal near infrared spectroscopy (FNIRS) has also been used to differentiate between species, age, sex, and reproductive status in wild and domestic animals [26]. Here, we hypothesized that FNIRS can be applied to detect biological differences between the feces of male and female giant pandas, between age classes, and among females of different reproductive states. If successful, this will be the first step in developing a powerful non-invasive tool for wildlife population surveys and management that may be applicable to pandas as well as a variety of threatened and endangered species.

MATERIALS AND METHODS

Whole fecal samples were collected opportunistically across seasons for 2 years from eight adult (4 male; 4 female) and four juvenile (2 male; 2 female) giant pandas housed in the United States between 2006 and 2007 (Table 1). This included 2 adults at the Memphis Zoo, 2 adults and 1 juvenile at Zoo Atlanta, 2 adults and 2 juveniles at the San Diego Zoo, and 2 adults and 1 juvenile at the Smithsonian's National Zoo. Schaller et al. [4] defines a juvenile as <4 yrs old and in the process of being weaned, while an adult animal is >5 years old and is characterized by increased levels of testosterone for males or a demonstrated prominent estrus cycle for females. One of the adult females was pregnant during a portion of the study and gave birth to a live cub. Bamboo diets differed between institutions (Table 1), likely resulting in a range of fecal chemical composition. For instance, different bamboos fed to captive giant pandas in our study group included *Phyllostachys nuda*, *Pseudososa japonica* (Memphis), *Phyllostachys aurea*, *Bambusa oldhami*, *B. ventricosa* (San Diego), *Arundinaria gigantea*, *P. aureosulcata*, and *P. nigra henon* (Zoo Atlanta), and *P. bissetii* (National Zoo, Washington DC). Additionally, giant pandas seasonally select different bamboo parts consuming mostly culm (and shoots if available) in the spring and leaves throughout the summer, fall and winter [27].

Table 1. Bamboo species offered and mean daily consumption for captive giant pandas (*Ailuropoda melanoleuca*) housed in US zoos.

Institution	Giant panda			Mean daily consumption (%)		
Bamboo species fed[a]	ID No.	Age class	Sex	Biscuit	Produce	Bamboo
San Diego Zoo	371	Ad	F[b]	8.3	4.4	87.3
2–6	415	Ad	M	4.7	3.5	91.8
	596	juv	F	3.8	3.3	92.9
	563	juv	M	3.0	2.6	94.3
National Zoo	473	Ad	F	3.1	4.6	92.3
6, 8, 14	458	Ad	M	6.2	5.3	88.5
	595	juv	M	3.7	3.1	93.2
Memphis Zoo	507	Ad	F	2.1	1.6	96.3
5, 6, 8, 9, 12–14	466	Ad	M	1.5	2.2	96.3
Zoo Atlanta	452	Ad	F	3.4	3.5	93.1
1, 6–8, 10, 11, 13, 14	461	Ad	M	4.1	3.5	92.4
	649	juv	F	4.5	3.0	92.4

[a]Bamboo species fed: 1) *Arundinaria gigantea*, 2) *Bambusa oldhami*, 3) *B. ventricosa*, 4) *B. vulgaris vittata*, 5) *Phyllostachys aurea*, 6) *P. aureosulcata*, 7) *P. bambusoides*, 8) *P. bissetii*, 9) *P. glauca*, 10) *P. nigra* 'black', 11) *P. nigra* 'Henon', 12) *P. nuda*, 13) *P. rubromarginata*, 14) *Pseudosasa japonica*.
[b]Adult female pregnant during some period of the study, and which subsequently gave birth.
doi:10.1371/journal.pone.0038908.t001

https://doi.org/10.1371/journal.pone.0038908.t001

Zoo staff at the respective institutions collected giant panda fecal samples year-round during normal daily care activities. Samples were placed in clean plastic bags, sealed, and labeled with the panda ID, time, and date. Fecal samples were immediately frozen after collection and sent to the Grazingland Animal Nutrition Lab at Texas A&M University (College Station, TX) in sealed styrofoam coolers. For this initial study, we were not able to control for species of bamboo fed, nor were zoo staff asked to select for specific fecal boli representing different plant part consumption (i.e. bamboo leaf versus culm); instead samples were collected randomly. Thus, our original goal was to determine if physiological discriminant equations could be developed from all samples collected without accounting for diet preferences such as: (1) those that were artificially created by the keepers who selected the bamboo for feeding; (2) samples collected by the keepers with different quantities of plant parts or supplemental food items used during training; and (3) seasonal plant part selection by the bears themselves. However, our subsequent field experience with wild and captive pandas caused us to re-examine our original intent and to also analyze a large (n=105) subset of fecal samples that originated from diets consisting of predominately bamboo leaf.

Fecal samples were processed as previously described by Lyons and Stuth [22]. In brief, whole samples were dried in a forced-air oven at 60°C for 24 hours and subsequently blended and ground in an Udy Mill to pass uniformly through a 1 mm screen for greater homogeneity in particle size [28]. Prior to analysis, ground samples were re-dried at 60°C for at least 3 hours and put in a desiccator for one hour to stabilize sample temperature and moisture. Ground samples were then manually packed in sample cups with a quartz cover glass at a consistent level and compression. Initially, each sample was scanned using a Foss NIRS Systems 6500 Spectrometer (Foss North America, Eden Prairie, MN, USA) with spinning cup attachment. Measurements of reflectance were made over the visible and near infrared ranges (400 to 2500 nm). Samples were later returned to the Memphis Zoo and spectra (350 to 2500 nm) obtained again using an ASD FieldSpec3 portable Spectrophotometer (Analytical Spectral Devices, Boulder, CO, USA) for subsequent analysis.

Spectra were grouped by sex (Experiment 1), age class (Experiment 2), or female reproductive status (Experiment 3) for the appropriate discriminations. Pregnancy was determined by the respective zoo staff using a combination of behavioral observations, hormone level monitoring [29], confirmation by ultrasound [30],[31],[32],[33] and ultimately, birth of a cub. Retrospectively, there were two dietary factors that we hypothesized may impact our ability to predict sex, age, or reproductive status. The first factor was the species of bamboo being offered to the pandas at each institution but we did not achieve a testable distribution of this factor in our initial design. The second, and dominant factor, was the type of plant part consumed (leaf, culm/stalk, or shoot). We thus controlled for plant part in a subsequent analysis by selecting fecal samples derived from a bamboo leaf-dominated (>90%) diet. Such samples are easily determined visually or by spectroscopic analysis in the visible range (~400 to 100 nm; Vance, unpublished data). These "leaf only" samples were used to generate an additional male:female discriminant calibration (Experiment 4).

Discrimination between groups was accomplished for Experiments 1 through 3 using WinISI II v. 1.04a software, that utilizes the two-block partial least squares method [28]. This method predicts a set of indicator variables that are assigned to the calibration spectra as follows: samples associated with group A (e.g., male) are labeled {1, 2} in the algebraic matrix and conversely, samples associated with group B (e.g., female) are {2, 1}. "Unknown" samples predicted using the resulting calibration will be assigned to the higher predicted indicator variable of the pair. We used the software default criteria in that a predicted indicator value greater than 1.5 was required for a "correct" determination of sex, age class, or pregnancy status. For example, if the predicted value of sample X is {1.3, 1.7}, the sample is associated with group B. The strength of the group membership increases the closer the value of the respective indicator variable is to 2.0. Experiment 4 calibrations were generated using Grams version 9.1 from Thermo Fisher Scientific.

All discriminant equations were developed using log 1/reflectance (~400 to 2500 nm) spectra with a mathematical pre-treatment of a second order derivative and scatter correction. Calibrations to classify samples by sex were made using 118 fecal samples collected from adult males and 121 fecal samples from adult females. A similar discrimination was developed using 50 juvenile male samples and 43 samples from juvenile females. In order to weight age classes in the calibration equally, the age class discrimination was created using 100 adult samples randomly selected from the total of 239, versus 93 juvenile panda samples. A limited number of fecal samples were available from the pregnant female (n=8) for comparison with adult females that were not pregnant. To again weight the calibration equivalently, 10 samples from within the adult, not-pregnant female population were randomly selected for this comparison.

Prediction model validation was accomplished in stages. First, "leave one out" cross validation [34] was performed on an entire calibration set.

Secondly, we randomly selected and removed 25% of the calibration set, and then used the remaining 75% to create a new discriminant calibration with which to predict the respective validation samples. This calibration using 75% of each respective sample set will be referred to as a "reduced" calibration. Due to low sample size, cross validation on the full set was employed with the female reproductive status calibration and then 10 randomly selected samples from not-pregnant adult females, not used in the original calibration, were used as a validation set. Next, (on the male:female calibration set only) we performed a "round robin" series in which 3 zoos were used as the calibration set and the fourth zoo used as a validation set. In this exercise the percent of samples making up the validation set ranged from 7% in the case of San Diego Zoo to 67% in the case of Memphis Zoo, rather than an across the board 25%.

A final evaluation was performed in that discrimination was attempted between samples within a broad classification (i.e., sex or age class) that were randomly assigned to an arbitrary group irrespective of any biological significance. This random selection and subsequent group assignment is not to be confused with selection of random samples serving as validation sets for particular calibrations based on "biological" traits; or with similar steps taken to create equally weighted calibrations. The same validation steps (i.e. cross validation and 75% calibration versus 25% validation split) were applied to evaluate the effectiveness of calibrations based on these random "non-biological" groupings. Differences in proportion of fecal spectra that were correctly identified within groups (biological or arbitrary) as compared to a 50% chance of success were determined using Chi-square procedures [35].

RESULTS

Experiment 1: Male:Female discrimination

The original (i.e. all samples, not classified by plant part) adult male:female discrimination calibration was moderately successful with approximately 80% grouped correctly ($P<0.01$) for both sexes within the full calibration set (Table 2). Prediction success declined for both males and females after the removal of validation samples. Within the reduced calibration set, samples were 73% ($P<0.01$) and 65% ($P<0.05$) correctly predicted for females and males, respectively. Predicting the withheld samples using the reduced calibration equation resulted in 53% of the females ($P>0.1$) and 72% of the males ($P<0.05$) predicted correctly. When samples from each unique combination of 3 zoos were used to develop adult male:female calibrations, success rate averaged 72% for males and 80% for females ($P<0.01$; Table 3). Validation of these calibrations using the 4th respective zoo in a "round-robin" fashion, however, resulted in correct identifications of only ~40%.

Table 2. Application of near infrared spectroscopy of feces to discriminate between adult male and female captive giant pandas (*Ailuropoda melanoleuca*) housed in US zoos.

	Calibration[b]		Validation[c]	
Discriminant Model[a]	Correct M	Correct F	Correct M	Correct F
M vs. F 100%	78 (92/118)**	81 (98/121)**	NA	NA
M vs. F 75%	65 (58/89)*	73 (66/91)**	72 (21/29)†	53 (16/30)
	Correct A	Correct B	Correct A	Correct B
Random 100%	43 (51/120)	58 (69/119)	NA	NA
Random 75%	49 (44/90)	54 (48/89)	37 (11/30)	60 (18/30)

[a]Results are reported as: % correct identifications (number correct/group total). Within each discriminant model group, calibrations were developed using either the entire calibration set (100%), or with a reduced set (75%) after removing a randomly selected 25% of samples to be used as a validation set. A, B = random group.
[b]Refers to the prediction of group membership for each sample in the calibration set itself.
[c]Refers to the prediction of group membership for each sample in the validation set. NA = not applicable, i.e. there were no validation samples removed from the 100% calibration.
†Within a group, percentage of correct identifications differ from 50% ($P<0.1$).
*Within a group, percentage of correct identifications differ from 50% ($P<0.05$).
**Within a group, percentage of correct identifications differ from 50% ($P<0.01$).
doi:10.1371/journal.pone.0038908.t002

https://doi.org/10.1371/journal.pone.0038908.t002

Table 3. Effect of zoo on the ability of near infrared spectroscopy of feces to discriminate between adult male and female captive giant pandas (*Ailuropoda melanoleuca*).

	Calibration[b]		Validation[c]	
Discriminant Model[a]	Correct M	Correct F	Correct M	Correct F
Atlanta	78 (78/100)**	79 (78/99)**	28 (5/18)	32 (7/22)
National	74 (78/106)**	82 (89/109)**	25 (3/12)	67 (8/12)
San Diego	75 (82/110)**	81 (91/113)**	38 (3/8)	38 (3/8)
Memphis	63 (24/38)	81 (34/42)**	81 (65/80)**	33 (26/79)

[a]The discriminant model is labeled after the *validating* zoo, i.e. if the Atlanta Zoo samples are the validating set, then the calibration set is made up of samples from the remaining zoos.
[b, c]There are a total of 239 samples. The proportion of calibration to validation samples varies with each validation, i.e. Atlanta (83:17%), National (90:10%), San Diego (93:7%) and Memphis (33:67%).
**Within a group, percentage of correct identifications differ from 50% ($P<0.01$).
doi:10.1371/journal.pone.0038908.t003

https://doi.org/10.1371/journal.pone.0038908.t003

Male versus female was also discriminated within the juvenile age class (Table 4). In the full calibration, successful identifications were 98% for males (P<0.01) and 93% for females (P<0.01). Results from the reduced calibration were 92% and 91% for male and female samples respectively (P<0.01). Validation samples were correctly identified by applying the reduced calibration at 100% (male, (P<0.01) and 82% (female, (P<0.05).

Table 4. Application of near infrared spectroscopy of feces to discriminate between juvenile male and female captive giant pandas (*Ailuropoda melanoleuca*) housed in US zoos.

	Calibration[b]		Validation[c]	
Discriminant Model[a]	**Correct M**	**Correct F**	**Correct M**	**Correct F**
M vs. F 100%	98 (49/50)**	93 (40/43)**	NA	NA
M vs. F 75%	92 (35/38)**	91 (29/32)**	100 (12/12)**	82 (9/11)†
	Correct A	**Correct B**	**Correct A**	**Correct B**
Random 100%	51 (24/47)	50 (23/46)	NA	NA
Random 75%	51 (18/35)	65 (22/34)	42 (5/12)	50 (6/12)

[a]Results are reported as: % correct identifications (number correct/group total). Within each discriminant model group, calibrations were developed using either the entire calibration set (100%), or with a reduced set (75%) after removing a randomly selected 25% of samples to be used as a validation set. A, B = random group.
[b]Refers to the prediction of group membership for each sample in the calibration set itself.
[c]Refers to the prediction of group membership for each sample in the validation set. NA = not applicable, i.e. there were no validation samples removed from the 100% calibration.
†Within a group, percentage of correct identifications differ from 50% (P<0.1).
**Within a group, percentage of correct identifications differ from 50% (P<0.01).
doi:10.1371/journal.pone.0038908.t004

https://doi.org/10.1371/journal.pone.0038908.t004

Experiment 2: Age class discrimination

For age class discrimination, adult samples in the full calibration were predicted 85% correctly (P<0.01) while the juveniles were predicted at 89% (P<0.01), Table 5). Calibration success remained essentially the same for age class discrimination after the validation set removal with 88% of adult samples and 83% of juvenile samples correctly predicted (P<0.01) within this reduced calibration model. Predicting the withheld samples using the new calibration resulted in similar results, as 88% of the adults and 87% of the juvenile samples were classified correctly (P<0.01).

Table 5. Application of near infrared spectroscopy of feces to discriminate between adult and juvenile captive giant pandas (*Ailuropoda melanoleuca*) housed in US zoos

	Calibration[b]		Validation[c]	
Discriminant Model[a]	**Correct Ad**	**Correct juv**	**Correct Ad**	**Correct juv**
Ad vs. juv 100%	85 (85/100)**	89 (83/93)**	NA	NA
Ad vs. juv 75%	88 (66/75)**	83 (58/70)**	88 (22/25)**	87 (20/23)*
	Correct A	**Correct B**	**Correct A**	**Correct B**
Random 100%	57 (55/97)	35 (34/96)*	NA	NA
Random 75%	56 (41/73)	42 (30/72)	38 (9/24)	50 (12/24)

[a]Results are reported as: % correct identifications (number correct/group total). Within each discriminant model group, calibrations were developed using either the entire calibration set (100%), or with a reduced set (75%) after removing a randomly selected 25% of samples to be used as a validation set. A, B = random group.
[b]Refers to the prediction of group membership for each sample in the calibration set itself.
[c]Refers to the prediction of group membership for each sample in the validation set. NA = not applicable, i.e. there were no validation samples removed from the 100% calibration.
*Within a group, percentage of correct identifications differ from 50% ($P<0.05$).
**Within a group, percentage of correct identifications differ from 50% ($P<0.01$).
doi:10.1371/journal.pone.0038908.t005

https://doi.org/10.1371/journal.pone.0038908.t005

Experiment 3: Female reproductive status discrimination

Prediction results for the two reproductive categories within the full calibration set were 75% of pregnant ($P>0.1$) and 70% of not-pregnant ($P>0.1$) samples predicted correctly (Table 6). Prediction of the 10 validation samples from not-pregnant females was different ($P=0.06$) in that 8 were correctly identified as being not-pregnant while 2 were misidentified.

Table 6. Application of near infrared spectroscopy of feces to discriminate between pregnant and not pregnant female captive giant pandas (*Ailuropoda melanoleuca*) housed in US zoos.

	Calibration[b]		Validation[c]	
Discriminant Model[a]	**Correct Preg**	**Correct Not**	**Correct Preg**	**Correct Not**
Pregnant vs. Not	75 (6/8)	70 (7/10)	NA	80 (8/10)[†]
	Correct A	**Correct B**	**Correct A**	**Correct B**
Random	56 (5/9)	33 (3/9)	NA	NA

[a]Results are reported as: % correct identifications (number correct/group total). Preg = pregnant, not = not pregnant, A, B = random group.
[b]Refers to the prediction of group membership for each sample in the calibration set itself.
[c]Refers to the prediction of group membership for each sample in the validation set. NA = not applicable, i.e. there were no validation samples removed from the calibration, or in the case of pregnant validation samples, none were available.
[†]Percentage of correct versus incorrect identifications differ ($P<0.1$).
doi:10.1371/journal.pone.0038908.t006

https://doi.org/10.1371/journal.pone.0038908.t006

Experiment 4: Male:Female discrimination within "leaf only" calibration set

Using the "leaf only" calibrations, 93% of male and 95% of female samples were correctly identified ($P<0.01$) within the full calibration set (Table

7). Compared to Experiment 1, prediction success (93%) did not decline ($P<0.01$) after the removal of validation samples. Predicting the withheld samples using the reduced calibration equation resulted in 87% of the males ($P<0.01$) and 100% of the females ($P<0.01$) predicted correctly.

Table 7. The effect of using fecal samples derived from "leaf only" bamboo diets on the ability of near infrared spectroscopy to discriminate between adult male and female captive giant pandas (*Ailuropoda melanoleuca*).

	Calibration[b]		Validation[c]	
Discriminant Model[a]	**Correct M**	**Correct F**	**Correct M**	**Correct F**
M vs. F 100%	93 (62/67)**	95 (36/38)**	NA	NA
M vs. F 75%	93 (41/44)**	93 (27/29)**	87 (20/23)**	100 (9/9)**
	Correct A	**Correct B**	**Correct A**	**Correct B**
Random 100%	49 (26/53)	48 (25/52)	NA	NA
Random 75%	50 (20/40)	64 (25/39)	46 (6/13)	53 (7/13)

[a]Results are reported as: % correct identifications (number correct/group total). Within each discriminant model group, calibrations were developed using either the entire calibration set (100%), or with a reduced set (75%) after removing a randomly selected 25% of samples to be used as a validation set. A, B = random group.
[b]Refers to the prediction of group membership for each sample in the calibration set itself.
[c]Refers to the prediction of group membership for each sample in the validation set. NA = not applicable, i.e. there were no validation samples removed from the 100% calibration.
**Within a group, percentage of correct identifications differ from 50% ($P<0.01$).
doi:10.1371/journal.pone.0038908.t007

https://doi.org/10.1371/journal.pone.0038908.t007

Random Group Discrimination

Within the adult male and female samples, random calibration, using 120 samples assigned to Group A and 119 samples from Group B, was similar in success to the outcome of a coin toss (approximately 50%; ($P>0.1$) Table 2). The reduced random calibration samples were 49% and 54% correctly predicted ($P>0.1$) against the calibration equation for Groups A and B, respectively. Predicting the randomly withheld 25% with the 75% calibration equation resulted in 37% of Group A and 60% of Group B predicted correctly ($P>0.1$). Similar results were observed for the other "random" calibrations (Tables 3 to 7).

DISCUSSION

This study indicates that FNIRS could be applied to discriminate between different physiological groups of giant pandas, i.e. sex, age, and female reproductive status. Further validation of this technology should demonstrate its potential as a non-invasive method for wild panda management and population monitoring. However, as indicated by the "round robin" exercise, diet will clearly have an impact on our ability to predict physiological traits in the giant panda and future calibrations will need to take diet, and specifically plant part, into consideration. For example, while strong discriminations were generated for adult male: female when only leaf-based samples were used, calibration performance may have improved if bamboo species had been consistent. This scenario is unlikely for pandas in captive facilities; however, in the wild, giant pandas across certain mountain ranges (e.g. Qinling in Shaanxi Province) will all consume the same species of bamboo seasonally, which would facilitate calibration development.

Walker et al. [36] found that although spectral differences in domestic goat (*Capra hircus*) feces were affected by age, breed, and sex; the dominant factor was diet. The giant pandas in our study consumed a mix of bamboo leaf, culm, and to a lesser extent diet supplements (i.e. fruit or "biscuits"). Additionally, diets across zoos were from different bamboo species. Table 1 indicates that the animals from the National Zoo consumed a subset of the species consumed by those at Zoo Atlanta. In Figure 1, we present the average fecal spectra from adult pandas at these respective zoos, not accounting for diet selection. A comparison of difference spectra is illustrated in Figure 2. A discriminant equation developed using the samples from the Atlanta (n=47) and National (n=64) zoos correctly predicted 9 of 10 and 8 of 10 validation samples from each zoo respectively (P<0.1). Consistent with the findings of Walker et al. [36] the separation between fecal spectra collected at the two zoos (i.e. diet) is greater than that observed between the two sexes. In our "round-robin" exercise, successful identification of fecal samples ranged from 32 to 67% for females and from 25 to 81% for males. Although diets were from different bamboo species, fecal spectra were reasonably similar as indicated by Mahalanobis distance [37] values. These values for each individual zoo›s validation spectra were 2.4±0.2 (Atlanta), 1.9±0.1 (Memphis), 1.6±0.1 (National), and 1.7±0.2 (San Diego). Values for Mahalanobis distance greater than 3.0 are typically used to identify spectra as outliers from the calibration set [38]. Misidentified samples were largely those consisting of mostly bamboo culm rather than leaf, and or animals consuming higher proportions of diet supplements (author's personal observation). Predictions of male versus female were still moderately successful when all zoos/diets were combined in the original calibrations.

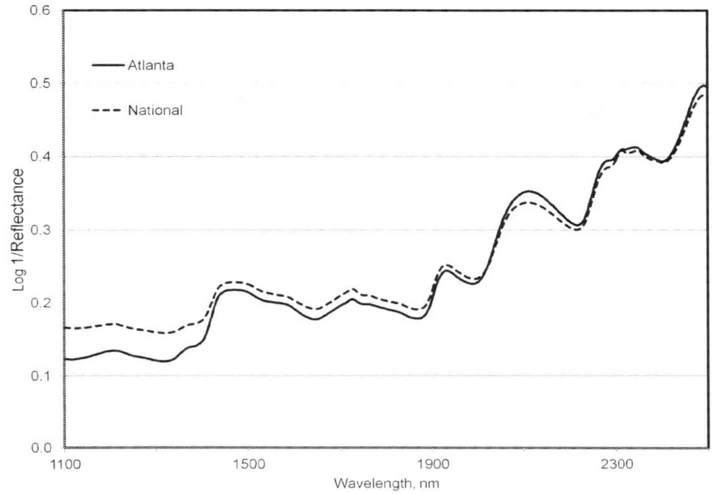

Figure 1. Average fecal near infrared spectra (log 1/reflectance, derivative=0, gap=0) from adult pandas at two different US zoos.

https://doi.org/10.1371/journal.pone.0038908.g001

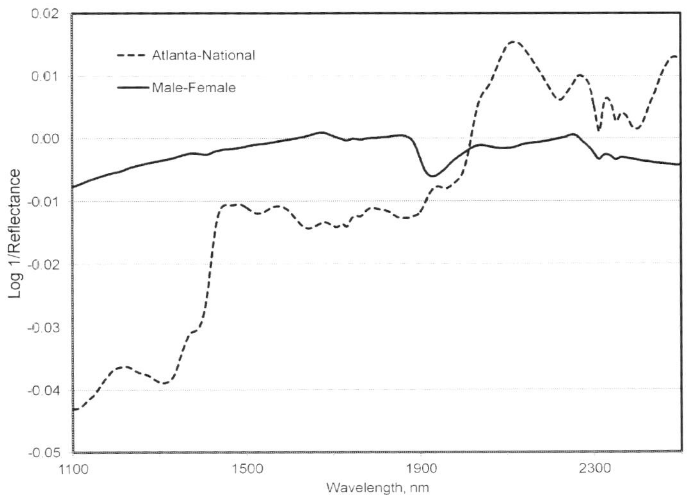

Figure 2. Fecal near infrared difference spectra (log 1/reflectance, derivative=0, gap=0) from adult M-F pandas versus all pandas from two zoos.

https://doi.org/10.1371/journal.pone.0038908.g002

Interestingly though, our subsequent experience with pandas in the wild combined with the previous discussion on the effects of diet on fecal NIR spectra, and more specifically the contribution made by plant part (leaf versus culm) leads us to conclude that: 1) bamboo plant part has a greater effect on fecal NIR spectra than bamboo species, 2) fecal samples in the wild can easily be identified visually as consisting of predominately leaf or culm, and 3) that using fecal samples derived from a leaf-dominated diet could result in greater discriminant ability via FNIRS, as compared to that observed with "mixed" samples. We therefore conducted the additional "leaf only" analysis to examine the potential effect of seasonal diet choices afforded pandas in the wild. The greater success rate for these latter discriminations as compared to the "mixed diet" calibrations indicates that application of NIRS as described here may prove to be more applicable in the wild than in captivity. Overall, this initial study on captive giant pandas demonstrated the potential for using FNIRS to distinguish between physiological classes in an ursid species.

Previous studies with captive cervids have shown that fecal spectra from males and females of the same species were significantly different. Our data indicate that we could determine giant panda sex on average 80% of the time for "whole" diets and greater than 90% of the time for leaf-based diets, which is similar to previous reports for pastured red deer (*Cervus elaphus*) and fallow deer (*Dama dama*) [39]. Interestingly, the calibrations for male:female determination in giant pandas are stronger than those reported for pen-fed white-tailed deer at approximately 70% (Osborn, personal communication). Application of FNIRS to discriminate males from females may not work in all herbivores as the technology has to date proven unsuccessful for the eastern grey kangaroo (*Macropus giganteus*; Foley and Billing, unpublished data). Percent correct determinations for male and female kangaroos were similar to those observed for random groupings in that study. Although male:female and reproductive status discrimination in livestock has been reported [26], research with cattle [22] found that there was no effect of physiological status of female cattle on FNIRS predictions of diet quality. Taken together, these conflicting results suggest there may be location and species-specific differences in the ability of FNIRS to discriminate between males and females.

For illustrative purposes, Figure 3 displays average FNIRS spectra from 3 of these aforementioned studies in comparison with the giant panda. Visual inspection of absorbance spectra for these species does not indicate obvious spectral differences between male and female except for those from fallow deer which also coincides with the greatest discriminant ability reported. When one then examines average difference spectra between male and female at 1100 to 2500 nm for fecal samples from these species; subtle departures become more evident (Figure 4). Again, the difference spectrum from fallow deer exhibits the greatest overall spectral separation and except for the water absorption region around 1900 nm, the kangaroo spectrum exhibits the least separation. Less intuitive is the observation that while the difference between male and female panda spectra appears to be less than that for white-tailed deer, we observed slightly better discriminant ability in the former. Mathematical range and variation, as well as sample number and the number of different individual animals sampled must of course be taken into consideration; therefore, a definitive determination of the ability of FNIRS to discriminate sex in various herbivores will require an experiment designed to incorporate these factors.

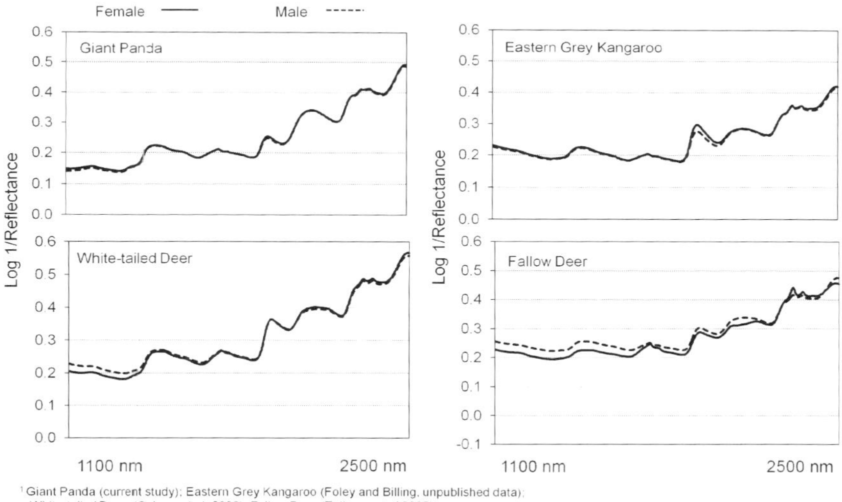

Figure 3. Average fecal near infrared spectra (log 1/reflectance, derivative=0, gap=0) from male and female animals of four different herbivore species.

https://doi.org/10.1371/journal.pone.0038908.g003

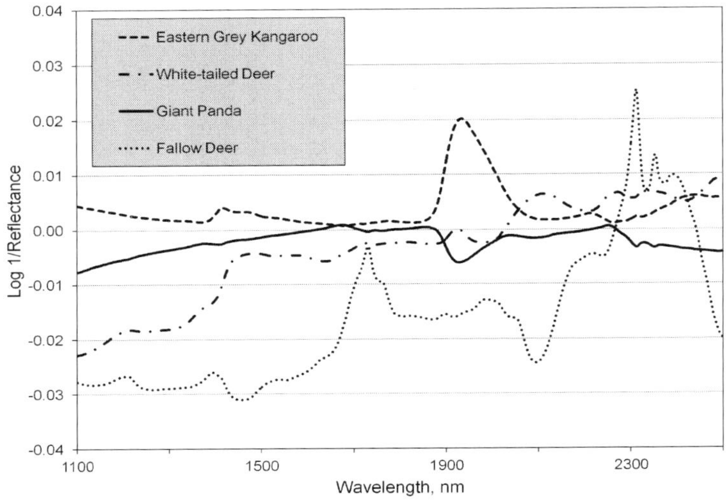

[1] Giant Panda (current study); Eastern Grey Kangaroo (Foley and Billing, unpublished data); White-tailed Deer, (Osborn et al. 2002); Fallow Deer (Tolleson et al 2005).

Figure 4. Fecal near infrared difference (male – female) spectra (log 1/reflectance, derivative=0, gap=0) from four different herbivore species.

https://doi.org/10.1371/journal.pone.0038908.g004

The average age class discrimination results were similar to the average values for males versus females on whole diets (not separated by plant part) and indicate that FNIRS can successfully differentiate between juvenile and adult giant pandas. It is unclear why we observed greater discrimination for juvenile animals compared to adult animals in our study. Reasons for the discrepancy in male:female predictions between adult and juvenile animals may be related to the number of juveniles available for the study compared to adult animals (4 versus 8, respectively) as the sex ratios were identical. As more animals are added to the calibration library for juveniles we will have a better understanding of whether FNIRS can also be used to identify sex in younger animals.

Both adults and juveniles discriminated with a reasonably high prediction rate and this is most likely due to physical rather than chemical differences in the samples, i.e. chewing and processing of bamboo. In their comparison of red and fallow deer, Tolleson et al. [39] proposed that differences in fecal spectra between sexes might be due to such factors as diet selection, retention time, or dentition. An additional explanation of spectral differences between juvenile and adult feces may be hormonal differences resulting from a lack of sexual maturity reached in the juveniles. Pandas do not reach sexual maturity until they are at least 4.5 years old [4]. Hormone levels, in particular steroids, and associated reproductive behaviors displayed by breeding adults that exhibit a normal estrus are different from juvenile pandas [30], [40]. These hormonal differences resulting from sexual maturity, or lack thereof, may be reflected in spectral differences between the fecal samples of various age classes. We should be clear to state that we do not ascribe potential differences in fecal spectra to the ability to measure hormones in minute amounts via FNIRS but rather that the presence of these hormones are likely exerting an effect on ingestive behavior, metabolism, absorption, etc [41], that could be manifest in fecal chemistry and thus, fecal spectra.

The results for our pregnant versus not-pregnant discrimination were similar to our average prediction rates for both sex and age-class. Both pregnant and not-pregnant states were predicted equally well in the full calibration set. Although certainly encouraging, the success of these particular discriminations should be viewed with caution considering our low sample size and number of individuals. Prediction of a validation set drawn from the population of samples collected from not-pregnant females was successful (80%). There are of course a limited number of pregnant pandas in the U.S. each year to obtain samples from. Historically, detection of pregnancy in female pandas has not been possible until the last three weeks prior to birth (using ultrasound) primarily due to female pandas experiencing embryonic diapause in addition to pseudo-pregnancy, even if not bred, which closely resemble the hormone profiles of true pregnancies [31],[32],[33],[40]. Early pregnancy detection would be an extremely valuable tool for captive management as well as monitoring females in the wild. More research is needed to track spectral changes of feces throughout the full reproductive cycle and pregnancy to determine whether we will be able to apply this technique in the field.

When sample libraries were reorganized using random number generators to confirm our categorical spectral separations and calibrations, the percentages for each discrimination were approximately 50%, or the same as a coin toss. This random calibration analyses lends additional strength to our findings that we are in fact identifying specific spectral differences in the physiological parameters we defined for giant panda feces, i.e. the results are due to biology rather than math. Here we demonstrate for the first time that FNIRS is a potential tool for giant panda management. Current calibration equations are robust enough for field testing although steps need to be taken to account for seasonal diet influences and plant part preference by the pandas [27]. An important next step is to acquire fecal samples from wild and captive pandas in China. Research to evaluate the effectiveness of FNIRS to monitor nutrition and physiology of free-ranging ranging pandas is underway. If successful, all three of these discriminations (sex, age, and pregnancy status) will be extremely valuable for field research and assessing wild population demographics. Being able to non-invasively determine the pregnancy status of a female giant panda also has an important implication for captive populations. By having a more accurate idea of the current demographics of wild panda populations, reserve managers can more effectively tailor their conservation plans for habitat management.

ACKNOWLEDGMENTS

We thank the zoo staff at the Memphis Zoo, San Diego Zoo, Smithsonian's National Zoo, and Zoo Atlanta that collected fecal samples that were integral to the study. We thank the USDA/ARS Biophotonics Initiative (No. 58-6402-3-0120) for providing Dr. Carrie Vance's time on the project. F. Falcone at the Memphis Zoo and L. Gerber with the University of Arizona provided assistance with the final preparation of the manuscript. We would also like to thank the Grazingland Animal Nutrition Lab at Texas A&M University for the use of their near infrared reflectance spectroscopy hardware and software for the duration of the research.

Author Contributions

Conceived and designed the experiments: EW AK RH JS DT. Performed the experiments: EW AK RH CV. Analyzed the data: EW DT CV. Wrote the paper: EW AK DT CV.

REFERENCES

1. IUCN (2010) IUCN Red List of Threatened Species. Available: http://www. iucnredlist.org. Accessed 2011 Jan 22.
2. State Forestry Administration of China. (2006) The Third National Giant Panda Census in China. Beijing: Science Press.
3. Breeding Strategy Recommendations and Summary of the Status of the Giant Panda Captive Population November (2010) Chengdu, China.
4. Schaller GB, Hu J, Pan W, Zhu J (1985) The giant pandas of Wolong. Chicago: University of Chicago Press. 298 p.
5. Zhan X, Li M, Zhang Z, Goossens B, Chen Y, et al. (2006) Molecular censusing doubles giant panda population estimate in a key nature reserve. Curr Biol 16: R451–R452.
6. Zhu L, Zhan X, Wu H, Zhang S, Meng T, et al. (2010) Conservation implications of drastic reductions in the smallest and most isolated populations of giant pandas (Report). Conserv Biol 24:1299–1306.
7. Pan W, Wang LYD, Wang H, Lu Z, Zhu X (2004) Future survival of giant pandas in the Qinling Mountains of China. In: Lindburg DG, Baragona K, editors. Giant pandas: Biology and conservation. Berkeley University of California Press, pp. 81–87.
8. He W, Lin L, Shen F, Zhang W, Zhang Z, et al. (2008) Genetic diversities of the giant panda (Ailuropoda melanoleuca) in Wanglang and Baoxing Nature Reserves. Conserv Genet 9:1541–1546.
9. Lifeng Z, Xiangjiang Z, Tao M, Shanning Z, Fuwen W (2010) Landscape features influence gene flow as measured by cost-distance and genetic analyses: a case study for giant pandas in the Daxiangling and Xiaoxiangling mountains.
10. BMC Genet 11:72–82.
11. Hu Y, Zhan X, Dunwu Q, Fuwen W (2010) Spatial genetic structure and dispersal of giant pandas on a mountain-range scale. Conserv Genet 11:2145–2155.
12. Stuth J, Jama A, Tolleson D (2003) Direct and indirect means of predicting forage quality through near infrared reflectance spectroscopy. Field Crops Res 84:45–56.
13. Lawler IR, Aragones L, Berding N, Marsh H, Foley W (2006) Near-

infrared reflectance spectroscopy is a rapid, cost-effective predictor of seagrass nutrients. J Chem Ecol 32:1353–1365.

14. Moore BD, Foley WJ (2005) Tree use by koalas in a chemically complex landscape. Nature 435:488–490.
15. Woolnough AP, du Toit JT (2001) Vertical zonation of browse quality in tree canopies exposed to a size-structured guild of African browsing ungulates. Oecologia 129:585–590.
16. Woolnough AP, Foley WJ (2002) Rapid evaluation of pasture quality for a critically endangered mammal, the northern hairy-nosed wombat (Lasiorhinus krefftii). Wildlife Res 29:91–100.
17. Wiedower E, Hansen R, Bissell H, Ouellette R, Kouba A, et al. (2009) Use of near infrared spectroscopy to discriminate between and predict the nutrient composition of different species and parts of bamboo: application for studying giant panda foraging ecology. J Near Infrared Spec 17:265–273.
18. Foley WJ, McIlwee A, Lawler I, Aragones L, Woolnough AP, et al. (1998) Ecological applications of near infrared reflectance spectroscopy – a tool for rapid, cost-effective prediction of the composition of plant and animal tissues and aspects of animal performance. Oecologia 116:293–305.
19. McIlwee AM, Lawler IR, Cork SJ, Foley WJ (2001) Coping with chemical complexity in mammal-plant interactions: near-infrared spectroscopy as a predictor of eucalyptus foliar nutrients and of the feeding rates of folivorous marsupials. Oecologia 128:539–548.
20. 19. Wallis IR, Foley WJ (2003) Validation of near-infrared reflectance spectroscopy to estimate the potential intake of eucalyptus foliage by folivorous marsupials. Aust J Zool 51:95–98.
21. Andre´ J, Lawler IR (2003) Near infrared spectroscopy as a rapid and inexpensive means of dietary analysis for a marine herbivore, dugong (Dugong dugon). Mar Ecol Prog Ser Vol.257: 259–266.
22. Volesky JD, Coleman SW (1996) Estimation of botanical composition of esophageal extrusa samples using near infrared reflectance spectroscopy. J Range Manag 49:163–166.
23. Lyons RK, Stuth JW (1992) Fecal NIRS equations for predicting diet quality of free-ranging cattle. J Range Manag 45:238–244.
24. Walker JW, Clark DH, McCoy SD (1998) Fecal NIRS for Predicting

Percent Leafy Spurge in Diets. J Range Manag 51:450–455.

25. Dixon R, Coates D (2009) Review: Near infrared spectroscopy of faeces to evaluate the nutrition and physiology of herbivores. J Near Infrared Spec 17:1–31.
26. Foley WJ (2009) Near infrared reflectance spectroscopy in ecological studies of plant-animal interactions. Spectrosc Europe 21:6–9.
27. Tolleson DR (2010) Fecal NIRS: What else, what next? In: Walker J, Tolleson D, editors. Shining light on manure improves livestock and land management. pp. 82–95.
28. Hansen RL, Carr MM, Apanavicius CJ, Jiang P, Bissell HA, et al. (2010) Seasonal shifts in giant panda feeding behavior: relationships to bamboo plant part consumption. Zoo Biol 29:470–483.
29. 28. Martens H, Martens M (2001) Multivariate analysis of quality: An introduction. John Wiley & Sons Ltd., Chichester; New York. 445 p.
30. Willis EL, Kersey DC, Durrant BS, Kouba AJ (2011) The Acute Phase Protein Ceruloplasmin as a Non-Invasive Marker of Pseudopregnancy, Pregnancy, and Pregnancy Loss in the Giant Panda. PLoS One 6: e21159.
31. Swaisgood RR, Zhou X, Zhang G, Lindburg DG, Zshang H (2003) Application of behavioral knowledge to conservation in the giant panda. International J Comp Psychol 16:65–84.
32. Kersey DC, Wildt DE, Brown JL, Snyder RJ, Huang Y, et al. (2010a) Unique . biphasic progestagen profile in parturient and non-parturient giant pandas (Ailuropoda melanoleuca) as determined by faecal hormone monitoring. Reproduction 140:183–193.
33. Kersey DC, Wildt DE, Brown JL, Snyder RJ, Huang Y, et al. (2010b) Endocrine milieu of perioestrus in the giant panda (Ailuropoda melanoleuca), as determined by non-invasive hormone measures. Reproduction Fert Develop 22:901–912.
34. Sutherland-Smith MP, Morris J, Silverman S (2004) Pregnancy detection and fetal monitoring via ultrasound in a giant panda (Ailuropoda melanoleuca). Zoo Biol 23:449–461.
35. Stone M (1974) Cross-validatory choice and assessment of statistical predictions (with discussion). J Roy Statistical Society: Ser B 36:111–147.

36. Steel RGD, Torrie JH (1980) Principles and Procedures of Statistics. McGrawHill Book Co., New York.
37. Walker JW, Campbell ES, Lupton CJ, Taylor Jr CA, Waldron DF, et al. (2007) Effects of breed, sex, and age on the variation and ability of fecal near-infrared reflectance spectra to predict the composition of goat diets. J Anim Sci 85:518–526.
38. Mahalanobis PC (1930) On tests and measures of groups divergence I. Journal of the Asiatic Society of Bengal 26:541.
39. Shenk JS, Westerhaus MO (1991) Population definition, sample selection, and calibration procedures for near infrared reflectance spectroscopy. Crop Sci. 31:469–474.
40. Tolleson DR, Randel RD, Stuth JW, Neuendorff DA (2005) Determination of sex and species in red and fallow deer by near infrared reflectance spectroscopy of the faeces. Small Ruminant Res 57:141–150.
41. Mainka SA, Cooper RM, Mao L, Guanlu Z (1990) Urinary Hormones in two juvenile female giant pandas (Ailuropoda melanoleuca). J Zoo Wildlife Med 21:334– 341.
42. Black HE (1988) The effects of steroids upon the gastrointestinal tract. Toxicol Pathol 16:213–223.

CHAPTER 17

Physiological Stress and Refuge Behavior by African Elephants

David S. Jachowski[1], Rob Slotow[2], and Joshua J. Millspaugh[1]

[1]Department of Fisheries and Wildlife Sciences, University of Missouri, Columbia, Missouri, United States of America,
[2]Amarula Elephant Research Programme, School of Life Sciences, University of Kwa-Zulu-Natal, Durban, South Africa

ABSTRACT

Physiological stress responses allow individuals to adapt to changes in their status or surroundings, but chronic exposure to stressors could have detrimental effects. Increased stress hormone secretion leads to short-term escape behavior; however, no studies have assessed the potential of longer-term escape behavior, when individuals are in a chronic physiological state.

Citation: Jachowski DS, Slotow R, Millspaugh JJ (2012) Physiological Stress and Refuge Behavior by African Elephants. PLoS ONE 7(2): e31818. https://doi.org/10.1371/journal.pone.0031818

Such refuge behavior is likely to take two forms, where an individual or population restricts its space use patterns spatially (spatial refuge hypothesis), or alters its use of space temporally (temporal refuge hypothesis). We tested the spatial and temporal refuge hypotheses by comparing space use patterns among three African elephant populations maintaining different fecal glucocorticoid metabolite (FGM) concentrations. In support of the spatial refuge hypothesis, the elephant population that maintained elevated FGM concentrations (iSimangaliso) used 20% less of its reserve than did an elephant population with lower FGM concentrations (Pilanesberg) in a reserve of similar size, and 43% less than elephants in the smaller Phinda reserve. We found mixed support for the temporal refuge hypothesis; home range sizes in the iSimangaliso population did not differ by day compared to nighttime, but elephants used areas within their home ranges differently between day and night. Elephants in all three reserves generally selected forest and woodland habitats over grasslands, but elephants in iSimangaliso selected exotic forest plantations over native habitat types. Our findings suggest that chronic stress is associated with restricted space use and altered habitat preferences that resemble a facultative refuge behavioral response. Elephants can maintain elevated FGM levels for ≥6 years following translocation, during which they exhibit refuge behavior that is likely a result of human disturbance and habitat conditions. Wildlife managers planning to translocate animals, or to initiate other management activities that could result in chronic stress responses, should consider the potential for, and consequences of, refuge behavior.

INTRODUCTION

In responding to real or perceived threats, vertebrates initiate a physiological stress response that has broad implications if stress levels are maintained at a high level (i.e., chronic) [1]. The production of stress hormones is a key physiological step in balancing the expenditure of energy, and facilitates the ability of an individual to survive exposure to a stressor [2], [3]. While this response is effective in the presence of short-term stressors, chronic levels of stress can result in various pathological dysfunctions, including an increase in blood glucose, or the inhibition of reproduction, immune function, or growth [1], [4]. Therefore, while short-term releases of stress hormones help a vertebrate adapt to its surroundings, over extended periods of time, chronic release of hormones should be minimized to reduce deleterious effects [2].

Vertebrates limit chronic exposure to stressors through three kinds of facultative behavioral responses [5]: (1) the individual exhibits escape behavior away from the perturbation; (2) the individual remains in the area, but identifies and uses a refuge to avoid the perturbation; and (3) the individual identifies and uses a refuge, but will move outside the refuge during periods of non-disturbance. Many studies have focused on short-term escape behavior away from disturbances [5], [6]. The latter two kinds of responses have received considerably less attention. Previous studies have suggested that use of refugia typically is temporary, and that normal space use continues once the disturbance passes [5], [7]. However, to our knowledge, there has been no research to evaluate if long-term use of refugia is likely to occur if the animal does not adjust to the source of perturbation, and maintains a chronic physiological state.

Descriptions of wildlife use of "refuges" or "refugia" are increasingly widespread in ecology and conservation biology. In the ecological literature, refugia frequently are defined by fine-scale spatial responses of animals to perturbations [8], [9], [10]. While particular behaviors and space use patterns have been reported as refuge behavior, little is known about the facultative process behind those observations. Initiation of refuge behavior is an active process involving an external cue (i.e. the stressor), internal physiological response, and active movement and selection of refugia [11]. The extent to which physiological state influences the timing and duration of refuge behavior is poorly understood, despite its potential importance in predicting when, where, to what extent, and for how long refuge behavior will occur.

The refuge behavior of African elephants (*Loxodonta africana*) is relatively well documented through long-term behavioral studies. Elephants are long-lived with high cognitive ability for spatial memory [12] that allows them to adapt space use patterns based on the location of resources [13], boundaries [14], or past experiences [15]. Behavioral observations suggest that elephants exhibit at least two facultative behavioral responses indicative of spatial and temporal refuge behavior. Firstly, humans have restricted elephant movements, and fragmented habitat, through the creation of real (e.g., electric fences) or perceived (e.g., human land use and disturbance) boundaries [16]. In response, elephants have restricted space use patterns and have identified, used, and rarely occurred outside of protected areas or refugia [17]. Secondly, in addition to restricting movements spatially, space use can be modified temporally to avoid areas during periods of disturbance [18], [19]. This pattern of spatio-temporal refuge behavior allows elephants to reoccupy habitats when humans are absent [20], [21].

In South Africa, where elephants are being reintroduced to relatively small fenced reserves, there is a particular need to consider the potential for refuge behavior. Elephants have been translocated for reintroduction into over 58 reserves in South Africa [22]. The process of translocation is well established and designed to be as unobtrusive to the animals as possible [23], but still results in an elevated physiological stress response for up to 30 days post-release [24], [25]. However, little is known about the potential for longer-term stress responses in elephants following translocation [26], despite the need to understand how they habituate to their new surroundings, and if they exhibit aberrant behavior that poses a risk to elephants, other animals and people [23]. To facilitate acclimatization, it has been suggested that managers provide "refuge areas" to allow translocated elephants freedom from harassment [27]. Thus, there is interest in identifying when and where refuge behavior occurs, to mitigate potential human-elephant conflict.

In this study, we evaluated spatial and temporal hypotheses of refuge behavior in elephants by comparing space use patterns among three restored elephant populations. These populations maintained different levels of physiological stress, including one with chronic levels. Under the spatial refuge hypothesis, where individuals restrict space use when stress hormone levels are elevated, we expected elephant populations that were chronically stressed to avoid disturbance by exhibiting restricted space use patterns. Therefore, we examined two metrics: home range size, and the proportion of the area used by elephant family groups in each reserve. Under the temporal refuge hypothesis, where individuals temporally alter their use of space when stress hormone levels are elevated, we expected elephant family groups in a state of chronic stress to restrict their use of space during the day, when human disturbance existed, and increase their use of space at night, when disturbance ceased. We tested support for the temporal refuge hypothesis by evaluating whether elephant family group home range sizes were smaller during the day than at night, whether family groups used the same areas during the day and night, and whether seasonal resource selection differed between night and day. By comparing these metrics across elephant populations in different physiological states, we were able to link stress with refuge behavior.

RESULTS

From 2000 to 2006, we collected and assayed 709 fecal samples from elephant populations in the three reserves included in this study (Phinda Private

Game Reserve n=195; iSimangaliso Wetland Park n=366; Pilanesberg National Park n=148). Fecal glucocorticoid metabolite concentrations were significantly higher for elephants in iSimangaliso than for elephants in the other two reserves ($F_{2, 708}$=80.17, P<0.0001) (Figure 1). Elephants in iSimangaliso consistently had FGM concentrations around 50 ng/g, indicative of a chronic stress response [25], [28]. In comparison, elephants in Phinda and Pilanesberg had relatively moderate FGM concentrations (25–35 ng/g), typical of baseline levels in elephants [25], [28] (Figure 1). Across all reserves, FGM values were 20% higher in the dry season than in the wet season ($F_{1, 705}$=23.20, P<0.0001). We observed differences in FGM levels among years ($F_{5, 700}$=2.79, P=0.0167). However, annual differences primarily occurred in FGM concentrations of elephants in Phinda; FGM levels of elephants in iSimangaliso were consistently elevated across all years (Figure 1).

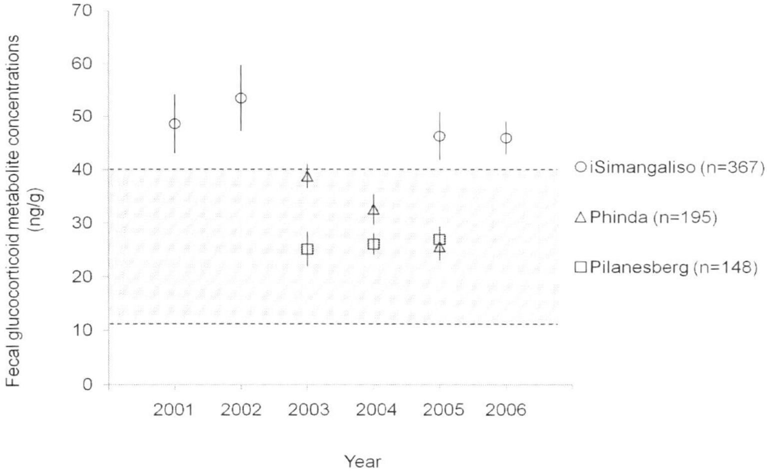

Figure 1. Fecal glucocorticoid metabolite values of elephants in each reserve.

Average (with 95% confidence intervals) fecal glucocorticoid metabolite (FGM) concentrations (in dry weight ng/g) for each year samples were collected. Basal FGM concentrations for elephants (15–40 ng/g) are shaded grey.

https://doi.org/10.1371/journal.pone.0031818.g001

In support of the spatial refuge hypothesis, from 2004 to 2007 elephants in iSimangaliso maintained smaller home ranges and used a smaller proportion of the reserve compared to elephants in the other two populations.

Despite iSimangaliso being slightly larger (602 km^2) than Pilanesberg (560 km^2) (Figure 2), elephant home range size was twice as large in Pilanesberg than in iSimangaliso ($F_{2, 52}$=48.45, P<0.0001). Within all reserves, home range size was consistent across years ($F_{4, 52}$=1.66, P=0.1744), but on average 65 km^2 larger during the wet as opposed to the dry season ($F_{1, 52}$=18.47, P<0.0001). When scaled in proportion to the total area available within the reserve, elephant home ranges in iSimangaliso occupied 20% less of the available area ($\bar{x}$=0.35, SE=0.04, range=0.13–0.56), than in the similarly-sized Pilanesberg ($\bar{x}$=0.55, SE=0.03, range=0.17–0.74), and 43% less than in the smaller Phinda (180 km^2) ($\bar{x}$=0.78, SE=0.02, range=0.63–0.98) (Figure 2). Elephants utilized more of the available area during each season in the relatively small Phinda reserve than in the other two reserves ($F_{2, 52}$=49.29, P<0.0001) (Figure 3). Similar to home range size, scaled home ranges were consistent across years ($F_{4, 52}$=2.02, P=0.1059), but on average elephants utilized 9% more of the reserve during the wet as opposed to the dry season ($F_{1, 52}$=21.14, P<0.0001).

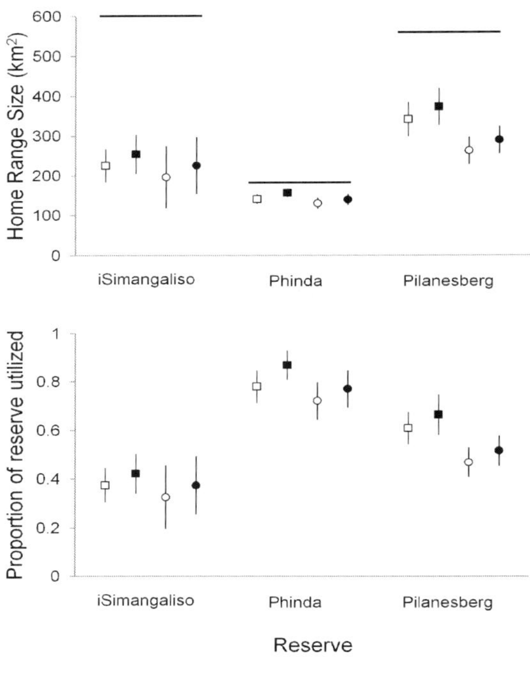

Figure 2. Home range size of elephants in each reserve.

Average (with 95% confidence interval) home range size (km^2) during the wet (squares) and dry (circles) seasons (top graph). Horizontal lines indicate the size of each reserve. The bottom graph depicts the average (with 95% confidence interval) proportion of each reserve occupied by elephant home ranges. Solid symbols represent mean average home range sized based on utilization distributions (UDs) calculated from nighttime locations and hollow symbols represent UDs calculated from daytime locations.

https://doi.org/10.1371/journal.pone.0031818.g002

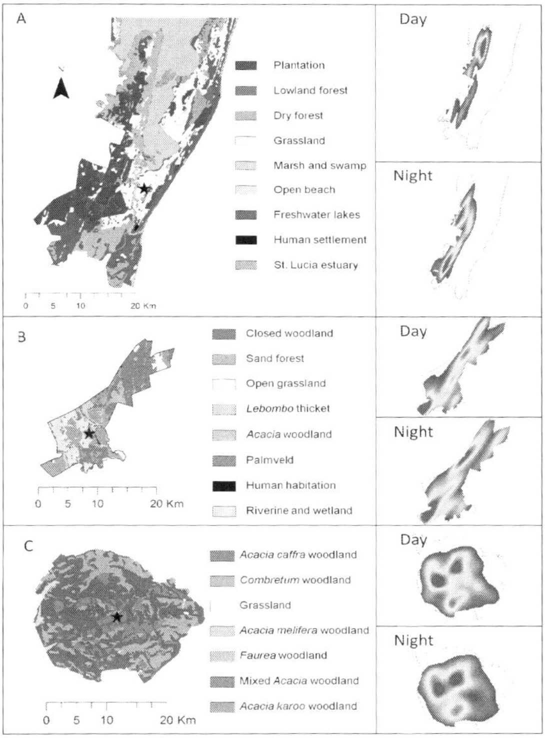

Figure 3. Space use patterns by elephants in each reserve.

The distribution of habitat types within iSimangaliso Wetland Park (A), Phinda Private Game Reserve (B), and Pilanesberg National Park (C). The star within each reserve represents the location of the boma (or preconditioning enclosure) that was also the initial release site of elephants. Inset on the right are 95%fixed kernel seasonal utilization distributions (UDs) for a select adult female elephant in each of our study areas based on daytime (top) and nighttime (bottom) locations. Areas in red within the UD represent areas of high intensity use, which fade to blue in areas of low use, and reserve boundaries are demarcated by solid lines. Space use was restricted and differed between day and night at iSimangaliso Wetland Park (A), compared to Phinda Private Game Reserve (B) and Pilanesberg National Park (C).

https://doi.org/10.1371/journal.pone.0031818.g003

We found mixed support for our temporal refuge hypothesis. We found no difference between day and night home range sizes of elephants within any reserve (iSimangaliso, $F_{6, 16}$=0.20, P=0.9706; Pilanesberg, $F_{12, 48}$=0.27, P=0.9921; Phinda, $F_{10, 18}$=0.39, P=0.9324) (Figure 2). However, across reserves, we observed significantly less day vs. night space use overlap in iSimangaliso compared to Pilanesberg and Phinda (Figures 3 and 4), at both the home range ($F_{2, 47}$=7.52, P=0.0015) and core area ($F_{2, 46}$=8.26, P=0.0009) scales. In iSimangaliso, we observed 66.6% overlap in daytime and nighttime space use at the home range scale and 55% space use overlap at the core area scale (Figure 4). By contrast, we observed 7–10% more overlap in daytime and nighttime space use in Pilanesberg and Phinda at the home range scale, and 8–10% more at the core area scale. The amount of day-night space use overlap did not differ by season (home range, $F_{1, 47}$=0.42, P=0.5225; core area, $F_{1, 46}$=0.04, P=0.8346) or year of investigation (home range, $F_{1, 47}$=1.74, P=0.1573; core area, $F_{1, 46}$=1.32, P=0.2751).

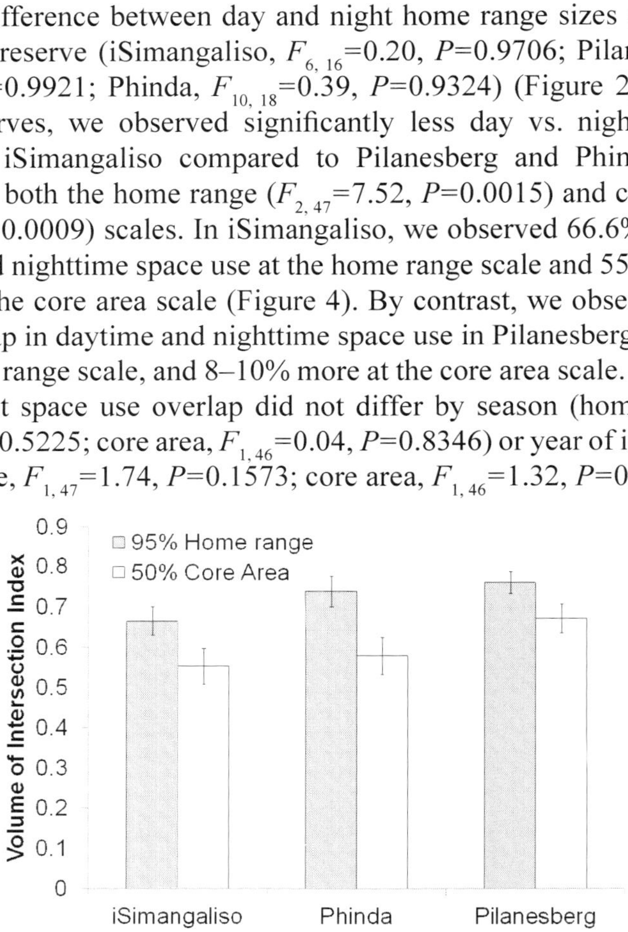

Figure 4. Day vs. night space use overlap by elephants in each reserve.

Mean (with 95% confidence interval) volume of intersection index scores for elephant based on comparisons between day and night home range (grey) and core area (white) space use patterns. Volume of intersection index statistic measures the amount of overlap between two utilization distributions. Index values range from 0 to 1, where higher scores indicate a higher degree of overlap.

https://doi.org/10.1371/journal.pone.0031818.g004

In terms of resource selection patterns, in iSimangaliso, with the exception of dry forest, elephants selected forest plantation over all other habitat types. This pattern was consistent across seasons (Pillai's Trace=1.3258, $F_{21, 30}$=1.13, P=0.3713) and time of day (Pillai's Trace=1.4295, $F_{42, 78}$=0.58, P=0.9718), suggesting that elephants generally tended to select forest plantation in favor of most native habitat types regardless of time of day or their relative availability (Figure 5). In contrast, in Phinda and Pilanesberg where tree plantations were not present, elephants exhibited differing seasonal resource selection patterns that favored native forest habitats. In Phinda, we observed seasonal differences in resource selection (Pillai's Trace=2.0965, $F_{35, 80}$=1.65, P=0.0338), where elephants selected sand forest and closed woodland over all other habitat types during in the dry season, and selected *Acacia* woodland in the wet season. Similar to iSimangaliso, we did not observe differences in resource selection between day and night (Pillai's Trace=2.0032, $F_{70, 126}$=0.72, P=0.9328), and resource use did not consistently correspond with the relative availability of habitats within the reserve (Figure 5). In Pilanesberg, resource selection differed between seasons (Pillai's Trace=1.0712, $F_{36, 276}$=1.67, P=0.0128), but was consistent between day and night (Pillai's Trace=1.0270, $F_{72, 276}$=0.79, P=0.8812), similar to iSimangaliso and Phinda. Elephants in Pilanesberg tended to select *Combretum*, *Faurea*, and *Acacia caffra* woodland over other habitat types during both the wet and dry seasons, but varied in their selection of grassland and mixed *Acacia* woodland among seasons. Furthermore, in contrast to iSimangaliso and Phinda, resource selection more closely followed the relatively availability of habitats (Figure 5). Overall, despite the failure to observe temporal day vs. night differences in resource selection in iSimangaliso that would provide support for our temporal refuge hypothesis, the differences in resource selection patterns we observed among reserves provides further support for our spatial refuge hypothesis. In particular, selection of forest plantation and dry forest in favor of available habitat in iSimangaliso, regardless of season, suggests that restricted space use patterns are related to the avoidance of a particular area rather than to the availability of suitable habitat.

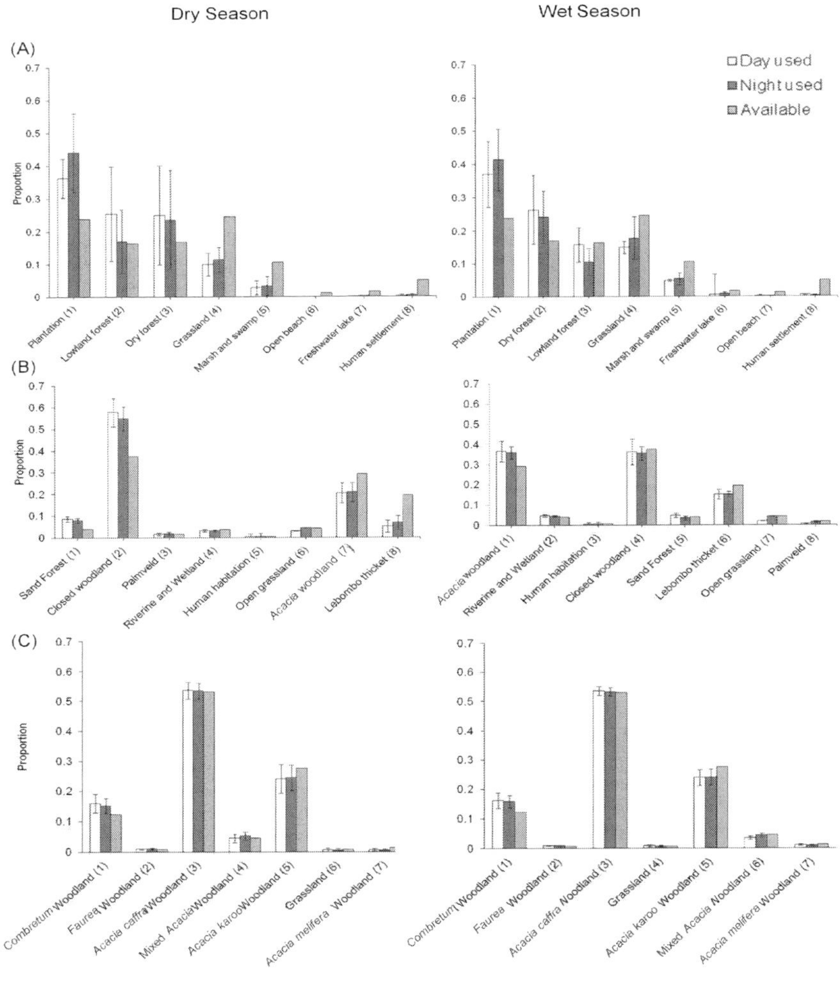

Figure 5. Compositional analysis of habitat use by elephants between day and night among reserves.

Mean (with 95% confidence intervals) weighted day and night time use (calculated by summing UD fixed kernel scores by habitat type), compared to availability of habitat types at the reserve level. Habitat names are followed by their compositional analysis rank (Aebischer et al. 1993). Graphs are separated by dry (left column) and wet (right column) season as well as by reserve in rows: iSimangaliso Wetland Park (A), Phinda Private Game Reserve (B), and Pilanesberg National Park (C).

https://doi.org/10.1371/journal.pone.0031818.g005

DISCUSSION

Our study suggests that chronic stress hormone concentrations are associated with restricted space use and altered habitat preferences that resemble a facultative refuge behavioral response. The elephant population in iSimangaliso displayed FGMs indicative of chronic stress and used a smaller portion of this reserve throughout the year. These results contrast with findings for other translocated populations with lower FGM concentrations, and other wild elephant populations [13], [29], [30], [31]. Restricted space use patterns indicative of refuge behavior have been documented for a variety of species, but few previous studies have linked the internal physiological status and selection of refugia [8], [32]. Our results suggest that if stressors are persistent and result in a chronic physiological state, populations will restrict space use and occupy refugia for an extended period of time.

Chronic stress response by elephants in iSimangaliso following translocation could be a consequence of delayed acclimatization. Previous studies on elephants have documented short-term elevations in FGM concentrations associated with poaching [33], hunting [34], fire [35], tourism [27], and translocation [24], [25]. Elephants selected for translocation to iSimangaliso exhibited baseline FGM concentrations prior to capture; however, FGM values did not return to baseline conditions within 30 days after the translocation event as found in previous studies of FGM responses to translocation of working elephants [24] and elephants allowed to navigate to their original home range [25]. One potential explanation is that it simply takes an extended period of time for wild elephants to acclimatize. For example, those elephant populations in our study with lower FGM concentrations were in reserves where initial translocations occurred 10–20 years prior to the initiation of the study, so it is possible that >6 years is required for physiological acclimatization following translocation to a new environment.

The spatial refuge behavioral response of elephants that we observed in iSimangaliso is potentially a consequence of avoiding the area associated with translocation and release. Because previous behavioral research has shown that there are sometimes long-term sociological and behavioral effects of traumatic events on elephants [36], [37], a persistent stress response could be attributed to the experience of a population or family group with the process of capture and translocation. Elevated stress responses to translocation have been reported with subsequent dispersal away from the release site for multiple species [4], including elephants [24]. Similarly, upon translocation to iSimangaliso, elephants were released in the Eastern Shores section, but quickly dispersed to the Western Shores section. All three separately introduced elephant family groups have subsequently resided in the latter section for 6 years post-release. This suggests that, given elephants were released on the Eastern Shores section, they could be avoiding the location associated with a translocation, a highly stressful event [4]. This avoidance following dispersal might be compounded by the presence of Lake St. Lucia, which could act as a barrier to movement between the two sections. However, during our study elephant family groups easily traversed the lake, crossing it 20 times to visit the Eastern Shores section for short periods (typically 24–48 hours) before returning to the Western Shores section.

The timing and frequency of human disturbances within iSimangaliso also could influence the refuge behavior pattern we observed. Wild elephant populations, similar to most wildlife, avoid areas associated with persistent interactions with humans [17]. The Eastern Shores section of iSimangaliso is open to the public and receives a consistently high level of tourism visitation, a factor known to elicit a physiological response in elephants [27]. By contrast, the Western Shores section is closed to the public, yet contains forest plantations that occasionally have a high amount of human disturbance by plantation workers, but which is localized to a particular stand. It is likely that elephants make trade-offs between relative risks associated with human disturbances within their environment. That is, elephants in iSimangaliso might utilize forest plantations, which are only intermittently visited by humans, and which occur in a matrix of native forest habitat that provides opportunities for the animals to escape disturbance, in favor of the Eastern Shores section, which more consistently receives human disturbance.

In addition to human disturbance, restricted foraging by elephants in iSimangaliso in dry forests and tree plantations could have influenced chronic FGM levels. Elevated elephant FGM concentrations might be related to nutritional stress and overall diet quality, where FGM concentrations are inversely related to the amount of nitrogen present in their diet [38]. The restricted space use patterns we observed in iSimangaliso, where elephants selected for and likely primarily consumed browse in dry forest and non-native tree plantations, likely further limited access to high quality forage regardless of season, and might have contributed to elevated FGM levels in that reserve. Therefore, in addition to potential human disturbance stressors, the impact of nutritional stress on chronic FGM concentrations is an area in need of research in translocated elephant populations.

The differences we observed in elephant space use patterns among populations did not correspond with our current understanding of how reserve shape, competition, and resource selection could restrict elephant space use. Elephants within fenced reserves have been shown to avoid areas in proximity to boundary fences [14], thus the shape of the reserves could influence elephant movement and space use. However, elephants in the most elongated and narrowest reserve (Phinda) utilized the highest portion (78%) of their reserve. By comparison, elephants in the round reserve (Pilanesberg), where we would expect less of an edge effect, utilized 55% of their reserve. African elephant family groups establish age- or size-related dominance hierarchies [39], which could result in competition between elephants that restricts space use patterns when populations are highly abundant and/or resources are limited [40]. While individual family groups could have exhibited greater competition and avoidance in iSimangaliso than the other two reserves, in general we would expect greater competition in reserves with higher elephant density [40]. In contrast, the reserve that exhibited the most restricted space use patterns contained the lowest elephant density (iSimangaliso, 0.04 elephants per km^2) compared to the other two reserves (Phinda, 0.51 elephants per km^2; Pilanesberg, 0.32 elephants per km^2). The distribution of dry native forest and tree plantations within iSimangaliso, which were primarily limited to the Western Shores where we observed restricted spaces use patterns, suggests that these habitats could be limiting and selected over grasslands habitats that dominated the Eastern Shores. However, elephants generally are considered habitat generalists [41], [42]. Further, given that grasslands likely contained higher quality forage most similar to the donor site (Kruger National Park; [43]), particularly during the wet season [44], we feel selection of tree plantations and dry forests of the Western Shores is more likely due to elephants avoiding open areas (i.e. grasslands) and sources of human disturbance than nutritional attributes that typically drive habitat-related patterns in elephant movement [44].

The differences we observed in space use overlap between day and night in iSimangaliso suggest that elephants might slowly be adapting temporal refuge behavior in response to human disturbance. Despite restricting their use of space to the Western Shores, elephants in iSimangaliso continue to exhibit a state of chronic physiological stress. If utilizing forest plantation is a persistent stressor to elephants in iSimangaliso, under the concept of allostasis an individual or population should continue to adapt and change its behavior to minimize the likelihood of encountering stressors [2]. However, elephants tend to be slow in developing novel movement patterns in response to changes in their environment [15]. Therefore, given variation in the timing and location of disturbances in forest plantations, developing reliable movement patterns is likely difficult and elephants might only slowly be adapting to stressors in the Western Shores. It also is possible that the elephants have finer-scale refuge behavior that we were unable to detect at the scale of our analysis, such as avoidance of roads during periods of peak use by plantation workers. Future research is needed to evaluate if elephants in iSimangaliso continue to maintain an elevated physiological state, and if they modify their fine-scale spatial and temporal behavior over time.

Chronic stress and elephant refuge behavior could have a number of potential short and long-term consequences to elephant health, human safety, tourism, ecosystem processes, and biodiversity. Two months following the initial release of elephants in iSimangaliso, mortality of an 8 month-old male calf occurred, likely as a result of stressors associated with translocation and long, continuous movement of the family group post-release. This incident suggests that chronic stress is likely to be a problem for young animals, and that providing refugia to limit continuous movements could reduce the risk of future mortalities. The history of human deaths caused by elephants in the reserves included in this study, while anecdotal, suggests that chronic stress and refuge behavior might be linked to incidences of elephant aggression toward humans. In iSimangaliso, despite closure of the Western Shores to the public, elephants in a single family group have killed two reserve workers. Also, in Phinda, within 3–5 years after introduction, a female elephant killed a human. In Pilanesberg, by contrast, although there have been a number of elephant attacks on humans and one person has been killed, all attacks were by male elephants of which most if not all were in musth. Given that it is comparatively less common for female elephants to be aggressive [45], and stress associated with socially disruptive events like translocation have previously been associated with incidences of elephant aggression [37], [45], our findings collectively suggest that chronic stress and refuge behavior following translocation are at least loosely linked to elephant aggression toward humans. Refuge behavior by elephants also limits their tourism value. Elephants frequently are seen by tourists in Pilanesberg and Phinda, but rarely are seen in iSimangaliso, where the opportunity for viewing elephants was one of the primary reasons for their reintroduction. Finally, the repeated use of refugia by elephants over an extended period of time could lead to extensive habitat modification [46] and potentially to loss of biodiversity [47]. In the case of iSimangaliso, refuge behavior also could exacerbate the damage to commercially valuable trees in forest plantations.

Accounting for refuge behavior has important implications to our understanding of elephant space use. Seasonal variation in the spatial distribution of resources, primarily forage and water availability [13], as well as social interactions [48] and the shape of fenced reserves [14], are known to be key drivers of elephant space use. In addition to these factors, physiological state could influence space use and resource selection patterns. For example, in Pilanesberg and Phinda, elephants generally used resources in proportion to their availability (Figure 5). In contrast, elephants in iSimangaliso exhibited restricted spaces use patterns and selected forest plantations on the Western Shores in favor of native habitats. This does not rule out the possibility that elephants in Pilanesberg and Phinda exhibited refuge behavior over shorter periods of time or that they identified areas as refugia. It is likely that elephants in Phinda and Pilanesberg identified refugia that allowed them to recover following exposure to a stressor. For example, Woolley et al. [35] documented that, following a catastrophic fire event in Pilanesberg, elephants exhibited a short-term elevation in stress hormone levels and moved to the northern portion of the reserve, which is designated as a "wilderness zone" closed to tourists. Thus, the availability of refugia, when needed, is likely critical to successfully avoiding development of a chronic physiological state.

The identification of refugia is particularly important in South Africa, where elephants are increasingly restricted to fenced reserves [49]. The use of fences in South Africa generally has been effective at limiting elephant movements and potential human-elephant conflict [50], [51]. However, our findings suggest that issues of human-elephant conflict and refuge behavior within fenced reserves need to be addressed. One potential solution to this problem is to identify areas that can serve as refugia for elephants (such as wilderness zones as in Pilanesberg) and limit human disturbance in those areas. Alternatively, managers might identify specific areas and periods when refugia are needed, similar to the current concept of virtual fences used to mitigate human-elephant conflict [51]. For example, in iSimangaliso, where an individual female in each family group is monitored with a Global Positioning System collar linked to a cellular phone network, managers are using real-time elephant movement data in combination with computer technology based on geospatial maps, to send a notification message to one or more cell phones any time a collared elephant moves into a pre-determined zone, such as across a reserve border [51]. Similarly, if reserve managers are able to identify refugia spatially or predict via movement patterns when elephants are exhibiting refuge behavior, they could limit human disturbance to that area for a period of time, and potentially provide corridors into or among refugia to mitigate the risk of chronic stress and potentially dangerous human-elephant interactions.

In summary, managers considering the translocation or reintroduction of wildlife should consider the possibility of chronic stress and potential consequences of refuge behavior. Chronic stress is common following wildlife translocation, and has been associated with reproductive failure, increased predation risk, disease risk, and movement away from the release site [4], [52]. Our results suggest that chronic stress is associated with refuge behavior in translocated elephants, and we predict that it is likely to occur as a common facultative response in other species following translocation. Thus, future efforts to predict when, where, and to what extent wildlife populations will exhibit refuge behavior could likely be improved by an understanding of their physiological response.

MATERIALS AND METHODS

Ethics Statement

The collection of elephant fecal samples and field observation techniques were approved by the Animal Ethics Sub-committee of the University of KwaZulu-Natal Ethics Committee (permit reference 012/09/Animal).

Study Areas

We selected three reintroduced elephant populations in South Africa, which were each contained by electrified boundary fences: Pilanesberg Game Reserve (25°8′–25°22′S, 26°57′–27°13′E), iSimangaliso Wetland Park (28°49′–27°55′S, 32°68′–32°22′E), and Phinda Private Game Reserve (27°92′–27°68′S, 32°44′–32°20′E). Most individual elephants within our three study sites were translocated from Kruger National Park, or were the offspring of such animals [53]. Exceptions were 10 individuals at Phinda Private Game Reserve brought in 1993 from Gonarhezou in Zimbabwe [23], and six individuals (two from US captive populations, two from Namibia, and two from Mabula Game Reserve) that were released in Pilanesberg National Park [54]. Pilanesberg National Park (hereafter referred to as Pilanesberg), located in the North West Province, is 560 km^2 in size and is composed of hilly terrain containing a mix of open grasslands and closed *Acacia* and broad-leaf bushveld [29]. We classified habitats based on seven major vegetation types in the park [14], [55]: (1) *Acacia caffra* woodland, (2) *A. karoo* woodland, (3) *A. mellifera* woodland, (4) *Combretum* woodland, (5) *Faurea* woodland, (6) mixed *Acacia* woodland, and (7) grassland. Fifty-eight male and 37 female elephants were reintroduced from 1981 to 1998, primarily from Kruger National Park [56]. In 2004 there were at least 16 family groups [57], and by 2009 there were approximately 180 individual elephants in the park (S. Dell, Pilanesberg National Park, personal communication).

Phinda Private Game Reserve (hereafter referred to as Phinda), located in the KwaZulu-Natal Province, is 180 km^2 in size and contains a range of habitats that include sweet lowveld bushveld, Natal low bushveld, and coastal bushveld [58]. We used existing land use and vegetation maps created by Noel van Rooyen and Simon Morgan for reserve management to classify habitats into eight categories: (1) *Acacia* woodland, (2) human habitation, (3) open grassland, (4) closed woodland, (5) riverine and wetland, (6) sand forest, (7) Lebombo thicket, and (8) palmveld. Managers released 54 orphan elephants in 1992–1994 and 3 mature adult bulls in 2003 [15]. In 2009, there are at least five family groups, and the total population in 2010 was estimated to be 93 individuals (T. Burke, Phinda Private Game Reserve, personal communication).

The iSimangaliso Wetland Park is located on the eastern coast in KwaZulu-Natal Province. It is 602 km^2 in size and is composed of the Eastern Shores section (273 km^2) bordered by fencing to the north and south, by the Indian Ocean to the east and the estuary of Lake St. Lucia to the west; and the Western Shores section (329 km^2) bordered by Lake St. Lucia to the east and electrified fence along its other boundaries. We used existing vegetation and land use maps created by Noel van Rooyen for park management to classify iSimangaliso into eight major habitat types: (1) tree plantations (composed of either *Eucalyptus globulus* or *Casuarina equisetifolia*), (2) dry forest, (3) lowland forest, (4) marsh and swamp, (5) freshwater lake, (6) grassland, (7) human settlement, and (8) open beach. We did not consider the estuarine Lake St. Lucia as available habitat in our analysis. The reintroduction of elephants to iSimangaliso was initiated in 2001 with the translocation of a 24 elephants (15 females and 9 males) from Hluhluwe-iMfolozi Park (originally from Kruger National Park), and in 2002 and 2003 with two additional family groups directly from Kruger National Park.

Stress hormone data

From 2000 to 2006, we sampled FGM concentrations of elephants in each of the three reserves. In the field, fecal samples were collected opportunistically by trained employees of the reserves or by the University of KwaZulu-Natal. On average, samples from Phinda, Pilanesberg, and iSimangaliso were collected within 30 min, 10 hrs, and 20 hrs respectively. Across all reserves, time between deposition and collection for all samples used was <72 hrs, and similar to other FGM-based studies on elephant [25], [34], [59]. We recorded the approximate age of the sample as well as the location of collection, but were unable to consistently identify which individual or family group deposited the sample. Samples for laboratory analysis were collected by opening, and taking a portion from the center of the bolus [25], [34]. After collection, samples were immediately treated with a 2% acetic acid solution and frozen for shipment [60]. In the laboratory, samples were stored at −80°C, freeze-dried, ground, and sifted through a stainless steel mesh. We extracted FGM from the feces using corticosterone I^{125} radioimmunoassay kits (MP Biomedicals, Costa Mesa, CA) following validated and established protocols [28]. Inter-assay variation for 11 assays was 7.3% and average intra-assay variation was 3.9%.

We conducted a nested analysis of variance (ANOVA) to determine if significant differences occurred in mean FGM concentrations of elephants among the three reserves, and if differences within reserves occurred between years. In addition, we evaluated if FGM concentrations followed a pattern of variation between seasons (wet and dry) similar to that seen in previous studies [61], [38]. We partitioned data into annual wet and dry seasons based on rainfall patterns for our study areas, where the wet season occurred from November to April, and the dry season occurred from May to October [30], [54].

Location Data

From 2004 to 2007, GPS collars were placed on a single adult female individual in each of 14 family groups (iSimangaliso n=3, Phinda n=5, Pilanesberg n=6). Because adult female elephants live in cohesive family units, we assumed that GPS collars deployed on adult female elephants capture the movements of an entire family group [17]. All collars were programmed to record elephant locations at predetermined intervals (ranging from 30 min to 12 hrs depending on the individual elephant) and to transmit coordinates by Global System for Mobile Communications

(GSM) cell phone signal or satellites to a ground station where they were stored on a master computer [15]. We omitted locations in Pilanesberg from September 2005 to September 2006 due to a catastrophic fire that altered elephant space use patterns [35]. We also omitted locations in Phinda prior to September 2005 due to removal of a section of fence at that time that allowed for expansion of the reserve [15]. While locational data were not validated, location error was relatively low (<50 m) based on evaluations of similar GPS collars on elephants [13], [14].

Analysis of Space Use Patterns

In analyzing elephant space use, we first wanted to identify distinct, biologically meaningful time intervals among which we could compare space use patterns by family groups over time. Elephant space use patterns consistently vary between two annual seasons based on rainfall (i.e. wet and dry seasons) [13], [29]. Given that elephants at the donor site (Kruger National Park) also exhibit distinctive wet and dry season movement patterns [62], we predicted that translocated elephants at our study sites would similarly exhibit seasonal movement patterns [30], [31].

We developed seasonal utilization distributions (UDs) [63] to estimate space use for each season during which an elephant continuously wore a GPS collar. Between 2004 and 2007, within each season we captured ≥300 locations ($\bar{x}$=303.51, SE=8.33, range=90–370) of elephants separated by 12±2 hours in each of our three study sites. We represented space use by each elephant family group during each season by creating 95% fixed kernel UDs using the plug-in method of bandwidth selection [64]. Because elephant space use is limited by hard boundaries (i.e. electric fences) at each reserve, we trimmed each UD by the reserve boundary and standardized the remaining UD value so that cell values in each UD summed to 1.0.

To evaluate the spatial refuge hypothesis, that elephant family groups with high FGM levels exhibit restricted space use, we compared the proportion of a reserve utilized by elephants among reserves. Because each reserve was completely fenced around the entire perimeter (except for portions of iSimangaliso bordered by lake or ocean) and fences created an edge effect influencing elephant movement [14], home range size was likely influenced by reserve size [30]. Therefore, because reserves were different sizes (180 km^2 to 602 km^2), we evaluated elephant space use based on the percent of the reserve occupied by the UD contour in addition to home range size estimates. We evaluated the home range size and proportion of the reserve utilized by each family group during each season for normality and compared among reserves, family groups, years, and seasons using a nested factorial ANOVA. In the ANOVA, reserve, year and season were fixed effects, elephant family group was nested within reserve, and home range size or proportion of the reserve utilized was the dependent variable.

To evaluate the temporal refuge hypothesis, that elephants with elevated FGM levels exhibit different behavioral patterns in day vs. night, we compared day home range size to night home range size within each reserve. Given that tourist game drive traffic and, in the case of iSimangaliso, forestry operations, primarily occur during daylight, we hypothesized that there might be differences in day and night space use by elephants. We categorized locations into day or night separately for each season and computed UDs for each family group using procedures described above. We defined day as between 0800 and 1900, and night as between 2100 and 0600. We omitted locations between 1900 and 2100 and 0600 and 0800 due to seasonal variations in the time of sunrise and sunset, and because some guided tourism viewing occurs during those periods. We computed home range sizes for both day and night within each season for each elephant, and evaluated if there were significant differences in home range size between day and night within each reserve individually using a factorial nested ANOVA. In the ANOVA, elephant family group was a fixed effect, season was nested within elephant family group, time (in terms of UDs based on daytime or nighttime locations) was nested within season and elephant family group, and home range size was the dependent variable.

To determine the extent to which elephants used the same area by day as by night, we evaluated space use overlap by individual family groups between day and night within each season using a volume of intersection (VI) analysis [65], [66]. The VI index measures overlap in space use between two UDs (as distinct from polygon overlap). Volume of intersection scores range from 0–1, where a VI score of 1 indicates perfect overlap of the UDs. Therefore, we interpreted higher VI scores as evidence of the repeated use of space between day and night. To account for potential day-night variation in highly utilized areas, we computed VI scores for both the home range scale of 95% fixed-kernel UDs, and core area scale of 50% fixed-kernel UDs [67]. We log-transformed VI scores and used a nested ANOVA to test the null hypothesis that no difference occurred in VI scores among reserves [68]. In the ANOVA, reserve, year and season were fixed effects, elephant family group was nested within reserve, and the VI score for comparing day vs. night space use was the sampling unit.

Analysis of resource selection

We assessed resource selection by elephants in each reserve using a weighted compositional analysis [69]. We utilized the 95% fixed-kernel UD for each day and night period and summed UD values for each habitat type. We divided the summed UD values for each habitat type by total UD score to get weighted proportional use of each habitat type by an elephant. We substituted 0.5% for 0 for all non-used habitats [70] and subtracted log-transformed use data from log-transformed availability data (at the reserve level) for each elephant at each sampling interval to calculate the difference in log-ratios [69], [71]. We evaluated if overall selection occurred using Wilk's lambda statistic to test if the mean vector of log-ratio differences differed from a vector of zeros, and when selection occurred, we ranked habitats based on their relative utilization [71]. We tested for effects of season (wet vs. dry) and time of day (day vs. night) on log-ratio difference values for each habitat type in each reserve using a nested multivariate analysis of variance (MANOVA) [72]. In the MANOVA, elephant family group was the fixed effect, season was nested within elephant family group, time of day was nested within season and elephant family group, and the log-ratio differences were the sampling unit.

ACKNOWLEDGMENTS

We thank the staff at iSimangaliso Wetland Park, Pilanesberg National Park and Phinda Private Game Reserve who provided habitat data and assisted in the collection of fecal samples, and in the collaring of elephants. We thank and acknowledge the following for collecting dung samples: Tarryne Burke, Heleen Druce, Samantha Janse van Rensburg, Leah-Anne Woolley, Chantal Dickson, Sibusiso Mfeka and Grant Burdon. We are grateful to D. Atwill of Columbia, Missouri, USA, for introductions leading to collaborative research projects between the University of Missouri and the University of KwaZulu-Natal. We thank R.J. Woods, B.E. Washburn and employees of the University of Missouri who completed stress hormone assays.

Author Contributions

Conceived and designed the experiments: DJ JM RS. Performed the experiments: JM RS. Analyzed the data: DJ JM. Contributed reagents/materials/analysis tools: RS JM. Wrote the paper: DJ RS JM.

REFERENCES

1. Romero LM. Physiological stress in ecology: lessons from biomedical research. Trends in Ecology and Evolution. 2004;19:249–255.
2. McEwen BS, Wingfield JC. The concept of allostasis in biology and biomedicine. Hormones and behavior. 2003;43:2–15.
3. Romero LM, Butler LK. Endocrinology of stress. International Journal of Comparative Psychology. 2007;20:85–95.
4. Dickens MJ, Delehanty DJ, Romero LM. Stress: An inevitable component of animal translocation. Biological Conservation. 2010;143:1329–1341.
5. Wingfield JC, Romenofsky M. Corticosterone and facultative dispersal in response to unpredictable events. Ardea. 1997;85:155–166.
6. Breuner CW, Greenberg AL, Wingfield JC. Noninvasive corticosterone treatment rapidly increases activity in Gambel's white-crowned sparrows. General and Comparative Endocrinology. 1998;111:386–394.
7. Astheimer LB, Buttemer WA, Wingfield JC. Interactions of corticosterone with feeding, activity and metabolism in passerine birds. Ornis Scandinavica. 1992;23:355–365.
8. Elliot JM. Pools as refugia for brown trout during two summer droughts: trout responses to thermal and oxygen stress. Journal of Fish Biology. 2000;56:938–948.
9. Riegl B, Piller WE. Possible refugia for reefs in times of environmental stress. International Journal of Earth Science. 2003;92:520–531.
10. Ultsch GR. The ecology of overwintering among turtles: where turtles overwinter and its consequences. Biological Reviews. 2006;81:339–367.
11. Bruener CW, Hahn TP. Integrating stress physiology, environmental change, and behavior in free-living sparrows. Hormones and Behavior. 2003;43:115–123.
12. Hart BL, Hart LA, Pinter-Wollman N. Large brains and cognition: Where do elephants fit in? Neuroscience and Biobehavioral Reviews. 2008;32:86–98.
13. Loarie SR, Van Aarde RJ, Pimm SL. Fences and artificial water affect African savannah elephant movement patterns. Biological Conservation. 2009;142:3086–3098.

14. Vanak AT, Thakar M, Slotow R. Do fences create an edge-effect on the movement patterns of a highly mobile mega-herbivore? Biological Conservation. 2010;143:2631–2637.
15. Druce HC, Pretorius K, Slotow R. The response of an elephant population to conservation area expansion: Phinda Private Game Reserve, South Africa. Biological Conservation. 2008;141:3127–3138.
16. Hoare RE, Du Toit JT. Coexistence between people and elephants in African Savannas. Conservation Biology. 1999;13:633–639.
17. Douglas-Hamilton I, Krink T, Vollrath F. Movements and corridors of African elephants in relation to protected areas. Naturwissenchaften. 2005;92:158–163.
18. Theuerkauf J, J-drzejewski W, Schmidt K, Gula R. Spatiotemporal segregation of wolves from humans in Białowieża forest (Poland). The Journal of Wildlife Management. 2003;67:707–716.
19. Hebblewhite M, Merrill E. Modeling wildlife-human relationships for social species with mixed-effects resource selection models. Journal of Applied Ecology. 2008;45:834–844.
20. Graham MD, Douglas-Hamilton I, Adams MW, Lee PC. The movement of African elephants in a human-dominated land-use mosaic. Animal Conservation. 2009;12:445–455.
21. Wedge PH, Rowland ED, Thompson BG, Batruch N. Use of acoustic tools to reveal otherwise cryptic responses of forest elephants to oil exploration. Conservation Biology. 2010;24:1578–1585.
22. Garaï ME, Slotow R, Carr RD, Reilly B. Elephant reintroductions to small fenced reserves in South Africa. Pachyderm. 2004;37:28–36.
23. Grobler DG, van Altena JJ, Malan JH, Mackey RL. Elephant translocation. In: Scholes RJ, Mennell KG, editors. Elephant management: A scientific assessment for South Africa. Johannesburg, South Africa: Wits University Press; 2008. pp. 241–246.
24. Viljoen JJ, Ganswindt A, du Toit JT, Langbauer WR. Translocation stress and faecal glucocorticoid metabolite levels in free-ranging African savanna elephants. South African Journal of Wildlife Research. 2008;38:146–152.
25. Millspaugh JJ, Burke T, Van Dyk G, Slotow R, Washburn BE, et al. Stress response of working African elephants to transportation and safari adventures. Journal of Wildlife Management. 2007;71:1257–1260.

26. Poole JH, Granli P. Signals, gestures, and behavior of African elephants. In: Moss CJ, Croze H, Lee PC, editors. The Amboseli elephants: A long-term perspective on a long-lived mammal. University of Chicago Press, Chicago, Illinois; 2011. pp. 109–127.
27. Pretorius Y. Stress in the African elephant of Mabula Game Reserve, South Africa. 2004. Master of Science thesis, University of Natal, Durban, South Africa.
28. Wasser SK, Hunt KE, Brown JL, Cooper K, Crokett CM, et al. A generalized fecal glucocorticoid assay for use in a diverse array of nondomestic mammalian and avian species. General and Comparative Endocrinology. 2000;120:260–275.
29. Ngene SK, Van Gils H, van Wieren SE, Rasmussen H, Skidmore AK, et al. The ranging patterns of elephants in Marsabit protected areas, Kenya: the use of satellite-linked GPS collars. African Journal of Ecology. 2010;48:386–400.
30. Shannon G, Page B, Slotow R, Duffy K. African elephant home range and habitat selection in Pongola Game Reserve, South Africa. African Zoology. 2006;41:37–44.
31. Shannon G, Page B, Duffy K, Slotow R. The ranging behavior of a large sexually dimorphic herbivore in response to seasonal and annual environmental variation. Austral Ecology. 2010;35:731–742.
32. Kauffman MJ, Varley N, Smith DW, Stahler DR, MacNulty DR, et al. Landscape heterogeneity shapes predation in a newly restored predator-prey system. Ecology Letters. 2007;10:690–700.
33. Gobush KS, Mutayoba BM, Wasser SK. Long-term impacts of poaching on relatedness, stress physiology, and reproductive output of adult female African elephants. Conservation Biology. 2008;22:1590–1599.
34. Burke T, Page B, van Dyk G, Millspaugh J, Slotow R. Risk and ethical concerns of hunting male elephant: Behavioural and physiological assays of the remaining elephants. PLoS One. 2008;3:e2417. [PMC free article]
35. Woolley L, Millspaugh JJ, Woods RJ, van Rensburg SJ, Mackey RL, et al. Population and individual elephant response to a catastrophic fire in Pilanesberg National Park. PLoS One. 2008;3:e3233. [PMC free article]

36. Bradshaw GA, Schore AN, Brown JL, Poole JH, Moss CJ. Elephant breakdown. Nature. 2005;433:807.
37. Bradshaw GA, Shore AN. How elephants are opening doors: developmental neuroethology, attachement and social context. Ethology. 2007;113:426–436.
38. Woolley L, Millspaugh JJ, Woods RJ, van Rensberg SJ, Page BR, et al. Intraspecific strategic responses of African elephants to temporal variation in forage quality. Journal of Wildlife Management. 2009;73:827–835.
39. Archie EA, Morrison TA, Foley CAH, Moss CJ, Alberts SC. Dominance rank relationships among wild African elephants, Loxodonta africana. Animal Behaviour. 2006;71:117–127.
40. Wittemyer G, Getz WM, Vollrath F, Douglas-Hamilton I. Social dominance, seasonal movements, and spatial segregation in African elephants: a contribution to conservation behavior. Behavioral Ecology and Sociobiology. 2007;61:1919–1931.
41. Laws RM. Elephants as agents of habitat and landscape change in East Africa. Oikos. 1970;21:1–15.
42. Owen-Smith N. Megaherbivores: the influence of very large body size on ecology. Cambridge University Press, Cambridge, UK; 1988.
43. Smit IPJ, Grant CC, Whyte IJ. Landscape-scale sexual segregation in the dry season distribution and resource utilization of elephants in Kruger National Park, South Africa. Diversity and Distributions. 2007;13:225–236.
44. Codron J, Lee-Thorp JA, Sponheimer M, Codron D, Grant RC, et al. Elephant (Loxodonta africana) diets in Kruger national Park, South Africa: Spatial and landscape differences. Journal of Mammalogy. 2006;87:27–34.
45. Twine W, Magome H. Interactions between elephants and people. In: Scholes RJ, Mennell KG, editors. Elephant management: A scientific assessment for South Africa. Wits University Press, Johannesburg, South Africa; 2008. pp. 206–240.
46. Skarpe C, Aarrestad PA, Andreassen HP, Dhillion SS, Dimakatso T, et al. The return of the giants: ecological effects of an increasing elephant population. Ambio. 2004;33:276–282.
47. Kerley GIH, Landman M, Kruger L, Owen-Smith N, Balfour D, et al. Effects of elephants on ecosystems and biodiversity. In: Scholes RJ,

Mennell KG, editors. Elephant management: A scientific assessment for South Africa. Wits University Press, Johannesburg, South Africa; 2008. pp. 84–145.

48. Mutinda H, Poole JH, Moss CJ. Decision making and leadership in using the ecosystem. In: Moss CJ, Croze H, Lee PC, editors. The Amboseli elephants: a long-term perspective on a long-lived mammal. University of Chicago Press, Chicago, Illinois; 2011. pp. 246–259.
49. Hayward MW, Kerley GIH. Fencing for conservation: restriction of evolutionary potential or a riposte to threatening processes? Biological Conservation. 2009;142:1–13.
50. Grant CC, Bengis R, Balfour D, Peel M, Davies-Mostert W, et al. Controlling the distribution of elephants. In: Scholes RJ, Mennell KG, editors. Elephant management: A scientific assessment for South Africa. Johannesburg, South Africa: Wits University Press; 2008. pp. 329–369.
51. Slotow R. Fencing for purpose: a case study of elephants in South Africa. In: Somers MJ, Hayward MW, editors. Fencing for conservation: restriction of evolutionary potential or a riposte to threatening processes? Springer-Verlag, New York, New York; 2012. pp. 91–104.
52. Teixeira CP, De Azevedo CS, Mendl M, Ciprest CF, Young RJ. Revisiting translocation and reintroduction programmes: the importance of considering stress. Animal Behaviour. 2007;73:1–13.
53. Slotow R, Garaï ME, Reilly B, Page B, Carr RD. Population dynamics of elephants re-introduced to small-fenced reserves in South Africa. South African Journal of Wildlife Research. 2005;35:23–32.
54. Burke T. The effect of human disturbance on elephant behavior, movement dynamics and stress in a small reserve: Pilanesberg National Park. 2005. Master of Science thesis, University of KwaZulu-Natal, Durban, South Africa.
55. Brockett BH. Woody vegetation states and community descriptions reference guide for Pilanesberg National Park, Bophuthatswana National Parks and Wildlife Management Board, South Africa Version 1.0. Johannesburg, South Africa; 1993.
56. Slotow R, van Dyk G, Poole J, Klocke A. Older bull elephants control young males. Nature. 2000;408:425–426.
57. Shannon G, Page BR, Mackey RL, Duffy KJ, Slotow R. Activity budgets and sexual segregation in African elephants (Loxodonta africana). Journal of Mammalogy. 2008;89:467–476.

58. Low AB, Robelo AG. Vegetation of South African, Lesotho and Swaziland. 1996. Department of Environmental Affairs and Tourism, Pretoria, South Africa.
59. Ahlering MA, Millspaugh JJ, Woods RJ, Western D, Eggert L. Elevated levels of stress hormones in crop-raiding male elephants. Animal Conservation. 2010;14:124–130.
60. Millspaugh JJ, Washburn BE, Milanick MA, Slotow R, van Dyk G. Effects of heat and chemical treatments on fecal glucocorticoid measurements: implications for sample transport. Wildlife Society Bulletin. 2003;31:399–406.
61. Foley CAH, Papageorge S, Wasser SK. Noninvasive stress and reproductive measures of social and ecological pressures in free-ranging African elephants. Conservation Biology. 2001;15:1134–1142.
62. De Knegt HJ, van Lengevelde F, Skidmore AK, Delsink A, Slotow R, et al. The spatial scaling of habitat selection by African elephants. Journal of Animal Ecology. 2010;80:270–281.
63. van Winkle W. Comparison of several probabilistic home-range models. Journal of Wildlife Management. 1975;54:322–330.
64. Gitzen RA, Millspaugh JJ, Kernohan BJ. Bandwidth selection for fixed-kernel analysis of animal utilization distributions. Journal of Wildlife Management. 2006;70:1334–1344.
65. Seidel KD. Statistical properties and applications of a new measure of joint space use for wildlife. University Press, University of Washington, Seattle, Washington; 1992.
66. Millspaugh JJ, Gitzen RA, Kernohan BJ, Larson MA, Clay CL. Comparability of three analytical techniques to assess joint space use. Wildlife Society Bulletin. 2004;32:148–157.
67. Barg JJ, Jones J, Robertson RJ. Describing breeding territories of migratory passerines: suggestions for sampling, choice of estimator, and delineation of core areas. Journal of Animal Ecology. 2005;74:139–149.
68. Millspaugh JJ, Brundige GC, Gitzen RA, Raedeke KJ. Elk and hunter space-use sharing in South Dakota. Journal of Wildlife Management. 2000;64:994–1003.
69. Millspaugh JJ, Nielson RM, McDonald LL, Marzluff JM, Gitzen RA, et al. Analysis of resource selection using utilization distributions. Journal of Wildlife Management. 2006;70:384–395.

70. Bingham RL, Brennan LA. Comparison of Type 1 error rates for statistical analyses of resource selection. Journal of Wildlife Management. 2004;68:206–212.
71. Aebischer NJ, Robertson PA, Kenward RE. Compositional analysis of habitat use from animal radio-tracking data. Ecology. 1993;74:1313–1325.
72. Smith RA, Jennings NV, Robinson A, Harris S. Conservation of European hares Lepus europaeus in Britain: is increasing habitat heterogeneity in farmland the answer? Journal of Applied Ecology. 2004;41:1092–1102.

CHAPTER 18

Haematology and Plasma Biochemistry of Wild Black Flying-Foxes, (Pteropus alecto) in Queensland, Australia

Lee McMichael[1,2], Daniel Edson[2], Amanda McLaughlin[2], David Mayer[3], Steven Kopp[1], Joanne Meers[1], Hume Field[4]

[1] School of Veterinary Science, The University of Queensland, Gatton, Qld 4343, Australia

[2] Biosecurity Queensland, Department of Agriculture, Fisheries and Forestry, Brisbane, Qld 4108, Australia

[3] Department of Agriculture Fisheries and Forestry, Brisbane, Qld 4103, Australia

[4] EcoHealth Alliance, New York, NY 10001, United States of America

ABSTRACT

This paper establishes reference ranges for hematologic and plasma biochemistry values in wild Black flying-foxes (*Pteropus alecto*) captured

Citation: McMichael L, Edson D, McLaughlin A, Mayer D, Kopp S, Meers J, et al. (2015) Haematology and Plasma Biochemistry of Wild Black Flying-Foxes, (Pteropusalecto) in Queensland, Australia. PLoS ONE 10(5): e0125741. https://doi.org/10.1371/journal.pone.0125741.

in South East Queensland, Australia. Values were found to be consistent with those of other *Pteropus* species. Four hundred and forty-seven animals were sampled over 12 months and significant differences were found between age, sex, reproductive and body condition cohorts in the sample population. Mean values for each cohort fell within the determined normal adult reference range, with the exception of elevated levels of alkaline phosphatase in juvenile animals. Hematologic and biochemistry parameters of injured animals showed little or no deviation from the normal reference values for minor injuries, while two animals with more severe injury or abscessation showed leucocytosis, anaemia, thrombocytosis, hyperglobulinemia and hypoalbuminemia.

INTRODUCTION

Pteropid bats (*Pteropodidae*: *Pteropus*), colloquially known as flying-foxes, are ecologically important for the nocturnal pollination and long range seed dispersal of Australian plant species [1]. Species of bats, including flying-foxes have also been identified as a source of a number of emerging pathogenic viruses of animal and human health significance, including Hendra virus [2–9]. A range of physiological and ecological factors, including physiological stress associated with reproduction and sub-optimal nutrition are reported to constitute risk factors for viral infection in flying-foxes [10–12].

The study of hematologic and biochemical values has been employed to describe the health status of many wildlife species. Ruykys *et al*., compared the health status of wild and captive populations of warru (*Petrogale lateralis*); Clarke *et al*., assessed the health of threatened western ringtail possums (*Pseudocheirus occidentalis*) prior to translocation; and Pacioni *et al*., assessed the population health of woylie (*Bettongia penicillata ogilbyi*) [13–15]. Similar approaches will be useful in assessing the health of Australian flying-foxes in order to understand and meaningfully interpret population health in ecological and epidemiological contexts. Establishing hematologic and biochemical reference ranges for wild Australian flying-foxes may help to identify indicators of poor nutritional status, disease and the effect of environmental stressors.

Current literature on haematology and biochemistry of Australian flying-foxes is minimal and published reference ranges have previously been based on small numbers (between 4 and 60) of captive or wild caught animals, including opportunistic sampling of injured flying-foxes [16–19].

This study presents hematologic and plasma biochemistry reference values for clinically normal adult wild-caught *Pteropus alecto*, commonly known as the Black flying-fox, and compares physiological, age-based and injured cohorts within the population.

MATERIALS AND METHODS

Animals, Ethics And Study Sites

Blood samples were collected from 447 individual *P. alecto*, during 7 catching events from June 2012 to May 2013, at a peri-urban parkland roost at Boonah in South East Queensland, Australia, as a component of a larger epidemiological study (reported elsewhere). Fieldwork was conducted under the Queensland Department of Agriculture, Fisheries and Forestry Animal Ethics Committee Permit SA 2011/12/375, and Department of Environment, Heritage and Protection Scientific Purposes Permits WISP05810609 and WISP14100614.

Animal Capture

Bats were captured pre-dawn in mist nets (18m wide x 8m deep) hoisted between two 20m fibre glass masts. Each bat was immediately removed from the net, placed in a cotton bag and allowed to hang calmly prior to processing. Animals were anaesthetised (under veterinary supervision) using the inhalation agent Isoflurane and medical oxygen [20]. A basic physical examination was conducted, identifying any obvious lesions or abnormalities of the skin, bones, abdomen, thorax, mouth and nervous system. Age class was estimated by morphometric measurements of forearm length (mm), weight (g), and presence of secondary sexual characteristics [21]. Adult males were distinguished from sub-adult males on the basis of fully developed penis and testes. Adult females were distinguished from sub-adult females on the basis of worn, elongated nipples, indicating that they had suckled at least once in their lifetime. Juveniles (< 12 months old) were classified on their smaller size and rudimentary development of sexual characteristics. Sex and body condition (assessed primarily through palpation of the pectoral muscle mass and associated prominence of sternal carinum), were recorded for each bat. Pregnancy was determined by gentle abdominal palpation and lactation status by expression of milk from the teats. All bats were marked by painting hind limb claws with coloured acrylic lacquer to avoid short-term resampling. All blood samples were collected within 6 hours of capture. Two mL of blood was collected from the propatagial (cephalic) vein and dispensed into a 1.3 mL Lithium Heparin blood tube (Sarstedt 2269201) and a 0.5 ml EDTA tube (Microtainer 5974). Blood glucose concentration was measured at the time of bleeding using an ACCU-CHEK Performa glucometer (Roche Diagnostics GmbH). After anaesthesia, each bat was monitored until conscious, and haemostasis at the venepuncture site confirmed prior to being placed into a bag and allowed to recover for at least 30 minutes before release at the capture site.

Haematology and Biochemistry Analysis

All samples were kept chilled prior to and during shipment to the Biosecurity Sciences Laboratory in Brisbane. EDTA whole blood samples were shipped daily to a commercial pathology laboratory, (Queensland Medical Laboratories (QML)), in Brisbane for complete blood counts using a Sysmex XE-2100 Automated Haematology System. Lithium heparin plasma was collected within 6 hours of bleeding, following separation by centrifugation, and were either shipped daily on ice, or stored at minus 80°C until thawing and shipment to QML for processing using a Siemens ADVIA 2400 Chemistry System. On occasion, there was insufficient sample volume to conduct all hematologic or biochemical tests. Therefore, the sample size for adult animals for reference range calculation varied for different parameters, ranging from 253–293 animals (Table 1). A total of 385 total blood samples were submitted for haematology and 364 plasma samples submitted for plasma biochemistry. When platelet aggregation was identified on blood smears absolute platelets counts were not conducted. All values were reported in standard international (SI) units.

Table 1. Haematological and plasma biochemistry reference ranges for clinically normal wild-caught adult *P. alecto*

Parameter	n	Mean	95% CI	Range	Ref Range (n)
Hb (g/L)	293	163.49	140.37–186.60	136.00–218.00	143.40–182.18 (280)
RCC ($x10^{12}$/L)	293	9.13	7.85–10.41	7.60–12.60	8.01–10.19 (282)
Hct	293	0.47	0.41–0.53	0.39–0.62	0.42–0.52 (286)
MCV (fL)	293	51.79	47.02–56.56	45.00–58.00	47.34–56.12 (284)
MCH (pg)	293	17.97	16.44–19.50	16.00–21.00	16.53–19.31 (286)
MCHC (g/L)	293	347.03	329.38–364.68	318.00–388.00	331.82–361.30 (282)
Plat ($x10^9$/L)	293	367.39	189.32–545.45	106.00–719.00	217.54–499.14 (276)
WBC ($x10^9$/L)	293	5.96*	2.95–12.03	2.50–22.00	2.38–10.01 (277)
Neut ($x10^9$/L)	293	3.57*	1.64–7.74	0.90–12.50	1.29–6.18 (278)
Lymp ($x10^9$/L)	293	1.72*	0.50–5.90	0.30–12.10	0.58–5.11 (279)
Mono ($x10^9$/L)	293	0.14*	0.03–0.62	0.00–1.20	0.03–0.50 (282)
Eos ($x10^9$/L)	293	0.06*	0.00–2.33	0.00–4.59	0.00–2.17 (291)
Baso ($x10^9$/L)	293	0.00*	0.00–0.01	0.00–0.01	0.00–0.01 (293)
Na (mmol/L)	260	138.39	133.50–143.27	132.00–147.00	134.27–142.00 (256)
K (mmol/L)	260	4.02*	2.68–5.76	2.50–9.80	2.84–4.93 (250)
Cl (mmol/L)	260	104.99	98.23–111.74	92.00–114.00	99.07–111.22 (259)
Bicarb (mmol/L)	260	14.68	10.38–18.97	9.00–21.00	10.72–18.34 (252)
Anion gap (mmol/L)	260	22.75*	17.39–29.22	15.00–38.00	17.70–27.14 (248)
Ca (mmol/L)	258	2.41	2.14–2.68	1.98–2.82	2.11–2.71 (247)
P (mmol/L)	258	1.4	0.62–2.19	0.50–2.50	0.66–2.11 (253)
Glucose (mmol/L)	993	6.8*	4.3–10.7	3.5–21.4	4.5–10.0 (955)
AST (U/L)	258	76.83*	25.21–174.05	23.00–502.00	28.85–135.71 (246)
ALT (U/L)	258	15.84*	6.56–38.28	< 10.00–52.00	9.02–30.06 (235)
ALP (U/L)	259	407.46*	156.90–877.52	69.00–1289.00	102.01–697.82 (247)
Protein (g/L)	257	65.51	54.05–76.98	50.00–82.00	55.27–75.11 (243)
Albumin (g/L)	255	36.12	29.57–42.67	27.00–44.00	30.03–41.73 (244)
Globulin (g/L)	255	29.34	22.32–36.35	20.00–40.00	22.97–35.10 (245)
Alb/Glob Ratio	255	1.25	0.95–1.54	0.80–1.70	0.93–1.54 (241)
Urea (mmol/L)	260	1.36*	0.25–7.57	< 0.50–7.40	0.25–7.57 (260)

CK (U/L)	254	466.28*	40.07–1577.35	57.00–9552.00	52.83–857.82 (237)
Triglyc (mmol/L)	253	0.19*	0.05–0.74	< 0.10–1.20	0.05–0.61 (244)
GGT (U/L)	259	-	-	< 5.00–19.00	-
Creatinine (μmol/L)	260	-	-	< 20.00–181.00	-
Bilrubin (μmol/L)	260	-	-	< 2.00–8.00	-
Chol (mmol/L)	253	-	-	< 0.50–1.40	-

*Natural log transformation.

Statistical Analysis

Reference ranges for hematologic and plasma biochemical measurements were calculated using XLSTAT (Version 2008.6.08) from 376 adult animal samples, excluding 42 animals with observed injury or clinical signs of disease. Normality of the distributions of variables was tested using the Fisher test for skewness and Shapiro-Wilk normality test. Variables that were not normally distributed were transformed using natural log, and rechecked for normality using the Shapiro-Wilk test. Arithmetic means and 95% confidence intervals (plus or minus 1.96 standard deviations) are presented for normally distributed variables, and geometric means and asymmetrical 95% confidence intervals are presented for those variables that had non-normal distributions. Values lying outside the mean plus or minus 1.96 standard deviations were excluded and the normal hematologic and plasma biochemistry adult reference ranges reported as the recalculated mean plus or minus 1.96 standard deviations for each variable [22, 23].

Effects of age, sex, body condition and reproductive status on hematologic and plasma biochemical parameters were assessed. Where values were reported lower than the detectable limit of the assay, half of the lowest detectable concentration was assigned and used in the comparative analyses. Data that proved to be positively skewed with heterogeneous variance, were transformed using the natural log (ln). Each hematologic or biochemical variable were subjected to an unbalanced generalised linear model [24], under the normal or log normal distribution as appropriate for continuous variables and the binomial distribution and logit link for binary variables, using GenStat [25]. Adjusted means and standard errors were estimated for each variable (standardised for age + sex + month + body condition + pregnancy + lactation). The residual plots for most variables proved to be approximately normal. The quoted mean levels were directly back-transformed from the ln-scale, and reported as geometric means. Adjusted means and 95% confidence intervals are reported for each variable where groups showed significant differences. Post-hoc t-tests between means were conducted for specific contrasts of the mating male cohort, using the individual standard errors for each mean.

RESULTS

Haematology and Biochemistry Reference Ranges

Hematologic and plasma biochemistry normal reference ranges for adult animals are presented in Table 1. Reference ranges for glucose were calculated from 993 animals that were involved in the greater study. Ten percent or less of the values for triglycerides, urea and ALT were lower than the lowest detectable limit of the assay. Approximately 90% of values for bilirubin were reported as <2 μmol/L, 70% of values for cholesterol reported as <0.5 mmol/L, 50% of values for GGT reported as <5 U/L and 25% of values for creatinine reported as <20 μmol/L. Reference ranges were not calculated for those variables with more than 10% of the values lower than the lowest detectable limit of the assay and the data are shown as the range of values obtained.

Cohort Comparisons

The majority of mean values for each cohort of significant difference fell within the demonstrated normal adult reference range. Statistical comparison of the mean values for each of the hematologic and biochemical variables, between the age cohorts, juveniles, sub-adults and adults showed significant differences for 17 parameters (Table 2). A significant increase in haemoglobin, red cell count, haematocrit and mean corpuscular haemoglobin concentration (MCHC) was observed with increasing maturity. Total leukocyte counts were highest in sub-adults and lowest in adults, neutrophil counts were highest in sub-adults and lowest in juveniles and lymphocyte and monocyte counts decreased with increasing maturity where juveniles had highest counts and adults the lowest. Aspartate aminotransferase (AST) and alanine aminotransferase (ALT) levels were lowest in sub-adults and highest in juveniles. ALP and phosphorus levels decreased with increasing maturity, where ALP levels for juveniles were highest, and higher than the normal reference range established for adult animals. Total protein, globulin and creatinine values increased with maturity while the albumin to globulin ratio and triglycerides decreased.

Table 2. Mean haematological and plasma biochemistry values for age cohorts with significant differences

Parameter	Mean (95% CI)			P
	Adult	Subadult	Juvenile	
Hb (g/L)	162.9 (161.4–164.4)	159.7 (154.2–163.9)	145.9 (137.3–153.1)	<0.001
RCC (x10^{12}/L)	9.10 (9.02–9.20)	8.96 (8.70–9.28)	8.41 (7.92–8.86)	<0.001
Hct	0.47 (0.465–0.473)	0.46 (0.446–0.471)	0.43 (0.402–0.444)	<0.001
MCHC (g/L)	347.7 (346.7–348.4)	347.5 (344.7–350.3)	341.0 (337.6–346.4)	0.018
WBC (x10^9/L)*	6.19 (4.13–8.25)	9.18 (6.99–11.37)	8.23 (5.94–10.52)	<0.001
Neut (x10^9/L)*	3.70 (3.50–3.91)	4.60 (3.85–5.48)	2.75 (2.06–3.67)	0.001
Lymp (x10^9/L)*	1.83 (1.67–2.01)	3.03 (2.25–4.10)	3.97 (2.44–6.45)	<0.001
Mono (x10^9/L)*	0.18 (0.00–0.31)	0.24 (0.00–0.61)	0.43 (0.00–1.00)	0.005
Creatinine (μmol/L)*	31.31 (30.11–34.50)	31.03 (23.13–37.08)	23.38 (16.74–44.97)	<0.001
AST (U/L)*	62.61 (58.85–66.62)	51.06 (41.06–63.50)	91.65 (58.21–144.32)	0.007
ALT (U/L)*	15.66 (14.78–16.59)	13.82 (11.25–16.98)	22.04 (14.37–33.82)	0.042
ALP (U/L)*	399.41 (373.90–426.67)	554.46 (440.54–697.85)	850.95 (716.81–985.09)	<0.001
Protein (g/L)*	65.04 (64.26–65.83)	61.31 (59.03–63.69)	60.46 (55.81–65.50)	<0.001
Globulin (g/L)*	29.02 (28.50–29.55)	26.87 (25.25–28.59)	24.12 (21.18–27.47)	<0.001
Alb/Glob*	1.24 (1.21–1.26)	1.29 (1.20–1.38)	1.48 (1.28–1.71)	0.004
P (mmol/L)*	1.36 (1.33–1.39)	1.66 (1.54–1.79)	1.83 (1.56–2.14)	<0.001
Triglyc (mmol/L)*	0.19 (0.18–0.20)	0.20 (0.17–0.24)	0.36 (0.26–0.50)	<0.001

Haematology: adult n = 293, sub-adult n = 39, juvenile n = 18. Plasma biochemistry: adult n = 306, sub-adult n = 36, juvenile n = 13.

* Natural log transformation.

Statistical comparison of the mean values for each of the variables between the sexes showed significant differences for 16 parameters (Table 3). Mean corpuscular haemoglobin (MCH), MCHC, platelet counts, neutrophil counts, glucose, triglyceride and cholesterol levels were higher in females than males. Males reported higher ALT, ALP, albumin, albumin to globulin ratio, creatinine, calcium, phosphorous, sodium and chloride levels compared to females.

Table 3. Mean haematological and plasma biochemistry values for male and females with significant differences

Parameter	Mean (95% CI)		P
	Male	Female	
MCH (pg)	17.80 (17.53–18.09)	18.10 (17.68–18.38)	0.011
MCHC (g/L)	347.20 (344.32–348.43)	348.00 (346.32–351.28)	0.01
Platelets (x10⁹/L)	359.20 (336.28–381.96)	380.00 (349.20–415.70)	0.013
Neut (x10⁹/L)*	3.67 (3.11–4.32)	3.84 (3.36–4.39)	0.007
Na (mmol/L)*	138.24 (137.03–140.30)	137.82 (135.80–139.32)	0.001
Cl (mmol/L)*	105.32 (103.59–107.08)	104.17 (102.33–106.04)	<0.001
Creatinine (µmol/L)*	38.47 (32.27–45.88)	31.98 (26.50–38.59)	0.003
Glucose (mmol/L)*	6.54 (6.17–6.95)	6.86 (6.39–7.37)	0.003
ALT (U/L)*	17.65 (14.98–20.80)	13.65 (11.73–15.89)	0.007
ALP (U/L)*	475.33 (436.59–517.50)	346.54 (316.08–379.93)	<0.001
Albumin (g/L)*	35.94 (35.22–36.69)	35.30 (34.48–36.15)	0.018
Alb/Glob*	1.29 (1.22–1.36)	1.19 (1.13–1.26)	0.003
Ca (mmol/L)*	2.46 (2.41–2.51)	2.36 (2.31–2.41)	0.031
P (mmol/L)*	1.43 (1.35–1.51)	1.34 (1.26–1.42)	0.05
Chol (mmol/L)*	0.27 (0.24–0.30)	0.45 (0.39–0.46)	<0.001
Triglyc (mmol/L)*	0.12 (0.11–0.14)	0.31 (0.27–0.35)	<0.001

Haematology: male n = 205, female n = 181. Plasma biochemistry: male n = 176, female n = 179.

*Natural log transformation.

Statistical comparison of the mean values for each of the variables between the reproductive cohorts of mating males, pregnant females and lactating females compared with non-reproducing adult cohorts, showed significant differences for 17 parameters (Table 4). A significant decrease in haemoglobin, red cell count, haematocrit, MCH and mean corpuscular volume (MCV) was observed in mating males. Mating males also had decreased globulin, triglyceride levels but increased platelet counts, creatinine and urea levels. Lymphocyte, monocyte, eosinophil counts, total protein, albumin, globulin, phosphorous, triglyceride and creatinine levels, and anion gap were all decreased in lactating females while calcium levels increased with lactation. Pregnant females had increased total protein, albumin, globulin, calcium and urea levels but decreased platelet counts.

Table 4. Mean haematological and plasma biochemistry values for reproductive cohorts with significant differences

Parameter	Mean (95% CI)					
	Mating male	**P**	**Pregnant female**	**P**	**Lactating female**	**P**
Hb (g/L)	160.90 (157.43–164.40)	<0.001	-	-	-	-
Hct	0.45 (0.45–0.46)	<0.001	-	-	-	-
MCV (fL)	50.00 (49.20–50.80)	0.015	-	-	-	-
MCH (pg)	17.66 (17.40–17.92)	0.001	-	-	-	-
Platelets (x1012/L)	389.40 (360.22–418.60)	0.045	345.80 (270.46–421.10)	0.021	-	-
Lymph ($x10^9/L$)*	-	-	-	-	1.60 (1.10–2.35)	<0.001
Mono ($x10^9/L$)*	-	-	-	-	0.11 (0.07–0.19)	0.035
Eos ($x10^9/L$)*	-	-	-	-	0.06 (0.03–0.14)	0.004
Anion gap (mmol/L)*	-	-	-	-	21.65 (20.27–23.12)	0.029
Urea (mmol/L)*	2.13 (1.63–2.77)	0.001	1.89 (1.07–3.33)	0.013	-	-
Creatinine (μmol/L)*	58.91 (50.80–68.32)	0.001	-	-	28.65 (21.75–37.72)	0.01
Protein (g/L)*	-	-	66.02 (62.62–69.61)	0.015	61.37 (58.66–64.22)	<0.001
Albumin (g/L)*	-	-	36.60 (34.64–38.66)	0.03	34.40 (32.83–36.04)	0.002
Globulin (g/L)*	28.08 (26.95–29.25)	0.042	29.28 (26.90–31.88)	0.039	26.71 (24.85–28.71)	0.002
Ca (mmol/L)*	-	-	2.49 (2.40–2.59)	0.004	2.41 (1.08–1.53)	0.006
P (mmol/L)*	-	-	-	-	1.28 (1.08–1.53)	0.007
Triglyc (mmol/L)*	0.12 (0.10–0.15)	0.027	-	-	0.14 (0.10–0.21)	<0.001

Haematology: mating male n = 30, lactating female n = 59, pregnant female n = 49. Plasma biochemistry: mating male n = 35, lactating female n = 54, pregnant female n = 59.

*Natural log transformation.

Statistical comparison of the mean values for each of the variables between the body condition cohorts poor, fair and good body condition, showed significant differences for 11 parameters (Table 5). A significant increase in haemoglobin, red cell count, haematocrit, total protein, albumin, globulin, calcium, anion gap and triglycerides was observed with improving body condition. Platelet, total leukocyte and eosinophil counts and cholesterol levels decreased with improving body condition.

Table 5. Mean haematological and plasma biochemistry values for body condition score cohorts with significant differences

Parameter	Mean (95% CI)			P
	Poor	Fair	Good	
Hb (g/L)	157.5 (154.8–160.6)	161.4 (160.3–163.1)	165.4 (163.0–168.2)	<0.001
RCC ($x10^{12}$/L)	8.87 (8.74–9.08)	9.04 (8.98–9.14)	9.21 (9.06–9.37)	0.001
Hct	0.45 (0.45–0.46)	0.47 (0.46–0.47)	0.48 (0.47–0.48)	<0.001
Platelets ($x10^{12}$/L)	389.80 (362.70–416.90)	372.40 (361.51–383.29)	355.1 (331.55–378.65)	0.044
Anion (mmol/L)*	21.98 (21.24–22.74)	22.38 (22.07–22.68)	22.76 (22.09–23.45)	0.018
Protein (g/L)*	62.99 (61.58–64.43)	64.59 (63.95–65.23)	66.22 (64.94–67.53)	<0.001
Albumin (g/L)*	34.61 (33.78–35.45)	35.62 (35.29–35.96)	36.67 (35.92–37.44)	<0.001
Globulin (g/L)*	27.99 (26.99–29.03)	28.73 (28.32–29.15)	29.49 (28.59–30.42)	<0.001
Ca (mmol/L)*	2.35 (2.31–2.39)	2.41 (2.39–2.43)	2.47 (2.43–2.50)	<0.001
Chol (mmol/L)*	0.36 (0.32–0.42)	0.35 (0.33–0.37)	0.33 (0.30–0.37)	0.041
Triglyc (mmol/L)*	0.17 (0.14–0.21)	0.19 (0.18–0.21)	0.21 (0.18–0.25)	0.012

Haematology: poor n = 38, fair n = 121, good n = 43. Plasma biochemistry: poor n = 33, fair n = 123, good n = 41.

*Natural log transformation.

Injured Animal Analysis

Hematologic and biochemical variables were analysed for 11 injured animals and compared to the established normal reference ranges. Values from 9 animals with recent minor or healing injuries, including abrasions, lesions, abscesses, and in one case a missing eye, showed little or no deviation from either hematologic or biochemistry reference values. An adult female with a torn anal sphincter returned an elevated neutrophil count (8.9 $x10^9$/L), while all other variables were within the normal ranges. The most severe injury observed was in a sub-adult female, which had extensive abscessation extending 7 cm along the dorsum of her spine, including deep 3 cm pockets in the axilla and ventral neck region, likely resulting from a raptor attack. Chest X-rays revealed significant opacity over the left lung field, and thoracocentesis confirmed the presence of a serosanguinous effusion, suggesting lung abscessation. This animal showed decreased haemoglobin concentration (132 g/L), red cell count (7 $x10^{12}$/L) and haematocrit (0.38), consistent with mild anaemia, while platelets (887 $x10^9$/L), white blood cells (19.5 $x10^9$/L), neutrophils (7.2 $x10^9$/L), lymphocytes (10.7 $x10^9$/L) and monocytes (1.6 $x10^9$/L) were all elevated. It also had elevated globulins (36 g/L), concomitant with a decrease in albumin (28 g/L), resulting in a decreased albumin:globulin ratio (0.8). Due to a poor long-term prognosis, the animal was euthanased.

DISCUSSION

This manuscript presents the first large study to establish baseline health data for a wild Australian *Pteropus* species, presenting haematological and plasma biochemical reference ranges for *P. alecto*. It also describes variation in these values related to changes in physiological and life cycle status. *P. alecto* is a known host of Hendra virus and, as bats have been identified as a source of a number of emerging pathogenic viruses of animal and human health significance, this work is relevant to "One Health" research and to the conservation and management of *Pteropus* species.

This study focussed on one flying-fox roost in South East Queensland (SEQ) considered representative of a dynamic regional population of flying-foxes. This contention is supported by observations of flying-fox roosts suggesting a potential 10% daily population turnover (Welbergen, personal communication), telemetry studies showing that *P. alecto* move frequently and up to 450 km between roost sites (Edson, personal communication), and genetic studies showing that *P. alecto* on mainland Australia are panmictic, suggesting that the mating between individuals is not influenced by any environmental, hereditary, or social interaction [26].

The normal reference ranges determined in this study represent values from clinically normal animals assessed by physical examination, identifying obvious lesions or abnormalities of the skin, bones, abdomen, thorax, mouth and nervous system. We excluded those animals with clinical signs of illness, in addition to outlying values. Accounting for underlying issues not presenting with clinical signs, such as infection with known or unknown viral, bacterial or parasitic agents, was beyond the scope of the study and thus their effect on the reference ranges is not established.

Heard *et al*., [27] concluded that plasma can successfully be used for biochemical analysis in flying-foxes, but may result in higher globulin and creatinine kinase values and lower potassium values compared to serum analysis. Plasma biochemical analysis was chosen in preference to serum analysis in this study, as the lithium heparin plasma matrix can also be used in a variety of other assays. The aforementioned findings associated with the use of the plasma matrix were considered in the interpretation of the findings.

Day *et al*., [28] found that values for glucose and chloride in heparinised blood samples decrease over storage time, while potassium and phosphorus levels increase with time. To avoid these issues, all lithium heparin blood samples were chilled immediately after collection, plasma separated by centrifugation and collected within 6 hours, and blood glucose measurements taken at the time of bleeding. The glucose measurements are considered to be putative fasting measurements as bleeding routinely occurred at least 2 hours post capture. However, we recognise that there may be variation in blood glucose levels dependent on the time between capture restraint and the time of sampling, due to a typical physiological gluconeogenesis response to capture stress [29].

Heard *et al*., [30] found that both short term physical restraint and isoflurane anaesthesia of *Pteropus hypomelanus* were associated with changes in a number of hematologic and biochemical variables, but found the overall magnitude of the effect was greater for the physically restrained animals. No stress leukogram response was observed with either restraint method. Given the animal welfare and human safety benefits, we used isoflurane anaesthesia to facilitate the collection of blood samples.

The mean hematologic and plasma biochemistry values and normal adult reference ranges reported here for *P. alecto* are consistent with previous reports for *Pteropus* species, with the exception of mean leukocyte, neutrophil and lymphocyte counts [19, 13, 31, 32, 33, 34]. In some *Pteropus* species, higher mean total leukocyte counts have been reported, usually associated with a higher mean lymphocyte count than we report in *P. alecto*. An approximate 2:1 ratio of mean neutrophil to lymphocyte counts was found for *P. alecto* in this study, which is similar to findings for *P. poliocephalus*, *P. scapulatus*, *P. conspicillatus*, *P. giganteus*, *P. voeltzkowi*, *P. vampyrus*, and *P. rodricensis*, while lymphocyte counts for *P. melanotus*, *P. pumilus*, and *P. hypomelanus*, were equal to or exceeded neutrophil counts [19, 23, 31, 32, 33, 34].

While all of the above species' mean leukocyte counts fell within established reference ranges for *Pteropus* species, care must be taken when interpreting hematologic data as reflective of a 'normal' population. It is not clear whether the differences among *Pteropus* species' mean leukocyte counts reported in other studies are due to physiologic or environmental differences, small sample size, sampling biases or different immune system stimulations indicative of physiological stress or disease status. The use of biochemical values as indicators of nutritional and health status of a population must also be considered with care and appropriate statistical analyses, as they may be confounded by large environmental variation [13]. We assert that for this study the large number of samples, exclusion of sick and injured individuals, year round sampling (encompassing all reproductive states and seasonal effects) and exclusion of outlying values prior to reference range calculation, make a robust normal reference range for *P. alecto*.

Comparison of the normal reference ranges established in this study for *P. alecto* with those of a number of other mammalian species demonstrates that the former are biologically plausible. Erythrocyte parameters are relatively consistent between flying-foxes, common brushtail possums, dogs and rats. Leukocyte parameters, where the predominant leukocytes present are neutrophils, are similar between flying-foxes, brushtail possums, canines and cats, but dissimilar to rats and cattle where lymphocytes are the predominant circulating leukocyte in healthy individuals [35, 36]. *P. alecto* biochemical parameters are similar to those of dogs, but differ with respect to liver enzymes, urea, creatinine and cholesterol values. This is plausibly due to the high protein and fat diet of carnivores compared to the low protein and fat diet of flying-foxes, which are predominantly blossom and fruit feeders [24].

As sex-specific seasonal changes were demonstrated by Welbergen [37] highlighting the difference in timing of maximal reproductive effort between sexes of *P. poliocephalus*, this study sought to investigate the possible effect of life stage by analysing differences between partitioned groups for sex, age and reproductive status. Care must be taken in interpreting the results if group size is small. Ruykys *et al*., [13] suggest a target size of 40 values to establish reliable comparisons. By this definition, we had small cohorts for mating males, immature animals and animals in poor body condition. With respect to immature animals and animals in poor body condition, the small data sets were due to the low numbers of these individuals captured and the decision on welfare grounds not to take volumes of blood exceeding 1% of body weight. The small data set for mating males was due to the short peak mating season, with only a limited number of animals able to be sampled during that period.

While differences were found for some parameters between age, sex and reproductive cohorts in the study, and we discuss these with respect to clusters of hematologic and biochemical parameters, the majority of mean values for each cohort of significant difference fell within the demonstrated normal adult reference range. The biological significance of these statistically significant findings should not be over interpreted, and it is likely that some of the observed differences are of little or no clinical relevance [38].

The increased ALP and phosphorous reported in young animals is expected where bone growth is occurring. AST and ALT levels were also significantly increased in juvenile animals. While ALT activity tends to be relatively specific to hepatic parenchymal cells, AST is found in a wide range of tissues, including the liver, heart, brain, skeletal muscle and erythrocytes. Whilst elevated AST can be an indicator of capture myopathy [14] a concomitant elevation in creatine kinase (CK) is also expected in such situations and was not seen in this study. The magnitude of difference between age cohort levels of AST was relatively modest, wherein situations of tissue insult, this enzyme would be expected to elevate several-fold, and hence it is most likely that the higher AST and ALT values in the juvenile cohort were representative of physiologic status and altered enzyme activity related to growth.

Total protein, albumin and globulin levels for immature animals tended to be low-normal, consistent with the expected higher physiological demand during growth. The erythrocyte panel demonstrated significant increases with increasing maturity, suggesting that the increasing trend of red cell indices represents not only a greater circulating red cell mass, but also a greater red cell content of haemoglobin. The lower leukocyte counts observed in adult flying-foxes is not unexpected given the mature status of their immune system. Higher absolute lymphocyte numbers observed in juveniles are a common finding in mammals of many species [31]. Absolute lymphocytosis can occasionally be significant in young animals due to excitement or immune stimulation which would be a natural response to the capture process [23], and are also reflective that the young animal is still adapting immune responses to its environment. Stress leukograms were an uncommon finding in this study with only two individual animals exhibiting lymphopenia in combination with neutrophilia.

The low lymphocyte, monocyte and eosinophil counts for lactating females raises the possibility that sex and reproductive status differences between variables could be related to different hormone profiles and behavioural characteristics. Although sex differences in red cell counts are common in animals, with males usually having higher values [14], we found no significant differences between sexes, apart from lower values in mating males which may indicate the high demand of mating. Total protein, albumin and globulin levels for lactating females tended to be low-normal, consistent with the expected higher physiological demand during reproductive life cycles. Higher albumin levels, often reported in lactating females as a function of hydration status [14] were not observed in the study. Lower levels of glucose were reported for males which also had high normal levels of both creatinine and urea during peak mating, possibly associated with dehydration or catabolism of muscle protein [14]. Females also had high normal levels of urea during pregnancy and lactation, also likely to be associated with these physiologically demanding life cycles.

The decreased level of triglycerides in lactating females and mating males may indicate higher physiological demand and a depletion of energy stores, supported by the demonstrated increasing triglyceride levels with improving body condition. Conversely cholesterol levels were found to decrease with improving body condition. This raises questions on the importance of lipid metabolism in flying-foxes that consume a predominantly low fat diet. Decreased levels of total protein, albumin and globulin levels were reported for animals in poor condition which is consistent with other studies of *Pteropus* species [24, 39] and may be due to a lower proteinaceous plane of nutrition. The erythrocyte panel demonstrated significant increases with improving body condition, consistent with previously reported findings that haemoglobin and red cell count are positively associated with general body condition and survival [13].

The presence of only a small number of injured or ill animals examined in this study is normal within a wild population, as seriously injured or sick wild animals would likely succumb to death by predation or natural attrition. While the cohort was small, this did afford the opportunity to compare hematologic and biochemical parameters between clinically normal individuals used to establish normal reference ranges, and those with evidence of injury. While values reported in animals with recent, minor or healed injuries or lesions, demonstrated little or no deviation from either normal hematologic or biochemistry reference values, when lesions were more significant and showed signs of infection, leukocyte counts were significantly elevated, demonstrating the expected normal mammalian response to insult and infection.

This study has advanced knowledge and understanding of the hematologic and plasma biochemistry values in *P. alecto*. While sample collection occurred at a single roost in southeast Queensland, observational, telemetry and genetic studies demonstrate the connectivity of *P. alecto* across their mainland Australian range, and thus the relevance of our findings to the species as a whole. Temporal analyses of changes in hematologic and biochemical parameters are necessary to investigate if there are seasonal changes or environmental stressor effects. Further investigations into correlation of parameters and disease prevalence may be important in establishing the parameters for monitoring disease and risk of disease in *P. alecto* populations.

ACKNOWLEDGMENTS

We acknowledge funding from the National Hendra Virus Research Program supported by the Queensland, New South Wales and Australian governments, and PhD scholarship funding (Lee McMichael) from the (past) Australian Biosecurity CRC for Emerging Infectious Diseases and the University of Queensland APA. We would like to thank the Queensland Centre for Emerging Infectious Diseases (QCEID) personnel for assistance in field collections and laboratory analyses, in particular, Lauren Goldspink, Miranda Vidgen, Alice Broos, Debra Melville and Craig Smith; staff of the Royal Society for the Prevention of Cruelty to Animals for diagnostic tests; DAFF regional field officers Perry Jones and Paul Willett, Scenic Rim Regional Council environmental officer Andrew McLaughlin and the Boonah police station officers and facilities with assistance in facilitating the field work.

AUTHOR CONTRIBUTIONS

Conceived and designed the experiments: LM DE HF. Performed the experiments: LM DE AM HF. Analyzed the data: LM DE AM DM SK. Contributed reagents/materials/analysis tools: LM DE. Wrote the paper: LM DE AM SK JM HF.

REFERENCES

1. Churchill S (2008) *Australian* Bats. Allen & Unwin, Sydney, Australia.
2. Murray K, Selleck P. Hooper P, Hyatt A, Gould A, Gleeson L, et al. (1995) A Morbillivirus that caused fatal disease in horses and humans. Science 268(5207):94–97. pmid:7701348
3. Chua KB, Bellini WJ, Rota PA, Harcourt BH, Tamin A, Lam SK, et al. (2000) Nipah virus: a recently emergent deadly paramyxovirus. Science 288(5470):1432–1435. pmid:10827955
4. Halpin K, Young PL, Field HE, Mackenzie JS (2000) Isolation of Hendra virus from pteropid bats: a natural reservoir of Hendra virus. J Gen Virol 81:1927–1932. pmid:10900029
5. Li W, Shi K, Yu M, Ren W, Smith C, Epstein JH, et al. (2005) Bats are natural reservoirs of SARS-like coronaviruses. Science 310(5748): 676–679. pmid:16195424
6. Calisher CH, Childs JE, Field HE, Holmes KV, Schountz T (2006) Bats: Important reservoir hosts of emerging viruses. Clinical Microbiology Reviews 19(3):531–545. pmid:16847084
7. Pourrut X, Souris M, Towner JS, Rollin PE, Nichol ST (2009) Large serological survey showing cocirculation of Ebola and Marburg viruses in Gabonese bat populations, and a high seroprevalence of both viruses in Rousettus aegyptiacus. BMC Infect Dis 9:159. pmid:19785757
8. Barr JA, Smith C, Marsh GA, Field H, Wang LF (2012) Evidence of bat origin for Menagle virus, a zoonotic paramyxovirus first isolated from diseased pigs. J Gen Virol 93:2590–2594. pmid:22915696
9. Ge XY, Li JL, Yang XL, Chmura AA, Zhu GG, Epstein JH, et al. (2013) Isolation and characterization of a bat SARS-like coronavirus that uses the ACE2 receptor. Nature https://doi.org/10.1038/nature12711
10. Field HE (2005) The ecology of Hendra virus and Australian bat lyssavirus. PhD thesis, The University of Queensland, Brisbane, Australia.
11. Plowright RK, Field HE, Smith C, Divljan A, Palmer C, Tabor G, Daszak P, et al. (2008) Reproductive and nutritional stress are risk factors for Hendra virus infection in little red flying foxes (*Pteropus scapulatus*). Proc Royal Soc 275(1636):861–869. pmid:18198149
12. Breed AC, Breed MF, Meers J, Field HE (2011) Evidence of endemic Hendra virus infection in flying-foxes (*Pteropus conspicillatus*)—Implications for disease risk management. PloS ONE 6(12):1–7.

13. Ruykys , Rich B, McCarthy P (2012) Haematology and biochemistry of warru (Petrogale lateralis MacDonnell Ranges race) in captivity and the wild. Aust Vet Journal 90(9):331–340. pmid:22928680
14. Clarke J, Warren K, Calver M, de Tores P, Mills J, Robertson I (2013) Hematologic and serum biochemical reference ranges and assessment of exposure to infectious diseases prior to translocation of the threatened Western ringtail possum (*Pseudocheirus occidentalis*). J Wildl Dis 49(4):831–840. pmid:24502710
15. Pacioni C, Robertson ID, Maxwell M, van Weenen J, Wayne AF (2013) Hematologic characteristics of the woylie (*Bettongia penicillata ogilbyi*). J Wildl Dis 49:4 816–830.
16. Wightman J, Roberts J, Chaffey G, Agar NS (1987) Erythrocyte biochemistry of the Grey-headed fruit bat (Pteropus poliocephalus). Comp Biochem Physiol 88B(1):305–307. pmid:3677608
17. Agar NS, Godwin IR (1992) Erythrocyte metabolism in two species of bats: common bent-wing bat (*Miniopterus schreibersii*) and red fruit bat (*Pteropus scapulatus*). Comp Biochem Physiol 101B:9–12.
18. O'Brien GM, Endean CK (2001) Normal blood values for Pteropus poliocephalus and P. scapulatus. Australasian Bat Society Newsletter 16:10.
19. Olsson AR (2002) Normal haematological and serum biochemical values and the effects of anaesthesia and parasites in the Spectacled flying-fox, *Pteropus conspicillatus*. MSc thesis, James Cook University of North Queensland, Australia.
20. Jonsson NN, Johnston SD, Field H, De Jong C, Smith C (2004) Field anaesthesia of three Australian species of flying fox. Vet Rec 154:664. pmid:15198316
21. Epstein JH, Prakash V, Smith CS, Daszak P, McLaughlin AB, Meehan G, et al. (2008) Henipavirus infection in fruit bats (*Pteropus giganteus*), India. Emerg Infect Dis 14(8): 1309–1311. pmid:18680665
22. Gascoyne SC, Bennett PM, Kirkwood JK, Hawkey CM (1994) Guidelines for the interpretation of laboratory findings in birds and mammals with unknown reference ranges—plasma biochemistry. Vet Rec 134: 7–11. pmid:8128571
23. McLaughlin AB, Epstein JH, Prakash V, Smith CS, Daszak P, Field HE, et al. (2007) Plasma biochemistry and haematological values for wild-caught flying foxes (*Pteropus giganteus*) in India. J Zoo Wildl Med 38(3):446–452. pmid:17939354

24. McCullagh P, Nelder JA (1989) Generalized Linear Models (2nd ed.). Chapman and Hall, London.
25. GenStat (2013) GenStat for Windows, Release 15.3. VSN International Ltd., Oxford.
26. Webb NJ, Tidemann CR (1996) Mobility of Australian Flying-Foxes, Pteropus spp. (Megachiroptera): Evidence from Genetic Variation. Proc Royal Soc 263 (1369): 497–502. pmid:8637931
27. Heard DJ, Ruiz MM, Harr KE (2006) Comparison of serum and plasma for determination of blood biochemical values in Malaysian flying foxes (*Pteropus vampyrus*). J Zoo Wildl Med 37(3):245–248. pmid:17319121
28. Day RL, Heard DJ, LaBlanc D (2001) The effect of time at which plasma separation occurs on biochemical values in small island flying foxes (*Pteropus hypomelanus*). J Zoo and Wildl Med 32(2):206–208.
29. McMichael LA, Edson D, Field H (2014) Measuring physiological stress in Australian flying-fox populations. EcoHealth 11: 400–408. pmid:24990534
30. Heard DJ, Huft VJ (1998) The effects of short-term physical restraint and isoflurane anaesthesia on haematology and plasma biochemistry in the island flying fox (*Pteropus hypomelanus*). J Zoo Wildl Med 29(1):14–17. pmid:9638618
31. Heard DJ, Whittier DA (1997) Hematologic and plasma biochemical reference values for three flying fox species (Pteropus sp.) J Zoo Wildl Med 28(4): 464–470. pmid:9523641
32. Clark P (2004) Haematology of Australian Mammals. CSIRO Publishing, Australia.
33. Hall J, Rose K, Smith C, De Jong C, Phalen D, Austen J, et al. (2013) Health assessment of the Christmas Island Flying fox (*Pteropus melanotus natalis*). J Wildl Dis 50(3):447–458. pmid:24807172
34. International Species Information System (2014) Available: www.isis.org
35. Wells RMG, Jones A, Clout MN, Sarre SD, Anderson RK (2000) Seasonal effects on the haematology and blood chemistry of wild brushtail possums, *Trichosurus vulpecula* (Marsupialia: Phalangeridae) in New Zealand. Comp Haem Int 10: 68–73.
36. Thrall MA (2004) Veterinary Haematology and Clinical Chemistry. Lippincott Williams and Wilkins, USA.

37. Welbergen JA (2011) Fit females and fat polygynous males: seasonal body mass changes in the grey-headed flying fox. Oecologica 165: 629–637. pmid:21153744
38. Thrusfield M (1997) Veterinary Epidemiology. 2 ed. Oxford: Blackwell Science.
39. Hossain MB, Islam MN, Shaikat AH, Yasin MG, Hassan MM, Islam SKMA, et al. (2013) Biochemical profile of wild-captured Indian flying fox (Pteropus giganteus) in Bangladesh. Bangl J Vet Med 11(1): 2308–0922.

CHAPTER 19

Serum Biochemistry Panels In African Buffalo: Defining Reference Intervals and Assessing Variability Across Season, Age and Sex

Claire E. Couch[1], Morgan A. Movius[2], Anna E. Jolles[1,2], M. Elena Gorman[2], Johanna D. Rigas[3], Brianna R. Beechler[2]

[1] Oregon State University, Department of Integrative Biology, Corvallis, Oregon, United States of America,

[2] Oregon State University, College of Veterinary Medicine, Corvallis, Oregon, United States of America,

[3] Utah State University, School of Veterinary Medicine, Logan, Utah, United States of America

Citation: Couch CE, Movius MA, Jolles AE, Gorman ME, Rigas JD, Beechler BR (2017) Serum biochemistry panels in African buffalo: Defining reference intervals and assessing variability across season, age and sex. PLoS ONE 12(5): e0176830. https://doi.org/10.1371/journal.pone.0176830.

ABSTRACT

Serum biochemical parameters can be utilized to evaluate the physiological status of an animal, and relate it to the animal's health. In order to accurately interpret individual animal biochemical results, species-specific reference intervals (RI) must be established. Reference intervals for biochemical parameters differ between species, and physiological differences including reproductive status, nutritional resource availability, disease status, and age affect parameters within the same species. The objectives of this study were to (1) establish RI for biochemical parameters in managed African buffalo (*Syncerus caffer*), (2) assess the effects of age, sex, pregnancy, and season on serum biochemistry values, and (3) compare serum biochemistry values from a managed herd to a free-ranging buffalo herd and to values previously published for captive (zoo) buffalo. Season profoundly affected all biochemistry parameters, possibly due to changes in nutrition and disease exposure. Age also affected all biochemical parameters except gamma glutamyl transferase and magnesium, consistent with patterns seen in cattle. Sex and reproductive status had no detectable effects on the parameters that were measured. The biochemical profiles of managed buffalo were distinct from those observed in the free-ranging herd and captive buffalo. Biochemical differences between buffalo from captive, managed, and free-ranging populations may be related to nutritional restriction or lack of predation in the context of management or captivity. The reference intervals provided in this study, in addition to the seasonal and age-related patterns observed, provide a foundation for health investigations that may inform management strategies in this ecologically and economically important species.

INTRODUCTION

Accurate methods for monitoring wildlife health are important for wildlife management, and for predicting threats to human and domestic animal health. Serum biochemistry parameters are often used to detect physiological abnormalities in animals that may reflect disease, but typical values for these parameters vary between populations, age classes, and across season. Thus, serum biochemistry values can only be used to measure wildlife health if typical serum biochemistry values for the population of interest are known.

Serum biochemistry parameters can vary significantly depending on individual age, sex, diet, and reproductive status. For instance, neonates and juveniles are known to have different biochemical parameters compared to adults, related to bone growth, an immature immune system, and distinct metabolism of nutrients [1–7]. Additionally, sex-related differences are often attributed to reproductive cycles or behavioral difference between the sexes [1,2]. Findings vary between species and studies [1,7] which may be due to different management systems or environments, or due to methodological discrepancies in sampling and storage procedures [8].

Season affects resource availability and reproductive cycles, leading to changes in nutrition, immunity, and behavior of foraging animals, which can result in significant effects on disease dynamics [9,10]. These seasonal patterns can lead to systemic biochemical fluctuations, such as changes in protein [11] and mineral [12] concentrations. It is important for disease monitoring purposes to be aware of normal seasonal variation in physiology. As shown by Perez et al, the use of statistical models can help to distinguish between the effects of disease and other seasonal effects on physiological parameters [13].

Environmental conditions such as nutrition, predation, and disease patterns may also differ between captive and wild populations, which may be reflected in variations of the serum biochemical parameters. Differences in biochemistry parameters between captive and free-ranging populations have been demonstrated for several wildlife species [14–17]. These differences are likely due to species-specific responses to capture and captivity, ranging from zoos to managed game parks. Additionally, free ranging and captive animals often have dramatically different diets, which can be reflected in serum levels of essential minerals such as calcium and phosphorus [18–19]. Disease is more intensively controlled in captive animals than in free-living populations, and differences in disease management strategies could contribute to altered biochemical parameters [20]. As such, it is crucial to consider the complex influences of environment factors on the physiological status of wildlife when interpreting biochemical parameters for wild animals.

African buffalo (*Syncerus caffer*) are valuable game animals that contribute to the economies of many countries in sub-Saharan Africa, notably through ecotourism, big game hunting, and game meat. On the other hand, they are natural reservoirs for several diseases that are important at the wildlife-livestock interface, including foot-and-mouth disease virus (FMDV), theileriosis, and brucellosis [21]. Although biochemical parameters have not previously been compared between free-ranging and captive buffalo, Beechler et al [22] demonstrated that hematological parameters vary between these two groups. Establishing reference intervals

(RI) for buffalo biochemical parameters provides a valuable tool for detecting changes in herd health, which could have important implications for management decisions. For instance, if nutritional parameters such as magnesium or phosphorous are decreased in captive buffalo, it could indicate that captive management methods need to be altered. If enzymes related to muscle damage are elevated, it may indicate capture stress and perhaps the need to seek alternative capture methods. Prior to this work, RI for African buffalo have been available for hematologic parameters [22] but not for biochemical parameters. Limited biochemical ranges have been published by Species360, which synthesizes opportunistically collected samples in zoos worldwide [23], but sample size in this database is too small to calculate RI. The objectives of this study were to (1) establish RI for serum biochemical parameters in managed African buffalo, (2) assess the effects of age, sex, pregnancy and season (early dry through late wet) on serum biochemistry values, and (3) compare the serum biochemistry values between a managed herd, a free-ranging herd, and previously published values for captive (zoo) buffalo.

METHODS

African buffalo included for this study (n = 418) were located within Kruger National Park (KNP), a 19,000 km^2 preserve located in northeastern South Africa which contains approximately 30,000 free-ranging buffalo [24]. Two separate populations were sampled for this study, and we also reference Species360 values for 20 captive buffalo for comparison [23]. The majority of our analyses were performed on a managed herd contained within a 900 hectare enclosure in the central area of KNP. Approximate size of this herd at any given time was between 50–65 due to births and deaths, and total number of animals sampled throughout the study period from the managed herd was 101. The managed herd was captured and sampled at 2–3 month intervals between February 2014 and August 2015 as part of an ongoing study on foot-and-mouth disease dynamics in buffalo. Four seasons were assigned based on rainfall patterns in KNP [25] with November-January designated early wet season, February-April designated late wet season, May-July designated early dry season and August-October designated late dry season (Table 1). Average rainfall and average temperature in the study area are highest during the early wet season (261 cm, 26 C), and decrease through the late wet (198 cm, 25.2 C) and early dry (26 cm, 18.8 C) before increasing again slightly during the late dry (62 cm, 21.4 C) [26]. A subset of data from this managed herd was used to generate RI, and evaluate the effects of season, age, sex and pregnancy on the biochemical parameters listed in Table 2. A second analysis compared the values obtained from the managed herd to samples obtained from an age-targeted cohort of 317 female free-ranging African buffalo followed for 4 years and captured and sampled every 6 months from 2008–2012 in southern KNP as part of a study on bovine tuberculosis [27]. All buffalo in both herds were marked so that individuals could be followed over time.

Table 1. Sample collection details. The number of samples collected from each age group across four seasons in a herd of managed buffalo contained in a 900 hectare enclosure within KNP

Season	Capture Month & Year	Calves <1 year	Subadults 1–5.5 years	Adults 5.5–15 years	Geriatrics >15 years
Early Dry	• June 2014 • June 2015	42	26	58	14
Late Dry	• August 2014 • August 2015 • October 2014	49	43	83	20
Early Wet	• December 2014	12	13	24	5
Late Wet	• February 2014 • February 2015	23	27	42	13

Table 2. Serum biochemistry parameters. Eleven biochemistry parameters commonly used in diagnostics were measured in this study and used to generate reference intervals

Parameter	Unit of Measurement	Description	Chemical Methods [34]	Relevant Citations
Albumin, Alb	g/dL	Protein generated by the liver that can change with liver disease, nutrition status, and hydration status	Dye binding with Bromcresol green.	[38, 39]
Globulins, Glob	g/dL	Proteins that function as carriers, enzymes, complement, and antibodies. May change with inflammation and hydration status	Calculated from total protein—albumin	
Total Protein, TP	g/dL	Albumin plus globulins	Protein forms a complex with copper in the presence of hydroxide, from which absorbance is measured.	

Alkaline Phosphatase, ALP	U/L	Enzyme produced in the liver and bone. Changes can result from liver and bone disease.	Measures rate at which ALP hydrolyzes ρ-Nitrophenyl Phosphate	[5, 39]
Aspartate Aminotransferase, AST	U/L	Enzyme produced in the liver and muscle. A change can result from liver and muscle damage.	Measures rate of NAD+ formation due to oxidation of product formed when AST catalyzes reaction of L-aspartate + α-ketoglutarate.	
γ-Glutamy ltransferase, GGT	U/L	Enzyme produced primarily in the liver and sensitive to biliary disease	GGT catalyzes the transfer of a glutamyl group from L- γ-glutamyl-3-carboxyl-4-nitroanilide to Gly-gly. Rate of 3-carboxy-4-nitroaniline production is measured.	
Creatine kinase, CK	U/L	Enzyme specific to muscle tissue. An increase signifies muscle damage.	Dephosphorylates Creatine Phosphate. Changes in absorbance due to NADPH production.	[34,39]
Blood Urea Nitrogen, BUN	g/dL	A byproduct of protein metabolism that is filtered by the kidneys	A coupled reaction of Urea+H_2O. Rate of NAD^+ formation is directly proportional to amount of Urea in sample.	[5, 39]
Calcium, Ca	mg/dL	Most common mineral in the body, needed for muscle contraction, bone and tooth formation, cardiac function, and blood clotting.	Binds with Arsenazo III to form calcium-dye complex	[5, 39, 40]
Magnesium, Mg	mg/dL	Mineral needed for muscle and nerve function, development of bone and teeth	Mg activates hexokinase to phosphorylate glucose. Mg concentration measured by increase in NADPH absorbance	[39]
Phosphorous, P	mg/dL	Mineral needed for muscle and nerve function, development of bone and teeth	A coupled reaction forms NADH and other products from sucrose and P. Amount of NADH formed is measured.	[5, 39]

Managed Herd Sample Collection

Buffalo for this study were captured in Northern KNP during the early 2000's and moved to an enclosure near Satara camp (Fig 1). The 900-hectare enclosure includes 50–65 buffalo and other herbivores and small mammalian predators typical of the ecosystem. Large herbivores (e.g. rhinoceros, elephants) and large predators (e.g. lions, leopards) are absent from the enclosure. Water is available to the buffalo at a natural pan (a seasonal pond) and a man-made water reservoir. Buffalo graze and breed as they would in the wild. The current study spans the period from February 2014 to August 2015. Data from a total of 8 captures taken every 2–3 months are included in this analysis. There were a total of 494 samples collected from 101 individual buffalo. Distribution of age categories (see demographic methods for details) by season is shown in Table 1. To sample the buffalo, they were herded into a capture corral, separated into groups of 4–10 animals, and sedated for data collection. Buffalo that could not be herded into the corral were darted individually from a helicopter.

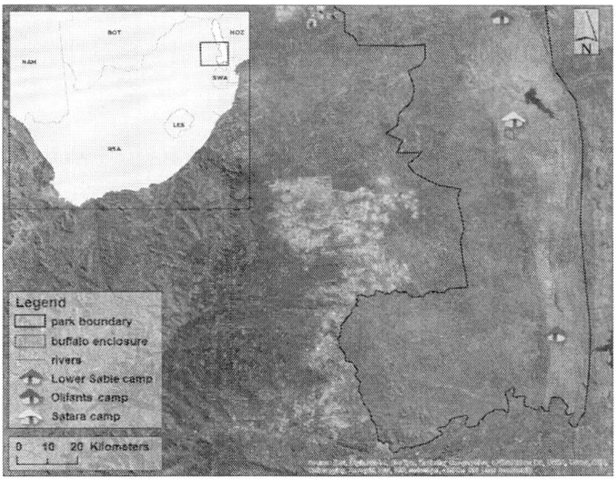

Figure 1. The location of African buffalo used in this study. The enclosure is located in the central portion of KNP near Satara and experiences only 500mm of rain per year, with the majority occurring during the summer months. Total size of the enclosure is 900 hectares, and is enclosed in a double fence to exclude large predators. The buffalo captured as part of the free-ranging herd were initially located in the south-eastern portion of the park centered around Lower Sabie, which experiences similar rainfall to the managed herd. The free-ranging herd was allowed to disperse as normal over the 4 years of the study which resulted in individuals being re-captured throughout the southern portion of the park, as far north as Olifants and spanning the geographic area the managed herd is maintained in [25].

Sedation was achieved with 7–10 ug etorphine hydrochloride and 0.04–0.07 mg azaperone per kg body weight for each buffalo. Blood was collected via jugular venipuncture with an 18-gauge needle directly into vacutainer tubes without additives within 5–30 minutes of sedation, and stored on ice for transport back to the laboratory. At the laboratory (approximately 6–8 hours post collection), blood was centrifuged at 5000g for 10 minutes and serum pipetted off the cellular layer into sterile microcentrifuge tubes and stored at -80°C until analysis. All animal work for this study was approved by the institutional animal care and use committee at Oregon State University, ACUP #4478 and at Kruger National Park.

Free-Ranging Herd Sample Collection

Between June 2008 and August 2012, an age-targeted cohort (median age 3.2 years, range 1–15 years) of female African buffalo were captured every 6 months for 4 years in KNP as part of a study on bovine TB [27–31]. Two hundred animals were originally captured from 3 locations spread across the Southern portion of KNP in 2008. Buffalo were re-captured and blood collected every 6 months until 2012. Over the course of the study, 117 animals died and were replaced with similarly aged animals for the duration of the study, resulting in 1750 total samples from 317 individuals. Blood was collected and serum obtained as described for the managed herd. This study was approved by the animal care and use committees at Oregon State (ACUP 3267), University of Montana (AUP 027-05VEDBS-082205), and Kruger National Park.

Demographic Parameter Collection

For both studies, age in years was calculated from incisor emergence and tooth wear [32], while sex was determined visually. Pregnancy status was determined by rectal palpation [22] by an experienced veterinarian and animals were recorded as either pregnant or not. Body condition was assessed by palpation of four regions (ribs, hips, spine, and base of tail) each of which are scored on a scale of 1–5 and then averaged across the four regions. This method has been shown to correlate with kidney fat index and total hematocrit [32].

Serum Biochemistry Panel Protocol

Frozen serum samples were thawed and stored at 4°C for 1–3 days prior to analysis using an Abaxis Vetscan VS2 (Abaxis Inc., Union City, CA, USA) chemistry analyzer and the large animal profile (Abaxis SKU500-023). This machine is maintained by Veterinary Wildlife Services in KNP and is serviced twice a year [33]. The machine is self-calibrating and is equipped with intelligent quality control (iQC) to verify chemistry, optics, and electronic functions of the analyzer during each run. Chemical methods, including sensitivity and specificity for each parameter in the large animal profile, are described by the manufacturer [34]. Parameters measured are described in Table 2. The parameters reported here are stable during transport and storage as we described, with the exception of CK [35]. The half-life of cattle CK is only 4–8 hours [36], but it is known to be stable at -80C [37]. Since our samples were not frozen for several hours after capture, CK values in buffalo serum may be underestimated here.

Reference Intervals

RI were calculated from the managed herd instead of from the free-ranging herd for two reasons: (1) this herd comprised both male and female buffalo across all age groups (calf to geriatric), whereas sampling from the free-ranging herd was age- and sex-targeted; and (2) because the free-ranging herds were not sampled evenly across season. RI were not calculated for the Species360 (captive) buffalo because there were only 20 individuals in the database. RI were calculated using Reference Value Advisor in Microsoft Excel [41] and were calculated following the NCCLS and Clinical and Laboratory Standards Institute (CLSI) guidelines. One randomly selected sample was included in the analysis for animals captured more than one time. Since adults had had significantly different values than calves, RI were calculated using only individuals greater than 5.5 years of age. To restrict our analysis to apparently healthy animals, only individuals with a body condition above 2.5 and that survived for at least 2 more months after capture were included. In conjunction with reference range calculation, we assessed the data for any outliers using the Tukey method [42, 43]. Any identified outliers were removed prior to generating reference rages. The final sample sizes reported exclude these outliers. Due to small sample size, we used a nonparametric robust method on box-cox transformed data to calculate reference intervals [42], similar to the robust approach described by Horn et al [43]. Parametric methods were not used due to small sample size [44]. General linear mixed models (GLMM) demonstrated no significant differences between sexes; therefore, RI were calculated for adults and not separated into male and female intervals. Separate reference intervals were calculated for the wet and dry seasons because of the seasonal variability demonstrated in the GLMM results (Table 3).

Table 3. Reference intervals for serum biochemistry parameters of managed adult buffalo

Parameter	**Adult Wet Season (n = 29 animals)**	**Adult Dry Season (n = 40 animals)**
Albumin (g/dL)	3.12–4.54	3.37–4.67
Globulin (g/dL)	3.55–6.25	2.52–5.12
Total protein (g/dL)	7.49–10.15	6.6–9.29
Albumin:Globulin	0.542–1.180	0.761–1.854
Alkaline phosphatase (ALP, U/L)	31.1–240.8	24.3–257.1
Aspartate aminotransferase (AST, U/L)	124.2–279.2	135.6–429.9
γ-Glutamyltransferase (GGT, U/L)	9.2–19.9	6.9–20.6
Creatine kinase (U/L)	0–693.4	233–2926.1
Urea nitrogen (BUN, g/dL)	3.7–15.8	5.9–24.5
Calcium (mg/dL)	7.12–9.41	6.99–9.5
Magnesium (mg/dL)	1.82–4.54	2.07–3.7
Phosphorus (mg/dL)	4.02–9.47	3.09–8.16
Calcium:Phosphorus	0.792–2.066	0.953–2.678

Statistical Analysis for Effects of Season, Pregnancy, Age and Sex on Biochemistry Parameters

GLMM were used to evaluate the effects of age (linear), sex (binomial), season (ordinal), and pregnancy (binomial) on each parameter. Multiple samples per individual were used for this analysis, and animal ID was included as a random effect. Separate models were used for each biochemistry parameter, with the concentration of the parameter as the response variable, age, sex, and season as fixed effects, and animal ID as a random effect. For all models, age was rescaled by centering and dividing by 2 standard deviations to remove the scale effect [45]. Gaussian error distribution was assumed. Akaike's information criterion (AIC) was used to estimate the relative quality of each model, and then select the model that minimizes loss of information [46]. Marginal R^2 values were estimated for each of the final models using the MuMIn package [47] following the protocol proposed by Nakagawa et al [48]. Many biochemical parameters vary non-linearly with age such that the parameters tend to be highest or lowest in animals of intermediate age groups [49]. These unimodal associations of focal parameters with age can be approximated by fitting models that include both age and age squared (age^2) as explanatory variables. As such, we considered effects of both age and age^2 on our focal parameters, and included age^2 if it resulted in a model with a significantly better fit based on an ANOVA F-test. Statistical significance was defined as $p \leq 0.05$. Statistical analyses were performed in R, using the lmerTest and lme4 packages [50–51].

Comparison of Free-Ranging Herd and Managed Herd Values

Effects of management (free-ranging herd vs managed herd) were evaluated using separate GLMM that accounted for age and season. Only females greater than 3 years of age were compared between the two herds since the free-ranging herd was an age and gender targeted cohort of buffalo. Age was rescaled for graphing purposes by dividing by two standard deviations to make all the x-axes equivalent. Values for captive (zoo) buffalo from Species360 were obtained, but they are not directly compared due to the small sample size and unknown background of these individuals.

RESULTS

Reference Intervals

RI for the managed herd were generated for adults during the wet and dry seasons (Table 3). There are large differences in the RI by season, notably AST, GGT, total protein (TP), BUN, CK, and albumin:globulin ratio (Table 3).

Significance of Age, Sex, Pregnancy Status, and Season

Sex and pregnancy status did not correlate significantly with any biochemical parameter. For this reason, these variables were not included in the final models. Seasonal variation significantly affected all parameters except GGT (Table 4, Fig 2), although none of the seasonal averages were outside the RI established in this study. Age and/or age^2 correlated significantly with all biochemical parameters except magnesium, GGT and BUN (Table 4, Fig 3). TP, globulin, ALP, and phosphorus show nonlinear responses with age while albumin, ALP, AST, CK and calcium respond linearly (Table 4, Fig 3). The percent of variance explained by individual differences is provided in Table 4 (marginal R^2).

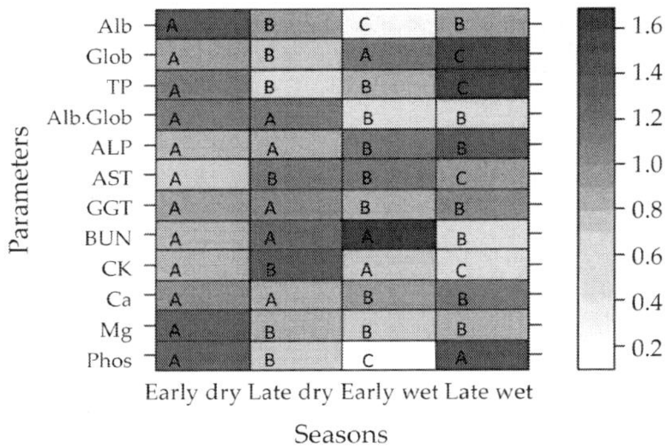

Figure 2. Managed African buffalo serum biochemistry parameters across seasons. Within each parameter, shared letters indicate no significant difference across seasons. The heat map shows the percent difference between seasonal average value and overall average for each parameter, with red being very high concentration and white being very low concentration. All parameters were centered and transformed so that color varies on the same scale for each parameter.

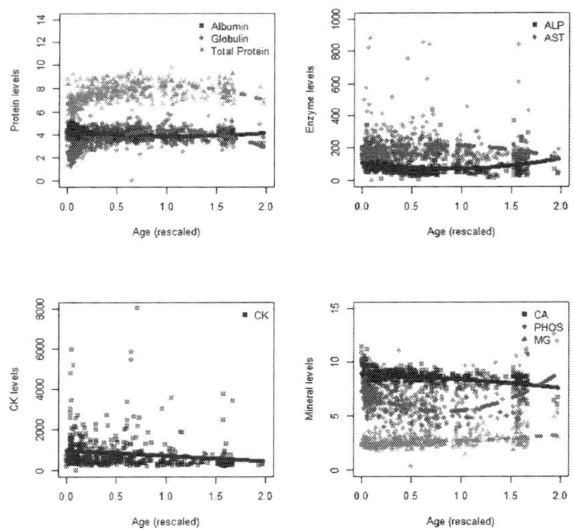

Figure 3. Effects of age on serum biochemistry parameters. Serum biochemistry parameters are only shown here if age or age squared are significant model parameters. The line represents the fitted line from the model while the points

are each individual data point. CK is shown on a separate graph due to scale. Age has been rescaled (see Methods).

Table 4. Effects of demographics and season in general linear mixed models for serum biochemistry values in managed buffalo

Parameter	Age		Age Squared		Season	Marginal R^2
	Estimate	Significance	Estimate	Significance		
Albumin	-0.77	***	0.37	**	*	0.17
Globulin	3.79	***	-1.79	***	*	0.40
Total protein	3.02	***	-1.43	***	*	0.37
Albumin:Globulin	-2.26	***	1.08	***	*	0.34
Alkaline phosphatase	-1.23	***	0.63	***	*	0.12
Aspartate aminotrans-ferase	0.43	*	-0.28	*	*	0.05
γ-Glutamyltransferase	NS	NS	0.098	0.29	NS	0.17
Creatine Kinase	-0.32	***	NA	NA	*	0.03
Blood Urea Nitrogen	NS	NS	NS	NS	*	0.37
Calcium	-0.38	***	NA	NA	*	0.18
Magnesium	NS	NS	NS	NS	*	0.18
Phosphorous	-4.39	***	2.60	***	*	0.02

Herd Comparisons

Although only the managed herd was used to generate reference intervals, data from the free-ranging herd were used to compare the two populations (managed vs. free-ranging) of buffalo. GLMM comparing individual serum biochemical values demonstrated that herd management (managed vs. free-ranging) was a significant variable in all parameters except for ALP and calcium after accounting for age, sex, and season (Table 5). Values for captive animals were obtained from Species360 and are included for reference, but were no direct comparisons were drawn with Species360 values because information on animal age and health status were not reported. Although statistical analysis including Species360 values was not possible, some of the parameters that show increase or decrease between managed and free-ranging buffalo follow the same trend in zoo animals. For instance, albumin decreases from free ranging wild a nimals to the managed herd, and is lower in zoo animals than managed animals. An inverse correlation exists in globulins and BUN, with the highest values in Species360 animals, followed by managed, then by free-ranging animals. Other parameters including calcium, CK, GGT, AST and ALP do not demonstrate a linear relationship from wild to managed to Species360 groups.

Table 5. General linear mixed model results comparing wild and managed herds, and mean serum biochemistry values for wild, managed, and captive African buffalo (+/- standard error of the mean). The captive values are from Species360 and are listed in italics as the data was not collected during this project, but is listed for comparison as the only prior information on these values in African buffalo. Asterisks indicate level of significance (* ≤0.05, **≤0.01, ***≤0.001) and parameters with statistically significant differences between managed and wild are bolded

Parameter	GLM results		Mean values (+/- SEM)		
	Estimate	P-value	Wild	Managed	*Captive* [23]
Albumin	0.72	$<2e^{-16}$*	4.6 +/- 0.02	3.89 +/- 0.37	*2.1 +/- 0.35*
Globulins	-1.22	0.00061*	2.29 +/- 0.07	3.57 +/- 0.89	*4.9 +/- 0.26*
Total protein	-1.21	0.03*	6.2 +/- 0.15	7.47 +/- 0.045	*7 +/- 0.21553*
Alkaline phosphatase	0.19	0.15	97.44 +/- 2.59	89.2 +/- 3.10	*190 +/- 30.67*
Aspartate aminotransferase	-0.40	$4.03e^{-9}$***	114.2 +/- 1.62	198.5 +/- 4.92	*187 +/- 21.46*
γ-Glutamyltransferase	-3.46	$1.12e^{-5}$***	12.21 +/- 0.11	17.43 +/- 0.74	*15 +/- 5.20*
Creatine kinase	-0.56	$1.50e^{-8}$***	330.1 +/- 7.78	731.5 +/- 46.34	*397 +/- 72.95*
Blood urea nitrogen	-0.431	$<2e^{-16}$***	8.43 +/- 0.14	13.66 +/- 0.36	*20 +/- 1.11*
Calcium	0.38	0.10	8.28 +/- 0.051	8.47 +/- 0.04	*9.1 +/- 0.19*
Magnesium	0.21	0.012*	2.78 +/- 0.02	2.69 +/- 0.08	*2.05 +/- 0.35*
Phosphorous	0.76	0.00015***	6.41 +/- 0.037	5.96 +/- 0.09	*2.05 +/- 0.44*

DISCUSSION

In this study, reference intervals were established for 11 serum biochemistry parameters in a herd of African Buffalo kept in a 900ha enclosure within KNP ("managed"). Comparisons were made with a free-ranging population in the same park. Effects of demographic parameters and season on the biochemical parameters were evaluated. Many of the biochemical parameters differed significantly between the managed and free-ranging herds, and also differed from values previously published for captive animals [23]. The provided reference intervals, and the comparative differences observed in this study will be useful for interpreting diagnostic tests.

Season consistently influenced all biochemistry parameters measured, with the exception of GGT. This finding illustrates the central role of environmental factors in mediating the physiology of wild animals. However, variation was not extreme enough to cause means for biochemical parameters in the wet season to fall outside of the RI generated for the dry season, or vice versa. Our results suggest that the observed variation reflects naturally fluctuating physiological shifts that may be due to seasonal variability in nutrition, water availability or disease exposure. Our findings are similar to a study by Yokus et al [12] which found that, in domestic cattle, magnesium and phosphorus levels increased in the spring in response to availability of lush pasture, while total protein levels were higher in the summer due to decreased hydration status.

In African buffalo, there is demonstrated seasonality in disease exposure [52, 53], altered immune defense [54] and altered host behavior due to foraging requirements [55] and reproductive timing [56, 57]. This may explain the seasonal variability in immunoglobulin levels and resultant decreased albumin: globulin ratio. Globulin is strongly associated with immune stimulation and inflammation, while albumin is a negative acute phase protein that decreases in response to inflammation, therefore this ratio is often used as an indicator of recent disease exposure. The decreased albumin: globulin ratio during the wet season could be attributed to either increased disease exposure, or alteration in immune status due to reproductive timing.

Nutritional intake and water availability is greatest at the end of the wet season and the beginning of the dry season, corresponding with increased body condition scores in the buffalo. Body condition scores then decrease throughout the dry season, and may reach conditions near starvation at the end of the dry or early wet season [58]. Seasonal variation was also seen in the physiological parameters reflecting nutritional intake with albumin, phosphorous and magnesium being highest in the wet season. Poor body condition and subclinical dehydration may also affect animals' response to stress [59]. Dry season conditions result in hypoperfusion of tissues which may predispose them to injury and increase tissue damage during capture events, as reflected in the high levels of CK and AST during the dry season. In addition, decreased hydration during the dry season likely affects renal perfusion causing mild azotemia, as suggested by high blood urea nitrogen concentrations when water was less available.

Somewhat surprisingly, there were no observable effects of sex and reproductive status on any of the biochemistry parameters measured. This latter finding is in contrast to prior studies in other free-ranging ungulates such as fallow deer and collared peccaries, where reproductive status affected ALP and AST, respectively [60, 61]. The lack of a direct effect of sex on biochemical parameters in this study may be due to the sample population, which consisted predominantly of adult females with immature calves and subadults, and only few sexually mature, adult males. Additionally, because females are typically only pregnant once every two years and are lactating in the interim to support calves, there may be too few non-lactating or non-pregnant females for comparison. Moreover, comparing differences between stages of pregnancy (early, mid, late) and/or stages of lactation may lead to altered results. This population did not have enough individuals to assess for this pattern.

In contrast to sex and reproductive status, variation of serum biochemical parameters was consistently demonstrated between different animal ages, although all averages remained within the normal reference intervals. These results demonstrate dynamic shifts in what may be considered normal depending on individual physiology and environment, and are important factors to consider when evaluating an individual biochemical profile. Some of the age-related changes found in this study were predictable based on previous research. Animals are born with underdeveloped immune systems and hepatic function which is reflected by their lower globulins, total protein, and urea nitrogen concentrations. Due to normal bone growth during the first year of life neonates typically have higher phosphorous and ALP concentrations relative to adult animals [62–64]. Several age related changes noted in this study were somewhat surprising, including increased albumin and CK concentrations. Albumin was highest in calves, which may be due to reduced hydration as they compete with larger animals for water access. CK was highest in calves and decreased linearly with age, suggesting that calves experience more acute muscle damage compared to adults during capture events, which may be due to increased trauma during capture and sedation.

Several differences in serum biochemistry values were observed between populations. Specifically, albumin, albumin: globulin ratio, phosphorous, and magnesium were elevated in the wild versus the managed population and globulins were lower in wild animals relative to the managed herd. African buffalo in KNP freely move about in fission fusion societies [65] with dispersal being common [25] and little genetic substructure [66], so it is unlikely that the differences observed between managed and wild herds are due to genetic variation. Additionally, GPS data from the free-ranging herd demonstrates that they ranged broadly throughout the southern portion of the park, and their ranges overlapped with the location of the managed herd [25], so differences in capture location are unlikely to explain differences in serum biochemistry. Biochemical differences may be a result of inadequate nutrition in the managed herd possibly due to reduced ability to migrate to better forage when fenced and debilitation due to increased disease exposure from higher population density [67, 68]. These findings could have implications for management on small game reserves where animals are maintained on limited acreage. AST, GGT, CK, along with urea nitrogen were all significantly elevated in the managed population. Differences in enzyme concentrations may be driven by underlying nutritional differences or other differences in management or capture methods. Values obtained from captive zoo animals [23] differ from the managed and wild herds, and the trends appear consistent with increasing disease and decreasing health in contained populations. However, it is important to remember that the Species360 database contains only opportunistically collected samples, which may be taken when the animal is immobilized to a health problem. Overall, the comparison between the free-ranging, managed, and captive values demonstrates that significant differences in biochemical parameters exist between buffalo populations depending on their management level, and highlights the importance of having reliable RI when assessing the health of an individual or herd.

Results from this study provide RI for a number of important biochemical parameters in African buffalo, and demonstrate the importance of changing environments and management techniques on the physiology of a herd. Moving forward, these RI can be used as a baseline to establish which parameters are the most sensitive and accurate indicators of individual health. This will provide an important tool for evaluating herd health and identifying at risk populations, and adds valuable information to design management strategies to improve overall husbandry.

SUPPORTING INFORMATION

S1 Table. ANOVA F-test results for each of the final models used to evaluate the effects of season and age on serum concentrations of biochemistry values in managed African buffalo. The F-test compares the model fit with the null model (no predictors other than the intercept) and gives the sum of squared residuals (SSR), numerator degrees of freedom (NumDF), denominator degrees of freedom (DenDF), F values, and probability based on an F distribution (Pr) for each model parameter

Biochemistry Value	Age					Age squared					Season				
	SSR	NumDF	DenDF	F	Pr	SSR	NumDF	DenDF	F	Pr	SSR	NumDF	DenDF	F	Pr
Albumin	1.0430	1	118.24	14.278	0.0002486 ***	0.5777	1	118.24	14.278	0.0057929 **	8.9085	3	388.92	40.650	< 2.2e-16 ***
Globulins	34.43	1	117.47	125.683	< 2.2e-16 ***	18.722	1	111.53	68.342	3.206e-13 ***	39.505	3	402.19	48.071	< 2.2e-16 ***
Albumin: Globulins	8.4368	1	122.13	112.974	< 2.2e-16 ***	4.7165	1	118.39	63.158	1.248e-12 ***	4.8707	3	388.75	21.741	4.896e-13 ***
Total protein	24.839	1	109.14	92.114	4.441e-16 ***	13.547	1	102.69	50.237	1.771e-10 ***	45.119	3	400.71	55.772	< 2.2e-16 ***
Alkaline phosphatase	2.0525	1	162.26	22.957	3.716e-06 ***	1.2835	1	159.32	14.355	0.0002142 ***	4.8838	3	406.47	18.208	4.193e-11 ***
Aspartate aminotransferase	0.8248	1	79.81	5.3975	0.02271 *	0.868	1	72.08	5.6804	0.01979 *	5.0376	3	392.86	10.9887	6.066e-07 ***
γ-Glutamyltransferase	0.32004	1	95.05	2.9075	0.09143	0.12591	1	85.27	1.1438	1.1438	0.59435	3	412.68	1.7998	0.14656
Creatine kinase	4.917	1	76.06	13.760	0.0003929 ***	NA	NA	NA	NA	NA	33.825	3	399.52	31.551	< 2.2e-16 ***
Blood urea nitrogen	0.1253	1	88.66	0.2550	0.6148	0.0001	1	76.29	0.0001	0.9903	12.2965	3	426.49	8.3444	2.107e-05 ***
Calcium	10.2834	1	85.06	42.986	4.058e-09 ***	NA	NA	NA	NA	NA	2.6712	3	399.34	3.722	0.01158 *
Phosphorous	59.594	1	115.98	45.624	6.024e-10 ***	51.239	1	108.39	39.227	7.777e-09 ***	179.178	3	410.65	45.725	< 2.2e-16 ***
Calcium:Phosphorous	10.741	1	127.65	6.7181	0.010656 *	11.383	1	111.80	7.119	0.008757 **	16.473	3	445.13	3.4344	0.016962 *

ACKNOWLEDGMENTS

This work would not have been possible without the assistance of Veterinary Wildlife Services at KNP for capture assistance, especially Peter Buss & Markus Hofmeyer. We also would like to acknowledge Michele Miller for allowing us to use her abaxis VS2 to perform the chemistry panels. We would like to thank Vanessa Ezenwa for allowing us to use the samples from her free-ranging buffalo study. Robert Spaan, Johannie Spaan, Hannah Tavalire, Caroline Glidden, Brian Dugovich, Danielle Sisson, Henri Combrink, Katherine Forssman and countless others were instrumental in the capture work.

Author Contributions

1. **Conceptualization:** BRB AEJ CC MM EG.
2. **Data curation:** CC MM.
3. **Formal analysis:** BRB CC.
4. **Funding acquisition:** AEJ BRB.
5. **Investigation:** BRB AEJ MM.
6. **Methodology:** AEJ BRB MM.
7. **Project administration:** AEJ BRB.
8. **Resources:** AEJ BRB JR EG.
9. **Supervision:** AEJ BRB.
10. **Validation:** JR EG.
11. **Visualization:** CC BRB AEJ.
12. **Writing – original draft:** CC MM BRB.
13. **Writing – review & editing:** BRB AEJ MM CC JR EG.

REFERENCES

1. Borjesson DL, Christopher MM, Boyce WM. Biochemical and hematologic reference intervals for free-ranging desert bighorn sheep. *J Wildl Dis*. 2009;36(2): 294–300.
2. Casas-Díaz E, López-Olvera JR, Marco I, Mentaberre G, Lavín S. Hematologic and biochemical values for Spanish ibex (*Capra pyrenaica*) captured via drive-net and box-trap. *J Wildl Dis*. 2008;44(4): 965–72. 10.7589/0090-3558-44.4.965
3. Bush M, Smith EE, Custer RS. Hematology and serum chemistry values for captive dorcas gazelles: variations with sex, age and health status. *J Wildl Dis*. 1981;17(1): 135–43.
4. English AW, Lepherd EE. The haematology and serum biochemistry of wild fallow deer (*Dama dama*) in New South Wales. *J Wildl Dis*. 1981;17(2): 289–95.
5. Doornenbal H, Tong AK, Murray NL. Reference values of blood parameters in beef cattle of different ages and stages of lactation. *Can J Vet Res*. 1988;52(1): 99
6. Tsuchiya N, Harada Y, Taki M, Minematsu S, Maemura S, Amagaya S. Age-related changes and sex differences on the serum chemistry values in Sprague-Dawley rats—I. 6–30 weeks of age. *Exp Anim*. 1995;43(5): 671–678.
7. Marler RJ. Some hematologic and blood chemistry values in two herds of American bison in Kansas. *J Wildl Dis*. 1975;11(1): 97–100.
8. Maceda-Veiga A, Figuerola J, Martinez-Silvestre A, Viscor G, Ferrari N, Pacheco M. Inside the redbox: applications of haematology in wildlife monitoring and ecosystem health assessment. *Sci Total Environ*. 2015;514:322–32. 10.1016/j.scitotenv.2015.02.004
9. Altizer S, Dobson A, Hosseini P, Hudson P, Pascual M, Rohani P. Seasonality and the dynamics of infectious diseases. *Ecol Lett*. 2006;9(4): 467–484. 10.1111/j.1461-0248.2005.00879.x
10. Jolles AE, Beechler BR, Dolan BP. Beyond Mice and Men: Environmental Change, Immunity and Infections in Wild Ungulates. *Parasite Immunol*. 2015;37(5): 255–66. 10.1111/pim.12153
11. Huber D, Kusak J, Žvorc Z, Rafaj RB. Effects of sex, age, capturing method, and season on serum chemistry values of brown bears in Croatia. *J Wildl Dis*. 1997;33(4): 790–94. 10.7589/0090-3558-33.4.790

12. Yokus B, Cakir UD. Seasonal and physiological variations in serum chemistry and mineral concentrations in cattle. *Biol Trace Elem Res*. 2006;109(3): 255–66. 10.1385/BTER:109:3:255
13. Perez JM, Serrano E, Soriguer RC, Gonzalez FJ, Granados JE, Cano-Manuel FJ, et al. Distinguishing disease effects from environmental effects in a mountain ungulate: seasonal variation in body weight, hematology, and serum chemistry among Iberian ibex (Capra pyrenaica) affected by sarcoptic mange. *J Wildl Dis*. 2015;51:148–156. 10.7589/2014-01-008
14. Broughton H, Govender D, Shikwambana P, Chappell P, Jolles AE. Bridging gaps between zoo and wildlife medicine: establishing normal reference ranges for free-ranging African lions (Panthera leo). *J Zoo Wildl Med*, in press
15. Moen R, Rasmussen JM, Burdett CL, Pelican KM. Hematology, serum chemistry, and body mass of free-ranging and captive Canada lynx in Minnesota. *J Wildl Dis*. 2010;46(1): 13–22. 10.7589/0090-3558-46.1.13
16. Schmidt DA, Barbiers RB, Ellersieck MR, Ball RL, Koutsos EA, Griffin ME, et al. Serum chemistry comparisons between captive and free-ranging giraffes (Giraffa camelopardalis). *J Zoo Wildl Med*. 2011;42(1): 33–39. 10.1638/2009-0218.1
17. Constable P, Hinchcliff K, Demma N, Callahan M, Dale B, Fox K, et al. Serum biochemistry of captive and free-ranging gray wolves (Canis lupus). *J Zoo Wildl Med*. 1998; 29(4): 435–440.
18. Miller M, Weber M, Valdes EV, Neiffer D, Fontenot D, Fleming G, et al. Changes in serum calcium, phosphorus, and magnesium levels in captive ruminants affected by diet manipulation. *J Zoo Wildl Med*. 2010;41(3), 404–408. 10.1638/2009-0001.1
19. Sullivan K, Van Heugten E, Van Heugten KA, Poore MH, Dierenfeld ES, Wolfe B. Analysis of nutrient concentrations in the diet, serum, and urine of giraffe from surveyed North American zoological institutions. *Zoo Biol*. 2009;29(4): 457–469
20. Miller LD, Thoen CO, Throlson KJ, Himes EM, Morgan RL. Serum biochemical and hematologic values of normal and Mycobacterium bovis-infected American Bison. *J Vet Diagn Invest*. 1989;1(3): 219–22. 10.1177/104063878900100304

21. Michel AL and Bengis RG. The African buffalo: a villain for inter-species spread of infectious diseases in Southern Africa. *Onderstepoort J Vet Res*. 2012;79(2): 453 10.4102/ojvr.v79i2.453
22. Beechler BR, Jolles AE, Ezenwa VO. Evaluation of hematologic values in free-ranging African buffalo (*Syncerus caffer*). *J Wildl Dis*. 2009;45(1): 57–66. 10.7589/0090-3558-45.1.57
23. Zoological Information Management System [Internet]. Bloomington (MN): Species360. [Updated: 2014 January 27; cited 2015 July 1]. http://www.species360.org
24. Cross PC, Heisey DM, Bowers JA, Hay CT, Wolhuter J, Buss P, et al. Disease, predation and demography: assessing the impacts of bovine tuberculosis on African buffalo by monitoring at individual and population levels. *J Appl Ecol*. 2009;46(2): 467–75.
25. Spaan RS. Dispersal behavior in African buffalo (Syncerus caffer): trade-offs between nutritional resources and disease exposure [MS Thesis]. Corvallis:Oregon State University; 2016.
26. South African National Parks [Internet]. Pretoria (South Africa). Average monthly and seasonal temperatures (°C) of the Kruger National Park; 2006 [cited 2017 March 11]. https://www.sanparks.org/parks/kruger/conservation/scientific/weather/rainfall/2006/temps_rainfall_averages.pdf
27. Ezenwa VO, Jolles AE. Opposite effects of anthelmintic treatment on microbial infection at individual versus population scales. *Science*. 2015;347(6218): 175–77. 10.1126/science.1261714
28. Beechler BR, Bengis R, Swanepoel R, Paweska JT, Kemp A, Vuren PJ, et al. Rift valley fever in Kruger National Park: do buffalo play a role in the inter-epidemic circulation of virus?. *Transbound and Emerg Dis*. 2015;62(1): 24–32.
29. Gorsich EE, Ezenwa VO, Jolles AE. Nematode—coccidia parasite co-infections in African buffalo: epidemiology and associations with host condition and pregnancy. *Int J Parasitol Parasites Wildl*. 2014;3(2): 124–34. 10.1016/j.ijppaw.2014.05.003
30. Henrichs B. Hidden dynamics: community ecology and costs of parasite infection in African buffalo (Syncerus caffer) [MS Thesis]. Corvallis: Oregon State University; 2014.
31. Budischak SA, Hoberg EP, Abrams A, Jolles AE, Ezenwa VO. 2015. A combined parasitological molecular approach for noninvasive

characterization of parasitic nematode communities in wild hosts. *Mol Ecol Resour*. 2015;15(5): 1112–119. 10.1111/1755-0998.12382

32. Ezenwa VO, Jolles AE, O'Brien MP. A reliable body condition scoring technique for estimating condition in African buffalo. *Afr J Ecol*. 2009;47(4): 476–81.
33. *Abaxis VetScan Operator's Manual*. Union City, California: Abaxis, Inc; 2009
34. *VetScan Large Animal Profile Product Information*. Union City, California: Abaxis, Inc; 2001.
35. Stockham SL & Scott MA. *Fundamentals of veterinary clinical pathology*. 2nd Ed Victoria, Australia: Blackwell Publishing; 2008.
36. Braun JP, Lefebvre H, Bezille P, Rico AG, Toutain PL. Creatinine Kinase in cattle: a review. *Rev Med Vet-Toulouse*. 1995;146(10): 615–622
37. Kale VP, Patel SG, Gunjal PS, Wakchaure SU, Sundar RS, Ranvir RK, et al. Effect of Repeated Freezing and Thawing on 18 Clinical Chemistry Analytes in Rat Serum. *J Am Assoc Lab Anim Sci*. 2012;51(4): 475–478.
38. Alberghina D, Giannetto C, Vazzana I, Ferrantelli V, Piccione G. Reference intervals for total protein concentration, serum protein fractions, and albumin/ globulin ratios in clinically healthy dairy cows. *J Vet Diagn Invest*. 2011;23(1):111–4. 10.1177/104063871102300119
39. Cozzi G, Ravarotto L, Gottardo F, Stefani AL, Contiero B, Moro L, et al. Short communication: Reference values for blood parameters in Holstein dairy cows: Effects of parity, stage of lactation, and season of production. *J Dairy Sci*. 2011;94(8):3895–901. 10.3168/jds.2010-3687
40. Lincoln SD, Lane VM. Serum ionized calcium concentration in clinically normal dairy cattle, and changes associated with calcium abnormalities. *J Am Vet Med Assoc*. 1990;197(11):1471
41. Geffré A, Concordet D, Braun JP, Trumel C. Reference Value Advisor: a new freeware set of macroinstructions to calculate reference intervals with Microsoft Excel. *Vet Clin Pathol*. 2011;40(1): 107–112. 10.1111/j.1939-165X.2011.00287.x
42. Tukey JW. *Exploratory data analysis*. Reading: Addison-Wesley Pub; 1977.
43. Horn PS, Pesce AJ, Copeland BE. A robust approach to reference interval estimation and evaluation. *Clin Chem*. 1998;44: 622–31.

44. Le Boedec K. Sensitivity and specificity of normality tests and consequences on reference interval accuracy at small sample size: a computer-simulation study. *Vet Clin Pathol.* 2016;45:648–656. 10.1111/vcp.12390
45. Gelman A. Scaling regression inputs by dividing by two standard deviations. *Stat Med.* 2008;27(15): 2865–873. 10.1002/sim.3107
46. Akaike H. Information theory as an extension of the maximum likelihood principle. Second international symposium on information theory, Akademiai Kiado, Budapest. 1973:267–81.
47. Barton K. MuMIn: Multi-model inference; 2016. https://CRAN.R-project.org/package=MuMIn
48. Nakagawa S, Scheilzeth HA. A general and simple method for obtaining R2 from generalized linear mixed-effects models. *Methods Ecol Evol.* 2013;4:133–42
49. Rodgers JD, Gass GH. The Effect of Age on Serum Proteins in Mice. *Exp Geront.* 1983;18(1): 39–45.
50. Kuznetsova A, Brockhoff PR, Christensen RHB. LmerTest: tests in linear mixed effects models; 2016 [Cited 2017 Feb 10]. https://CRAN.R-project.org/package=lmerTest
51. Bates D, Maechler M, Bolker B, Walker S. Fitting linear mixed-effects models using lme4. *J Stat Softw.* 2015;67:1–48.
52. Henrichs B, Oosthuizen MC, Troskie M, Gorsich EE, Gondhalekar C, Beechler BR, et al. Within guild co-infections influence parasite community membership: a longitudinal study in African buffalo. *J Anim Ecol.* 2016;85(4): 1025–034. 10.1111/1365-2656.12535
53. Gorsich EE, Bengis RG, Ezenwa VO, Jolles AE. Evaluation of the sensitivity and specificity of an enzyme-linked immunosorbent assay for diagnosing brucellosis in African Buffalo (*Syncerus caffer*). *J Wildl Dis.* 2015;51(1): 9–18. 10.7589/2013-12-334
54. Beechler BR, Broughton H, Bell A, Ezenwa VO, Jolles AE. Innate immunity in free-ranging african buffalo (*Syncerus caffer*): associations with parasite infection and white blood cell counts. *Physiol Biochem Zool.* 2012;85(3): 255–64. 10.1086/665276
55. Ryan SJ, Jordaan W. Activity patterns of African buffalo *Syncerus caffer* in the Lower Sabie region, Kruger National Park, South Africa. *Koedoe.* 2005;48(2): 117–124.

56. Ryan SJ, Knechtel CU, Getz WM. Range and habitat selection of African buffalo in South Africa. *J Wildl Manag*. 2006;70(3): 764–776.
57. Sinclair ARE. *The African buffalo: a study of resource limitation of populations*. Chicago: University of Chicago Press; 1977.
58. Ryan SJ, Cross PC, Winnie J, Hay C, Bowers J, Getz WM. The utility of normalized difference vegetation index for predicting African buffalo forage quality. *J Wildl Manag*. 2012;76(7): 1499–1508.
59. Dayger CA, Cease AJ, Lutterschmidt DI. Responses to capture stress and exogenous corticosterone vary with body condition in female red-sided garter snakes (*Thamnophis sirtalis parietalis*). *Horm Behav*. 2013;64(4): 748–54. 10.1016/j.yhbeh.2013.09.003
60. Poljicak-Milas N, Slavica A, Janicki Z, Robic M, Belic M, Milinkovic-Tur S. Serum biochemical values in fallow deer (*Dama dama L.)* from different habitats in Croatia. *Eur J Wildl Res*. 2004;50(1): 7–12.
61. Lochmiller R and Grant W. Serum chemistry of the collared peccary (*Tayassu tajacu*). *J Wildl Dis*. 1984;20(2): 134–40.
62. Vahala J, Kaše F. Serum chemistry profiles for Lechwe waterbucks (*Kobus leche*): variations with age and sex. *Comp Biochem Physiol B*. 1993;106(1): 47–51.
63. Husakova T, Pavlata L, Pechova A, Hauptmanova K, Pitropovska E, Tichy L. Reference values for biochemical parameters in blood serum of young and adult alpacas (*Vicugna Pacos*). *Animal*. 2014;8: 1448–455. 10.1017/S1751731114001256
64. Gurgoze SY, Icen H. The influence of age on clinical biochemical parameters in pure-bred Arabian mares. *J Equine Vet Sci*. 2010;30(10): 569–74.
65. Cross PC, Lloyd-Smith JO, Getz WM. Disentangling association patterns in fission-fusion societies using African buffalo as an example. *Anim Behav*. 2005;69(2): 499–506.
66. Van Hooft P. Population genetics of the African buffalo: from ecology to evolution [PhD Dissertation]. Wageningen: Wageningen University. 2001.
67. Spears JW. Micronutrients and immune function in cattle. *Proc Nut Soc*. 2000;59: 587–94.
68. Hoste H, Torres-Acosta JF, Paolini V, Aguilar-Caballero A, Etter E, Lefrileux Y, Chartier C, et al. Interactions between nutrition and gastrointestinal infections with parasitic nematodes in goats. *Small Rumin Res*. 2005;60: 141-

SECTION 7: WILD ANIMALS

7.3. FISH

CHAPTER 20

Physiological State Influences the Social Interactions of Two Honeybee Nest Mates

Geraldine A. Wright[1], Joshua L. Lillvis[2], Helen J. Bray[3], Julie A. Mustard[4]

[1] Centre for Behaviour and Evolution, Institute of Neuroscience, Newcastle University, Newcastle upon Tyne, United Kingdom

[2] Neuroscience Institute, Georgia State University, Atlanta, Georgia, United States of America

[3] School of Biology, Newcastle University, Newcastle upon Tyne, United Kingdom

[4] School of Life Sciences, Arizona State University, Tempe, Arizona, United States of America

ABSTRACT

Physiological state profoundly influences the expression of the behaviour of individuals and can affect social interactions between animals. How

Citation: Wright GA, Lillvis JL, Bray HJ, Mustard JA (2012) Physiological State Influences the Social Interactions of Two Honeybee Nest Mates. PLoS ONE 7(3): e32677. https://doi.org/10.1371/journal.pone.0032677.

physiological state influences food sharing and social behaviour in social insects is poorly understood. Here, we examined the social interactions and food sharing behaviour of honeybees with the aim of developing the honeybee as a model for understanding how an individual's state influences its social interactions. The state of individual honeybees was manipulated by either starving donor bees or feeding them sucrose or low doses of ethanol to examine how a change in hunger or inebriation state affected the social behaviours exhibited by two closely-related nestmates. Using a lab-based assay for measuring individual motor behaviour and social behaviour, we found that behaviours such as antennation, willingness to engage in trophallaxis, and mandible opening were affected by both hunger and ethanol intoxication. Inebriated bees were more likely to exhibit mandible opening, which may represent a form of aggression, than bees fed sucrose alone. However, intoxicated bees were as willing to engage in trophallaxis as the sucrose-fed bees. The effects of ethanol on social behaviors were dose-dependent, with higher doses of ethanol producing larger effects on behaviour. Hungry donor bees, on the other hand, were more likely to engage in begging for food and less likely to antennate and to display mandible opening. We also found that when nestmates received food from donors previously fed ethanol, they began to display evidence of inebriation, indicating that ethanol can be retained in the crop for several hours and that it can be transferred between honeybee nestmates during trophallaxis.

INTRODUCTION

Homeostatic mechanisms for the maintenance of physiological state regulate many important behavioural processes in animals including foraging for and consuming food. Solitary animals search for food in a cycle that mirrors their nutritional state: hungry animals are more likely to forage and to move, and satiated ones are more likely to be quiescent after eating[1]–[4]. In contrast, in social insect societies such as bee or ant colonies, food seeking behaviour is largely carried out by specialist foragers who continually forage such that they collect food and return to share it with other colony members. Although adult bees feed directly from colony stores of pollen or honey, foraging worker honeybees use a portion of the food they collect or they are often given food from nestmates (e.g. nurses) rather than eating honey reserves (reviewed in Crailsheim, 1998)[5]. Within colonies, food sharing between adults is facilitated by behavioural cues such as begging[6]; as a result of food solicitation, trophallaxis, or food sharing, occurs between the begging ant or bee and the donor from whom it solicits food. When entire colonies are starved, begging between nestmates increases[7]–[9]. Food sharing is also an important way that nestmates communicate their own hunger state to other nestmates to aid in the regulation of foraging within the colony[7],[10]. Whether or not hunger state influences the dynamic of food sharing beyond an increase in food solicitation is not well-understood.

Invertebrate animals are often studied to understand how neural circuits give rise to behaviour. One of the main contributions of invertebrates has been in providing insight into the molecular mechanisms underlying the effects of human-abused drugs on the brain. In*Drosophila, C. elegans*and honeybees, motor function, learning and memory, communication, and reward seeking are all affected by exposure ethanol[11]–[18]. As models, they have been successful because the molecular targets of ethanol in neural circuits in mammals are also expressed in the insect nervous system[19].

While it is clear that intoxication with drugs such as ethanol leads to changes in individual behaviour, we know less about how the change in state that accompanies drug consumption or exposure influences social interactions among animals. For example, when animals consume ethanol, inebriation could lead to fewer social interactions and social isolation if inebriated individuals fail to communicate or interact appropriately. Intoxication with drugs or toxic substances could also affect food sharing behaviour in social insects if intoxicated individuals are unwilling to share food with others or unmotivated to eat as a result of consuming ethanol. Only a few studies to date have examined social interactions in intoxicated or inebriated insects[14],[20]–[22]. The experiments carried out to understand how ethanol influences social interactions in honeybees have reported conflicting results[21],[22].

Here, we investigated how hunger and inebriation state influence food sharing behaviour between two honeybee nestmates. Using a newly developed assay for measuring individual and social behaviours in pairs of honeybees, we aimed to identify whether the physiological state of food donors produced predictable patterns of reaction to begging behaviour by hungry nestmates. What distinguishes our experiments from all preceding studies of social interactions between honeybee nestmates is that we controlled the satiety state of all bees by restraining them and feeding them 1.0 M sucrose 24 h prior to the assay. At the start of the assay, the donors were fed defined amounts of sucrose solution to standardize each donor's satiety state. By doing this, we controlled variation in satiety state that allowed us to observe clear differences in the behaviour of hungry and fed bees. We expect that the hungry bees in our study were in a greater state of food deprivation, and that this control over state combined with the detailed observation of interactions between two individuals made it possible to distinguish the state-induced changes in all behaviours we recorded. Our study emphasizes the importance of carefully defining physiological state prior to measuring nestmate interactions. In a previous study of the influence of ethanol on honeybee motor function, we characterized inebriation using a suite of behaviours including grooming behaviour and the righting reflex. We measured motor function as before but also included behaviours that described social interactions between the two bees assayed. To identify whether inebriation state influenced the extent to which donors were willing to share food, we also measured the amount of food transferred to the hungry nestmate and observed receivers in order to identify whether ethanol received via trophallaxis had an effect on their behaviour.

METHODS

Subjects

The study was performed using honey bees (*Apis mellifera*) collected from indoor colonies during Oct-Dec 2004 and from outdoor colonies during June 2005 at the Rothenbuhler Honeybee Laboratory at Ohio State University. Returning foraging workers were captured from the colony entrance using glass vials. In the lab, they were subjected to cooling anaesthesia and restrained using standardised techniques[23]. Once harnessed, each subject was fed 18 μl of 2.0 M sucrose, and kept at room temperature for 18–24 h prior to being used in an experiment.

Behavioural Observations

Two hours prior to the observation, each bee was fed 9 μl of one of the following treatments: 1.0 M sucrose, 5% ethanol in 1.0 M sucrose, or 10% ethanol in 1.0 M sucrose. (The sucrose concentration was held constant across all treatments.) Just prior to the observation, bees that were designated as the 'donor' were given 20 μl 1.0 M sucrose; the 'receiver' bees were not fed again. In the 'no food' treatment, the donor was not given the 20 μl 1.0 M sucrose prior to the observation. Observational arenas were composed of 100×15 mm plastic Petri dishes (Figure 1). The lid of the petri dish was cut in half to allow for a cut-to-fit piece of plastic to be placed between the two halves of the lid to separate the two bees. Both bees were briefly cooled and the bee that would be the focus for the observation was marked on the thorax with white out. The bees were then weighed and placed either side of the divider; they remained separated for 10 min to allow them acclimate to their surroundings before the divider was removed and the 30 min observation period began. All observations occurred between 2–7pm by JLL using The Observer software (Noldus Information Technology). We classified behaviour as either 'individual' behaviour (as described in Maze*et al.*, 2006)[15]or social behaviour. Individual behaviours included the following: walking, grooming, stopped and upside down (fanning and flying behaviour, as reported in Maze*et al.*2006, were not reported or analysed here because they were very infrequently observed). Social behaviours included: antennation, proboscis extension/licking (begging), trophallaxis, and mandible opening. After the observation period, both bees were removed from the observation arena and reweighed to confirm that fluid had been transferred during trophallaxis. (Whether or not the bee was giving or receiving trophallaxis could not be easily determined, so these two behaviours were grouped into a single behavioural category).

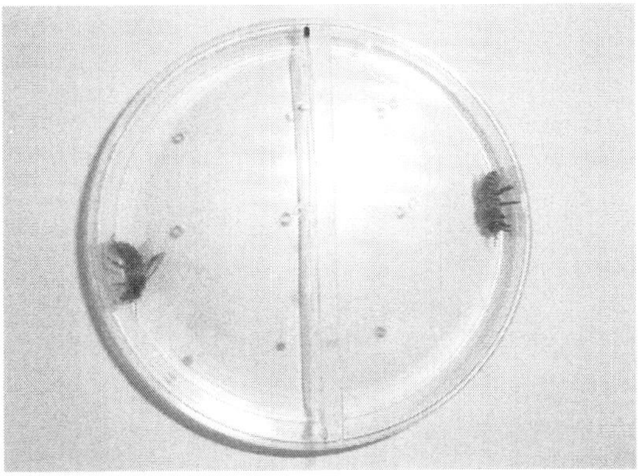

Figure 1. The observation arena. The bees were placed in the petri-dish separated by a plastic partition for 10 min prior to the experiment.

To determine whether the receiver bees obtained ethanol from the donors during trophallaxis, we also observed the behaviour of receiver bees when the donor had been given 20 μl of either 1.0 M sucrose or 1.0 M sucrose containing 5% ethanol prior to the observation. These observations were carried out as described above.

Statistical Analysis

All statistics were calculated using the software, SPSS. The observational data were compiled for 5 min time intervals and for lag sequential analysis using The Observer software and are reported in the fileData S1. We analysed the ontogeny of the behaviours in our 30 min interval for each behaviour using a repeated-measures general linear model (GLM) for the standardized amount of time engaged in the behaviour for each interval (as a percentage) and the mean bout duration as described in Maze*et al.*(2006) [15]. Bout frequency was analysed using repeated-measures Poisson Regression (PR)[24]. The behaviour of the receivers was analysed using multivariate ANOVA (MGLM). Multiple comparisons were calculated as least square differences (lsd) and were only reported as significant if they were less than P<0.05 after a Bonferroni correction. To identify whether it was possible to predict treatment using the behavioural data for the donor bees, we first reduced the dimensionality of the observation data using a principle components method for factor analysis[25]. The resulting scores for factors with eigenvalues greater than 1 were entered into a canonical discriminant analysis and used to classify the 4 donor treatments (sucrose, 5% ethanol, 10% ethanol, no food) in our experiment[25]. Differences in the weight of donors and receivers were calculated using ANOVA.

RESULTS

State Influences the Expression of Individual Motor Behavior in Donor Bees

A previous series of experiments in bees established that ethanol inebriation was characterized by a suite of changes in motor function that depended on the dose of ethanol bees had consumed[15]. Using this method, we measured the same variables for the donor and receiver bees in this assay (walking, stopped, grooming, upside down). For donors, we observed that in general the time spent walking increased at later intervals (Fig. 2A, interval main effect, GLM, $F_{1,63}=13.8, P<0.001$). However, on average, donor bees spent less time walking if they had been given an ethanol dose (main effect, GLM, $F_{3,63}=3.51, P=0.020$). The mean number of bouts of walking increased over time (Fig. 2B, main effect, Pois. Reg. $\chi_5^2=21.5, P=0.001$), but walking bout duration did not change over the observation period (Fig. 2C, main effect, GLM, $F_{1,58}=0.189, P=0.665$). Walking bout duration was on average longer for bees fed sucrose than for the other treatments (main effect GLM, $F_{3,58}$=5.58,*P*=0.002), but walking bout frequency was the same for all treatments (main effect, Pois. Reg. χ_3^2=4.47,*P*=0.215). In contrast, the amount of time spent stopped did not vary across the intervals (Fig. 2D, main effect, GLM, $F_{1,63}$=0.02,*P*=0.883), but it was affected by treatment (main effect, GLM, $F_{3,63}$=5.47,*P*=0.002). Stopped bout frequency clearly depended on whether or not the donors had been fed ethanol prior to the assay (Fig. 2E, main effect, Pois. Reg. χ_3^2=33.9,*P*<0.001), and donors fed 10% ethanol exhibited longer bout durations (Fig. 2F, main effect, GLM, $F_{3,58}$=3.54,*P*=0.020).

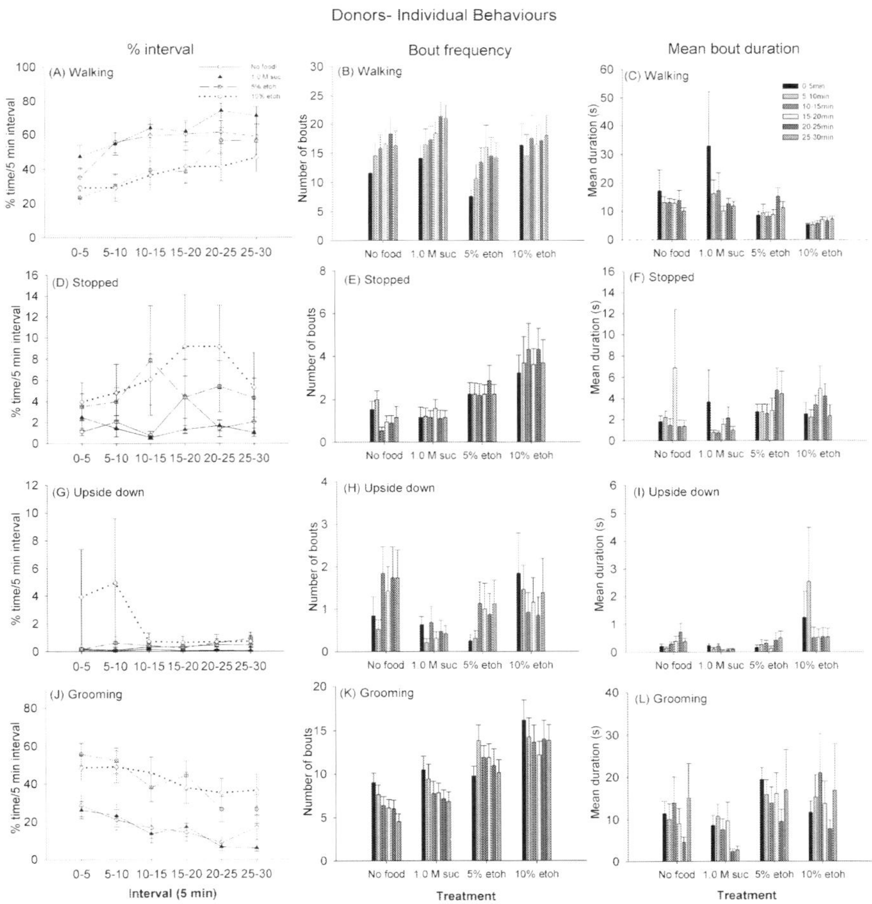

Figure 2. Hunger and inebriation state influence motor behaviour in donor honeybees. The time spent engaged in each of the behaviours, the mean bout frequency, and the mean bout duration are shown for each 5 min interval. (A–C) Bees spend less time walking if they had been pre-fed with a 1.0 M sucrose

solution containing ethanol; the ethanol treatments did not differ (P=0.381), nor did the hungry and fed bees (P=0.979). Bout duration and frequency were unaffected by treatment. (D–F) The amount of time spent stopped was greater for the donors pre-fed ethanol. The frequency of bouts of stopped behaviour was the same for the hungry and sucrose-fed bees (P=0.645), but the ethanol-fed bees had more frequent bouts of stopped behaviour than the controls (P<0.05 for all comparisons). Mean bout duration was not affected by treatment. (G–I) Ethanol treatment did not influence the total amount of time spent upside down (e.g. the ability to perform the righting reflex) but 10% ethanol-fed bees had longer and more frequent bouts of upside down behaviour (P<0.001 for all comparisons). Hungry bees were also more likely to perform bouts of upside down behaviour than the sucrose or 5% ethanol fed bees. (J–L) Donor bees fed ethanol spent more time grooming, and performed more bouts of grooming (P<0.05 for all comparisons) that lasted longer than hungry bees or those fed sucrose. N_{Hungry}=20, $N_{1.0\ M\ Suc}$=19, $N_{5\%\ etoh}$=17, and $N_{10\%\ etoh}$=13. ± SEM Note: Y-axis scale is not the same on all graphs.

Whether or not bees could right themselves (expressed as time spent upside down) and the time spent grooming were characteristic behaviours observed in inebriated bees[15]; bees fed doses of 25% or greater spend more time upside down, and bees fed 5–10% ethanol spend the most time grooming. In this study, we did not observe that donor bees fed ethanol spent more time upside down over the observation period (Fig. 2G, main effect GLM, $F_{3,63}$=1.53,P=0.216). However, the 10% ethanol-treated donors and the hungry donors had more frequent bouts of upside down behaviour those fed sucrose or 5% ethanol (Fig. 2H, main effect, Pois. Reg. χ_3^2=49.9,P<0.001). The 10% ethanol treated donors also had the longest bouts of upside down behaviour (Fig. 2I, main effect GLM, $F_{3,63}$=4.41,P=0.005). All of the bees spent more time grooming during the first 10 min of the observation, but the ethanol-treated donors were also more likely to spend time grooming during the entire assay (Fig. 2J, main effect, GLM, $F_{3,63}$=8.99,P<0.001). Ethanol-treated donors also exhibited more bouts of grooming (Fig. 2K, Pois. Reg. χ_3^2=33.8,P<0.001) that persisted for longer during each bout than either the hungry or sucrose-fed bees (Fig. 2 L, main effect GLM, $F_{3,58}$=312,P<0.001).

Hunger and inebriation affect social behavior in Donor Bees

We characterized social interactions in these experiments as antennation, begging (proboscis out to receive food), trophallaxis, and mandible opening. Antennation is the performance of contact chemoreception and, therefore, object recognition. The donor bees fed sucrose prior to the assay (the 1.0 M sucrose treatment and both ethanol treatments) generally spent more time antennating (Fig. 3A, main effect GLM, $F_{3,63}$=7.55,P<0.001) with more frequent bouts of this activity (Fig. 3B, main effect, Pois. Reg. χ_3^2=40.3,P<0.001). The sucrose-fed donors were more likely to spend time antennating than either the hungry bees (lsc,P<0.001) or either of the ethanol-fed bees (lsc, 5%:P=0.015; 10%:P=0.003). Time spent antennating was also greater during the first 15 min of the observation when the bees were first allowed to interact for all treatments (main effect, GLM, $F_{1,63}$=20.0,P<0.001). Mean bout duration was the same for all treatments (Fig. 3C, main effect, GLM, $F_{1,63}$=1.68,P=0.171).

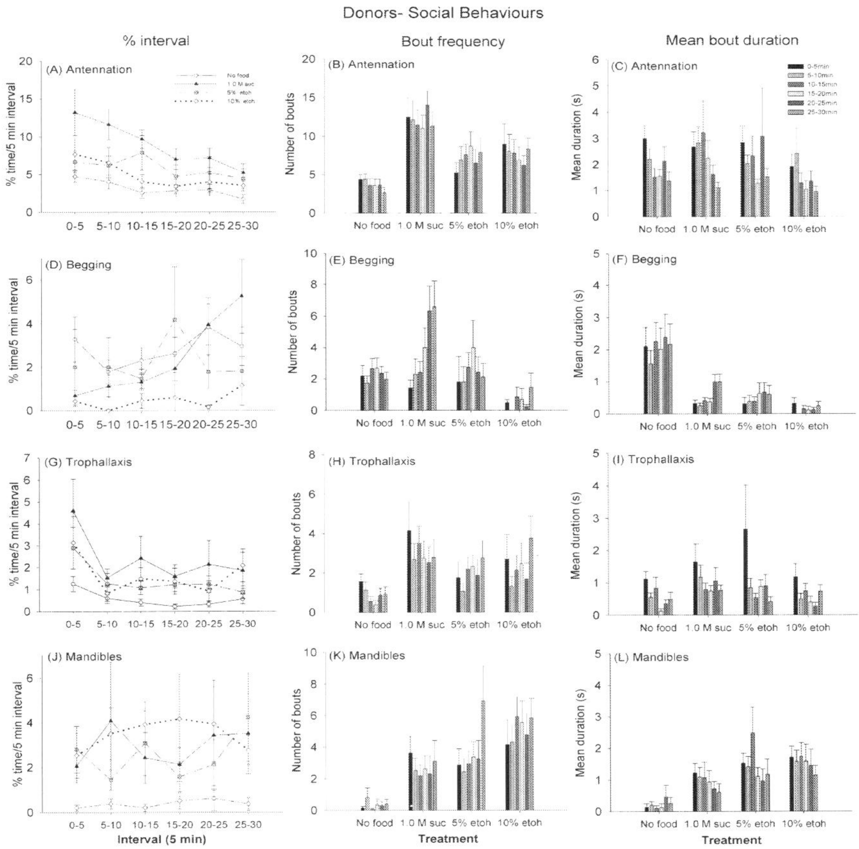

Figure 3. Hunger and inebriation state influence social behaviour. The time spent engaged in the each behaviour, the mean bout frequency, and the mean bout duration are shown for each 5 min interval. (A–C) Antennation was greatest at the start of the observation, and gradually decreased. The donors fed 1.0 M sucrose

spent more time antennating and had longer bouts of antennation than bees in any of the other treatment groups; the hungry bees spent the least time antennating, and had the shortest bouts of antennation (*P*<0.001 for all comparisons). (D–F) Begging behaviour (proboscis out) generally increased at later intervals for all treatments; the 10% ethanol-fed bees spent the least time begging. The hungry donors exhibited the longest bouts of begging. (G–I) Hungry bees were the least likely to spend time performing trophallaxis. They also had the least frequent and shortest bouts of trophallaxis. The ethanol treatments were not less likely than the sucrose fed bees to perform bouts of trophallaxis, nor were their bouts shorter (all comparisons,*P*>0.050). (J–L) The hungry bees were also least likely to exhibit mandible opening, but the other treatments did not differ in time spent performing mandible opening (all comparisons,*P*>0.050). Bout frequency and bout duration of mandible opening was greatest for the ethanol-fed donors (P<0.010). N_{Hungry}=19, $N_{1.0\ M\ Suc}$=19, $N_{5\%\ etoh}$=16, and $N_{10\%\ etoh}$=13. ± SEM Note: Y-axis scale is not the same on all graphs.

Begging and trophallaxis behaviour are the main ways that food is solicited and transferred among social insect nestmates. The amount of time spent begging depended both on whether bees were hungry or inebriated (Fig. 3D, main effect, GLM, $F_{3,63}$=4.11,*P*=0.010); the 10% ethanol-fed bees spent the least amount of time begging (*P*<0.05 comparisons with hungry and sucrose-fed bees). Donors in all of the other treatment groups begged more frequently than the 10% ethanol-fed bees (main effect, Pois. Reg. χ_3^2=10.5,*P*=0.014). The hungry bees had the longest bouts of begging (Fig. 3F, main effect, GLM, $F_{3,58}$=11.4,*P*<0.001). In fact, their bouts of begging were 2–3 times longer than any of the bees in the other treatments. All groups began to spend more time soliciting food during the last 10 min of the observation period (main effect, GLM, $F_{1,63}$=11.1,*P*=0.001).This increase in begging behaviour was most dramatic in the sucrose-fed donors, as the frequency of begging bouts increased significantly at later intervals (Fig. 2E) (interaction, Pois. Reg. χ_{14}^2=102,*P*<0.001).

Time spent performing trophallaxis, or food donation, was greatest at the start of the assay for all the donor bees (Fig. 3G, main effect, GLM, $F_{1,63}$=5.32,*P*=0.024); the bout duration was also longest at this time (Fig. 3 I, main effect, GLM, $F_{1,58}$=11.2,*P*=0.001). Trophallaxis bout frequency did not vary over the observation, (Fig. 3H, main effect, Pois. Reg. χ_5^2=9.20,*P*=0.102), but the sucrose and ethanol-fed donor bees were more likely than the hungry bees to perform bouts of trophallaxis (time spent, main effect GLM, $F_{3,63}$=3.53,*P*=0.020, bout frequency, main effect, Pois. Reg. χ_3^2=15.7,*P*=0.001). Inebriation state did not affect the time spent performing trophallaxis (lsc,P>0.050), and inebriated bees did not perform fewer bouts on average than the sucrose-fed bees (lsc,P>0.050).

Mandible opening is a form of aggression between social insect nestmates elicited to demand food[26]; in honeybees, it has also been shown to precede food donation[6]. The time spent displaying this behaviour was more or less constant over the entire observation period for all groups (Fig. 3J, interaction, GLM $F_{3,63}$=0.118,P=0.949), but the hungry donors were the least likely to display this behaviour of all of the treatment groups (lsc, P<0.050). The bout frequency of mandible opening increased for the inebriated bees at later intervals, but not for the hungry or sucrose-fed donors (Fig. 3K,2-way interaction, Pois. Reg χ_{15}^{2}=54.2,P<0.001). The ethanol-fed donors were also most likely to exhibit bouts of mandible opening (lsd,P<0.05 for all comparisons). Bout duration of mandible opening was approximately as long on average as bouts of begging or trophallaxis for all bees (i.e. 2–4 s,Fig. 3L). Bouts of mandible opening were longest in duration at the beginning of the assay, when the receiver bees were also most likely to be solicit food from the donor bee (main effect GLM, $F_{1,58}$=6.45,P=0.014). The sucrose and ethanol-fed bees exhibited longer bouts of mandible opening than the hungry bees (main effect GLM, $F_{3,58}$=10.2,P<0.001).

To identify whether mandible opening behaviour was a form of food offering behaviour, we performed a lag sequential analysis to look at the pattern of behaviour immediately following mandible opening. We specifically compared whether mandible opening was followed by trophallaxis or by a second bout of mandible opening behaviour. Using lag sequential analysis with a lag of 1 (i.e. the behaviour immediately following the focal behaviour) on the bouts of mandible opening, we found that hunger and inebriation did not influence the probability of performing mandible opening followed by trophallaxis (food offering behaviour) (Fig. 4,1-way ANOVA, $F_{3,142}$=0.597,P=0.618). On the other hand, inebriated bees, and in particular the 10% ethanol-fed bees, were more likely to repeatedly open and close their mandibles without performing trophallaxis (1-way ANOVA, $F_{3,142}$=2.82,P=0.041). (We also examined the probability that mandible opening was followed by the other social behaviours, antennation and begging. Treatment did not influence the probability of observing this sequence of behaviour for either variable, antennating: 1-way ANOVA, $F_{3,142}$=1.54,P=0.206; begging: 1-way ANOVA, $F_{3,142}$=1.88,P=0.136, data not shown).

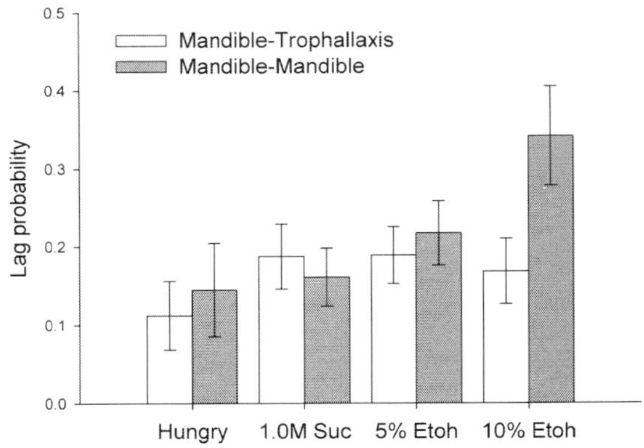

Figure 4. Lag sequential analysis of mandible opening behaviour of the donor bees revealed that the donors fed 10% ethanol were more likely to have bouts of mandible opening that were immediately followed by a second bout of mandible opening than bees receiving any of the other treatments. The probability that a mandible opening bout was followed by another mandible opening bout was the same for all treatments except the 10% ethanol-fed donors. N_{Hungry}=24, $N_{1.0\ M\ Suc}$=46, $N_{5\%\ etoh}$=46, and $N_{10\%\ etoh}$=27 (N refers to number of observations of mandible opening behaviour that were followed by antennation, begging, trophallaxis, or mandible opening in each group). ± SEM.

Social Behaviour Predicts Hunger and Inebriation State in Donor Bees

We tested whether it was possible to classify the satiety and inebriation state of the donor honeybees using the entire suite of behavioural variables that we measured. This analysis was performed using standardized amount of time spent per 5 min interval (mean percentage of time spent per interval). To do this, we first reduced the dimensionality of the data by performing a factor analysis (FA) on all the behavioural variables for each interval (6 intervals for 8 variables;Table S1). The factor scores generated for the 13 factors with eigenvalues over 1 were then entered into a canonical discriminant analysis (CDA).

The CDA produced two significant classification functions. The first function accounted for 50% of the variation in the data and separated donor bees fed sucrose solution containing ethanol from those fed nothing or 1.0 M sucrose (CDA, χ_{39}^{2}=77.5,P<0.001;Table 1). As indicated by the standardized discriminant function coefficients, this function was mainly derived from information from the first and second factors of the FA. These factors predominantly represented variation in the individual motor behaviours; for example, in F1, walking and grooming had strong factor loadings that were inversely correlated (as represented by the sign of the factor loading) and in F2, upside down behaviour was strongly represented, as indicated by the magnitude of the factor loading (Table S1). The 10% ethanol-fed bees, defined by the fact that they were less likely to walk, more likely to groom, and more likely to be upside down, were significantly separated from the bees that were not fed ethanol (No food and 1.0 M sucrose groups).

Table 1. Canonical Discriminant Analysis for All Behaviours

	Discriminant function		
	1	2	3
Eigenvalue	0.888	0.628	0.253
% variance	50.200	35.500	14.300
Wilks' Lambda	0.260	0.490	0.798
Standardized CDA Function Coefficients			
Factors			
F1	**0.814**	0.013	0.047
F2	**−0.550**	0.403	0.290
F3	−0.153	**0.756**	−0.073
F4	0.370	**0.638**	−0.164
F5	0.055	0.246	0.042
F6	−0.118	0.115	0.282
F7	0.060	0.194	0.298
F8	−0.193	0.153	0.000
F9	−0.013	0.119	**0.549**
F10	0.478	0.021	**0.563**
F11	0.182	−0.003	−0.005
F12	0.114	−0.051	0.362
F13	0.353	0.310	−0.338
Non-standardized Function Coefficients			
Treatment			
No food	−0.803	−0.958	−0.074
1.0 M sucrose	−0.826	1.117	−0.056
5% Ethanol	0.825	−0.009	0.711
10% Ethanol	1.237	0.026	−0.743

The second function accounted for 35.5% of the variation and significantly separated the donor bees fed 1.0 M sucrose from the hungry, unfed bees (CDA, $\chi_{24}{}^{2}$=40.9,P=0.017;Table 1). This function was largely represented by data from factors 3 and 4 in the PCA (Table S1); F3 was characterized by the fact that bees that exhibited mandible opening were also less likely to beg for food (factor 3) whereas F4 predominantly represented a correlation between trophallaxis and upside down behaviour (factor 4). Notably, in this function the ethanol-fed bees had similar unstandardized function coefficients, indicating that they had very similar behaviours.

To verify our conclusions above, we performed the FA-CDA analysis on the individual (walking, stopped, upside down, grooming) and social (antennation, begging, trophallaxis, mandibles open) behaviours separately. The analysis of the individual behaviours yielded one significant classification function that, as before, separated the ethanol-fed bees from those that did not receive ethanol (CDA, $\chi_{18}{}^{2}$=38.4,P=0.003;Table S2). Likewise, when the FA-CDA analysis was repeated on the social variables alone, we obtained one classification function that separated the hungry bees from those fed with 1.0 M sucrose prior to the experiment (CDA, $\chi_{23}{}^{2}$=43.1,P=0.003;Table S3).

Receivers Display Evidence of Ethanol-Induced Changes in Behaviour

In addition to observing state-based changes in the behaviour of donor bees, we also examined how the behaviour of receivers was affected when they interacted with either sucrose-fed or 5% ethanol-fed donors. In these observations, we specifically aimed to identify whether interacting with an ethanol-fed donor led to inebriation in the receiver. The receivers in both groups spent the same amount of time engaging in trophallaxis early in the observation (Fig. 5G, main effect, MGLM, $F_{1,126}$=0.260,P=0.611), indicating that food was probably transferred from the donor bee to the receiver. As expected if ethanol had been transferred to the receiver bees, the receivers that had been paired with a 5% ethanol-fed donor spent less time walking (Fig. 5A, main effect, main effect, MGLM, $F_{1,126}$=11.7,P<0.001) and more time grooming (Fig. 5D,main effect, MGLM, $F_{1,126}$=15.1,P<0.001) than those paired with a donor fed 1.0 M sucrose. Consistent with the behaviour of the donor bees, the amount of time spent upside down was not significantly greater in the receivers paired with the 5% ethanol donors (Fig. 5Cmain effect, main effect, MGLM, $F_{1,126}$=2.18,P=0.142), nor was the amount of time spent stopped (Fig. 5B, main effect, MGLM, $F_{1,126}$=0.180,P=0.672). The only social behaviour affected by whether or not a receiver was paired with a 5% ethanol donor was the amount of time spent antennating (Fig. 5E, interval×treatment, MGLM, $F_{5,126}$=3.31,P=0.008); time spent performing the other behaviours did not depend on the treatment (Fig. 5F,G,H, begging: $F_{5,126}$=1.89,P=0.171; trophallaxis: $F_{5,126}$=0.260,P=0.611; mandibles open: $F_{5,126}$=0.009,P=0.927). In general, the mean bout duration and the frequency of both individual and social behaviours were reflected in the trends seen in the percent of the time spent during the interval (Figure S1andS2).

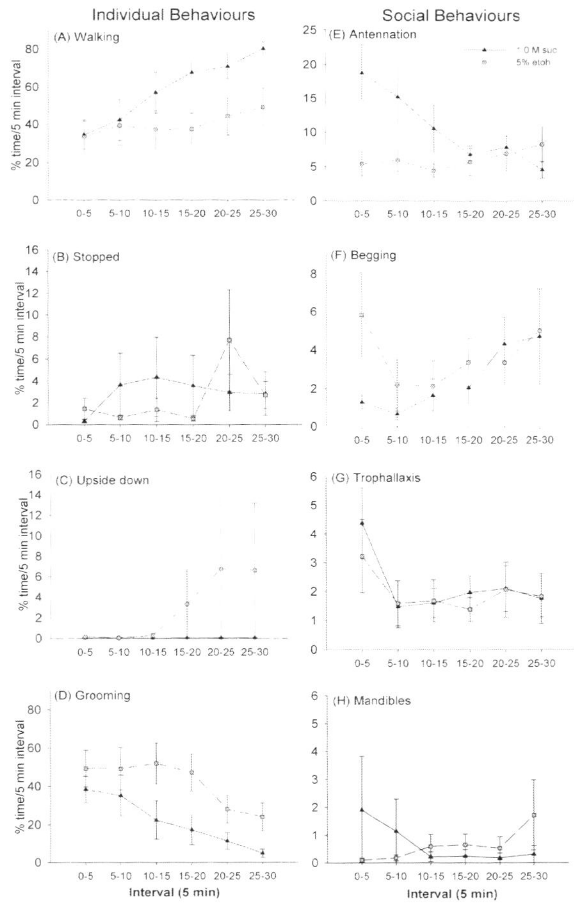

Figure 5. Receiver behaviour towards 1.0 M sucrose-fed and 5% ethanol-fed donors. (A–D) The receivers paired with 5% ethanol-fed donors walked less and groomed more often than the sucrose-fed donors (E–G). The only social behaviour that was affected by pairing with an ethanol-fed donor was anten-

nation: in this case, the receiver spent less time antennating the donor over the interval than the receivers paired with 1.0 M sucrose fed donors. $N_{1.0\ M\ Suc}=10$, $N_{5\%\ etoh}=13$. ± SEM Note: Y-axis scale is not the same on all graphs.

Food Transfer Takes Place during the Observation Period

To verify that the receivers actually received food during trophallactic interactions with the donors in all the experiments, we weighed both bees before and after the behavioural assay. All donors lost weight over the interval of the assay, and all receivers gained weight (Fig. 6, bee type main effect, GLM, $F_{1,123}=100, P<0.001$). The amount of weight lost by the donors depended on whether or not they had received food prior to the assay; the 'hungry' donors that had not been fed, lost weight (6.5 mg), but lost significantly less weight than donors that received food (1-way ANOVA, $F_{3,97}=5.78, P=0.001$). Whether or not a donor had been fed ethanol prior to the assay, however, did not influence the amount of weight lost by donors or the amount of weight gained by receivers (treatment main effect, GLM, $F_{2,123}=0.41, P=0.869$). We, therefore, conclude that treatment with ethanol did not affect the amount of food transferred from the donor to the receiver during trophallaxis. The amount of weight gained on average by the receivers (3 mg) was not equal to that lost by the donors (16 mg).

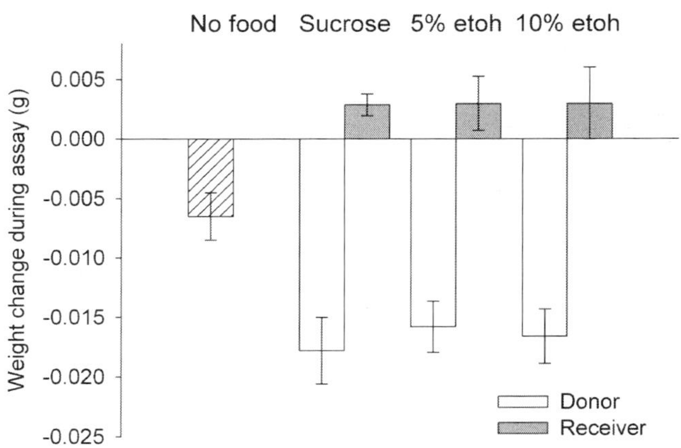

Figure 6. Ethanol-fed donors transferred the same amount of food on average as the sucrose-fed donors. Hungry honeybees, that had been starved 24 h prior to the assay, still transferred food, but significantly less food than those fed with 20 μl of 1.0 M sucrose prior to the assay (the 1.0 M sucrose, and 5% and 10% ethanol fed donors). $N_{Hungry}=34$, $N_{1.0\ M\ Suc\ Donors}=26$, $N_{1.0\ M\ Suc\ Receivers}=26$, $N_{5\%\ etoh\ Donors}=26$, $N_{5\%\ etoh\ Receivers}=26$ and $N_{10\%\ etoh\ Donors}=13$, $N_{10\%\ etoh\ Donors}=13$. ± SEM.

DISCUSSION

Our results clearly show that hunger and inebriation state influence social interactions between two honeybee nestmates. Compared to bees provided with sucrose without ethanol exposure, hungry bees were more likely to beg for food and less likely to communicate via antennation or share food. Inebriated bees were less likely to communicate via antennation or beg and more likely to repeatedly open their mandibles, potentially as a warning display to reject begging receivers. Our data also show that substances like ethanol that are retained in the crop can be transferred between honeybee nestmates via trophallaxis.

Hunger State and Social Interactions in Bees

Food is an important currency in social insect colonies, and the effect of hunger on social interactions has been studied previously to identify how the quality and quantity of food consumed by a donor influences its willingness to share with a hungry nestmate. For example, Farina and Nunez (1991)[27] used an assay with two nestmates similar to ours but that only measured trophallaxis behaviour and found that fully satiated bees were more willing to share food than partially fed or starved bees. In their assay, they showed that the sucrose concentration given to the donor had no effect on willingness to engage in trophallaxis of either donors or receivers.

Another study that examined the same social behaviours that we measured starved entire colonies of honeybees and observed the behaviour of the bees in the context of the hive.

This study only found that begging frequency increased between nestmates in hungry colonies; other social behaviours including trophallaxis were unaffected[8]. In our experiments by contrast, hunger state had a strong influence on all behaviours associated with social interactions between two nestmates. As seen in ants[9], hungry honeybees were more likely to perform long bouts of begging.

However, they were also less likely to give food via trophallaxis, to antennate, and to display mandible opening. The social behaviour of sucrose-fed bees, on the other hand, was characterized by antennation early in the observation period, food sharing, and a constant rate of mandible opening. These bees did not beg for food much in the beginning of the 30 min observation, but begged more later on, perhaps indicating that their hunger state was changing.

Inebriation State and its Influence on Social Behaviour

Mandible opening by honeybees during nestmate interactions has previously been interpreted as food offering behaviour that precedes trophallaxis with a nestmate[6],[8]. In contrast, studies of many insects including crickets[28], ants[29],[30], and bees[31], suggest that opening of the mandibles is a defensive posture that precedes a bite or sting attack when the animal recognizes an enemy[29],[31]. In sweat bees, mandible opening is also form of aggressive solicitation of food by dominant members of colonies[26], in which the dominant bee repeatedly opens and closes her mandibles and follows by nipping or biting. Non-aggressive food solicitation in this species is distinguished by the fact that it is preceded by antennation and then followed by extending the proboscis to beg for food[26]in a manner more similar to that described between honeybees[6].

In our study, mandible opening was not always followed by food offering behaviour. Hungry, sucrose-fed, and 5% ethanol-fed bees were equally likely to perform trophallaxis after opening their mandibles as they were to open their mandibles a second time. However, the bees fed 10% ethanol were 2 times more likely to perform bouts of mandible opening behaviour followed by a second bout of mandible opening than by trophallaxis. While we observed repeated mandible opening events, we did not observe nipping or biting between nestmates that might indicate that the inebriated bees were demanding food from the other bee in the arena. These bees were not motivated to seek food, as indicated by the fact that they were also the least likely to engage in food begging behaviour. This is corroborated by the observation in a previous study that bees fed 5% or 10% ethanol solutions were less motivated to eat[16].

The repeated mandible opening we observed may instead be a defensive posture on the part of the inebriated donor bee that perhaps indicates that it did not recognize the receiver bee as a nestmate or that it was unwilling to share food. Social insects antennate as a form of contact chemoreception which allows them to recognize nestmates and avoid or attack enemies[29],[32]and even facilitates food sharing[9]. We previously found that bees inebriated after feeding on 10% ethanol solutions had more difficulty distinguishing odours in an appetitive differential conditioning task in which they had to learn to differentiate odours associated with food from those that were not associated with a food reward[16]. Studies in*Drosophila*and*C. elegans*also suggest that ethanol affects olfactory processing in invertebrates[11],[33]. Here, inebriated donor bees were less likely to antennate other bees, and they might have also had more difficulty recognizing receiver's scent. For example, when ants do not recognize other ants as nestmates, they will not perform trophallaxis and instead open their mandibles as a form of warning display[29]. Although the inebriated bees were not on average less willing to engage in trophallaxis and transferred as much food as bees fed only sucrose, they did perform fewer bouts of trophallaxis and were less likely to beg for food from receivers. On average, the inebriated bees may have been less motivated to share food or to eat and thus were more likely to open their mandibles to reject begging from the receivers.

Inebriation State and its Influence on Trophallaxis

Ethanol inebriation reduces motor coordination, affects sensory function, and influences the motivation to feed and the ability to learn in honeybees[15],[16],[34], most likely via interaction with several neurotransmitters and intracellular signalling processes[35]. Our analysis confirmed that measurement of an individual's motor function could predict whether the donor bees in our study had been fed ethanol. Being able to perform social behaviours also relies on motor coordination, so one might expect that ethanol inebriation would affect whether or not these behaviours were performed; indeed, the ethanol-fed bees performed less antennation than those fed sucrose. Other behaviours, such as trophallaxis and begging occurred less frequently than observed in sucrose-fed or hungry honeybees.

The inebriated bees in our study did not spend less time performing trophallaxis than the sucrose-fed bees, and transferred as much food to the receiver bees as the sucrose-fed bees. Therefore, there are at least two possible reasons why the behaviour of receiver bees was affected by donor bees having consumed ethanol. First, it is possible that the difference in the behaviour of the receivers arose from the reduced social activity of donors fed ethanol. For example, the receivers paired with donors given 5% ethanol may have shown reduced antennation due to the reduced level of antennation by donors in the 5% ethanol group. Alternatively, our data suggest that the receivers paired with ethanol-fed bees obtained food that contained ethanol via trophallaxis, as the receivers began to exhibit behaviours expected from inebriated bees, such as more grooming and less walking. This indicates not only that ethanol must have remained in the crop for more than 2 h after feeding, but that donors were also willing to share potentially dangerous food with nestmates. Taken together, these data indicate that donor bees are willing to share food containing ethanol with nestmates, even when they themselves have been poisoned or inebriated. It also indicates that hungry receivers will solicit food from nestmates that do not actively engage in the social behaviours that precede trophallaxis like antennation.

Our assay is the first to report in detail the behaviour exhibited by a donor bee during its social interactions with a hungry nestmate. Our assay is unique from earlier studies of the behaviour of nestmates in that we report the detailed behaviour of donors after carefully defining the physiological state of both the donor and receiver. We anticipate that this assay will aid researchers in studies designed to quantify how stress, disease, or drugs influence social interactions in social insects such as bees on the scale of two individuals.

SUPPORTING INFORMATION

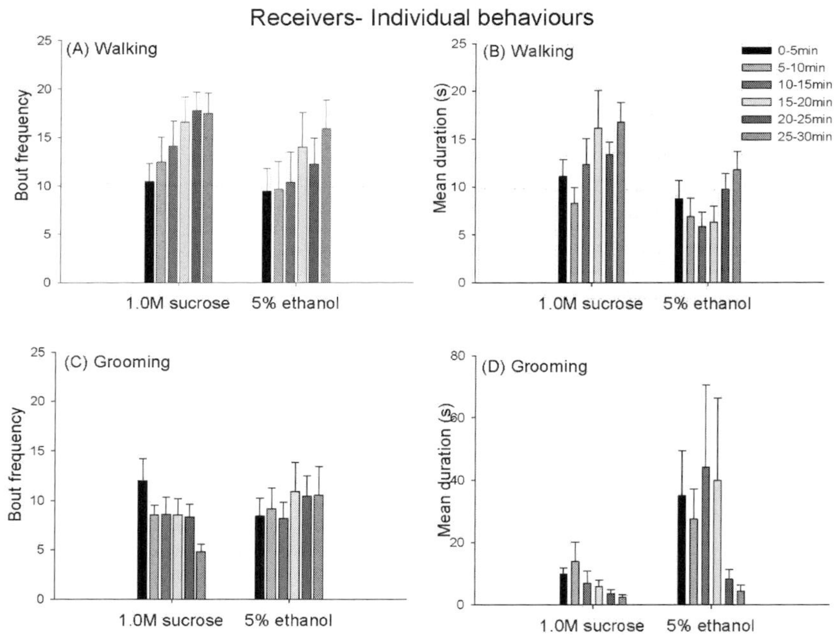

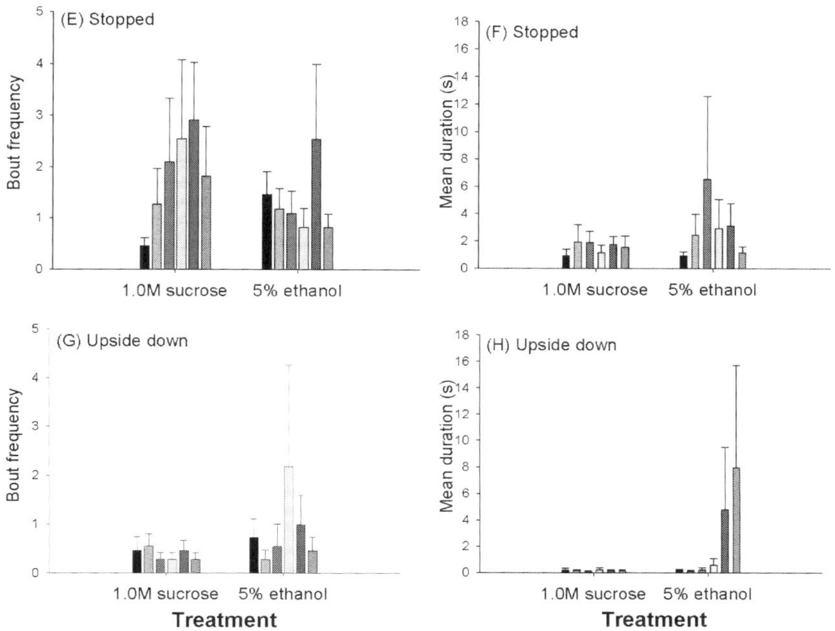

Figure S1. Individual behaviours of receivers towards 1.0 M sucrose-fed and 5% ethanol-fed donors.(A, B) The receivers performed more bouts of walking later in the interval (Pois. Reg. χ_5^2=12.3, p=0.031), but there was no effect of treatment (Pois. Reg χ_1^2=1.05, p=0.307). The average bout duration for the stopped behaviour did not change during the interval for the receivers (interval main effect (GLM), $F_{1,19}$=0.26, p=0.614), and was unaffected by treatment (treatment main effect (GLM), $F_{1,19}$=0.05, p=0.823). However, the number times that the receivers stopped generally became more frequent later in the interval for the receiver with the sucrose bee (interaction, Pois. Reg. χ_5=23.1, p<0.001). (C, D) Bouts of stopped behaviour depended on both interval and treatment (interaction, Pois. Reg. χ_5=23.1, p<0.001).The mean duration of bouts of stopped behaviour did not change during the interval for the receivers (interval main effect, GLM, $F_{1,19}$=0.26, p=0.614), and was not affected by treatment (treatment main effect, GLM, $F_{1,19}$=0.05, p=0.823). (E, F) During the observation, the bout duration of which the receiver was upside down did not change (interval main effect, GLM, $F_{1,19}$=0.96, p=0.340), and was not significantly influenced by the treatments (main effect of treatment, GLM, $F_{1,19}$=0.48, p=0.498). The receiver bee with the ethanol-donor turned upside down more frequently later in the interval, but this trend was not observed for the receiver interacting with the sucrose-donor (interaction: Pois. Reg. χ_5^2=22.6, p<0.001, interval: Pois. Reg. χ_5^2=12.5, p=0.029). Treatment did not affect how frequently the receivers went upside down (Pois. Reg. χ_1^2=0.961, p=0.327). (G,H) The bouts of grooming in the receiver became shorter over the interval (interval main effect, GLM, $F_{1,19}$=34.9, p<0.001), and were unaffected by dose (treatment main effect, GLM, $F_{1,19}$=2.66, p=0.119). Frequency of grooming bouts also decreased over time for the receiver (Pois. Reg. χ_5^2=15.1, p=0.010) but were not affected by treatment (Pois. Reg. χ_5^2=0.56, p=0.456).

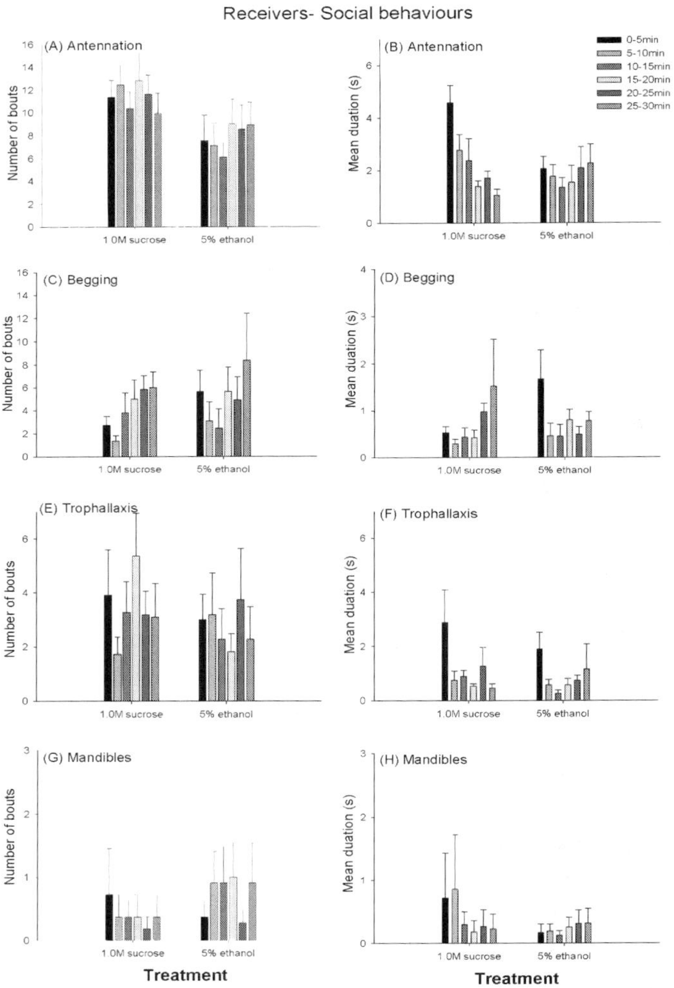

Figure S2. Social behaviour of receivers towards 1.0 M sucrose-fed and 5% ethanol-fed donors.(A,B) The number of bouts of antennation remained constant during the interval (Pois. Reg χ_5^2=6.65, p=0.248), and was not significantly affected by the treatment (Pois. Reg χ_1^2=3.31, p=0.069). Mean bout duration of antennation changed over the interval for the receiver (interval main effect, GLM, $F_{1,19}$=9.39, p=0.006); the receiver with the sucrose-donor had longer bouts of antennation at the start of the interval, whereas the receiver with the ethanol-donor had bouts of the same duration over the observation period (interaction main effect, GLM, $F_{1,19}$=9.36, p=0.006). (C, D) The receivers begged ('proboscis out') more frequently later in the interval (Pois. Reg., χ_5^2=16.1, p=0.006), but treatment did not affect the number of begging bouts of receivers (Pois. Reg., χ_1^2=0.35, p=0.554).The average bout duration of begging behaviour was constant over time in receivers (interval main effect, GLM, $F_{1,19}$=0.73, p=0.405), bouts were of the same duration for both treatments (treatment main effect, GLM, $F_{1,19}$=0.09, p=0.772). (E, F) The bout duration of trophallaxis by the receivers were generally of the same length over the interval (interval main effect, GLM, $F_{1,19}$=3.38, p=0.082), and were also unaffected by treatment (treatment main effect, GLM, $F_{1,19}$=1.36, p=0.258). The frequency of trophallaxis bouts in donors remained the same over the interval (Pois. Reg χ_5^2=4.82, p=0.438) once again this was not affected by dose (Pois. Reg χ_3=0.40, p=0.530). (G, H) For the receivers with the ethanol-donors, the bouts of mandible opening behaviour became more frequent later in the interval (Pois. Reg. χ_5^2=39.8, p<0.001), yet overall there was no effect of treatment on the number of bouts (Pois. Reg. χ_1^2=0.66, p=0.416).Bouts of mandible opening behaviour were generally of equal length throughout the observation (interval main effect, GLM, $F_{1,19}$=0.02, p=0.889), and were the same for receivers with the sucrose-donor and the ethanol-donor (treatment main effect (GLM), $F_{1,19}$=0.02, p=0.900).

Table S1. Factor Analysis for All Behavioural Variables

Factors													
	1	2	3	4	5	6	7	8	9	10	11	12	13
Eigen-values	9.621	5.260	4.714	4.441	2.990	2.225	1.972	1.613	1.529	1.374	1.255	1.189	1.094
% variance	20.044	10.958	9.822	9.251	6.230	4.635	4.108	3.361	3.186	2.863	2.614	2.477	2.278
Factor loading													
W1	**-0.480**	-0.088	0.217	-0.061	-0.054	0.222	-0.018	0.562	-0.052	-0.296	0.225	-0.184	0.259

W2	**-0.812**	0.091	0.052	-0.213	-0.038	-0.138	-0.104	0.073	-0.113	-0.239	-0.048	0.063	-0.016
W3	**-0.834**	0.040	0.150	-0.081	-0.024	-0.102	0.029	-0.165	-0.050	0.043	0.172	0.110	0.111
W4	**-0.868**	-0.015	0.168	-0.011	-0.030	-0.136	0.047	-0.111	-0.049	0.067	0.024	0.185	0.107
W5	**-0.814**	-0.062	0.227	-0.054	-0.211	-0.007	0.176	0.021	0.104	0.192	-0.018	-0.043	0.154
W6	**-0.711**	-0.076	0.313	-0.043	-0.172	0.021	0.227	0.141	0.003	0.341	-0.192	0.040	-0.044
S1	0.376	-0.062	-0.018	0.073	0.180	-0.395	0.560	0.147	0.153	-0.216	0.119	0.289	0.053
S2	0.508	-0.157	-0.044	-0.126	0.298	-0.228	0.567	0.130	-0.003	-0.057	-0.005	-0.221	-0.158
S3	0.310	-0.097	-0.195	-0.066	-0.023	-0.006	0.560	0.139	-0.246	0.288	-0.155	-0.371	-0.070
S4	0.456	-0.065	-0.267	-0.114	-0.067	0.207	0.013	0.142	-0.468	0.008	-0.216	-0.109	0.253
S5	0.572	-0.046	-0.081	0.044	0.001	0.070	0.027	-0.096	-0.424	-0.010	-0.139	0.332	0.140
S6	0.317	0.051	-0.337	-0.022	0.272	-0.092	0.501	-0.033	-0.038	-0.012	0.113	0.115	0.366
U1	-0.225	**-0.640**	-0.286	**0.545**	0.111	0.073	-0.081	-0.008	0.042	-0.040	-0.116	-0.131	0.001
U2	-0.224	**-0.628**	-0.299	**0.574**	0.113	0.066	-0.067	-0.018	0.071	-0.029	-0.105	-0.098	0.009
U3	-0.331	**-0.522**	-0.364	**0.633**	0.152	0.009	-0.009	-0.059	0.044	-0.065	-0.074	-0.002	-0.004
U4	-0.354	**-0.668**	-0.351	**0.441**	0.080	-0.003	-0.031	-0.079	0.039	-0.023	-0.032	0.027	-0.022
U5	-0.307	**-0.596**	-0.372	**0.502**	0.053	0.097	0.068	0.026	0.025	0.041	0.002	0.015	-0.055
U6	-0.314	**-0.467**	-0.297	**0.378**	-0.042	0.004	0.190	-0.013	0.035	0.108	0.194	0.183	0.064
G1	**0.630**	-0.013	-0.043	-0.023	-0.286	-0.169	0.055	-0.481	0.106	0.232	-0.110	0.021	-0.076
G2	**0.838**	0.008	-0.036	0.157	-0.214	0.148	0.019	-0.038	0.047	0.255	0.175	-0.058	0.071
G3	**0.840**	-0.087	-0.144	0.065	-0.072	0.085	-0.260	0.087	0.069	-0.014	-0.104	0.045	0.065
G4	**0.860**	-0.088	0.008	0.105	-0.088	-0.006	0.000	0.100	0.230	0.031	-0.040	0.042	-0.157
G5	**0.852**	-0.083	-0.172	0.040	-0.026	-0.060	-0.165	0.056	0.064	-0.074	0.028	0.055	0.161
G6	**0.709**	-0.039	-0.302	-0.042	0.081	-0.103	-0.353	-0.064	0.167	-0.209	0.129	0.015	0.138
A1	-0.143	0.399	0.346	0.376	0.266	0.122	0.026	-0.227	-0.208	-0.018	-0.186	0.090	0.206

A2	-0.097	0.456	0.271	0.342	0.013	0.321	0.093	-0.166	0.245	-0.098	-0.305	0.032	0.275
A3	0.146	0.201	0.280	0.273	0.195	0.271	0.263	0.081	0.468	-0.203	0.120	0.199	0.011
A4	-0.061	0.247	0.367	0.192	-0.036	0.304	0.095	-0.351	0.408	-0.007	-0.004	-0.356	0.113
A5	0.243	0.190	0.319	0.339	-0.126	0.537	0.228	0.243	0.106	0.009	0.070	-0.002	-0.191
A6	0.287	0.195	0.245	0.462	-0.032	0.552	0.017	0.165	-0.255	0.096	-0.065	0.180	-0.138
P1	-0.146	0.255	**-0.497**	-0.261	0.549	0.244	-0.083	-0.022	-0.013	0.048	0.090	-0.077	0.095
P2	-0.154	0.266	**-0.475**	-0.286	0.563	0.243	-0.036	0.073	0.152	0.247	-0.018	0.043	0.148
P3	-0.332	0.326	**-0.393**	-0.163	0.279	-0.007	-0.033	0.015	0.138	0.352	-0.044	0.211	-0.351
P4	-0.153	0.345	**-0.532**	-0.153	0.554	0.239	0.025	-0.080	0.024	0.208	0.109	-0.022	0.149
P5	-0.180	0.411	**-0.425**	-0.031	0.355	0.002	-0.128	0.042	0.049	-0.273	-0.051	-0.060	-0.365
P6	-0.209	0.625	**-0.123**	0.367	0.230	0.019	-0.006	-0.005	-0.235	-0.208	-0.055	-0.192	-0.222
T1	-0.044	0.377	-0.029	**0.522**	0.096	-0.356	-0.237	0.201	0.016	0.336	-0.016	-0.021	0.175
T2	0.130	0.315	0.037	**0.292**	-0.041	-0.447	-0.212	0.264	0.296	0.276	-0.210	-0.090	0.138
T3	-0.008	0.517	0.075	**0.586**	0.087	-0.301	-0.013	0.176	-0.061	-0.062	-0.198	0.086	0.009
T4	0.033	0.404	0.035	**0.491**	-0.102	-0.227	0.007	-0.167	-0.191	0.015	0.552	-0.073	0.022
T5	0.058	0.541	0.040	**0.595**	0.121	-0.346	0.106	0.100	-0.088	-0.114	-0.162	0.025	-0.029
T6	0.092	0.274	0.025	**0.543**	-0.030	0.039	0.005	-0.350	-0.245	0.090	0.389	-0.208	-0.083
M1	0.188	-0.117	**0.394**	0.171	0.163	0.130	-0.189	0.365	-0.033	0.331	0.321	0.253	-0.108
M2	0.151	-0.284	**0.551**	-0.104	0.534	-0.229	0.046	-0.107	-0.023	0.010	0.020	0.017	-0.054
M3	0.210	-0.453	**0.547**	0.073	0.474	0.009	-0.191	-0.012	-0.149	0.036	0.010	0.110	-0.017
M4	0.147	-0.307	**0.519**	-0.010	0.367	-0.105	-0.213	0.190	-0.016	0.110	0.052	-0.368	0.107
M5	0.146	-0.274	**0.621**	-0.068	0.571	-0.175	-0.019	-0.091	0.025	0.019	-0.008	-0.155	-0.024
M6	0.233	-0.310	**0.578**	-0.061	0.391	0.132	0.001	-0.274	-0.108	-0.020	-0.160	0.146	-0.058

W = walking, S = stopped, G = grooming, U = upside down, A = antennate, P = proboscis out (begging), T = trophallaxis, M = mandibles open. Each number indicates the interval (1=0-5min, 2=5-10min, 3=10-15min, 4=15-20min, 5=20-25min, 6=25-30min). The top portion of the table indicates the eigenvalues and % variance contributed by each factor. The bottom half of the table indicates the loading for each variable on each factor. Factor loadings in bold indicate the behaviours with the greatest contribution to each factor for the first 4 factors.

Table S2. Canonical Discriminant Analysis for Individual Motor Behaviours

	Discriminant function		
	1	2	3
Eigenvalue	0.631	0.121	0.027
Percentage variance	81.000	15.600	3.500
Wilks' Lambda	0.532	0.868	0.974
Factors			
F1	**0.798**	0.040	0.213
F2	**0.681**	0.445	0.102
F3	0.112	-0.262	0.229
F4	-0.467	**0.781**	0.329
F5	0.024	-0.159	0.363
F6	0.183	0.294	**-0.792**
Treatment			
No food	-0.515	0.178	-0.203
1.0M sucrose	-0.880	-0.159	0.189
5% Ethanol	0.831	-0.437	-0.050
10% Ethanol	0.856	0.506	0.130

The standardized coefficients in bold indicate the factors from the factor analysis that contributed the most to classification of the treatments by each of the functions (all factors contributed, but the criterion for emphasis was a coefficient greater than ± 0.5). The sign and magnitude of the non-standardized coefficients indicate how the functions classified bees according to treatment. The order of the functions indicates the distance in similarity between the treatments; the line indicates the most likely split between groups made by the classification function. Numbers in grey indicate neutral groups midway between the groups split by the secondary functions.

Table S3. Canonical Discriminant Analysis for Social Behaviours

	Discriminant function		
	1	2	3
Eigenvalue	0.567	0.204	0.080
% variance	66.600	24.000	9.400
Wilks' Lambda	0.491	0.769	0.926
Factors			
F1	-0.010	**0.621**	-0.009
F2	**0.970**	-0.058	-0.167
F3	0.270	0.251	0.309
F4	-0.105	-0.392	-0.342
F5	-0.049	-0.423	**0.666**
F6	0.138	0.427	0.476
F7	0.242	-0.418	0.322
Treatment			
No food	-1.005	0.247	-0.101
1.0M sucrose	0.936	0.429	-0.160
5% Ethanol	0.127	-0.107	0.462
10% Ethanol	0.157	-0.801	-0.241

The standardized coefficients in bold indicate the factors from the factor analysis that contributed the most to classification of the treatments by each of the functions (all factors contributed, but the criterion for emphasis was a coefficient greater than +0.5). The sign and magnitude of the non-standardized coefficients indicate how the functions classified bees according to treatment. The order of the functions indicates the distance in similarity between the treatments; the line indicates the most likely split between groups made by the classification function. Numbers in grey indicate neutral groups midway between the groups split by the secondary functions.

Data S1. Data for Donors and Receivers

Beenumber	Treatment (1=nothing, 2=sucrose, 3=5% etoh, 4=10% etoh)	Interval	walking	stopped	upsidedown	grooming	antennation	proboscis	trophallaxis	mandibles
1	4	1	46.54	1.43	1.04	35.78	3.65	1.23	7.22	1.72
1	4	2	3.63	18.57	0.34	58.52	2.73	0	4.4	11.81
1	4	3	16.75	6.2	0	65.54	0	0	0	10.92
1	4	4	7.43	10.22	0	48.02	7.45	0	0.12	26.75
1	4	5	39.18	0.62	0.33	30.9	8.01	0.59	0.15	20.22
1	4	6	35.86	0.85	0.55	53.41	3.72	0.37	0	5.25
2	4	1	6.53	4.36	0	80.11	2.48	0	5.04	1.48
2	4	2	0	2.33	0	94.28	3.39	0	0	0
2	4	3	0	6.12	0	82.9	10.42	0.56	0	0
2	4	4	1.79	32.57	0	65.64	0	0	0	0
2	4	5	1.06	11.64	0	86.19	1.11	0	0	0
2	4	6	0	1.35	0	98.65	0	0	0	0
3	4	1	24.05	2.25	0	52.86	19.32	0	0	1.52
3	4	2	43.97	1.54	0	25.96	25.6	0	0	2.94
3	4	3	40.36	9.39	0	38.15	9.5	0	0	2.44
3	4	4	59.2	4.6	0	26.35	9.7	0	0	0
3	4	5	40.9	4.25	0	32.56	18.28	0	0	4.01
3	4	6	64.31	3.29	0	16.71	6.75	0.13	1.17	7.64
4	4	1	10.23	0	0	70.12	13.01	0	0	6.63
4	4	2	18.6	0.38	0.26	59.81	8.16	0	0.81	11.97
4	4	3	40.64	0.33	0.65	44.09	3.17	0	4.31	6.81
4	4	4	19.56	1.03	0	68.31	4.23	0	2.03	4.85
4	4	5	34.65	3.71	0	39.19	5.62	0	4.37	12.46
4	4	6	34.58	1.41	0	52.9	4.34	0	2.4	4.37
5	4	1	43.57	0.49	45.16	2.72	0.99	0	5.41	1.66
5	4	2	35.82	0	60.58	2.48	0.67	0	0	0.45
5	4	3	55	3.66	7.99	20.69	3.4	0	1.03	8.23
5	4	4	80.8	0	5.92	8.14	1.94	0	0	3.2

5	4	5	83.08	0.42	6.99	7.82	1.1	0	0	0.59
5	4	6	75.26	0	5.07	11.11	2.93	0	3.27	2.36
6	4	1	64.2	0.17	0.09	33.09	2.01	0	0	0.44
6	4	2	56.65	1.2	0.12	34.55	1.05	0	0.1	6.34
6	4	3	70.91	0	0	20.29	0.68	0	0	8.12
6	4	4	77.89	0.75	0.82	11.52	1.02	0	0	8.01
6	4	5	64.01	1.88	0	26.4	2.35	0	0	5.35
6	4	6	77.52	0.12	0	14.43	3.39	0	0	4.55
7	4	1	63.59	8.82	0.43	21.67	0	0	0	0
7	4	2	76.21	2.12	0.81	8.61	0	0	0	0
7	4	3	65.33	1.71	0.52	14.34	1.67	0	2.33	2.05
7	4	4	75.62	1.06	0.66	4.85	0	0	2.73	0.65
7	4	5	81.77	0.77	0.71	4.58	0	1.18	0	0
7	4	6	83.05	0.21	3.02	1.57	0.1	0.5	1.2	0.32
8	4	1	4.96	0	0	77.62	17.12	0	0.3	0
8	4	2	2.71	1.04	0.59	77.71	17.21	0	0.47	0.27
8	4	3	7.91	0.55	0	82.7	2.61	0	2.96	3.26
8	4	4	35.98	1.59	0	54.89	3.38	0	1.97	2.19
8	4	5	22.18	0.9	0	73.95	1.87	0	0	0.77
8	4	6	2.04	13.06	0	76.24	3.76	0	2.05	2.85
9	4	1	1.48	0.33	0	70.76	13.42	2.68	10.64	0.69
9	4	2	6.62	1.27	0	84.23	7.03	0	0.85	0
9	4	3	58.99	9.16	0	15.43	8.69	4.52	2.77	0.36
9	4	4	65.72	2.46	0	7.94	9.48	7.66	6.75	0
9	4	5	35.23	15.31	0	2.21	1.43	0	0.68	0
9	4	6	51.34	3.19	0	14.05	6.17	12.08	10.01	0
10	4	1	26.84	5	0	58.48	5.49	0	0.71	3.48
10	4	2	25.35	0.17	0.37	63.09	5.15	0	0	3.09
10	4	3	2.12	7.32	0	89.64	0.37	0	0	0.55
10	4	4	2.51	60.99	0	36.04	0	0	0	0.45
10	4	5	0	49.7	0	49.03	0.9	0	0	0
10	4	6	28.68	2.69	0	53.47	8.68	1.65	1.14	2.87
11	4	1	27.22	5.09	0	39.83	2.95	0.44	11.53	12.94
11	4	2	24.19	1.21	0	63.87	4.7	0	2.63	3.41
11	4	3	16.39	2.08	0	66.67	3.4	0.7	4.99	5.77
11	4	4	19.24	1.92	0	66.73	4	0	4.3	3.58

11	4	5	21.33	4.41	0	64.34	2.28	0	3	4.64
11	4	6	49.07	0	0	39.89	3.11	0	4.24	3.68
12	4	1	52.46	0	4.3	19.88	19.52	1.16	0	1.87
12	4	2	79.92	0	1.26	8.69	8.32	0	0	1.46
12	4	3	90.25	0	0.25	2.53	3.91	0	0	2.27
12	4	4	88.62	0.11	0.92	4.21	1.01	0	0	3.32
12	4	5	93.89	0.21	0.83	0.7	0.84	0	0	2.17
12	4	6	93.79	0.09	0.73	0.46	1.86	0	0.14	2.01
13	4	1	8.03	23.67	0	67.43	0.33	0	0	0.49
13	4	2	5.52	32.81	0	54.87	1.41	0	1.57	3.83
13	4	3	8.69	32.53	0	53.42	4.22	0	1.07	0.06
13	4	4	5.28	1.92	0	89.58	2.12	0	0	1.1
13	4	5	23.13	25.09	0	38.87	7.91	0.28	3.9	0.83
13	4	6	13.67	43	0	41.14	0.45	0.15	1.58	0
14	3	1	44.08	0.54	0	42.45	12.39	0	0	0
14	3	2	94.46	0	0	3.08	1.05	0	0.35	0.84
14	3	3	96.89	0	0	1.93	0.18	0	0.21	0.21
14	3	4	86.54	0	0	6.91	1.51	0.94	0.95	2.61
14	3	5	98.55	0	0	0.53	0	0	0	0
14	3	6	82.21	0	0	3.66	2.99	6.74	1.04	1.79
15	2	1	40.04	0	0	47.86	4.28	0.47	4.7	2.65
15	2	2	83.68	0	0	7.31	3.61	0	4.25	1.15
15	2	3	90.96	0.12	0	2.95	1.9	1.11	2.17	0.79
15	2	4	88.64	0	0	1.89	3.61	3.45	0.75	1.56
15	2	5	84.38	0	0	1.27	4.89	6.48	1.72	0.4
15	2	6	91.57	0	0	0.35	2.34	3.86	0.85	0
16	2	1	47.47	15.23	0	35.69	1.34	0	0	0.27
16	2	2	48.37	0.36	0	17.1	0.61	0	0	0.19
16	2	3	57.86	1.7	0	31.61	3.66	0	2.3	2.87
16	2	4	77.66	0.38	0	11.14	4.66	0	2.95	3.21
16	2	5	90.46	0	0	5.91	1.49	0	0.83	1.31
16	2	6	75.85	0.95	0	15.18	1.52	0	3.04	3.45
17	3	1	18.66	20.05	0	59.4	0.73	1.15	0	0
17	3	2	3.19	16.18	0	74.99	1.8	1.66	1.67	0.51
17	3	3	0.71	0.24	0	59.43	37.02	0	0	2.6
17	3	4	0	3.02	0	93.45	2.23	0	1.3	0

17	3	5	0	6.92	0	89.47	0.94	1.72	0.95	0
17	3	6	0	1.61	0	97.6	0.4	0	0	0.38
18	2	1	36.34	1.21	0	26.47	17.94	0.31	17.22	0.51
18	2	2	89.14	0	0	2.89	1.52	2.43	4.03	0
18	2	3	92.18	0	0	2.78	1.45	0.95	2.64	0
18	2	4	87.17	2.91	0	3.39	1.12	3.5	1.78	0
18	2	5	92.83	0	0	1.74	1.01	1.98	2.44	0
18	2	6	94.33	0	0	0.35	0.92	3.64	0	0
19	3	1	50.11	6.52	0.77	17.31	14.44	1.68	5.99	3.19
19	3	2	63.19	2.12	8	11.9	11.46	2.07	1.26	0
19	3	3	52.56	1.87	2.29	18.5	17.78	3.54	2.35	0
19	3	4	46.72	0.59	0.59	35.11	9.18	5.88	1.94	0
19	3	5	84.06	0.33	3.26	5.77	2.69	1.06	2.82	0
19	3	6	66.08	6.23	3.7	9.52	4.15	6.39	3.66	0
20	2	1	34.35	7.91	0	3.12	36.36	1.91	16.07	0.29
20	2	2	62.76	0.95	0.26	4.88	26.14	0.46	4.08	0.48
20	2	3	58.92	1.98	1.6	1.81	14.83	2.56	17.92	0
20	2	4	76.54	4.61	0	2.8	5.92	4.32	5.09	0
20	2	5	50.08	2.8	0	5.11	11.14	11.78	18.41	0
20	2	6	54.76	4.89	0	1.63	11.61	24.43	1.96	0
21	3	1	1.78	9.82	0.31	70.98	3.39	0	13.73	0
21	3	2	0.39	0.77	1.74	79.47	8.75	3.04	4.96	0
21	3	3	0.18	0.67	0	89.04	4.56	2.73	2.82	0
21	3	4	15.98	1.96	0	76.82	2.69	1.52	0.29	0.73
21	3	5	58.98	4.87	0	29.1	3.41	0.38	3.26	0
21	3	6	75.33	2.23	0	18.42	3.35	0.68	0	0
22	2	1	21.29	13.44	0	37.99	16.25	0	0	4.14
22	2	2	31.18	13.13	0	10.6	1.66	0	0	43.43
22	2	3	58.85	0.8	0	6.26	16.55	0	0	17.54
22	2	4	70.84	0.2	0.33	18.56	1.77	0	0	8.29
22	2	5	35.94	2.44	0	18.66	0.61	0.38	0	41.98
22	2	6	63.11	5.38	0	2.05	0	0	0	29.46
23	3	1	18.86	1.44	0	63.11	0.38	0	0	0
23	3	2	29.53	8.13	0.2	44.75	11.43	2.04	0.35	3.57
23	3	3	25.43	3.67	0.24	62.01	4.82	0	3.16	0.37
23	3	4	58.21	1.61	0	35.17	2.44	0	1.1	0.77

23	3	5	89.89	2.2	0	3.88	1.25	0	1.7	1.08
23	3	6	87.86	2.98	0	4.46	0.37	0	0.84	3.49
24	2	1	100	0	0	0	0	0	0	0
24	2	2	15.98	0.81	0	63.14	17.72	0	0.55	1.8
24	2	3	67.06	0	0	17.02	15.29	0	0.31	0.31
24	2	4	44.89	0.71	0	27.29	22.74	0	1.59	2.78
24	2	5	80.9	0	0	2.25	15.98	0	0	0.86
24	2	6	66.53	1.76	0	22.63	3.16	0	0.18	5.42
25	3	1	40.5	7.58	0	25.13	10.78	0	1.74	14.27
25	3	2	0	8.11	0	91.89	0	0	0	0
25	3	3	15.76	12.48	0	47.22	16.96	1.24	2.04	4.29
25	3	4	27.21	1.89	0	64.42	2.89	0	1.59	2
25	3	5	26.64	9.67	1.23	22.86	37.89	0	0	1.72
25	3	6	58.74	1.24	1.54	10.18	20.98	0	1.46	5.86
26	2	1	8.33	3.39	0	42.57	12.16	0.64	3.94	0
26	2	2	57.35	0.44	0	17.25	20.62	1.54	2.8	0
26	2	3	82.07	0.17	0.16	1.7	11.26	2.43	2.2	0
26	2	4	52.95	0	0	24.59	18.72	3.74	0	0
26	2	5	80.81	4.16	0	1.02	6.02	6.82	1.18	0
26	2	6	93.74	0	0	0	2.8	2.4	1.06	0
27	3	1	20.11	2.22	0.82	73.51	3.33	0	0	0
27	3	2	0	17.56	0	82.2	0	0	0	0
27	3	3	0.49	89.4	0	10.1	0	0	0	0
27	3	4	0	31.7	0	66.49	0	1.81	0	0
27	3	5	54.6	3.2	0.89	38.54	1.21	0	0	0.4
27	3	6	95.28	0.79	0.4	0.29	0.99	0	0.42	0
28	2	1	62.1	0	0.07	21.45	0.33	0	0	0
28	2	2	48.21	0.31	0	34.27	16.26	0	0.95	0
28	2	3	78.5	0	0	7.1	12.58	1.81	0	0
28	2	4	54.73	1.32	0	36.19	7.09	0	0.68	0
28	2	5	77.05	0.61	0	9.53	8.51	3.33	0.98	0
28	2	6	76.24	0.13	0	4.96	7.74	9.11	1.67	0
29	3	1	19.42	2.83	0.16	75.92	1.02	0.64	0	0
29	3	2	57.93	0.38	0.48	32.46	3.86	0.77	2.67	1.45
29	3	3	89.76	0.26	1.52	3.72	1.08	2.18	0.44	0
29	3	4	75.52	0.55	1.63	8.15	2.49	4.1	1.35	0

29	3	5	86.94	0	3.72	4.62	2.71	1.37	0	0
29	3	6	84.03	0.27	5.24	3.27	5.79	1.39	0	0
30	2	1	26.62	0	0.62	59.08	7.67	0	3.86	1.47
30	2	2	0	0.15	0	85.89	9.57	0.31	2.65	1.43
30	2	3	0	0.81	0	81.53	16.11	0.11	1.44	0
30	2	4	0	5.31	0	89.24	5.42	0.03	0	0
30	2	5	32.88	5.16	0.57	35.87	20.12	3.98	1.41	0
30	2	6	58.69	1.06	0.44	13.29	18.16	5	2.87	0
31	3	1	2.46	0.33	0	52.1	13.39	0	0	0.4
31	3	2	0.62	5.4	0	58.28	5.56	0	0	0
31	3	3	0	0.7	0	78.37	0.73	0	0.46	19.74
31	3	4	0	3.24	0	93.11	0	0	0	3.64
31	3	5	9.32	40.99	0	43.61	4.21	0	0	1.86
31	3	6	17.11	1.03	0	40.61	13.75	1.3	0.49	25.71
32	2	1	9.17	0	0	52.47	38.36	0	0	0
32	2	2	74.03	0	0	4.68	21.29	0	0	0
32	2	3	60.37	0.57	0	20.5	13.66	0	0.92	3.99
32	2	4	78.77	0.86	0	8.47	6.78	0.48	0	4.63
32	2	5	87.08	3.64	0.24	2.34	6.03	0.47	0	0.2
32	2	6	85.89	1.21	0	2.21	5.65	0.44	2.93	1.68
33	3	1	52.74	0.98	0.66	8.28	18.54	0.61	6.79	11.4
33	3	2	48.12	1.64	0	35.13	12.01	0.53	0	2.57
33	3	3	52.13	0.81	0	16.76	13.31	0	0	16.99
33	3	4	41.05	0	0	42.79	5.11	0	0.16	10.89
33	3	5	73.36	0.34	0	6.84	2.89	0	0	16.56
33	3	6	73.27	0.32	0	18.38	1.92	0	0.24	5.88
34	2	1	65.61	0	0.16	17.02	4.58	0	0	11.41
34	2	2	37.65	0	0	40.71	4.35	0.85	0.46	14.84
34	2	3	72.85	1.26	0	7.74	5.73	1.47	0	10.19
34	2	4	62.3	2.2	0	21.62	3.43	0	0	10.39
34	2	5	82.38	6.13	0	4.9	3.51	0	0	3.08
34	2	6	69.87	1.25	0	8.75	6.99	1.21	0.71	11.16
35	3	1	16.58	3.38	0	63.87	7.83	0	6.05	1.89
35	3	2	2.98	0	0	79.4	16.76	0	0	0.86
35	3	3	78.15	3.94	2.03	4.46	3.73	1.92	0	0.83
35	3	4	56.62	4.17	1.14	13.41	3	10.54	4.96	0

35	3	5	56.22	8.6	0.96	13.74	3.9	10.34	1.77	0
35	3	6	41.84	26.97	4.61	23.06	0.8	0.72	0	0
36	2	1	72.39	0	0.69	3.26	15.81	0.69	3.98	1.73
36	2	2	74.72	1.2	0.12	1.57	15.95	0.82	1.34	0.6
36	2	3	18.7	0.4	0	1.58	8.38	1.2	0.49	0
36	2	4	28.28	0.41	0	4.63	4.29	0.96	1.99	1.37
36	2	5	71.06	0.16	0	3.9	7.73	10.58	2.38	1.62
36	2	6	71.47	0.59	0	2.59	6.24	15.79	0.89	1.68
37	3	1	34.15	0.82	0	21.37	12.43	29.97	1.27	0
37	3	2	32.04	3.33	0	31.76	9.31	23.56	0	0
37	3	3	35.98	13.22	0	36.3	9.17	5.33	0	0
37	3	4	31.23	18.39	0	4.18	4.62	41.59	0	0
37	3	5	54.98	3.79	0	29.94	1.86	9.42	0	0
37	3	6	22.84	19.74	0	45.14	4.08	8.2	0	0
38	2	1	48	0	0	12.16	15.55	3.2	14.8	5.64
38	2	2	69.82	0	0	4.2	13.4	8.51	3.83	0
38	2	3	77.6	0.31	0.2	3.55	9.26	5.65	2.73	0
38	2	4	91.07	0.22	0	0.08	3.83	4	0.38	0
38	2	5	92.24	0	0	0.43	2.38	4.27	0	0
38	2	6	97.04	0	0.11	0.28	0.92	0.7	0	0
39	2	1	88.81	0	0.18	6.08	2.42	0	0.93	1.39
39	2	2	59.75	0	0	20.03	17.03	0.12	0.31	2.58
39	2	3	66.77	0	0	12.49	11.09	2.98	4.3	1.82
39	2	4	82.49	0	0	4.31	3.84	7.77	0	0
39	2	5	80.96	0	0	2.19	5.45	8.86	0	0
39	2	6	84.47	0	0	3.53	2.87	6.69	0	0
40	3	1	30.09	0.93	0	59.41	0.32	0	7.61	1.64
40	3	2	0.41	0	0	92.12	1.15	0	6.32	0
40	3	3	18.32	2.37	0	74.95	1.05	0	3.32	0
40	3	4	16.73	5.16	0	72.76	1.81	0	2.7	0.84
40	3	5	1.51	10.84	0	84.01	0.15	0	3.49	0
40	3	6	0.18	9.15	0	83.75	3.92	0	2.99	0
41	3	1	21.39	0.09	0	66.86	4.05	0	3.14	4.38
41	3	2	66.3	1.15	0	27.41	2.97	0	0	1.92
41	3	3	81.27	0	0.24	6.25	5.64	3.38	0.4	2.09
41	3	4	78.67	1.82	0.49	8.35	4.28	2.95	2.54	0

41	3	5	86.19	0	0	4.47	1.73	4.49	2.13	0
41	3	6	95.77	0	0	0.94	1.05	0.87	0.83	0
42	2	1	71.96	0.47	0	20.94	1.8	0	4.25	0.34
42	2	2	92.8	0	0	6.46	0.74	0	0	0
42	2	3	49.34	0	0	16.74	2.81	0	0.9	0.75
42	2	4	25.44	2.13	0.08	5.32	5.89	0	5.74	0
42	2	5	83.51	0	0.05	1.84	9.76	4.13	0.52	0
42	2	6	11.41	0	0	0.18	0.97	1.05	0.3	0
43	3	1	10.89	0	0	81.11	1.65	0	0	6.34
43	3	2	29.44	0	0	67.04	0.7	0.41	0	2.42
43	3	3	17.6	0.21	0	79.73	0.98	0	0.5	0.98
43	3	4	14.22	0.37	0	78.75	4.32	0	1.07	1.28
43	3	5	43.3	0	0	51.7	4.55	0.28	0	0.18
43	3	6	24.08	0	0.08	75.67	0	0	0	0
44	3	1	4.91	0	0	87.35	7.74	0	0	0
44	3	2	49.36	2.69	0	35.84	7.34	0	0	4.61
44	3	3	70.85	0	0.38	17.24	8.54	0	0	2.56
44	3	4	56.74	0	0	14.34	27.42	0	0	1.28
44	3	5	77.46	0	0	5.06	7.42	0	0	9.94
44	3	6	67.49	0	0	3.57	3.54	0	0	24.56
45	3	1	12.62	2.02	0	76.95	1.19	0	3.02	4.2
45	3	2	38.63	0.16	0	40.3	11.16	0	3.85	5.9
45	3	3	35.79	4.18	0.57	41.95	8.51	4.54	2.69	1.76
45	3	4	48.93	0	0	38.96	6.6	1.78	0.81	2.91
45	3	5	59.38	0.37	0.31	17.96	11.54	1.17	4.9	4.37
45	3	6	66.03	1.11	0.34	13.76	7.04	4.46	2.82	4.31
46	2	1	15.42	0	0.1	41.05	31.35	2.9	8.41	0.77
46	2	2	23.6	6.14	0	58.83	10.18	0.26	0.62	0.37
46	2	3	76.21	1.2	0.08	10.42	9.07	0	2.22	0.79
46	2	4	59.16	0.73	0.07	17.06	11.39	3.66	6.36	1.38
46	2	5	59.46	3.4	0	12.59	11.91	4.14	6.48	1.81
46	2	6	37.56	0	0.11	22	10.96	14.77	14.4	0
47	2	1	56.8	0.26	0.08	17.88	18.4	1.47	0	4.48
47	2	2	59.21	0.61	0.16	15.79	16.99	3.72	0.22	2.71
47	2	3	76.46	0.21	0.72	5.48	11.2	1.92	0.69	2.26
47	2	4	70.17	0.08	0.25	17.15	8.77	0.87	0	2.59

47	2	5	80.84	0.13	0.45	4.38	5.71	0	0.1	7.06
47	2	6	80.9	0.25	0.07	3.41	5.87	0.19	0.73	6.71
48	1	1	48.52	0.86	0	13.47	12.45	1.3225	3.475	0
48	1	2	82.57	0.35	0	2.53	4.265	1.1325	0.565	0
48	1	3	76.93	0.49	1.63	8.5	0.76	2.3475	0	0
48	1	4	91.03	0	1.56	0.9	0.835	0.57	0.185	0
48	1	5	87.48	0	3	0.62	1.395	0.46	0.695	0
48	1	6	68.74	0	2.42	12.21	0.765	3.4075	0	0
49	1	1	19.02	0.84	0	5.09	5.49	15.385	1.255	0
49	1	2	79.07	0.84	0.49	2.82	2.495	2.5875	0.41	0
49	1	3	73.03	0.33	0.55	4.56	2.785	2.8025	2.38	0
49	1	4	49.21	1.28	0	2.53	5.4	3.9	0	0
49	1	5	8.33	0	0	1.61	0.415	2.6275	0.87	0
49	1	6	67.29	2.82	0	2.56	2.215	4.745	1.955	0
50	1	1	2.89	0	0	0	0	0	0	0
50	1	2	36.69	7.23	0.09	5.33	1.015	8.925	0	0
50	1	3	54.94	0	0.09	1.72	2.03	9.795	0	0
50	1	4	73.86	0.13	0	0.75	0	6.3125	0	0
50	1	5	69.03	0.64	0.05	0.51	0	7.4425	0	0
50	1	6	68.68	0	0.19	0.75	0	7.2025	0	0
51	1	1	51.78	4.57	0	3.95	4.79	6.52	0.52	0
51	1	2	66.44	0	0	0.93	0.815	7.2775	0	0
51	1	3	79.52	0	0.24	0.35	0	4.7775	0	0
51	1	4	63.67	0.15	0.73	0.64	0.525	7.775	0	0
51	1	5	70.4	0.97	0.91	2.21	0.23	5.8225	0	0
51	1	6	65.25	18.85	1.22	0.22	0.31	3.0775	0	0
52	1	1	3.5	0	0	56.3	2.215	8.9425	0	0
52	1	2	54.61	8.55	0	17.8	0	4.76	0	0
52	1	3	2.39	3.92	0	73.15	1.325	4.4725	0	0
52	1	4	14.58	0	1.15	61.56	0.265	5.54 75	0	0
52	1	5	0	0	0	9.41	0.525	22.1 525	0.465	0
52	1	6	0.68	0	0	97.1	0	0.555	0	0
53	1	1	48.72	0	0	28.05	8.425	0.1225	2.95	0
53	1	2	9.99	0	0	57.75	15.515	0	0.615	0
53	1	3	35.88	0	0	36.18	12.915	0.2025	0.565	0.16

53	1	4	68.59	0.48	0	14.01	8.46	0	0	0
53	1	5	72.23	0	0.11	5.98	9.86	0.1425	0.695	0
53	1	6	71.83	0.16	0	7.72	4.725	0.9275	3.285	0
54	1	1	42.06	0.82	1.47	4.63	2.545	9.845	3.28	0
54	1	2	67.93	3.63	0.15	12.03	2.365	2.8875	0	0
54	1	3	68.75	0	1.79	22.33	1.115	1.2225	0	0
54	1	4	34.05	11.3	0	11.98	0.605	10.365	0	0
54	1	5	1.52	0	2	26.36	1.365	16.7575	0.185	0
54	1	6	1.04	0	0	40.24	0.31	13.2925	2.465	0
55	1	1	48.56	0	0	22.49	11.13	1.355	0.635	0
55	1	2	56.94	0	0	8.13	7.025	1.61	2.485	0
55	1	3	7.69	0	0	27.32	0	0.2125	1.94	0
55	1	4	45.64	1.51	0	17.25	13.35	0.3425	0.81	5.68
55	1	5	60.37	0	0.16	8.77	3.89	2.35	1.06	11.32
55	1	6	57.1	0	0	0.67	1.215	9.255	1.085	0
56	1	1	15.72	0	0	74.62	2.04	0.16	0	0
56	1	2	31.27	0	0	47.89	8.855	0.04	1.11	0.58
56	1	3	61.65	0	0	21.87	4.755	1.045	1.24	0
56	1	4	7.21	0	0	73.68	7.175	0.285	1.805	0
56	1	5	61.39	5.09	0	15.44	6.145	1.2475	0	0
56	1	6	53.81	0	0.43	5.69	0.94	8.8575	0	0
57	1	1	40.35	1.16	0.45	56.17	0.16	0	0	0.43
57	1	2	0	8.31	0	91.69	0	0	0	0
57	1	3	4.22	6.97	0	88.81	0	0	0	0
57	1	4	3.67	69.78	0	26.54	0	0	0	0
57	1	5	23.51	4.23	0.57	50.83	10.42	0	0	0
57	1	6	16.64	15.97	0	42.13	9.59	0	0.145	5.79
58	1	1	3.91	5.69	0	83.32	3.175	0.18	0	0
58	1	2	27.23	2.46	0	50.03	10.14	0	0	0
58	1	3	12.78	0	0	11.99	0.22	0.0875	1.075	0
58	1	4	26.1	2.83	0	37.71	3.755	0	0.26	0
58	1	5	54.62	2.22	0	22.2	7.755	0	2.395	0

58	1	6	7.74	1.87	0	11.43	1.785	0	1.34	0
59	1	1	31.94	0.41	0.23	62.3	2.185	0	0.28	0
59	1	2	76.45	4.38	0	12.91	2.16	0.1325	0.67	0
59	1	3	60.47	0.32	0.29	10.96	1.835	5.9275	0.105	0
59	1	4	55.5	0	0	8.94	0.75	8.285	0	0
59	1	5	58.33	0	0	5.97	0.325	8.2025	0	0
59	1	6	71.85	0	0	4.19	0.47	5.4925	0	0
60	1	1	37.46	1.4	0	8.44	2.43	10.97 75	1.585	0
60	1	2	85.68	0	0	0.18	0.93	2.6525	0.08	0
60	1	3	80.64	0	0.18	1.13	0.44	3.645	0.32	0
60	1	4	86.16	0	0	2.83	0.075	2.2075	0	0
60	1	5	85.26	0	0.3	3.44	0.14	2.41	0	0
60	1	6	91.21	0	0	1.18	0.375	1.045	0	0
61	1	1	27.87	2.44	0	38.76	5.155	4.8725	0.515	0
61	1	2	59.45	0.16	0.08	29.91	1.16	1.385	0.28	0
61	1	3	75.85	0	1.04	3.22	0.685	2.57	0	0
61	1	4	89.8	0	1.88	0.76	1.145	0.5375	0	0
61	1	5	85.07	0	0.84	1.63	0.73	2.2125	0	0
61	1	6	92.95	0	0.14	0	0.4	0.3775	0	0
62	1	1	15.25	0.55	0.12	6.09	4.485	1.35	2.975	0
62	1	2	20.3	1.25	0	12.03	2.12	0.76	0	0
62	1	3	68.67	0	0	3.86	3.555	4.92	0.225	0
62	1	4	84.11	0	0.2	3.59	2.065	1.3375	0	0
62	1	5	80.3	0	0.84	6.65	1.18	1.4475	0	0
62	1	6	91.93	0	0.34	2.54	0.415	0.5025	0	0
63	1	1	17.61	1.99	0	53.33	6.555	0.865	5.25	0
63	1	2	38.76	1.5	0	39.12	3.12	0.3225	4.355	4.38
63	1	3	85.05	1.08	0	4.51	1.91	1.385	0	0
63	1	4	64.51	0.63	0	8.61	5.495	3.8125	0	0
63	1	5	56.97	15.38	0	8.95	2.93	3.2075	0	0
63	1	6	2.65	1.47	0	92.64	0.1	0.33	0.86	0
64	1	1	50.53	0.1	0	8.42	1.475	3.4675	0	0
64	1	2	91.55	0.71	0	1.77	1.365	0.455	0	0
64	1	3	94.57	1.62	0	0.49	0.76	0.1675	0	0
64	1	4	91.67	0	0	0.24	0.925	0.7125	0	0

64	1	5	91.84	0	0.23	0.37	0.155	0	0	0
64	1	6	92.97	0	3.33	0	0	0	0	0
65	1	1	51.84	1.22	0.09	23.93	8.545	0.165	2.155	0.85
65	1	2	65.31	0.3	0.18	10.11	10.285	0	0.885	1.78
65	1	3	88.84	0	0.26	2.51	2.91	0	0.205	0.49
65	1	4	92.63	0	0.59	4.23	0.655	0	0	0
65	1	5	81.33	0	0.33	4.25	5.82	0	0.17	0.69
65	1	6	74.86	0	0.42	8.15	7.08	0	0	1.43
66	1	1	63.39	0	0	17.7	8.065	0	0	2.78
66	1	2	79.01	0	0	12.86	3.435	0	0	0.92
66	1	3	71.33	0.27	0	9.66	7.655	0	0	3.42
66	1	4	77.32	1.37	0	8.96	2.395	0	1.425	4.71
66	1	5	92.1	0	0	3.43	1.945	0	0	0.42
66	1	6	88.66	0	0	7.3	1.96	0	0	0.11
67	1	1	82.08	0	1.25	7	2.95	0.1625	0.385	0
67	1	2	82.52	1.02	0.75	4.09	3.115	0.665	0.55	0
67	1	3	83.66	0	0.37	2.06	5.53	0.475	0	0
67	1	4	86	0	0.59	2.5	2.5	0.605	0	0
67	1	5	88.46	0	0	0.64	3.07	0.1925	0	0
67	1	6	91.75	0	0.82	0.8	1.44	0	0	0

ACKNOWLEDGMENTS

The authors wish to thank Sue Cobey for access to her honeybees and Brian H. Smith for the use of his laboratory facilities at Ohio State University.

AUTHOR CONTRIBUTIONS

Conceived and designed the experiments: GAW JAM JLL. Performed the experiments: JLL. Analyzed the data: GAW HJB. Contributed reagents/materials/analysis tools: GAW. Wrote the paper: GAW HJB JLL JAM.

REFERENCES

1. 1.Barton-Browne L, Evans DR. Locomotor activity of the blowfly as a function of feeding and starvation.*Journal of Insect Physiology.*1960;4:27–37.
2. 2.Connolly KJ. Locomotor activity in Drosophila as a function of food deprivation.*Nature.*1966;209:224.
3. 3.Raubenheimer D, Gade G. Separating food and water deprivation in locusts: Effects on the patterns of consumption, locomotion and growth.*Physiol Entomol.*1996;21:76–84.
4. 4.Siegal PS, Steinberg J. Activity level as a function of hunger.*Journal of Comparative Psychology.*1949;42
5. 5.Crailsheim K. Trophallactic interactions in the adult honeybee (Apis mellifera L.).*Apidologie.*1998;29:97–112.
6. 6.Free JB. A study of the stimuli which release the food begging and offering responses of worker honeybees.*The British Journal of Animal Behaviour.*1956;4:94–101.
7. 7.Leoncini I, Le Conte Y, Costagliola G, Plettner E, Toth AL, et al. Regulation of behavioral maturation by a primer pheromone produced by adult worker honey bees.*P Natl Acad Sci USA.*2004;101:17559–17564.[PMC free article]
8. 8.Schulz DJ, Vermiglio MJ, Huang ZY, Robinson GE. Effects of colony food shortage on social interactions in honey bee colonies. *Insect Soc.*2002;49:50–55.
9. 9.Mc Cabe S, Farina WM, Josens RB. Antennation of nectar-receivers encodes colony needs and food-source profitability in the ant Camponotus mus.*Insect Soc.*2006;53:356–361.
10. 10.Schulz DJ, Huang ZY, Robinson GE. Effects of colony food shortage on behavioral development in honey bees.*Behav Ecol Sociobiol.*1998;42:295–303.
11. 11.Bettinger JC, McIntire SL. State-dependancy in C-elegans.*Genes Brain Behav.*2004;3:266–272.
12. 12.Davies AG, Bettinger JC, Thiele TR, Judy ME, McIntire SL. Natural variation in the npr-1 gene modifies ethanol responses of wild strains of C. elegans.*Neuron.*2004;42:731–743.
13. 13.Kaun KR, Azanchi R, Maung Z, Hirsh J, Heberlein U. A Drosophila model for alcohol reward.*Nat Neurosci.*2011;14:612-U103.

14. 14.Lee HG, Kim YC, Dunning JS, Han KA. Recurring Ethanol Exposure Induces Disinhibited Courtship in Drosophila.*Plos One.*2008;3
15. 15.Maze IS, Wright GA, Mustard JA. Acute ethanol ingestion produces dose-dependent effects on motor behavior in the honey bee (Apis mellifera).*Journal of Insect Physiology.*2006;52:1243–1253.
16. 16.Mustard JA, Edgar EA, Mazade RE, Wu C, Lillvis JL, et al. Acute ethanol ingestion impairs appetitive olfactory learning and odor discrimination in the honey bee.*Neurobiol Learn Mem.*2008;90:633–643.
17. 17.Rothenfluh A, Heberlein U. Drugs, flies, and videotape: the effects of ethanol and cocaine on Drosophila locomotion.*Curr Opin Neurobiol.*2002;12:639–645.
18. 18.Scholz HaM JA. Invertebrate Models of Alcoholism.*Current Topics in Behavioral Neuroscience.*2011In press.
19. 19.Wolf FW, Heberlein U. Invertebrate models of drug abuse.*J Neurobiol.*2003;54:161–178.
20. 20.Barron AB, Maleszka R, Helliwell PG, Robinson GE. Effects of cocaine on honey bee dance behaviour.*J Exp Biol.*2009;212:163–168.
21. 21.Bozic J, Abramson CI, Bedencic M. Reduced ability of ethanol drinkers for social communication in honeybees (Apis mellifera carnica Poll.).*Alcohol.*2006;38:179–183.
22. 22.Mixson TA, Abramson CI, Bozic J. The Behavior and Social Communication of Honey Bees (Apis Mellifera Carnica Poll.) under the Influence of Alcohol.*Psychol Rep.*2010;106:701–717.
23. 23.Bitterman ME, Menzel R, Fietz A, Schafer S. Classical-Conditioning of Proboscis Extension in Honeybees (Apis-Mellifera).*Journal of Comparative Psychology.*1983;97:107–119.
24. 24.Agresti A.*Categorical Data Analysis.*Hoboken, NJ: Wiley; 2002.
25. 25.Johnson RA, Wichern DW. 2007. Applied Multivariate Statistical Analysis Englewood Cliffs, NJ: Prentice Hall.
26. 26.Wcislo WT, Gonzalez VH. Social and ecological contexts of trophallaxis in facultatively social sweat bees, Megalopta genalis and M. ecuadoria (Hymenoptera, Halictidae).*Insect Soc.*2006;53:220–225.
27. 27.Farina WM, Nunez JA. Trophallaxis in the Honeybee, Apis-Mellifera (L) as Related to the Profitability of Food Sources.*Anim Behav.*1991;42:389–394.

28. 28.Judge KA, Bonanno VL. Male Weaponry in a Fighting Cricket.*Plos One.*2008;3
29. 29.Boulay R, Soroker V, Godzinska EJ, Hefetz A, Lenoir A. Octopamine reverses the isolation-induced increase in trophallaxis in the carpenter ant Camponotus fellah.*J Exp Biol.*2000;203:513–520.
30. 30.Guerrieri FJ, d'Ettorre P. The mandible opening response: quantifying aggression elicited by chemical cues in ants.*J Exp Biol.*2008;211:1109–1113.
31. 31.Bell WJ, Hawkins WA. Patterns of Intraspecific Agonistic Interactions Involved in Nest Defense of a Primitively Eusocial Halictine Bee.*J Comp Physiol.*1974;93:183–193.
32. 32.Stuart RJ, Herbers JM. Nest mate recognition in ants with complex colonies: within- and between-population variation.*Behav Ecol.*2000;11:676–685.
33. 33.Kong EC, Woo K, Li HY, Lebestky T, Mayer N, et al. A Pair of Dopamine Neurons Target the D1-Like Dopamine Receptor DopR in the Central Complex to Promote Ethanol-Stimulated Locomotion in Drosophila.*Plos One.*2010;5
34. 34.Abramson CI, Stone SM, Ortez RA, Luccardi A, Vann KL, et al. The development of an ethanol model using social insects I: Behavior studies of the honey bee (Apis mellifera L.).*Alcohol Clin Exp Res.*2000;24:1153–1166.
35. 35.Scholz H. Intoxicated Fly Brains: Neurons Mediating Ethanol-Induced Behaviors.*J Neurogenet.*2009;23:111–119.

SECTION 7: WILD ANIMALS

7.4. FISH

CHAPTER 21

Physiological and Behavioural Responses to Noxious Stimuli in the Atlantic Cod (Gadus Morhua)

Jared R. Eckroth[1], Øyvind Aas-Hansen[2], Lynne U. Sneddon[3], Helena Bichão[2], Kjell B. Døving[1]

[1] Department of Molecular Biosciences, University of Oslo, Oslo, Norway

[2] Norwegian Institute of Food, Fisheries and Aquaculture Research, Nofima, Tromsø, Norway

[3] Institute of Integrative Biology, University of Liverpool, Liverpool, United Kingdom

ABSTRACT

In the present study, our aim was to compare physiological and behavioural responses to different noxious stimuli to those of a standardized innocuous stimulus, to possibly identify aversive responses indicative of injury detection in a commercially important marine teleost fish, the Atlantic cod. Individual fish were administered with a noxious stimulus to the lip under

Citation: Eckroth JR, Aas-Hansen Ø, Sneddon LU, Bichão H, Døving KB (2014) Physiological and Behavioural Responses to Noxious Stimuli in the Atlantic Cod (Gadus morhua). PLoS ONE 9(6): e100150. https://doi.org/10.1371/journal.pone.0100150

short-term general anaesthesia (MS-222). The noxious treatments included injection of 0.1% or 2% acetic acid, 0.005% or 0.1% capsaicin, or piercing the lip with a commercial fishing hook. Counts of opercular beat rate (OBR) at 10, 30, 60, 90 and 120 min and observations of behaviour at 30 and 90 min post-treatment were compared with pre-treatment values and with control fish injected with physiological saline, an innocuous stimulus. Circulatory levels of physiological stress indicators were determined in all fish at 120 minutes post-treatment. All treatments evoked temporarily increased OBR that returned to pre-treatment levels at 60 minutes (saline, 0.005% capsaicin, hook), 90 minutes (0.1% acetic acid, 0.1% capsaicin), or 120 minutes (2% acetic acid), but with no significant differences from the control group at any time point. Fish treated with 0.1% and 2% acetic acid and 0.1% capsaicin displayed increased hovering close to the bottom of the aquaria and fish given 2% acetic acid and 0.1% capsaicin also displayed a reduced use of shelter. The only effect seen in hooked fish was brief episodes of lateral head shaking which were not seen pre-treatment or in the other groups, possibly reflecting a resiliency to tissue damage in the mouth area related to the tough nature of the Atlantic cod diet. There were no differences between groups in circulatory stress indicators two hours after treatment. This study provides novel data on behavioural indicators that could be used to assess potentially aversive events in Atlantic cod.

INTRODUCTION

The ability of animals to detect stimuli that can cause harm to their tissues is a universal feature termed nociception [1]. This sensory ability encompasses the neural processing of noxious stimuli and may include the induction of both physiological and behavioural responses but with the sensation of pain not necessarily being implied [2]–[4]. In contrast, pain is regarded as an aversive sensory and emotional experience, representing awareness by the animal of actual or potential tissue damage [2], [5]. In fishes, it has been argued that the neural processing of noxious stimuli may involve peripheral and central nociception but does not allow for pain perception, and that behavioural responses to noxious stimuli may represent an activation of nocifensive motor programs which does not involve conscious awareness [6], [7]. In contrast, other studies have demonstrated nociception and suggested the existence of affective states and potentially the ability for pain perception in teleost fishes (reviewed by e.g. [8]–[11]). Studies on the rainbow trout (*Oncorhynchus mykiss*), for example, have demonstrated A-delta and

C-fibre nociceptors [12]–[15] and shown that subcutaneous injections of acetic acid or bee venom to the lips caused enhanced opercular beat rate (OBR) and prolonged time to resume feeding [15]. The trout showed rocking body movements or rubbing its lips against the wall of the tank or in the gravel that were ameliorated by the use of an analgesic [14], [16]. Adverse changes in behaviour and physiology have also been observed in studies of goldfish (*Carassius aurata*) and Nile tilapia (*Oreochromis niloticus*) [17]–[19]. Furthermore, in common carp (*Cyprinus carpio*) and zebrafish (*Danio rerio*), some behavioural and physiological responses to noxious stimuli were similar to those seen in the rainbow trout, whereas other behaviours were found to differ between these species [20].

In the present study, our aim was to evaluate whether some well-characterized noxious stimuli elicit physiological and behavioural responses differing from those induced by the administration of a standardized innocuous stimulus, thus possibly identifying aversive responses associated with injury detection. Specifically, we have investigated if the administration of acetic acid, capsaicin or a commercial fishing hook to the lip of the fish induced physiological and/or behavioral responses differing from the responses seen when administering physiological saline, a standardized innocuous stimulus [15]. Atlantic cod (*Gadus morhua*) was chosen as a marine teleost model due to the large body of general knowledge on the biology of this species and the need for increased knowledge on the fish welfare of this species due to its commercial importance in North Atlantic fisheries and aquaculture. To our knowledge, this is the first study to specifically investigate physiological and behavioral effects of noxious chemicals in a marine teleost, including the first observation of possible nociception with capsaicin in fish, a frequently used stimulus in mammalian pain research [21]. In addition, we are not aware of previous studies where possible effects of the presence of a commercial fishing hook has been investigated and compared to the effects of more commonly used tests in nociception and pain research. In order to facilitate inter-specific comparisons, we employed methods and experimental design evaluated in previous fish studies [12], [15], [16], [22]–[24], including the standardized use of temporary acute anaesthesia during stimulus administration [25]. We have expanded on most previous studies of nociception in fish by including measurements of circulatory levels of commonly used primary and secondary stress indicators in fish, to identify possible differences between treatments as nociception and pain is inherently stressful [26]–[28].

MATERIALS AND METHODS

Ethics Statement

All experiments were conducted in accordance with the Norwegian Fish Welfare and Laboratory Animals legislation [29], [30], which adheres to the European Convention for the Protection of Vertebrates used for Experimentation and other Scientific Purposes [31]. The protocol was approved by the Committee on the Ethics of Animal Experiments of the University of Oslo (Permit Number: 3052).

Animals and Rearing Conditions

Juvenile Atlantic cod (275±66 g, 30.7±2.4 cm, n=42) were obtained from Tromsø Aquaculture Research Station, Tromsø, Norway. The experiments were performed at the University of Oslo, Norway, in the period 16. February –30. March 2011. Before experiments, cod were distributed in two stock tanks (217×90×93 cm) with a constant flow (720 $L \cdot h^{-1}$) of filtered aerated saltwater (Hydrotech, Filter type Hdt 501-1p; 3.2% salinity, Marinium Reef Sea Salt), and were fed an appropriate diet (3% of the body mass, 5 mm pellets – Aglo Norse Coldwater ekstra, Eximo A/S). The fish were kept at a water temperature of 9±1°C and a 12 h:12 h light:dark cycle. Each cod was given a minimum of two weeks to acclimate in the stock tanks before they were used in experiments.

Experimental Protocol

Cod were caught at random and placed individually in one of ten rectangular glass aquaria (approximately 150 L; 110×35×40 cm) with a flow (60 $L \cdot h^{-1}$) of aerated seawater (water temperature of 9±1°C and a 12 h:12 h light:dark cycle). All aquaria had a gravel substrate and were aerated via an air stone and tubing connected to an air pump. The aquaria had a dark coloured, rectangular, plastic shelter tube with one long side cut out for viewing (12×12×12 cm) placed on the bottom in the rear centre of the aquaria. All aquaria had yellow tape placed along the outer edges of the tank. The tape had black lines drawn on with 1 cm distances between each line. This provided a method to estimate swimming distance when analysing behaviours. To visually isolate the cod from one another and from other disturbance, opaque black panels on the sides separated the aquaria and dark fabric covered the front. Each fish was acclimated for seven days in its aquarium before the treatment trial commenced. On the seventh day treatment trials were commenced at the same time each day with one fish at 8:00 am and a second fish at 10:30am. A total of six treatment groups were used with six to eight cod in each group; control treatment 1 (Saline, N=7), treatment 2 (0.1% Acetic acid, N=7), treatment 3 (2.0% Acetic acid, N=7), treatment 4 (0.005% Capsaicin, N=6), treatment 5 (0.1% Capsaicin, N=8) and treatment 6 (Fishing hook, N=7). The chosen chemical stimuli and their concentrations were based on pilot studies and on existing literature on nociception and potential pain in other teleost fish species (*e.g.* acetic acid: [15], [20], [22]–[24]) and in studies of nociception and potential pain in human pain research (*e.g.* capsaicin: see [21], [32]–[34]). All treatments were applied to the lip of the fish, as this area has also been used in similar studies in other fish species (*i.e.* [15], [16], [20], [22]–[24], [35]) and nociceptors were identified on the lips of the rainbow trout [13]–[15]. Each fish underwent one treatment only and all treatments were randomly assigned. The procedure consisted of the single cod being quickly but carefully netted from its aquarium and into a 5 L anaesthetic bath (40 mg/L; metacaine (MS-222), Pharmaq AS, Oslo, Norway). Metacaine [36] was chosen as it has been used in comparable previous studies with freshwater fish, is comparable to other drugs used [14]–[16], [20] and has well-documented effects in the Atlantic cod including rapid induction and recovery times [37]. When deep plane anaesthesia was attained i.e. loss of tail reflex and lack of response to tail pinch, the designated treatment was applied. After treament the cod was returned to its aquarium. The time lapse from dip-netting to being returned to the aquarium was always less than 180 sec, and the fish regained equilibrium within 529±43 sec after reimmersion. For the fish assigned to treatment groups 1 to 5, 25 μl of the chemical solution was injected subcutaneously to the upper lip, and 25 μl was injected to the lower lip, always on the left side, 10 mm from the front center of the lips, using a sterile 250 uL N Syringe with cemented needle (725N 250 μl SYR (22s/2″/2), Hamilton Bonaduz AG, Switzerland). The needle was inserted into the subcutaneous connective tissue of the lip before the injection took place. For the fish assigned to treatment group 6, a standard recreational saltwater fishing hook (40 mm from eye to bend, 10 mm gape, straight hook eye w/cutting point, Mutu Circle Hooks, Australia) was inserted from the mouth opening side going through the thin elastic fold of the skin and then through the lower left subcutaneous tissue of the lip (*i.e.* in the same position as the injection treatments), and remained attached to the cod for the entire experiment. Following the 120 min post-treatment observation period, the cod was quickly dip-netted and killed by an overdose of metacaine (160 $mg.L^{-1}$), and a blood sample was taken. Time between cod being dipnetted and killed by metacaine was 180±30 sec. Finally, the fish was weighed (Top Pan Precision Balances, PGL 4001, Adam Equipment, Danbury CT, USA), standard length measured to 0.1 cm, and the site of injection/hook inspected to notice possible physical effects of the treatment.

Measurements of Opercular Beat Rate (OBR)

Twenty minutes (−20 to −14) before treatment and at at the time points 10 min (10–16), 30 min (30–36), 60 min (60–66), 90 min (90–96), and 120 (114–120) min after treatment, OBR was measured by counting the number of opercular beats per minute during three intervals (one minute between each interval) so an average OBR could be calculated for that observation period.

To ensure the cod were not disturbed, OBR was observed from a TV-screen in an adjacent room by means of a digital video camera placed in front of the aquarium (described below).

Behavioural Observations

Digital wide-angle lens video cameras (Sony Handycam, DCR-SR47 HDD, Toyko, Japan) were placed in front of the glass aquaria for counting opercular beat rates and recording behaviour.

The cameras were placed in front of tanks 18 hours prior to commencement of each trial to allow habituation. Each fish was continuously recorded from 30 min pre-treatment until 120 min post-treatment.

From these video recordings, behavioural data were either measured by the percentage of time the fish engaged in a behaviour or by the number of times a behaviour was observed according to a simple ethogram (Table 1). The ethogram contains the most common types of behaviours observed during pre and post treatments.

The behaviours were scored for the 15 min period prior to treatment (10–25 min) and from (30–45 min) and (90–105) min after the treatment. High reliability and validity of these measurements was tested and demonstrated by intra- observer reliability tests preformed on two of the videos (intra- $r^2=0.859$, $r^2=0.998$; Spearman Rank correlation) and inter-observer reliability tests performed on three of the videos chosen at random (inter- $r^2=1.000$; $r^2=0.967$, $r^2=0.994$; Spearman Rank correlation).

Table 1. Ethogram presenting the categorization of behaviors in the present study. Behavioural data are expressed as the percentage duration (%) or counts of episodes (#).

Category	Measurement	Description
Shelter	Duration (%)	Positioned with 75% of body under shelter located on bottom of aquaria
Hovering top	Duration (%)	Hovering and/or stationary in top 15% portion of water column
Hovering bottom	Duration (%)	Hovering and/or stationary on bottom 15% portion of water column
Swimming	Duration (%)	Swimming around entire tank, moving more than one body length per second
Head shaking	Count (#)	Episodes with a series of very rapid lateral movements of the head.

Measurement of Blood Parameters

A sample of blood (1.0–1.5 ml) was taken from the caudal vasculature using a heparinised sterile 2.0 ml syringe (BD Plastipak, UK) and 21G×1 ½ ″ hypodermic-needle (Braun, Switzerland). Whole blood was immediately analysed using hand-held meters evaluated for use in fish [38], [49], and frequently used in studies in the Atlantic cod (*e.g.* [40], [41]). Lactate levels were measured using Lactate Pro™ (Arkray, Kyoto, Japan) whereas hematocrit and concentrations of sodium (Na+), potassium (K+) and glucose were measured using an i-STAT Portable Clinical Analyzer (Abbot Point of Care Inc.; Princeton, NJ, USA). The remaining blood was placed on ice in a 4.0 ml lithium-heparinized vacuum tube (BD Vacutainer, Franklin Lakes, NJ, USA), centrifuged (8 min, 2700×g at 4.0°C, Beckman Coulter Avanti J-20XP Centrifuge, USA) and plasma frozen and stored at −20°C for later analysis of cortisol levels. Plasma cortisol was measured by means of radioimmunoassay (RIA) according to a protocol recently validated for Atlantic cod (Dr. H. Tveiten, personal communication Nofima AS, Muninbakken 9–13, NO-9291, Tromsø, Norway). Briefly, cortisol was extracted from 200 µl of plasma with 4 ml diethyl ether under vigorous shaking for 4 min. The aqueous phase was frozen in liquid nitrogen, whereas the organic phase was transferred to a glass tube, evaporated in a water bath at 45°C and then reconstituted by addition of 600 µl assay buffer and then assayed according to established protocols [42], [43]. The detection limit of the assay was 0.6 $ng{\cdot}ml^{-1}$.

Chemicals

Solutions were dissolved in saline (Ringers solution). The acetic acid solutions were made on the day of the experiments to ensure the solutions were not neutralized by the saline. The 2% acetic acid was made by diluting 200 μl of glacial acetic acid (laboratory grade, VWR International, Oslo, Norway) in 10 ml of Ringer's solution (pH=2.5). The 0.1% acetic acid was made by diluting 10 μl of glacial acetic acid in 10 ml of Ringer's solution (pH=3.2). Capsaicin was obtained from Sigma-Aldrich Norway AS (Oslo, Norway). To ensure solubility of capsaicin, the capsaicin solutions were prepared in one dimethyl sulfoxide (DMSO) (D5879-Sigma Aldrich Co, St. Louis, MO, USA,) to give a final concentration of 1% DMSO in Ringers. A stock solution of capsaicin was prepared by dissolving 10 mg capsaicin in 0.1 ml DMSO. The 0.1% capsaicin treatment solution was made by adding 10 μl of capsaicin stock solution in 100 μl of DMSO and diluting in Ringer's solution to give a total volume of 10 ml. The 0.005% capsaicin was made by adding 50 μl of the 0.1% capsaicin/Ringers solution in 105.5 μl of DMSO and diluting in Ringer's solution to give a total volume of 10 ml. Ringer's physiological solution was made by mixing distilled water with ($g{\cdot}L^{-1}$) NaCl 12.85, KCL 0.29, $MgCl_2$ 0.20, $CaCl_2$ 0.29, and Hepes 2.38; pH was adjusted to 7.2. All the chosen acetic acid and capsaicin concentrations used were tested on a limited number of cod in a preceding pilot study to ensure sufficent effectiveness. It was found that injection of 25 μl of the high concentrations of both acetic acid and capsaicin into the lip of Atlantic cod caused white spots which progressed into necrosis within a few days. Accordingly, it was concluded from these tests that the high concentrations of these chemicals were sufficiently high to act as noxious stimuli and that higher concentrations were not needed.

Data and Statistical Analysis

Statistical analyses ($p \leq 0.05$ considered statistically significant) were performed using SPSS version 19 software (IBM Corporation, Armonk, NY, USA). All data were tested for normal distribution using Kolmogorov-Smirnov test and homogenity of variances was tested using Levene's F test and those that met parametric assumptions were tested accordingly whereas those that did not were analysed using non-parametric statistics. To determine if there were any differences in the physiological parameters between the fish tested at 8:00 am and the fish tested at 10:30 am independent t-tests were conducted ($p \leq 0.05$). OBR data were tested for differences between treatment groups at each time-point (*i.e.* at 20 min prior to treatment application and at 10, 30, 60, 90 and 120 min following treatment application) using two-way repeated measure ANOVA with a Greenhouse-Geisser correction followed by post-hoc tests with a Bonferroni correction. Greenhouse-Geisser correction was used as it is conservative in incorrectly rejecting the null hypothesis. To determine differences in OBR between the before and after effects of each treatment, a repeated measures ANOVA was applied again using a Greenhouse-Geisser correction followed by post-hoc tests using the Bonferroni correction. Video recordings were analyzed for behaviours using Adobe Premiere Pro CS4.1 (Adobe Systems Incorporated, USA). Possible differences between and within treatment groups in the occurrence of the categorized behaviours (Table 1) were tested using a repeated measures general linear model (GLM) followed by post-hoc tests using the Bonferroni correction. Blood parameters were tested to determine any differences between treatment groups using one-way ANOVA for normally distributed data and Kruskal Wallis tests for non-normally distributed data.

RESULTS

Opercular Beat Rate (OBR)

OBR in the cod increased after treatment in all groups (Figure 1; Saline ($F(3, 16)=28.721$, $p<0.001$), 0.1% Acetic Acid ($F(2, 13)=43.434$, $p<0.001$), 2.0% Acetic Acid ($F(2, 15)=29.882$, $p<0.001$), 0.005% Capsaicin ($F(2, 9)=12.398$, $p<0.001$). 0.1% Capsaicin ($F(2, 14)=27.062$, $p<0.001$), Hook in lower lip ($F(3, 16)=19.994$, $p<0.001$). Post hoc tests with the Bonferroni correction showed that from 10 min post treatment onwards, the OBR steadily declined with time and recovered to approximately normal ($p>0.05$ difference from its pre-treatment level) between 60 min (Saline, 0.005% Capsaicin and Hook in lower lip), 90 min (0.1% Acetic acid and 0.1% Capsaicin) and 120 min (2% Acetic acid) post treatment (Figure 1; Table S1). At any given time point, there were no statistically significant ($p>0.05$) differences in OBR between the saline treatment group (control) and any of the other treatment groups (Figure 1; Table S1).

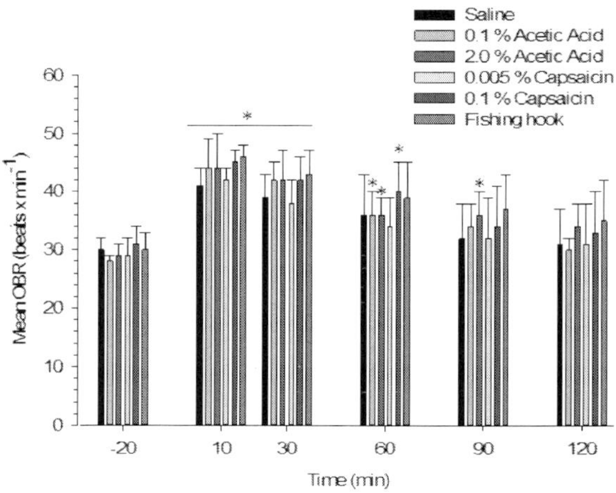

Figure 1. Opercular beat rates (beats×min^{-1}; mean OBR±S.E.) for each treatment group at 20 min pre-treatment (−20 min) and at 10, 30, 60, 90 and 120 min post-treatment (i.e. corresponding to the data presented in Table S1). There were no statistically significant differences ($p>0.05$) in OBR evidenced between treatment groups at any time-point (two-way repeated measures ANOVA with a Greenhouse-Geisser correction followed by post-hoc test using the Bonferroni correction). Asterisks (*) denote a statistically significant ($p\leq0.05$) within-group difference in OBR compared to its respective pre-treatment recording (one-way repeated measures ANOVA with a Greenhouse-Geisser correction followed by post-hoc test using the Bonferroni correction). N=7 fish per group except for 0.005% Capsaicin (N=6) and 0.1% Capsaicin (N=8).

Behavioural Observations

During the pre-treatment period all of the categorized behaviours (Table 1) were observed except for head-shaking. Two types of behaviors were significantly affected by treatments: hovering on the bottom of the aquaria ($F(5,108)$,=4.709 $p=0.001$) and use of shelter ($F(5,108)=2.427$, $p=0.050$); (Figure 2; Table S2). Post hoc tests (mean % of time±S.E.) for use of shelter using the Bonferroni correction showed that 2.0% Acetic acid significantly reduced percentage of time the sheltering behaviour was observed (1.6% ±1.6 at 30 min and 0.0±0.0% at 90 min) compared with 0.1% Acetic acid (39.8±16.7% at 30 min and 43.1±20.1% at 90 min) ($p=0.033$) and Saline (57.1±20.2% at 30 and 90 min) ($p=0.021$). 0.1% Capsaicin also decreased the % of time the sheltering behaviour was observed (16.0±12.5% at 30 min and 18.8±13.1% at 90 min) compared with 0.1% Acetic acid (39.8±16.7% at 30 min and 43.1±20.1% at 90 min) ($p=0.020$) and Saline (57.1±20.2% at 30 and 90 min) (p=0.013). Post hoc results (Bonferroni correction, mean % of time±S.E.) associated with hovering on the bottom demonstrated that 2.0% Acetic Acid significantly increased the percentage of time (47.8±14.0% at 30 min and 21.5±6.8% at 90 min) that cod were observed hovering on the bottom of the aquaria as compared with 0.005% Capsaicin (2.8 2.2% at 30 min and 10.8±7.0% at 90 min) (p=0.010) and Saline (2.0±1.9% at 30 min and 0.3±0.2% at 90 min) (p=0.001). Further 0.1% Capsaicin increased the percentage of time (28.7±14.0% at 30 min and 40.2±13.5% at 90 min) cod were observed hovering on the bottom of the aquaria as compared with 0.005% Capsaicin (2.8±2.2% at 30 min and 10.8±7.0% at 90 min) (p=0.02), Hook (19.2±8.0% at 30 min and 10.6±7.3% at 90 min) (p=0.021) and Saline (2.0±1.9% at 30 min and 0.3±0.2% at 90 min) (p<0.001). Also, 0.1% Acetic acid increased the mean percentage of time (34.3±12.6% at 30 min and 11.5±6.7% at 90 min) cod were observed hovering on the bottom of the aquaria compared with Saline (2.0±1.9% at 30 min and 0.3±0.2% at 90 min) (p=0.032). The analysis demonstrated that only the behaviour of hovering in the top of the tank was affected by time ($F_{(2, 108)} = 3.092$, p=0.050) with post hoc analysis showing that the total mean percentage of time this behavior was observed was significantly different between the pre-treatment observations (6.69±2.69 (mean % of time±S.E.)) and those at post 90 min (20.14±4.74 (mean % of time±S.E.); p=0.017) for all treatments. There was no interaction of time and treatment on any of the behaviors.

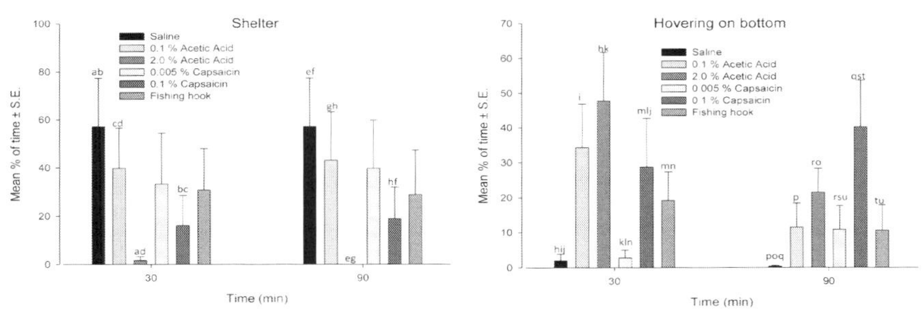

Figure 2. Occurrence of Shelter and Hovering on the bottom behaviours in Atlantic cod before and after saline, acetic acid, capsaicin, and fishing hook treatments. The data are expressed as mean percentage of time (%,±S.E.) the behaviour was displayed during 15 min segments at 30 min and 90 min after treatment. For each time point, identical letters denote a statistically significant (p≤0.05) difference between treatment groups (repeated measures GLM followed by post-hoc test using the Bonferroni correction). N=7 fish per group except for 0.005% Capsaicin (N=6) and 0.1% Capsaicin (N=8).

Only cod that had a hook inserted into the lower lip displayed episodes of head-shaking. Each episode lasted 1 - 2 seconds and consisted of a few very quick sideway movements of the head (the movements were too quick to be counted based on the video recordings). Head shaking episodes were not observed in any fish pre-treatment, but a total of 12 episodes (range 0–6 episodes per fish) were observed in five of the seven fish during the 30–45 min observation segment and again a total of 10 episodes (range 0–4 episodes per fish) in four of the same five fish during the 90–105 min observation segment. In all fish, the hook remained in place for the entire length of the experiment.

Blood Parameters

There were no statistically significant ($p \leq 0.050$) differences in physiological blood parameters (plasma cortisol and whole blood lactate, glucose, sodium, potassium and hematocrit) between treatment groups at termination of the experiment at 120 min post-treatment (Table 2).

Table 2. Whole blood parameters and plasma cortisol levels in Atlantic cod at 120 min post-treatment.

Treatment	Saline	0.1% Acetic acid	2.0% Acetic acid	0.005% Capsaicin	0.1% Capsaicin	Fishing hook
***Cortisol ($ng \cdot ml^{-1}$)**	54.4± 32.7	31.9±22.2	59.3±27.8	12.2±18.4	31.0±24.0	44.0±32.7
***Lactate ($mmol \cdot L^{-1}$)**	1.9±1.9	1.8±0.8	2.7±1.4	1.5±0.6	1.7±0.6	1.7±0.6
Glucose ($mmol \cdot L^{-1}$)	3.7±0.7	3.5±0.8	3.8±1.6	3.4±1.4	3.3±0.8	3.4±0.8
Na^+ ($mmol \cdot L^{-1}$)	160.9±4.5	155.1±10.0	159.9±2.4	158.5±4.0	158.8±9.0	159.0±4.0
***K^+ ($mmol \cdot L^{-1}$)**	3.6±0.6	3.5±0.6	3.7±0.6	3.2±0.6	3.6±0.7	3.4±0.4
Hematocrit (%)	20.0±2.5	17.9±3.4	21.9±2.5	19.0±3.4	17.8±3.5	18.6±3.0

DISCUSSION

Responses to Acetic Acid

Reduced pH of the extracellular fluid is associated with pathophysiological conditions such as inflammation, hypoxia and anoxia, and cutaneous injection of low pH solution causes prolonged activation of sensory afferents which may cause a sharp stinging pain in humans [44]. In fish, acetic acid is found to activate sensory nociceptive afferents [16] and this stimulus has been used in studies of rainbow trout (*O. mykiss*), zebrafish (*D. rerio*), common carp (*C. carpio*) and goldfish (*C. aurata*). In the present study, injection of 2% acid into the Atlantic cods lips reduced the amount of time spent under shelter and both 0.1% and 2% acetic acid increased the amount of time spent hovering near the bottom of the aquaria, as compared to saline-injected control. There were, however, no differences in circulatory stress indicators or opercular beat rate (OBR) at any time-point between saline-treated controls and acid-injected fish.

Although all treatments including saline control resulted in temporarily increased OBR, the injection of acetic acid was associated with a slight delay in this recovery to pre-treatment OBR-values with fish given 0.1% acid recovering in 90 minutes and fish given 2% acid taking in 120 minutes to recover, compared to 60 min in the saline control group.

In rainbow trout [5], [19], [22] and zebrafish [19] the injection of acetic acid into the lip resulted in an opercular beat rate (OBR) that was greater than saline controls and took significantly longer to return to normal values. In the cyprinids goldfish and carp, however, OBR was found not to be elevated above controls following the injection of acetic acid, even when much higher doses (5 and 10%) of acetic acid were used in the carp [17], [20]. Possibly, the lack of differences between control and noxiously stimulated Atlantic cod in OBR and circulatory stress parameters may be linked to consistent individual variation known to exist in both behaviour and stress physiology in a variety of species including fish [45]–[47]. Such differences have been recorded during acid injection in rainbow trout after noxious treatment [35] but has not been investigated in cod. Similarly, the effects of acid injection on the use of shelter has been investigated in rainbow trout and carp but with inconsistent results [14], [20], [23].

Interestingly, our cod did not display any changes in rapid swimming behaviour in the present study. This is in contrast to a prolonged decrease in swimming activity with acid injection in zebrafish and rainbow trout [20] and an immediate and vigorous increase in swimming activity in unanaesthetized rainbow trout and goldfish [17], [22], but corresponds to unaffected swimming in the common carp [20]. In line with this, the lack of differences in circulatory stress parameters in the present study corresponds to previous laboratory studies which have found that unstressed cod could be described as hovering or swimming slowly in the water column whereas stressed fish displayed fast swimming, jumping or laying still on the bottom [48]–[50].

In common carp, rainbow trout, and goldfish, the injection of acetic acid was associated with observations of anomalous behaviours such as sideways rocking body movements and/or rubbing of the affected areas against the tank wall or in the gravel [17], [19], [20]. No such behaviours were observed in Atlantic cod in this study or in zebrafish after injection of 5% acetic acid into the lips [22]. One cannot exclude, however, that this may reflect a species-specific difference with anomalous behaviours being avoided as they may attract the attention of predators [20], [51], and future studies could explore if anti-predator behaviour is disrupted in Atlantic cod during a noxious stimulation.

Responses to higher concentrations of noxious chemicals often elicit a greater response from the animals. In line with this, the more prolonged recovery of OBR, the reduced use of shelter and the increased time spent hovering in the bottom of the aquaria seen in cod given 2% acid compared to 0.1% acid may be expected. In our cod, the acetic acid as well as the capsaicin concentrations used were tested in a preceding pilot study where it was found that the high concentrations of these chemicals (but not the chosen low doses) caused white spots which progressed rapidly into necrosis, indicating that the high concentrations used were sufficiently high to act as noxious stimuli. In contrast, investigations using common carp and zebrafish have used much higher doses of acetic (5–10%) to elicit a behavioural response. Comparably, injections of 0.7% acetic acid in the cheek of goldfish was reported to cause tissue damage with the injection site turning white, and was associated with an immediate vigorous escape and rubbing behaviour which was reduced with the administration of morphine [17]. We cannot exclude the possibility, however, that higher concentrations may have elicited more pronounced responses in cod.

Although some physiological parameters were unaffected possibly due to time of sampling, the delayed recovery of OBR, the reduced use of shelter and the increased hovering in the bottom of the aquaria differentially induced by 0.1% and 2% acetic acid in our cod correspond to responses seen with acetic acid in rainbow trout, zebrafish and goldfish, possibly representing responses specific to the detection of injurous stimuli.

Responses to Capsaicin

Capsaicin was included in the present study because it is a known potent agonist for the vanilloid TRPV1 receptor (previously called VR1) and may produce dose-dependent long-lasting intense burning pain in humans [32], [33], [52], [53]. Recently, the presence of vanilloid-family receptors has been demonstrated in goldfish [54], zebrafish [54], [55] and the European sea bass, (*Dicentrarchus labrax*) [56] but capsaicin has not previously been explored as a nociceptive stimulus in fish. Of the two doses of capsaicin tested in the present study, only Atlantic cod which received the higher dose (0.1%) displayed significant effects, consisting of a reduced use of shelter, increased hovering near the bottom of the aquaria, and an indicated slight delay in the recovery of OBR to pre-treatment levels (90 minutes with 0.1% capsaicin compared to 60 minutes for 0.005% capsaicin and saline control). This demonstrates that 0.1% capsaicin is able to stimulate sensory afferents in the lip of the Atlantic cod. Although responsiveness to capsaicin may suggest the presence of TRPV1 receptors and a possible stimulation via C-fibre nociceptors [34], [57] this was not subject to investigation in the present study and therefore remains unknown. The presence of C-fibres has been documented in fish [13], [19] and similar histological and electrophysiological investigations are underway also in the Atlantic cod (Bichão *et al.*, unpublished data). The responses to capsaicin in our cod generally were mild but the responses observed with 0.1% capsaicin corresponded to that seen with acetic acid and may support the assumption of being specific to noxious stimulation.

Responses to a Fishing Hook

The insertion of a fishing hook through the lip of a fish undoubtedly causes tissue damage and should be regarded a noxious stimulus. In humans, punctate mechanical stimuli may cause the experience of sharp pain [58], [59]. In the present study, the only effect observed in hooked Atlantic cod was episodes of head shaking movements which were not seen pre-treatment or in the other treatment groups. This demonstrates that the presence of the hook is detected by the cod and induces a behavioural change, possibly seeking to get rid of it. The lack of other physiological and behavioural effects to the inserted hook suggests that the mechanical tissue damage *per se* may contribute little to reported angling-induced responses in this species, and that other factors such as pull force, physical exhaustion and stress responses to play, landing, air exposure and subsequent handling may be more influential on the physiology and behviour of angled fish [7], [60], [61].

In a previous laboratory study by Fernö and Huse [62], head-shaking episodes similar to those described in the present study were reported during repeated capture of wild-caught cod by a baited hook. In that study, the proportion of cod taking the hook was reduced with successive trials and it was concluded that aversive physical stimulation from the hook was the main negative reinforcement in this conditioning [63].

However, although Atlantic cod are found to display impressive associative learning capability and long time memory retention [64], the study by Fernö and Huse [62] found that individual Atlantic cod repeatedly encountered a baited hook over several days in two sets of experiments separated by almost 6 months [62], demonstrating a low hook avoidance in Atlantic cod compared to reports from some other freshwater species (e.g. [65], [66]). In line with this, the absence of observable responses other than head shaking to the presence of the hook in the present study may suggest a resiliency to hook damage in the Atlantic cod. Indeed, the almost complete absence of observable responses to punctate mechanical injury of the lip in the present study may be related to the eating habits of cod, which include species with hard or pointed components such as mussels, clams, whelks, brittle stars and crustaceans [67].

Notably, these observtions with puncture of the lip of the Atlantic cod may also correspond to a similar observation during puncture of the skin of the head of White Sea cod (*Gadus ogac*), that did not cause as strong responses when compared with stimulation of the fins and olfactory epithelium [68].

Use of Anaesthesia

In the present study, as in most previous studies on responses to noxious stimuli in fish (*e.g.* [15], [17], [20], [22], [53], [69]) noxious treatments were applied under acute anaesthesia so that the treatments could be applied to a precise area and to facilitate humane handling of the animals.

However, the use of anaesthesia will also provide temporary analgesia, thus masking acute effects of the tissue damage and possibly impeding longer term responses [16], [36]. In a study using un-anaesthetized rainbow trout, the injection of acetic acid into the lip caused an immediate and temporary loss of equilibrium in two of eight fish at a concentration of 2% and in seven of eight fish at a concentration of 5% [22]. Apart from this, effects on swimming behaviour and changes in OBR for the unanaesthetized rainbow trout were similar to those reported for anaesthetized rainbow trout [15]. This demonstrates that chemical stimuli may have the potential to remain in the tissue and continue stimulating sensory afferents well beyond the time of analgesic effects induced by the anaesthetic [23], [70].

The introduction of the fishing hook may have a similar effect, implying that any observed responses are not to the initial puncture of the tissue per se, but rather to the persisting mechanical damage and possible stimulations by movements of the hook in the awake fish. Although we cannot exclude the possibility that the anaesthetic used may have delayed responses more than 2 hours post-treatment, it has been demonstrated that the anaesthetic used did not influence commonly used behavioural parameters and that it may be possible to start behavioural observations even sooner than 30 min after anaesthesia [25].

Impact on Physiological Stress

Although the use of anaesthetic served to minimize stress and ensure rapid and standardised stimulus administration, the confinement, handling, anaesthesia and asphyxia may still initiate a stress response irrespective of treatment group [36], [71], [72]. In line with this, OBR is typically elevated with stress [28], [48], and a temporary post-treatment increase in this parameter was observed in all treatment groups in the present study. Surprisingly, no significant differences were found in the levels of circulatory stress indicators in our Atlantic cod two hours post-treatment. This suggests that none of the applied noxious stimuli may be considered significantly more stressful than the handling and administration of the innocuous control stimulus. Similarly, a general lack of differences in stress indicators between treatment groups was reported for Nile tilapia (*O. niloticus*) where OBR and circulatory stress parameters did not differ between tailfin-clipped fish and controls that were only handled [19]. For noxiously stimulated rainbow trout only acid injected fish that were held individually exhibited an elevation in plasma cortisol levels compared with controls. That was not observed in fish held in social groups although cortisol values were much higher in group housed fish [16], [37]. Because these studies all use terminal sampling at the end of the experiments and in the case of trout when all behavioural and physiological effects had subsided 3 hours after treatment, it is possible that the fish had recovered or adapted to the treatment.

Conclusions

This study has found that acetic acid, capsaicin and the presence of a fishing hook in the lip of Atlantic cod did affect behaviour when compared to the responses seen with the innocuous control treatment. Only head shaking was seen in the hook insertion group and whether this reflects a real resiliency to tissue damage in the mouth area remains to be tested. Given the diet of cod a lack of response to the hook may be related to the eating habits of this species. Still, the delayed recovery of OBR, the reduced use of shelter and the increased hovering behaviour induced by both acetic acid and capsaicin correspond to previous observations with the injection of acetic acid in rainbow trout, zebrafish and goldfish possibly representing indicators to assess injury detection and potentially aversive events in Atlantic cod. The lack of all these responses apart from head-shaking events in hooked fish possibly indicate that the fish are aware of the presence of the hook, and possibly seek to remove it from their lips. Future studies should explore noxious stimulation in the context of individual differences in stress coping styles and in a more realistic angling context and show the presence of nociceptors.

SUPPORTING INFORMATION

Table S1. Pre-treatment OBR (beats×min^{-1}; mean±S.E.M.), mean increase in post-treatment OBR (±S.E.M.), and the statistical significance (P-value) for the within-group comparison of post-treatment OBR with the corresponding pre-treatment value at each time point. Recovery time represents the time-point at which post-treatment OBR no longer differed significantly (p>0.05) from the pre-treatment measures (one-way repeated measures ANOVA with a Greenhouse-Geisser correction followed by post-hoc test using the Bonferroni correction). There were no statistically significant differences (p>0.05) in OBR evidenced between treatment groups at any time-point (two-way repeated -Geisser correction followed by post-hoc test using the Bonferroni correction). N=7 fish per group except for 0.005% Capsaicin (N=6) and 0.1% Capsaicin (N=8).

Treatment	Absolute pre-treatment OBR	Time point (min)	10	30	60	90	120	Recovery time (min)
Saline	30 ± 2	OBR	41 ± 3	39 ± 4	36 ± 7	32 ± 6	31 ± 6	60
		P value	**0.002**	**0.031**	0.054	0.967	0.998	
0.1 % Acetic acid	28 ± 1	OBR	44 ± 5	42 ± 3	36 ± 4	34 ± 4	30 ± 2	90
		P value	**0.002**	**0.001**	**0.017**	0.074	0.642	
2.0 % Acetic acid	29 ± 2	OBR	43 ± 6	42 ± 5	36 ± 3	36 ± 4	34 ± 4	120
		P value	**0.004**	**0.003**	**0.009**	**0.031**	0.407	
0.005 % Capsaicin	29 ± 3	OBR	42 ± 4	38 ± 5	34 ± 2	32 ± 3	31 ± 2	60
		P value	**0.001**	**0.027**	0.065	0.954	.978	
0.1 % Capsaicin	31 ± 3	OBR	45 ± 2	42 ± 4	40 ± 5	34 ± 7	33 ± 7	90
		P value	**0.001**	**0.001**	**0.011**	.934	.955	
Fishing hook	30 ± 3	OBR	46 ± 2	43 ± 4	39 ± 6	37 ± 6	35 ± 7	60
		P value	**0.002**	**0.004**	0.058	0.321	.977	

Table S2. Occurrence of Shelter and Hovering on the bottom behaviours in Atlantic cod before and after saline, acetic acid, capsaicin, and fishing hook treatments (i.e. corresponding to the data presented in Fig. 2). The data are expressed as mean percentage of time (%,±S.E.) the behaviour was displayed during 15 min segments prior to treatment administration (−20 min) and at 30 and 90 min after treatment administration. For each behaviour, identical letters denote a statistically significant (p≤0.05) difference between treatment groups at the same time point (repeated measures GLM followed by post-hoc test using the Bonferroni correction). N=7 fish per group except for 0.005% Capsaicin (N=6) and 0.1% Capsaicin (N=8).

Behaviour	Time (min)	Saline	0.1 % Acetic acid	2.0 % Acetic acid	0.005 % Capsaicin	0.1 % Capsaicin	Fishing hook
Shelter (%)	*- 20 to -5*	30.9 ± 17.9	55.1 ± 17.6	50.4 ± 18.3	34.2 ± 20.8	13.4 ± 12.4	34.0 ± 15.0
	+ 30 to + 45	57.1 ± 20.2**ab**	39.8 ± 16.7**dc**	1.6 ± 1.6**ad**	33.3 ± 21.1	16.0 ± 12.5**cb**	30.8 ± 17.1
	+ 90 to + 105	57.1 ± 20.2**ef**	43.1 ± 20.1**gh**	0.0 ± 0.0**eg**	39.7 ± 20.0	18.8 ± 13.1**hf**	28.8 ± 18.4
Hovering on bottom (%)	*- 20 to -5*	4.3 ± 2.8	11.2 ± 8.0	18.2 ± 8.0	7.7 ± 7.7	29.7 ± 12.8	15.3 ± 9.4
	+ 30 to + 45	2.0 ± 1.9**hij**	34.3 ± 12.6**i**	47.8 ± 14.0**hk**	2.8 ± 2.2**kl**	28.7 ± 14.0**mlj**	19.2 ± 8.0**m**
	+ 90 to + 105	0.3 ± 0.2**poq**	11.5 ± 6.7**p**	21.5 ± 6.8**ro**	10.8 ± 7.0**rs**	40.2 ± 13.5**qsn**	10.6 ± 7.3**n**

Acknowledgments

We are grateful to Lars Torgersen for maintenance and assistance of laboratory facilities; Asbjørn Vøllestad and Jack Thomson for helping us with statistical analyses and Tromsø Aquaculture Research Station for providing us with Atlantic cod.

Author Contributions

Conceived and designed the experiments: JRE ØAH LUS HB KBD. Performed the experiments: JRE KBD HB ØAH. Analyzed the data: JRE LUS HB ØAH KBD. Contributed reagents/materials/analysis tools: KBD ØAH LUS. Wrote the paper: JRE ØAH LUS HB KBD.

REFERENCES

1. Sherrington CS (1903) Qualitative differences of spinal reflex corresponding with qualitative difference of cutaneous stimulus,. *Journal of Physiology* 30: 39–46. [PMC free article]
2. IASP (2011) IASP Taxonomy. The International Association for the Study of Pain.
3. Kavaliers M (1988) Evolutionary and comparative aspects of nociception. *Brain Research Bulletin* 21: 923–931.
4. Loeser J, Treede R (2008) The Kyoto protocol of IASP Basic Pain Terminology. 137: 5.
5. 5. Molony V, Kent JE (1997) Assessment of acute pain in farm animals using behavioral and physiological measurements. *Journal of Animal Science* 75: 266–272.
6. Rose JD (2007) Anthropmomrphism and 'mental welfare' of fishes. *Diseases of Aquatic Organisms* 75: 139–154.
7. Rose JD, Arlinghaus R, Cooke SJ, Diggles BK, Sawynok W, et al. (2012) Can fish really feel pain? *Fish and Fisheries*: 1–37.
8. Braithwaite VA, Huntingford FA (2004) Fish and welfare: do fish have the capacity for pain perception and suffering? *Animal Welfare* 13: 87–92.
9. Chandroo KP, Duncan IJH, Moccia RD (2004) Can fish suffer?: perspectives on sentience, pain, fear and stress. *Applied Animal Behaviour Science* 86: 225–250.
10. Huntingford FA, Adams C, Braithwaite VA, Kadri S, Pottinger TG, et al. (2006) Current issues in fish welfare. *Journal of Fish Biology* 68: 332–372.
11. Sneddon LU (2011) Pain Perception in Fish. *Journal of Consciousness Studies* 18: 209–229.
12. Ashley PJ, Sneddon LU, McCrohan CR (2007) Nociception in fish: stimulus-response properties of receptors on the head of trout Oncorhynchus mykiss. *Brain Research* 1166: 47–54.
13. Sneddon LU (2002) Anatomical and electrophysiological analysis of the trigeminal nerve in a teleost fish, Oncorhynchus mykiss. *Neuroscience Letters* 319: 167–171.
14. Sneddon LU (2003a) The evidence for pain in fish: the use of morphine as an analgesic. *Applied Animal Behaviour Science* 83: 153–162.

15. Sneddon LU, Braithwaite VA, Gentle MJ (2003b) Do fish have nociceptors? Evidence for the evolution of a vertebrate sensory system. *Proc R Soc Lond* 270: 1115–1121.
16. Mettam JJ, Oulton LJ, McCrohan CR, Sneddon LU (2011) The efficacy of three types of analgesic drugs in reducing pain in the rainbow trout, Oncorhynchus mykiss. *Applied Animal Behaviour Science* 133: 265–274.
17. Newby NC, Wilkie MP, Stevens ED (2009) Morphine uptake, disposition, and analgesic efficacy in the common goldfish (Carassius auratus). *Canadian Journal of Zoology* 87: 388–399.
18. Nordgreen J, Garner JP, Janczak AM, Ranheim B, Muir WM, et al. (2009) Thermonociception in fish: Effects of two different doses of morphine on thermal threshold and post-test behaviour in goldfish (Carassius auratus). *Applied Animal Behaviour Science* 119: 101–107.
19. Roques JAC, Abbink W, Geurds F, van de Vis H, Flik G (2010) Tailfin clipping, a painful procedure: Studies on Nile tilapia and common carp. *Physiology & Behavior* 101: 533–540.
20. Reilly SC, Quinn JP, Cossins AR, Sneddon LU (2008) Behavioural analysis of a nociceptive event in fish: Comparisons between three species demonstrate specific responses. *Applied Animal Behaviour Science* 114: 248–259.
21. Julius D, Basbaum AI (2001) Molecular mechanisms of nociception. *Nature* 413: 203–210.
22. Newby NC, Stevens ED (2008) The effects of the acetic acid "pain" test on feeding, swimming, and respiratory responses of rainbow trout (Oncorhynchus mykiss). *Applied Animal Behaviour Science* 114: 260–
23. Sneddon LU (2003b) Trigeminal somatosensory innervation of the head of a teleost fish with particular reference to nociception. *Brain Research* 972: 44–52.
24. Sneddon LU, Braithwaite VA, Gentle MJ (2003a) Novel object test: examining nociception and fear in the rainbow trout. *The Journal of Pain* 4: 431–440.
25. Nordgreen J, Tahamtani FM, Janczak AM, Horsberg TE (2014) Behavioural Effects of the Commonly Used Fish Anaesthetic Tricaine Methanesulfonate (MS-222) on Zebrafish (Danio rerio) and Its Relevance for the Acetic Acid Pain Test. *PLoS ONE* 9: e92116.

26. Melzack R, Katz J (2006) Pain assessment in adult patients. In: McMahon SB, Koltzenburg, M., editor. Wall and Melzack's Textbook of Pain: Elsevier Ltd. 291–304.
27. Rose JD (2002) The neurobehavioral nature of fishes and the question of awareness and pain. *Reviews in Fisheries Science* 10: 1–38.
28. Wendelaar Bonga SE (1997) The stress response in fish. *Physiological reviews* 77: 591–625.
29. Ministry of Agriculture of Norway (1996) Regulations on experiments with animals. Available: http://wwwlovdatano/cgi-wift/ldles?doc=/sf/sf/sf-19960115-0023html Accessed 2010 Dec 12.
30. Ministry of Agriculture and Food of Norway (2010) Animal welfare act. Available: http://wwwlovdatano/all/nl-20090619-097html Accessed 2010 Dec 12.
31. Council of the European Union (1998) European Convention for the Protection of Vertebrate Animals used for Experimental and Other Scientific Purposes. Available: http://conventions.coe.int/treaty/en/treaties/html/123.htm Accessed 2014 May 25.
32. Caterina MJ, Schumacher MA, Tominaga M, Rosen TA, Levine JD, et al. (1997) The capsaicin receptor: a heat-activated ion channel in the pain pathway. *Nature* 389: 816–824.
33. Chung G, Jung SJ (2010) The transient receptor potential (TRP) family in pain and temperature sensation. In: Beaulieu P, Lussier, D, Porreca, F, Dickenson, A H., editor. Pharmacology of pain. Seattle: IASP Press. 327–346.
34. Holzer P (1991) Capsaicin: cellular targets, mechanisms of action, and selectivity for thin sensory neurons. *Pharmacological Reviews* 43: 143–201.
35. Ashley PJ, Ringrose S, Edwards KL, Wallington E, McCrohan CR, et al. (2009) Effect of noxious stimulation upon antipredator responses and dominance status in rainbow trout. *Animal Behaviour* 77: 403–410.
36. Ross LG, Ross B (2008) Anaesthetic and sedative techniques for aquatic animals. Oxford: Blackwell Publishing Ltd.
37. Zahl IH, Kiessling A, Samuelsen OB, Hansen MK (2009) Anaesthesia of Atlantic cod (Gadus morhua) – Effect of pre-anaesthetic sedation, and importance of body weight, temperature and stress. *Aquaculture* 295: 52–59.

38. Brown JA, Watson J, Bourhill A, Wall T (2008) Evaluation and use of the Lactate Pro, a portable lactate meter, in monitoring the physiological well-being of farmed Atlantic cod (Gadus morhua). *Aquaculture* 285: 135–140.
39. Gallagher AJ, Frick LH, Bushnell PG, Brill RW, Mandelman JW (2010) Blood Gas, Oxygen Saturation, pH, and Lactate Values in Elasmobranch Blood Measured with a Commercially Available Portable Clinical Analyzer and Standard Laboratory Instruments. *Journal of Aquatic Animal Health* 22: 229–234.
40. Foss A, Kristensen T, Åtland Å, Hustveit H, Hovland H, et al. (2006) Effects of water reuse and stocking density on water quality, blood physiology and growth rate of juvenile cod (Gadus morhua). *Aquaculture* 256: 255–263.
41. Karlsson A, Rosseland B, Massabuau J-C, Kiessling A (2011a) Pre-anaesthetic metomidate sedation delays the stress response after caudal artery cannulation in Atlantic cod (Gadus morhua). *Fish Physiology and Biochemistry*: 1–11.
42. Schulz R (1985) Measurement of five androgens in the blood of immature and maturing male rainbow trout, Salmo gairdneri (Richardson). *Steroids* 46: 717–726.
43. Tveiten H, Bjørn PA, Johnsen HK, Finstad B, McKinley RS (2010) Effects of the sea louse Lepeophtheirus salmonis on temporal changes in cortisol, sex steroids, growth and reproductive investment in Arctic charr Salvelinus alpinus. *Journal of Fish Biology* 76: 2318–2341.
44. McMahon SB, Bennett DLH, Bevan S (2006) Inflammatory mediators and modulators of pain. Textbook of Pain. S.B. McMahon, M. Koltzenburg ed. Philadelphia: Elsevier. 49–72.
45. Øverli Ø, Sørensen C, Pulman KGT, Pottinger TG, Korzan W, et al. (2007) Evolutionary background for stress-coping styles: Relationships between physiological, behavioral, and cognitive traits in non-mammalian vertebrates. *Neuroscience & Biobehavioral Reviews* 31: 396–412.
46. 46. Koolhaas JM, Korte SM, De Boer SF, Van Der Vegt BJ, Van Reenen CG, et al. (1999) Coping styles in animals: current status in behavior and stress-physiology. *Neuroscience & Biobehavioral Reviews* 23: 925–935.

47. Schjolden J, Winberg S (2007) Genetically determined variation in stress responsiveness in rainbow trout: behavior and neurobiology. *Brain, behavior and evolution* 70: 227–238.

48. Artigas ML, Skjæraasen JE, UtnePalm AC, Nilsen T (2005) Recovery from handling stress in Gadus morhua. *Journal of Fish Biology* 67: 384–391.

49. Herbert N, Steffensen J (2005) The response of Atlantic cod, Gadus morhua, to progressive hypoxia: fish swimming speed and physiological stress. *Marine Biology* 147: 1403–1412.

50. Johansen JL, Herbert NA, Steffensen JF (2006) The behavioural and physiological response of Atlantic cod Gadus morhua L. to short-term acute hypoxia. *Journal of Fish Biology* 68: 1918–1924.

51. Lima SL, Dill LM (1990) Behavioral decisions made under the risk of predation: a review and prospectus. *Canadian Journal of Zoology* 68: 619–640.

52. Schmelz M, Schmid R, Handwerker HO, Torebjörk HE (2000) Encoding of burning pain from capsaicin-treated human skin in two categories of unmyelinated nerve fibres. *Brain* 123: 560–571.

53. Simone DA, Baumann TK, LaMotte RH (1989) Dose-dependent pain and mechanical hyperalgesia in humans after intradermal injection of capsaicin. *Pain* 38: 99–107.

54. 54. Zimov S, Yazulla S (2004) Localization of vanilloid receptor 1 (TRPV1/VR1)-like immunoreactivity in goldfish and zebrafish retinas: Restriction to photoreceptor synaptic ribbons. *Journal of Neurocytology* 33: 441–452.

55. Mangos S, Liu Y, Drummond IA (2007) Dynamic expression of the osmosensory channel trpv4 in multiple developing organs in zebrafish. *Gene Expression Patterns* 7: 480–484.

56. Bossus M, Charmantier G, Lorin-Nebel C (2011) Transient receptor potential vanilloid 4 in the European sea bass Dicentrarchus labrax: A candidate protein for osmosensing. *Comparative Biochemistry and Physiology - Part A: Molecular & Integrative Physiology* 160: 43–51.

57. Caterina MJ, Leffler A, Malmberg AB, Martin WJ, Trafton J, et al. (2000) Impaired Nociception and Pain Sensation in Mice Lacking the Capsaicin Receptor. *Science* 288: 306–313.

58. Magerl W, Fuchs PN, Meyer RA, Treede R-D (2001) Roles of capsaicin-insensitive nociceptors in cutaneous pain and secondary hyperalgesia. *Brain* 124: 1754–1764.

59. Meyer RA, Ringkamp M, Campbell JN, Raja SN (2006) Peripheral mechanisms of cutaneous nociception. Wall and Melzack's Textbook of Pain. S.B. McMahon, M. Koltzenburg ed. London: Elsevier. 3–34.
60. Arlinghaus R, Cooke SJ, Lyman J, Policansky D, Schwab A, et al. (2007) Understanding the Complexity of Catch-and-Release in Recreational Fishing: An Integrative Synthesis of Global Knowledge from Historical, Ethical, Social, and Biological Perspectives. *Reviews in Fisheries Science* 15: 75–167.
61. Olsen RE, Næsje TF, Poppe T, Sneddon L, Webb J (2010) Risk Assessment of Catch and Release. *Norwegian Scientific Committee for Food Safety* 09: 1–79.
62. Fernö A, Huse I (1983) The effect of experience on the behaviour of cod (Gadus morhua L.) Towards a baited hook. *Fisheries Research* 2: 19–28.
63. Ashley PJ, Sneddon LU, McCrohan CR (2007) Nociception in fish: stimulus–response properties of receptors on the head of trout Oncorhynchus mykiss. *Brain Research* 1166: 47–54.
64. Nilsson J, Kristiansen T, Fosseidengen J, Fernö A, Bos R (2008) Learning in cod (Gadus morhua): long trace interval retention. *Animal Cognition* 11: 215–222.
65. Beukema JJ (1970a) Acquired hook-avoidance in the pike Esox lucius L. fished with artificial and natural baits. *Journal of Fish Biology* 2: 155–160.
66. Beukema JJ (1970b) Angling experiments with carp (Cyprinus carpio L.) II: Decreasing catchability through one-trial learning. *Netherlands Journal of Zoology* 20: 81–92.
67. COSEWIC (2003) Assessment and Update Report on Gadus Morhua (Atlantic Cod). Ottawa. 76 p.
68. Chervova LS (1997) Pain sensitivity and behavior of fishes. *Journal of Ichthyology* 37: 98–102.
69. Park KM, Max MB, Robinovitz E, Gracely RH, Bennett GJ (1995) Effects of intravenous ketamine, alfentanil, or placebo on pain, pinprick hyperalgesia, and allodynia produced by intradermal capsaicin in human subjects. *Pain* 63: 163–172.
70. Le Bars D, Gozariu M, Cadden SW (2001) Animal Models of Nociception. *Pharmacological Reviews* 53: 597–652.

71. Sneddon LU (2012) Clinical Anesthesia and Analgesia in Fish. *Journal of Exotic Pet Medicine* 21: 32–43.

72. Zahl I, Kiessling A, Samuelsen O, Olsen R (2010) Anesthesia induces stress in Atlantic salmon (Salmo salar), Atlantic cod (Gadus morhua) and Atlantic halibut (Hippoglossus hippoglossus). *Fish Physiology and Biochemistry* 36: 719–730.

SECTION 7: WILD ANIMALS

7.5. MARSUPIALS

CHAPTER 22

Physiological Stress in Koala Populations near the Arid Edge of Their Distribution

Nicole Ashley Davies[1], Galina Gramotnev[2], Clive McAlpine[2], Leonie Seabrook[2], Greg Baxter[2], Daniel Lunney[3,4], Jonathan R. Rhodes[2], Adrian Bradley[1]

[1] School of Biomedical Sciences, The University of Queensland, St Lucia, Queensland, Australia

[2] Landscape Ecology and Conservation Group, School of Geography, Planning and Environmental Management, The University of Queensland, St Lucia, Queensland, Australia

[3] Office of Environment and Heritage New South Wales, Hurstville, New South Wales, Australia

[4] School of Veterinary and Life Sciences, Murdoch University, Perth, Western Australia, Australia

Citation: Davies NA, Gramotnev G, McAlpine C, Seabrook L, Baxter G, Lunney D, et al. (2013) Physiological Stress in Koala Populations near the Arid Edge of Their Distribution. PLoS ONE 8(11): e79136. https://doi.org/10.1371/journal.pone.0079136.

ABSTRACT

Recent research has shown that the ecology of stress has hitherto been neglected, but it is in fact an important influence on the distribution and numbers of wild vertebrates. Environmental changes have the potential to cause physiological stress that can affect population dynamics. Detailed information on the influence of environmental variables on glucocorticoid levels (a measure of stress) at the trailing edge of a species' distribution can highlight stressors that potentially threaten species and thereby help explain how environmental challenges, such as climate change, will affect the survival of these populations. Rainfall determines leaf moisture and/or nutritional content, which in turn impacts on cortisol concentrations. We show that higher faecal cortisol metabolite (FCM) levels in koala populations at the trailing arid edge of their range in southwestern Queensland are associated with lower rainfall levels (especially rainfall from the previous two months), indicating an increase in physiological stress when moisture levels are low. These results show that koalas at the semi-arid, inland edge of their geographic range, will fail to cope with increasing aridity from climate change. The results demonstrate the importance of integrating physiological assessments into ecological studies to identify stressors that have the potential to compromise the long-term survival of threatened species. This finding points to the need for research to link these stressors to demographic decline to ensure a more comprehensive understanding of species' responses to climate change.

INTRODUCTION

Stressors are pervasive and include environmental and ecological disturbances that have profound effects on the ecology and evolution of organisms [1]–[3]. The sources of stress include biotic factors (predation, competition, social dynamics), extremes in physical factors (temperature, salinity) and climatic factors (drought, storms) [3], [4]. Shifts in the environment can be labile (i.e. eventually subside, such as cold, storms, heat and drought) or permanent, such as global climate change, urbanization, habitat degradation, pollution, with permanent changes requiring more than just temporary assimilation and they can result in changes in range, adaptations or local extinctions [5]. Stressors can impact on both physical and biotic components of an organism's environment and, depending on their pervasiveness, magnitude and frequency, can profoundly influence the fitness of individuals via costs to health, reproduction and survival [6]. Ultimately, stressors can have profound effects on population viability, species' distribution and extinction risk [7]–[9]. Unfavourable conditions trigger physiological responses that result in hypothalamic-pituitary-adrenal (HPA) axis activation and glucocorticoid secretion (cortisol in marsupials) by the adrenal cortex [10]–[13]. These glucocorticoid hormones affect physiological and behavioural traits that regulate the responses of vertebrates to stressors, and potentially offer a metric for evaluating species risk to global change [6]. Thus, physiological investigations of wild animals are vital to identify specific conservation concerns and identify threat status. Depending on the type of sample collected, they can also offer a potentially non-invasive way to monitor the efficacy of management strategies [14].

In response to a stressor, the HPA axis is activated, resulting in the increased secretion of glucocorticoids (primarily cortisol and corticosterone) from the adrenal gland (stress response) [15]–[19]. Glucocorticoid receptors are distributed widely in the body, thus glucocorticoid hormones can initiate a wide range of responses affecting behaviour, reproduction, growth, the immune system, and metabolism and energy allocation [20]–[23]. The stress response seeks to reduce the negative effects of a stressor and may cause temporary suppression or interruption of an individuals normal life history stage (such as delays in reproduction, development, migration) to redirect resources into an 'emergency' state to facilitate immediate survival [4], [24]–[27]. When an emergency life-history stage is triggered, it allows an individual to temporarily move away from the source of disturbance (adjustment) or endure it while adopting energy-saving physiological strategies (adaptation) that allow it to cope [24]. As a consequence, glucocorticoid secretion resulting from acute stress (e.g. predator attack) improves the fitness of an organism by mobilizing energy during the stressful situation [22], [28]–[31]. However, a stress response involving excessive glucocorticoid secretion, often termed allostatic overload or chronic stress, arises when an individual's HPA axis is challenged by excessive stressor exposure or pervasiveness resulting in fitness loss [32], [33]. Evidence from the biomedical literature shows that sustained, high levels of glucocorticoids can cause deleterious effects on the physiological health of an animal and lead to greater susceptibility to disease and reduced fecundity and survivorship [12], [13], [20], [34], [35]. Accordingly, variation in the release of glucocorticoids, as a measure of the stress response, is increasingly being used in ecological and conservation studies to infer the health of animals [23] and to enable us to understand how both natural (e.g. predators, conspecifics, weather) and imposed environmental challenges (e.g. climate change, relocation and habitat disturbance) impact on individuals and populations [36]–[40].

While the deleterious effects of chronic stress are theoretically well conceptualized from biomedical evidence, there are few empirical studies showing fitness loss or pathology and ensuing demographic consequences of allostatic overload in wild populations [6], [41]. Exceptions include a few studies on wild populations showing predation pressure causing indirect and negative glucocorticoid-mediated effects on individual fitness, resulting in decreased prey population growth [42]–[44]. Boonstra [41] contends that although chronic stress does occur in nature, the impacts of chronic stressors as laid out in the biomedical literature (i.e. pathology) do not occur in nature. Boonstra [41] argues that both demographic (the indirect effects on reproduction) and hormonal evidence indicate that although some wild populations may be chronically stressed by predators, there is no evidence of pathology or an HPA axis with an inability to respond. Boonstra [41] suggests that chronic stress is not the inevitable result of exposure to a persistent, severe stressor, but is part of the normal experience of many animals in nature and, although there are fitness costs, it is an evolved adaptive response to particular ecological and habitat pressures – i.e. trade-off reproduction for survival but not to such an extent that survival itself is at risk. Different views of the impacts of chronic stressors, as laid out in the biomedical literature relating to pathology and the impacts within natural populations, must be taken into consideration when assessing glucocorticoid concentrations from wild populations.

Non-invasive methods for measuring glucocorticoid metabolites in faeces and hair have become a widely accepted tool for gaining important information about an animal's endocrine status [12], [45]–[48]. Faecal glucocorticoid metabolite analysis to assess physiological stress has been validated using different methods and employed in studies of several marsupial species including the honey possum (*Tarsipes rostratus*) [49], tammar wallaby (*Macropus eugenii*) [50], Gilbert's potoroo (*Potorous gilbertii*) [51], the southern hairy-nosed wombat (*Lasiorhinus latifrons*) [52], the numbat (*Myrmecobius fasciatus*) [53] and the koala (*Phascolarctos cinereus*) [54], [55].

Unlike measurements of plasma and serum hormone concentrations that measure the actual hormone, non-invasive methods measure various metabolic end products of the hormone after it has been cleared from the circulation and extensively modified by bacteria in the gut [47], [56]–[59].

A specific hormone is typically broken down into various metabolites during this process [59], and it is these glucocorticoid metabolites that are measured in faeces (faecal glucocorticoid metabolites) and reflect the free (unbound) glucocorticoid fraction of total glucocorticoids [60], [61]. However, there are a number of caveats relating to sex, diet, metabolic rate and individual differences in hormone metabolite formation that are the source of complications that need to be considered for a meaningful interpretation of hormone metabolite data (for a review see [62], [63]). Despite these caveats, it has been demonstrated that faecal glucocorticoid metabolite levels can reflect both baseline glucocorticoid levels and an animal's stress response [61], making this non-invasive technique a valuable alternative to invasive blood sampling in field studies [63]. Faecal glucocorticoid measures also provide an integrated measure of fluctuating blood concentrations in the preceding few days, thus providing a stable measure of an individual's average hormone secretion [64]–[66]. Furthermore, faecal glucocorticoid measures can be obtained easily without handling or observing the animal, and thereby not elevating stress levels [12], [65], [67], [68].

As a consequence of a changing climate, distributional changes are occurring at the boundaries of geographic ranges (i.e. at the range limit) of many species, including boundary expansions at the leading edge of a range and contraction at the trailing edge [69]. Populations, in particular of folivores, at the trailing edge of their range, are likely to be most vulnerable to climate change through physiological stress associated with the decline in the nutrient richness of their food source [70], [71]. Trailing edge populations can be critical to the long-term survival of species because they may contain individuals that can adapt to changing climatic conditions [72]–[74].

The koala (*Phascolarctos cinereus*) is an arboreal, marsupial folivore, which feeds almost exclusively on a limited variety of *Eucalyptus*, *Corymbia* and *Angophora* species. It is widely distributed, with its range extending across 30 bioregions from tropical Queensland to temperate Victoria and South Australia. The koala was listed in May 2012 under the Commonwealth's *Environmental Protection and Biodiversity Conservation Act 1999* as threatened in Queensland, the Australian Capital Territory and New South Wales. Previous research in southwestern Queensland, near the arid edge of their range, has determined that its distribution is directly affected by climatic factors (high temperature and low rainfall), which supports the view that these western koalas are near their physiological limits [75], [76]. Therefore, western koalas can be expected to cope poorly with climate change.

Significant effects of season or weather conditions (e.g. temperature, humidity, food and water availability) on glucocorticoid concentrations have been shown for several mammalian species, with most studies reporting higher levels during harsher conditions in winter, extreme heat, or during the dry season [66], [77]–[82]. Furthermore, a number of studies have attributed population declines to the effects of chronic stress in free-living mammals [36], [83], [84]. Additionally, a study on a long-lived seabird, the black-legged kittiwake, found that initial increases in allostatic load can evoke increased foraging behaviour and consumption, triggered by increasing glucocorticoid levels, which resulted in the accumulation of energy stores [85]. However, the metabolic effects of increased glucocorticoid levels experienced under high allostatic loads lead to a reduction of energy stores (indicating high physiological stress), with observations of decreased fitness (increased mortality, decreased reproduction) with high glucocorticoid levels in this species [85]. Throughout the koala's distribution, droughts and heatwaves have been responsible for population crashes and local extinctions [71], [86]–[88]. In southwestern Queensland, there was an 80% decline in koala numbers between 1995 and 2009, with drought being the main contributor to the decline [88]. Climate change predictions for this region indicate that the intensity and frequency of droughts and heatwaves will increase, and moisture availability will decrease [89]–[91]. Therefore measuring glucocorticoid concentrations of western koala populations will help us understand how climatic variables impact these populations, and allow us to anticipate where problems will arise and thereby highlight specific conservation concerns [14].

The aim of this study was to investigate the effect of climate variables on the physiological stress of koala populations of southwestern Queensland to predict the potential response of these populations to future climate change. Koalas in this region are at the western limits of their geographic range and form a trailing-edge population [88]. We used non-invasive techniques to measure glucocorticoid levels. Mixed effects modeling was used to relate the current physiological conditions of koalas to current climate and environmental conditions. These relationships were then used to estimate potential physiological conditions based on predicted future climates.

MATERIALS AND METHODS

The approved protocol for this project was approved by the Animal Ethics Committee at The University of Queensland (GPA/603/08/ARC) and the

Environmental Protection Agency (now the Department of Environment and Resource Management) (Scientific Purposes Permit WISP05343008).

Study Area

This study was conducted in semi-arid, southwestern Queensland, comprising portions of the Mulga Lands bioregion, the Mitchell Grass Downs bioregion and the Brigalow Belt South bioregion (Figure 1). Annual average rainfall ranges from 750 mm in the east declining to 250 mm in the west. Rain falls mainly in summer and is highly variable [92]. The hottest month is January with mean monthly maximum and minimum temperatures of approximately 35°C and 19°C respectively. The coolest month is July with mean maximum and minimum temperatures of approximately 19°C and 4°C respectively [92].

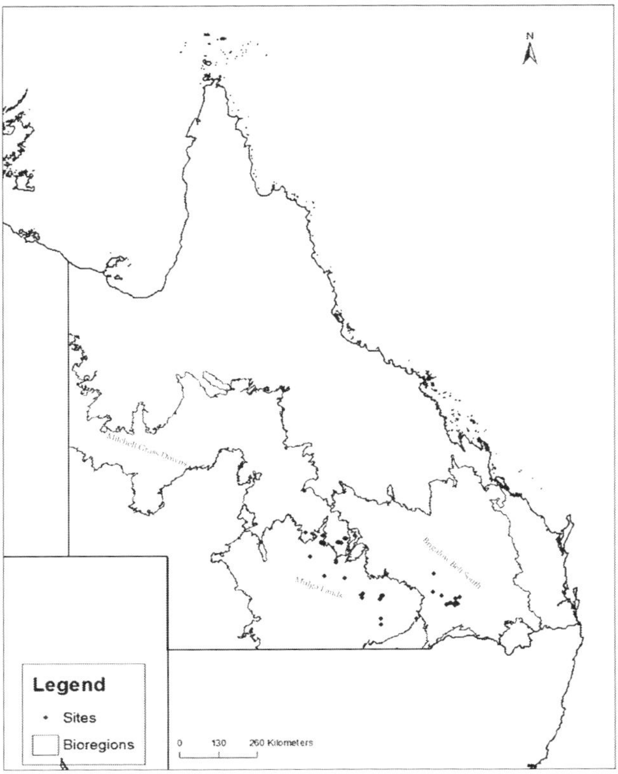

Figure 1. Study area within the Mulga Lands, Mitchell Grass Downs and Brigalow Belt South bioregions, Queensland, Australia. 'Sites' denote locations where fresh faecal pellets were collected.

The Mulga Lands bioregion is dominated by flat to undulating plains and low ranges supporting *Acacia aneura* (mulga) shrubland and low woodlands [93]. The woodlands that dominate waterways and associated floodplains are predominantly comprised of *Acacia* spp., *Eucalyptus populnea* (poplar box), *E. camaldulensis* (river red gum), *E. coolabah* (coolabah) and *E. orchophloia* (yapunyah) [93]. The western portion of the Brigalow Belt South bioregion is predominantly comprised of Acacia harpophylla (brigalow), Casuarina cristata (belah) and E. populnea open-forest woodland [93]. Riparian vegetation is dominated by E. camaldulensis, E. coolabah and E. largiflorens (black box) [94].

The Mitchell Grass Downs bioregion is dominated by treeless plains of Astrebla spp. (Mitchell grass) with some occasional ridges, rivers and gorges. Patches of low open woodland of E. coolabah occur across the region in association with low-lying plains and drainage lines [94].

Faecal Pellet Collection

Searches were conducted to collect fresh koala faecal pellets for cortisol analysis in 29 sites across the three bioregions (Mulga Lands 13 sites; Brigalow Belt South 11 sites; Mitchell Grass Downs 5 sites) (Figure 1) between August 2009 and June 2011. Morning (07 00–10 00 hours), afternoon (16 00–18 00 hours) and night (spotlighting, 19 00–04 00 hours) searches were conducted in a systematic manner along transects (1 km by 80 m) within riparian habitats. Previous research had identified riparian habitats as the primary resource used by koalas in southwestern Queensland [88], [95]. If a koala was sighted along a 1 km transect, then an extra 1 km was surveyed to locate additional koalas. Once a koala was located, sex was recorded and faecal pellets were collected by searching the entire area beneath the canopy of the tree in which the koala was located. Fresh, intact pellets were placed in clean containers using forceps and stored in the field in a liquid nitrogen dry shipper (MCE SC 20/12 V Vapour Shipper, In Vitro Technologies Pty. Ltd) for transport to the laboratory. The use of faecal cortisol analysis in the assessment of the activity of the HPA axis in freshly collected koala faecal pellets has previously been validated using the methods described below [54]. That research showed an excretion time lag with faecal cortisol metabolite concentrations corresponding to the release of glucocorticoids from the adrenal cortex at least 36 hours prior to deposition [54]. Other studies have also successfully assayed cortisol metabolites in koala faeces (with an ACTH challenge), with time lags of 24 and 48 hours in females and males respectively [55].

Extraction and Analysis

Faecal pellets from individual koalas were oven dried at 70°C for 10 hrs. The dried samples were crushed between two 10 mm PVC plates, passed through a 1 mm sieve to remove leaf particles, and collected into 5 ml flat bottom polypropylene screw capped vials. For each sample, 200 mg (±1 mg accuracy) of dry faecal powder was weighed into a 16×100 mm glass test tube. Borate buffer 2 ml (pH 6.5 0.1 M) was added to the dry powder, vortex mixed and then 50 μl of β glucuronidase (β-D-Glucuronoside glucuronosohydrolase, EC 3.2.1.31, Sigma Chemical Co. USA) containing approximately 4,000 units was added to each test tube to cleave glucuronide and sulfatase polar groups. Test tubes were then incubated for 4 hrs at 37°C on an orbital mixer. Redistilled diethyl ether 3 ml was added to each test tube and vortexed for 2 minutes, then the test tubes were allowed to stand for 2 minutes. The lower aqueous phase was frozen by briefly dipping the test tubes into liquid nitrogen and the supernatant ether containing the partitioned steroid was decanted into 12×75 mm glass test tubes and evaporated to dryness at 40°C in a hot block evaporator in a fume hood. The residue containing extracted steroid was re-dissolved in 200 μl of diluted zero cortisol calibration solution (Demeditec Saliva Free Cortisol Kit - diluted 1 10) and placed on an orbital mixer at 37°C for 60 minutes, followed by short, high-speed vortex (20 seconds). 100 μl of test samples, standards and controls were then pipetted into 96 well plates of Saliva Free Cortisol Kits (Demeditec) and run as per kit directions. The efficiency of the extraction process was progressively tracked by addition of 30,000 dpm ^{3}H-cortisol (1,2,6,7 ^{3}H cortisol 160 Ci/mmol Perkin-Elmer Life Sciences) and the final cortisol assay concentration was corrected for this efficiency. This assay allowed us to measure cortisol after first cleaving polar bonds in the major known metabolites (faecal cortisol metabolite: FCM), because it has been reported that there are no native, unmetabolized glucocorticoids present in faeces [60].

The specificity of the assay was: cortisol 100%, corticosterone 0.38%, cortisone 1.85%, 11-deoxycortisol 0.88%, prednisolone 9.89%, with all other steroids of similar structure <0.20%. Inter-assay variation was 5.4% CV, while intra-assay variation was 5.5% CV. Analytical sensitivity of the assay is 0.014 ng/ml or 4.2 pg per loaded sample. Serial dilutions of glucuronidase-treated koala faecal extracts run against Demeditec assay kit calibrator standards gave a satisfactory degree of parallelism for the assay. Assay data were analysed employing a four parameter logistic fit using MyAssays Analysis Software Solutions (www.myassays.com).

Explanatory Variables

The explanatory variables included sex, climate data, soil characteristics and location (Table 1). Climate data from the Bureau of Meteorology (1990–2011) were examined for stations closest to each site (within the same bioclimatic region and <30 km apart). Minimum and maximum average temperatures were considered for each sample corresponding to the day prior to sample collection to take into account the excretion time lag [54] (Table 1). Monthly rainfall was calculated because of time lags between rainfall and the response by vegetation [96]–[98]. Soil properties (e.g. total nitrogen and phosphorus) at each site were derived from geology mapping of Queensland (The Queensland Combined Soils dataset) at a scale of 1 250,000 to 1 200,000,000 (Table 1). Each site was also categorized by bioregion, catchment, basin, drought or post-flood event and season (Table 1).

Table 1. Explanatory variables used to examine variation in faecal cortisol concentrations

Variable	Units	Full description
Sex	Categorical	Sex of each koala from which faecal samples were collected.
Total Nitrogen	Kjeldahl	Soil total nitrogen content at each site - derived from geology mapping of Queensland (The Queensland combined soils dataset).
Phosphorous	Kjeldahl	Soil phosphorous content at each site - derived from geology mapping of Queensland (The Queensland combined soils dataset)
Annual rainfall	Millilitres	Average annual rainfall (1990–2011) for each site.
Rainfall 1–12 months	Millilitres	Total rainfall with monthly resolution at each site for 1–12 months prior to each faecal sample collected.
Temperature	Degrees Celsius	Average minimum and maximum temperatures at each site the day prior to sample collection.
Bioregion	Categorical	Bioregion (Mulga Lands, Mitchell Grass Downs, Brigalow Belt South)
Catchment	Categorical	Catchment (Mungallala, Warrego, Condamine, Moonie)
Basin	Categorical	Basin (Warrego, Moonie, Balonne-Condamine)
Drought or post-flood	Categorical	Categorized based on whether each sample was collected during the drought (2009) or post-flood event (2010 onwards).
Season	Categorical	Categorized based on whether each sample was collected during the koala breeding season (September to March) or non-breeding season (April–August).

Statistical Analysis

Statistical analyses were conducted using the Stata statistical software package [99] to determine the dependencies between the following variables: FCM concentration (ng/g) (dependent variable) and rainfall, temperature, soil nitrogen, soil phosphorus, sex, time of the year (month), and season (breeding/non-breeding) (independent variables – Table 1). As was demonstrated from the Shapiro-Wilk test for normality of data, the FCM (dependent) variable was not distributed normally ($p<0.001$). Therefore, the FCM variable was transformed logarithmically to achieve a normal distribution of the dependent variable. The explanatory variables were all standardized to have a zero mean and a standard deviation of one to allow comparison between the model parameter estimates.

Data were collected at 29 sites across the three bioregions. Therefore, we assumed a hierarchically structured population with random sampling of the regions, sites and subjects within the sites. Multilevel (three-level) mixed-effects linear regressions with random intercepts were used to assess the effect of each explanatory variable on koala FCM levels. The three levels within the considered model were associated with: (1) individual koala; (2) the site of data collection; (3) the region where the site was located. The random effects model determined the dependency of FCM levels on a particular rainfall variable (e.g., two month rainfall) for each site and/or region. Thus the random effects within the model are random intercepts and/or slopes for the respective regression lines of FCM levels versus a rainfall variable for each region and/or site. Though the model does not directly give the values of the random effects (random intercepts and/or slopes), their best linear unbiased predictions, also known as empirical Bayes predictions, can be calculated for both region and site levels [100].

RESULTS

All explanatory variables (Table 1), other than the covariates related to rainfall, were omitted from the considered model, as they had no significant impact on the levels of FCM ($p>0.1$). FCM concentration was assessed against rainfall for 1 to 12 months prior to sample collection (Table 2). The moving average dependence of the magnitudes of the standardized coefficients (Table 2) on the number of rainfall months is illustrated by Figure 2. Changes in FCM levels have a typical time lag of about 2–6 months after rain. This follows from the maximum of the curve in Figure 2, between ~2 and 6 months. Further, the points corresponding to 4, 5 and 6 months in Figure 2 show the effect size for the 2 months rainfall. This means that the effect size for months 4, 5 and 6 is enhanced by the effect size for the 2 months rain. Thus, the strongest impact on FCM levels comes from the rainfall between ~2–3 months prior to the collection time.

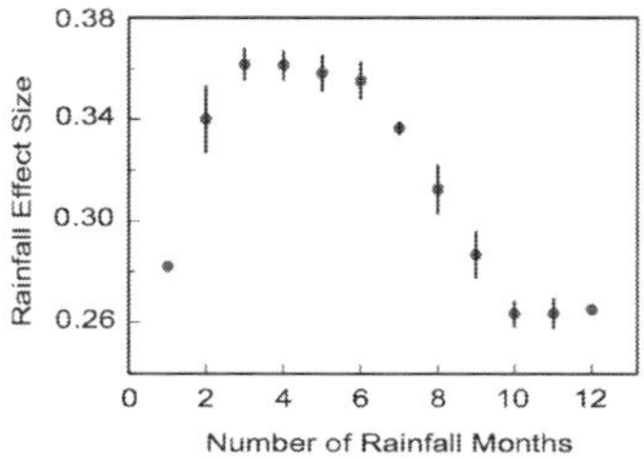

Figure 2. The moving average dependence of the effect size (standardized coefficients in Table 2) for the rainfall variable versus number of months with rain prior to data collection. The moving interval over which averaging is undertaken contains three consecutive values of the coefficients from Table 2. The first and the last points are added to this dependence directly from Table 2 (because these edge points are not included into the moving average dependence with the three-point moving interval). Error bars calculated as the errors of the mean within the moving interval are shown for each of the moving average points. Note, this figure shows the magnitude of the effect size, and the actual effect of rainfall on cortisol levels is negative.

Table 2. Statistical effects of rainfall for one to 12 months prior to sample collection on faecal cortisol concentration (Coefficients from standardized models (effect sizes))

Independent Variable	Coefficient	Standard error	P
Rainfall 1 month prior	−0.28	0.11	0.02
Rainfall 2 months prior	−0.391	0.095	<0.001
Rainfall 3 months prior	−0.347	0.096	<0.001
Rainfall 4 months prior	−0.348	0.097	<0.001
Rainfall 5 months prior	−0.39	0.11	<0.001
Rainfall 6 months prior	−0.34	0.10	0.001
Rainfall 7 months prior	−0.34	0.11	0.002
Rainfall 8 months prior	−0.34	0.11	0.001
Rainfall 9 months prior	−0.27	0.11	0.013
Rainfall 10 months prior	−0.26	0.10	0.013
Rainfall 11 months prior	−0.26	0.10	0.01
Rainfall 12 months prior	−0.27	0.10	0.01

Therefore, the model for the impact of rainfall on levels of FCM included two independent variables: rainfall for two months prior to the time of collection of the faecal pellets and rainfall for four months (3 to 6 months) prior to the time of collection. The variable of rainfall for months 3 to 6 prior to time of collection was included because it takes into account the decaying (but still extant) impact of rainfall during earlier months before the two-month period to the time of collection and its statistical significance (with the standardized coefficient being –0.25 and p=0.012) identifies the long period over which the effect is evident (Table 3).

Table 3. Statistical effects of rainfall 2 months prior and rainfall 3, 4, 5 and 6 months prior (combined) on faecal cortisol concentration (standardized)

Independent Variable	Coefficient	Z	P
Rainfall 2 months prior	−0.2647747	−4.11	0.000
Combined rainfall for months3, 4, 5 & 6 prior	−0.1715594	−2.51	0.012

We did not find significant, non-linear quadratic terms in the dependence of FCM levels on two-month rainfall. Further, since comparing the fit of the random intercept model to that of the multiple regression model, using the likelihood ratio test yielded p=0.01, we can thus reject the null hypothesis that the intercept is the same across all the regions and sites, and the random intercept model is valid and essential. At the same time, the likelihood ratio tests comparing the random slope model with the model that contains only random intercepts did not show a statistically significant difference for both the region or site levels (the p-values are close to 1 for all these tests). As a consequence, the consideration of random slopes was not necessary, so we limited our model only to the random intercepts.

The intra-class correlation, interpreted as a ratio of the variance of FCM levels due to different sites (or bioregions) to the total variance of FCM, determines the contribution of sites (or bioregions) to the total variance of FCM. That the intra-class correlation is equal to ~0.26 for the site level, and ~0.11 for the bioregion level means that the contribution of different sites to the variability of cortisol was significantly larger than that of different bioregions, indicating particularly strong clustering of the data with respect to the collection sites.

The dependence of the average FCM level on rainfall for the two-month period prior to the collection, adjusted for rainfall during the following months 3 to 6, is significant and demonstrates a monotonic decrease of FCM levels with increasing two-month rainfall (Figure 3). For a two-month rainfall of 400 mm, with the total rainfall for the previous 4 months being equal to its average value of 310 mm, the average FCM level is ~2.66 ng/g. If the two-month rainfall decreased from 400 mm to 50 mm, the FCM levels increased ~3 times to ~7.73 ng/g (Figure 3).

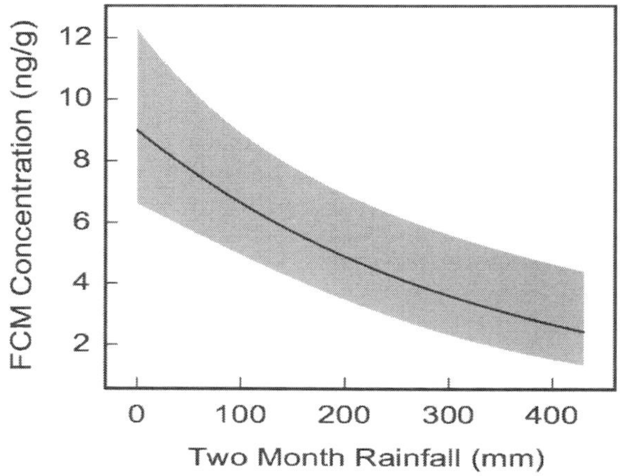

Figure 3. Faecal cortisol metabolite (FCM) concentration versus rainfall during the two months before the collection. The grey band shows the 95% prediction interval. The total rainfall for the previous 4 months (months 3 to 6) is assumed to equal its average value for that period of 310 mm.

The typical differences between the cortisol dependences for different bioregions are characterized by significantly lower levels of FCM in koalas at any level of rainfall (Figure 4). The model of the influence of rainfall on FCM concentration in koalas can account for ~10% of the total FCM variability. The relatively small fraction of the total FCM variability, explained by the rainfall variables, does not mean that this effect is insignificant. As demonstrated, the results are statistically significant with low *p*-values. Further, the contribution of the rainfall variables is comparable with the variability of FCM by region (also ~10%) and by site (~14%). This means that the other ~66% of FCM variability is due to other unexplained factors.

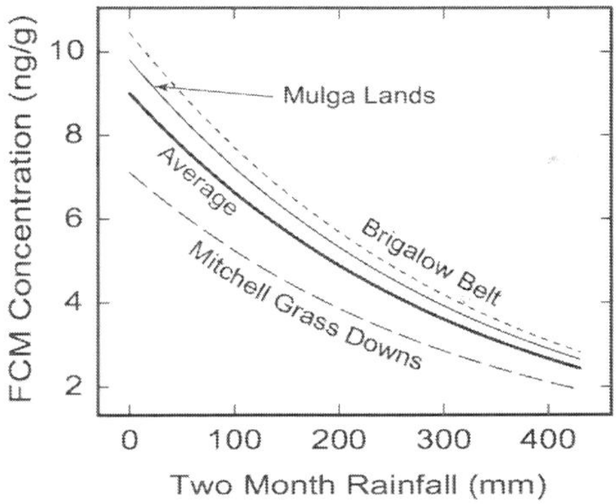

Figure 4. Regional variation in the relationship between faecal cortisol metabolite (FCM) concentration and rainfall during the two months before sample collection. The thick solid curve shows the cortisol levels averaged over all three bioregions combined (the same as the curve in Figure 3). The other three curves represent the average FCM levels for each of the bioregions: Brigalow Belt (small dashed curve), Mitchell Grass Downs (dashed curve), and Mulga Lands (thin solid curve). The total rainfall for the previous 4 months (months 3 to 6) is assumed to equal its average value of 310 mm.

To further assess the fit of the models, 12 quantile-quantile (Q-Q) plots of the standardized residuals were considered for rainfall during the periods from 1 month to 12 months prior to sample collection. These plots further demonstrated the validity of the considered models (Figure S1).

DISCUSSION

Chronic stress can reduce resistance to disease and affect population performance (including impacting negatively on survival and reproduction) [35], [101]–[105], thus understanding the physiological responses of populations exposed to environmental changes is critical to conservation. We show that higher FCM levels in koala populations near the arid edge of their range in western Queensland were associated with lower rainfall levels, consistent with physiological stress when moisture levels are low. These results are of particular concern to the continued survival of western koala populations, especially when a number of studies have attributed population declines of free-living mammals to the effects of chronic stress [36], [83], [84]. Droughts and heatwaves have been responsible for population crashes and local extinctions in a number of koala populations [71], [86]–[88]. Predictions indicate that the intensity and frequency of droughts and heatwaves will increase and moisture availability will decrease [89]–[91]. Our findings show that western koalas, at the trailing edge of their geographic range, will cope poorly with changes in rainfall patterns resulting from climate change.

Arid and semi-arid landscapes are temporally and spatially variable, heterogeneous ecosystems where ecological processes are primarily driven by climate and nutrients [106]. Previous research within these arid and semi-arid landscapes has determined that the distribution, density, habitat preferences and home range sizes of koalas are affected either by water availability (including leaf moisture) and/or rainfall [75], [76], [86], [107], [108]. The findings from the current research have also demonstrated the impact of rainfall on koala populations by showing that low rainfall negatively impacts on FCM levels. In particular, rainfall from the previous two months had the strongest influence on FCM levels. Time lags can occur between rainfall events and the subsequent response by vegetation, with previous studies showing that total rainfall over the preceding two months has a strong effect on vegetation growth [96]–[98]. This indicates that the influence of rainfall on FCM levels in koalas is likely to be caused by the improved leaf moisture and nutrition in the form of new growth resulting from the rainfall. The delay of the new growth after a rainfall causes the consequential time lag of about two months between the rainfall and the maximum reduction in FCM levels in koalas. Foliar moisture supplies most of the koala's water requirements. It has been proposed that in dry environments or during drought water, rather than leaf nutrients, influences tree selection by koalas [107], [109]. Therefore, rainfall two months prior is an essential determinant of leaf moisture content, and hence food quality, which in turn impacts on FCM concentrations of southwestern Queensland koalas. Results from this study are consistent with studies on African elephants, muriqui monkeys and spider monkeys [77], [79], [110]. These studies reported FCM concentrations significantly higher in the dry versus the wet season as food and/or water availability declined [77], [79], [110]. For example, FCM concentrations of free-ranging African elephants were significantly higher in the dry season as food, water availability and body condition declined and were inversely correlated with rainfall across seasons

[79]. The authors also noted that elephant diets change from high-quality grass in the wet season to less nutritious and more sparsely distributed bark and leaves in the dry season [111], [112]. Spider monkeys in conserved forests also show seasonal variation in faecal cortisol concentrations, with higher concentrations measured in the dry season compared to the wet season, when there is a reduction in the number of fruiting trees and hence food availability [110]. The same study also found that mean cortisol concentrations of spider monkeys living in fragmented habitat remained high throughout the year, indicating that individuals within these habitats may suffer from long-term stress [110]. In addition, serum cortisol was significantly higher in camels in drought-affected areas than camels in other areas [113].

When comparing cortisol concentrations in relation to prior rainfall, FCM concentrations of koalas from the Mitchell Grass Downs were below the combined bioregion average, whereas FCM concentrations of koalas from the Mulga Lands and Brigalow Belt South were above the average. The dominant soil type of sites within the Mitchell Grass Downs (Vertosol) has high water holding capacity and very high soil fertility compared to the dominant soils of the Mulga Lands (Kandosol) and Brigalow Belth South (Sodosol, Chromosol and Vertosol) bioregions [114]–[116]. The difference in soil water holding capacity and fertility among the regions could explain why the Mitchell Grass Downs koalas had lower FCM concentrations, i.e. trees within this region are likely to have higher leaf moisture and forage quality as a result of the higher water holding capacity and fertility of the soil. A previous study within southwestern Queensland found that there was a significant relationship between leaf moisture and soil types, although once soil types were divided into groups based on water availability, the authors suggested the relationship could be ascribed to water availability in sites [117]. Furthermore, in this study, soil properties had no significant impact on the levels of FCM. Therefore, although soil water-holding capacity and/ or fertility may explain the lower FCM concentrations of the Mitchell Grass Downs compared to the other two bioregions, rainfall two months prior is most likely the principal determinant of leaf moisture content and quality, regardless of soil type.

Significant effects of temperature on glucocorticoid concentrations have been shown for several mammalian species, with most studies reporting higher levels during harsher conditions in winter or during extreme heat [118], [119]. For example, a significant negative relationship was found between minimum ambient temperature and FCM levels of red deer [118]. In Marwari sheep, both hot ambient temperatures (maximum of 43–

46°C) and drought conditions resulted in a significant increase in serum cortisol [119]. However, we found no impact of minimum and maximum ambient temperatures on FCM levels of koalas. Behavioural adjustment and physiological adaptations of koalas living near the arid edge of their range help mitigate the physiological stress of extreme temperatures. In central Queensland, the temperature in trees used by koalas for shelter in summer can be around 2°C cooler than the ambient temperature [71], and in Gunnedah, in north-western New South Wales, tree choice was found to be mediated by temperature [120]. Furthermore, the pelt of the koala has the highest recorded insulation (including a high resistance to wind) for a marsupial [121]. The dorsal surface is more densely furred and less reflective of solar radiation than the ventral surface [121]. Degabriele [121] suggested that different postural adjustments, particularly in relation to wind direction, allow for insulation capabilities, and that the field water influx and low metabolic rate of koalas are likely to be adaptations to their arboreal, low-energy folivorous lifestyle that extends over a wide latitudinal range. However, throughout the koala's distribution, droughts and heatwaves have been responsible for population crashes and local extinctions [71], [86]–[88]. Although the metabolic and thermoregulatory adaptations of koalas help mitigate the effect of temperature on physiological stress, the studies to date identify that there is a threshold that, once crossed, cause population crashes. In this study, due to the relatively small number of samples collected at the lower and higher temperature extremes, this influence of temperature was not detected.

Variation in the physiological stress response, that influences the concentration of fecal GCMs, has been reported to reflect reproductive status or phase (i.e. oestrous, gestation, lactation) for a number of mammalian species [81], [122]–[124]. Glucocorticoids are also known to rise significantly near term in mammalian species because they trigger the cascade resulting in parturition [12], [125]. While these large changes in plasma cortisol occur in most eutherian mammals during gestation, the different endocrine control of parturition following the well known short gestation period in marsupials is not associated with any large changes in maternal plasma cortisol [126]. Consistently we did not detect differences between sexes or reproductive condition in this study. It is possible that during drought years, reproduction might have been suspended or reduced in southwestern Queensland koala populations, while as rainfall increases a normal pattern of reproduction might have been found. Indeed anecdotal evidence suggests such a scenario with only one mother and joey found during koala searches during the drought compared with numerous observations of mothers with joey's post-flood. However, we found the opposite with higher FCM during low rainfall periods (consistent with moisture stress) compared with periods of increased rainfall when females might be in reproductive condition. Narayan *et al* [55] reported that FCM levels of captive koalas significantly differed by sex, reproductive condition (lactating versus non-lactating koalas) and handling groups; however, *post hoc* comparisons revealed that although FCM levels were significantly different between the sexes and handling groups, there was no significant differences between the reproductive conditions. Narayan *et al* [55] also reported no significant difference in FCM levels between wild and captive koalas, with *post hoc* comparisons further supporting the suggestion that within each environment (captive or wild) there was a significant effect of sex but not reproductive condition on the levels of FCM. Sex differences in either plasma free cortisol or fecal cortisol were not apparent in a study on captive koalas by Davies *et al* [54], however, these animals were not lactating and not known to be pregnant at the time of the study. Because of the large difference between eutherian and metatherian reproductive endocrine control, the large changes in maternal plasma cortisol that occur in most eutherians during pregnancy do not occur during the short gestation in marsupials and fetal cortisol is more important for initiating parturition and enabling development and maturation of organs (in particular the lungs) in preparation for parturition [126]. Since this study measures a downstream excretory product of hepatic clearance of plasma free cortisol, not total cortisol, sex differences are likely to be small. This adds further support to the suggestion that rainfall, and the subsequent leaf moisture and/or leaf nutritional content, is the principal determinant of FCM levels of southwestern koala populations. However, future studies should address possible differences between sexes and reproductive status and to do this thoroughly, steroid partitioning in the blood plasma must be assessed enabling calculation of free cortisol.

Sustained high levels of glucocorticoids from chronic stress (e.g. from long-term climate change) can be deleterious to the physiological health of an animal and lead to greater susceptibility to disease, as well as reduced fecundity and survivorship [12], [13], [20], [34], [35]. While the deleterious effects of chronic stress are well conceptualized from biomedical evidence, there is some doubt whether or not these deleterious effects, especially relating to pathology, occur in nature [41], [127]. Boonstra [41] argues that both demographic (the indirect effects on reproduction) and hormonal evidence indicate that although some wild populations may be chronically stressed by predators, there is no evidence of pathology or an HPA axis with an inability to respond. Boonstra [41] suggests that chronic stress is not the inevitable result of a persistent, severe stressor, but is part of the normal experience of many animals in nature and, although there are fitness costs, it is an evolved adaptive response to particular ecological and habitat pressures. While this may be the case for predation stressors, there is no evidence in wild populations whether this adaptive response can be applied to other severe chronic stressors of a different nature, such as climate change or habitat loss and fragmentation. For species that have populations threatened by disease, chronic stress acting via well known effects on the immune system, can be expected to increase susceptibility to disease and threaten the long-term viability of populations. Koalas are subject to a range of diseases, which can lead either to death or infertility [128], [129]. Infectious diseases in koalas are recognised to threaten the long-term viability of populations in Queensland and New South Wales [129]. In northwestern New South Wales, the prevalence of the disease chlamydiosis was found to increase dramatically following intense stress from extreme hot weather (intense heatwaves) during a drought [87]. Following a review of studies of chlamydial disease and population decline, Lunney *et al.* [87] also linked chlamydiosis to koala populations under stress from habitat loss and fragmentation. Furthermore, Gordon [86] reported that on dry stretches of creek in southwestern Queensland koalas showed poorer health (poor condition, anemia, high tick loads) and suffered much higher mortality, i.e. population crashes, during drought compared to sites where trees were not affected by the drought (mainly on large permanent waterholes). We found low rainfall resulted in high cortisol levels, demonstrating that droughts have the potential to impact on the physiological stress of individual koalas as a result of reduced food and water availability. It is increasingly likely exposure to these environmental stressors will increase their susceptibility to disease. However, although there is anecdotal evidence reporting that stress and disease prevalence in various koala populations throughout its distribution are related, the link between chronic stress, disease and ensuing demographic impact needs detailed examination before any causal links can relate chronic stress to pathology. Our results provide new insights into the response of koalas to stressors, such as drought and heatwaves, and highlight the conservation concerns for the continued survival of western koala populations in light of predicted climate change.

Implications for Conservation

Drought and heatwaves have the potential to impact on physiological stress in individual koalas as these extreme climatic conditions can affect the quality of nutrients and moisture available in the diet of entire populations [130], [131]. Observed FCM concentration patterns in relation to rainfall levels strongly suggest that physiological stress in western koalas is being adversely affected by limited access to food and water. These western Queensland koala populations are at the semi-arid edge of their distribution, where drought and heatwaves have resulted in population declines [86], [88], which confirms that koalas in this semi-arid region are near their physiological tolerance limits. Climate change predictions for this region indicate that the intensity and frequency of droughts and heatwaves will increase, evapotranspiration will increase, and that foliar moisture will decrease [89]–[91]. This highlights conservation and management concerns for the continued survival of koala populations of western Queensland.

Investigating FCM levels at the trailing edge of a widely distributed species boundary, and how these levels vary with climatic factors, allows us to non-invasively identify changes likely to cause stress. It also allows us to identify the habitats that are least susceptible to climate change as priority sites for protection, as well as to facilitate management decisions to prevent further contractions in a species' distribution, and monitor the efficacy of management strategies. In southwestern Queensland, riparian habitat is the primary resource used by koalas [88], [95], [108], based on evidence of populations declining and contracting to riparian habitats during drought and heatwaves [86], [88]. It follows that since the moisture content of leaves in trees within riparian habitats would be higher, it makes them higher quality habitat for koalas. Considering that low rainfall, and hence low leaf moisture, resulted in higher FCM levels, it places even more importance on maintaining and improving the quality of riparian habitats for continued survival of these western koala populations, particularly during extreme weather. Historical land management practices have diminished riparian habitats from the silting of previously-permanent water-holes [86]. For the long-term conservation of western koala populations, fencing of riparian habitats to keep out feral animals and domestic stock should be considered. This will help to reduce the silting of waterholes and allow the recruitment of food trees. However, reduced grazing could also increase weed species, which could negatively impact on koalas and other native wildlife. This points to the need for close monitoring and control of weed infestations. Previous studies have also found water availability and rainfall to be the main factors defining habitat preference of koalas in arid and semi-arid regions [75], [107]. Maintaining access to freestanding water, such as farm dams, and planting favoured tree species, particularly those that tend to have higher leaf moisture content, such as *E. camadulensis* and *E. coolabah* [117]), close to dams would also increase the availability of high quality trees, as these trees are likely to have higher leaf moisture levels. Therefore, to ensure the continued survival of western Queensland koala populations in extreme weather, the most important actions for koala conservation within these semi-arid landscapes are the maintenance of the quality and quantity of riparian habitats, expanding the availability and accessibility of freestanding water, such as around farm dams, and providing specific trees as food resources.

CONCLUSION

Ecological studies are increasingly using the release of glucocorticoids as a measure of the stress response in animals. Using non-invasive methods to measure the glucocorticoid metabolites in faeces, we found that koalas had higher FCM levels during low rainfall periods. An analysis of mechanisms for the impact of rainfall on FCM levels in koalas is beyond the scope of this paper, but one mechanism could be through the new growth and higher foliar moisture content of eucalypt leaves during periods of rainfall. This would provide koalas with an abundance of food and water causing reductions of their cortisol levels. These results provide new insights into the effect of droughts and heat waves on physiological stress levels and highlight conservation concerns for the continued survival of western Queensland koala populations in the face of predicted climate change. Our results demonstrate the importance of integrating a physiological dimension into ecological studies to ensure a more comprehensive understanding of species' responses to habitat loss, particularly riparian habitat, and the increasing threat of relentless climate change.

SUPPORTING INFORMATION

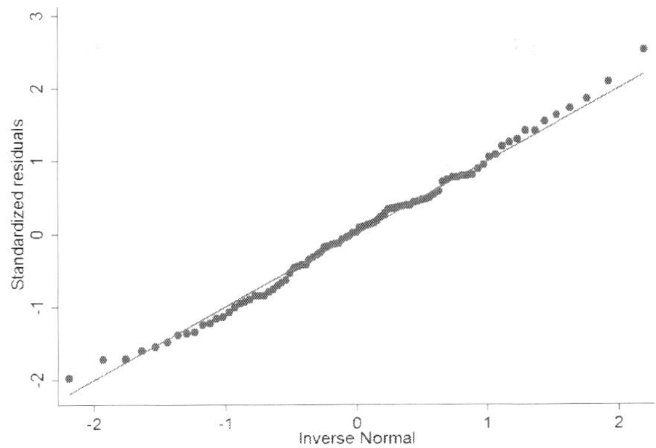

(a)

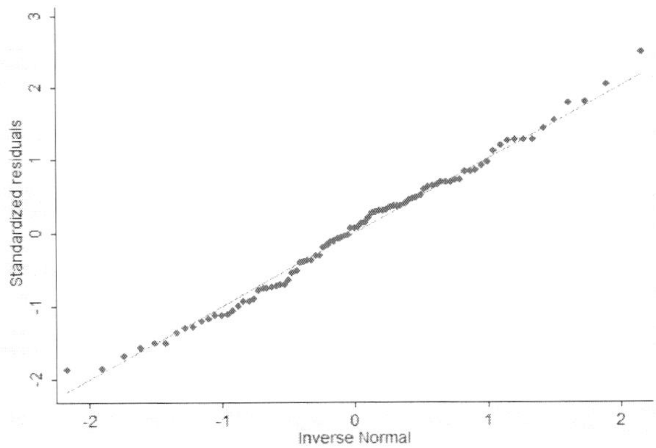

(b)

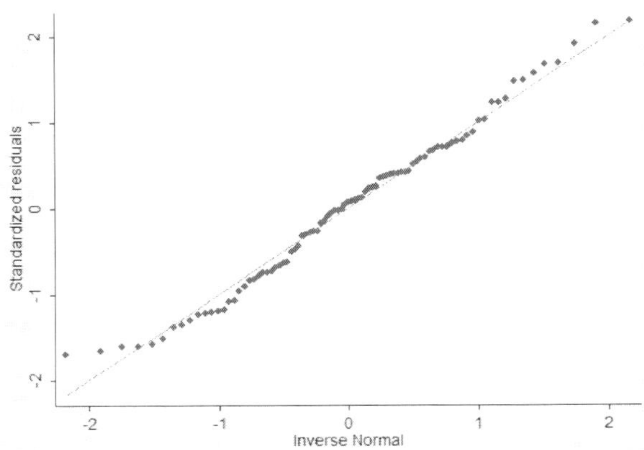

(c)

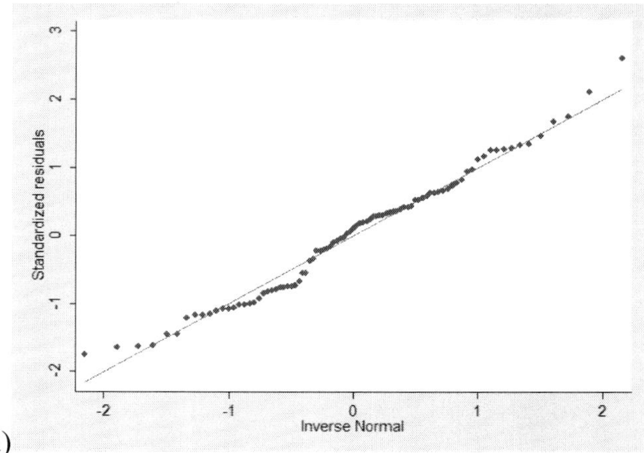

Standardized residuals
Inverse Normal
-2
-1
0
1
2
3

(d)

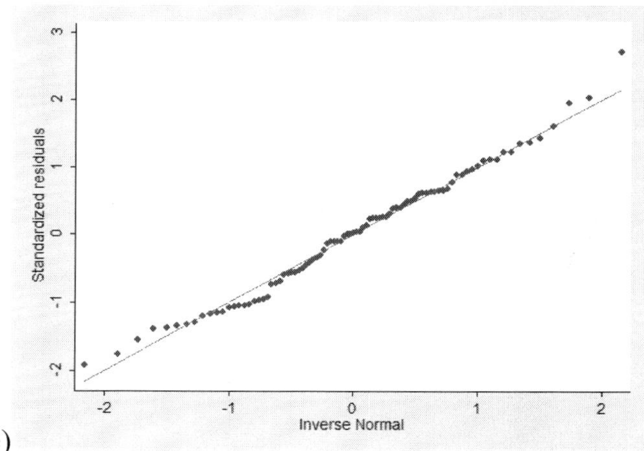

Standardized residuals
Inverse Normal
-2
-1
0
1
2
3

(e)

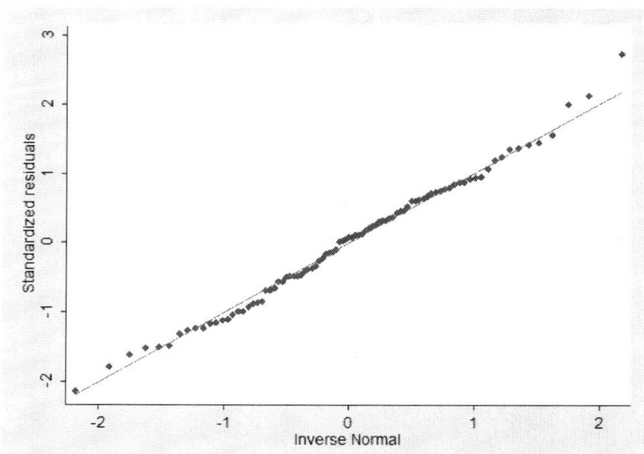

Standardized residuals
Inverse Normal
-2
-1
0
1
2
3

(f)

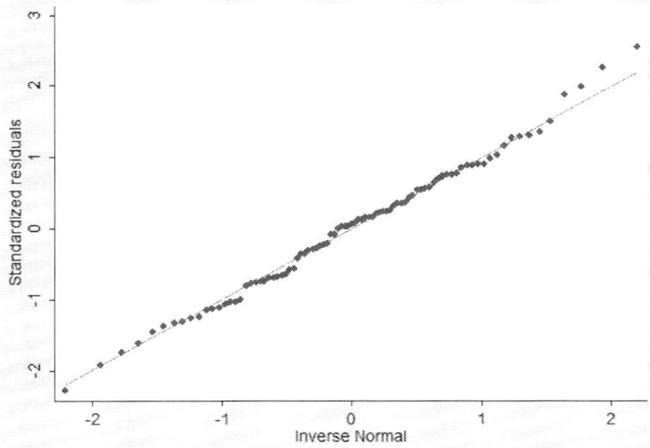
Standardized residuals
Inverse Normal

(g)

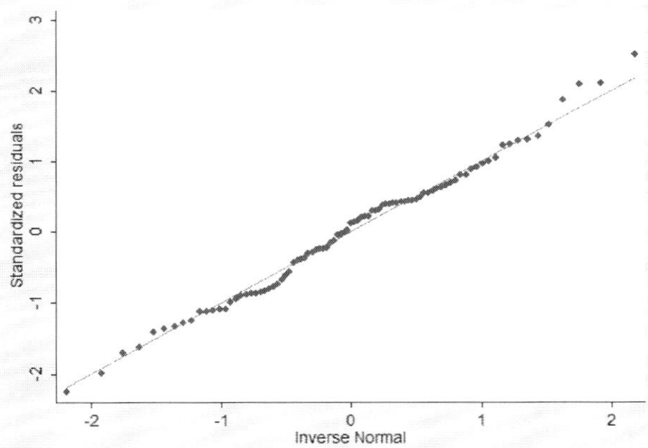
Standardized residuals
Inverse Normal

(h)

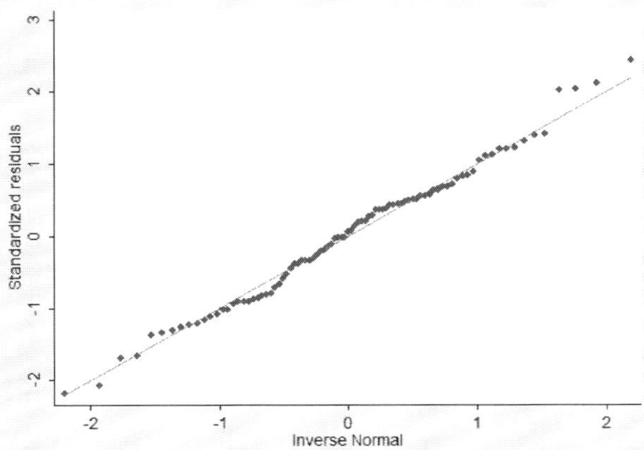
Standardized residuals
Inverse Normal

(i)

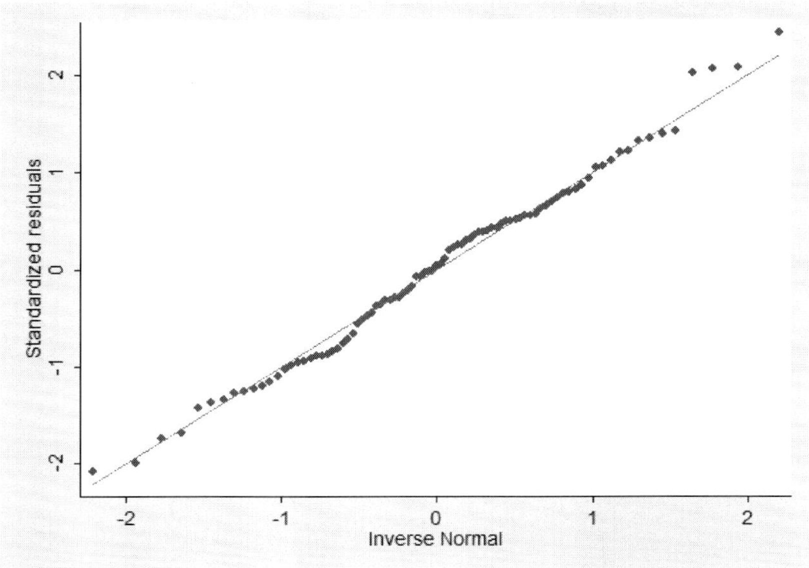

Standardized residuals
-2
-1
0
1
2
Inverse Normal

(j)

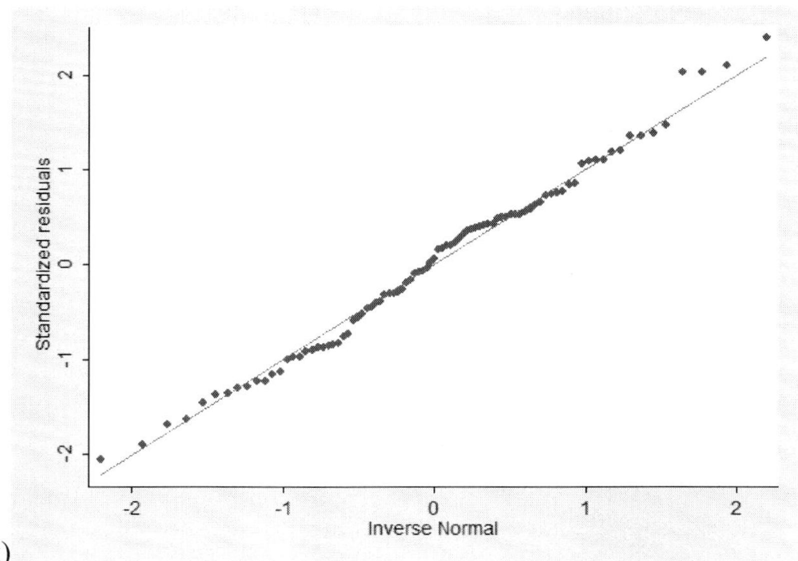

Standardized residuals
-2
-1
0
1
2
Inverse Normal

(k)

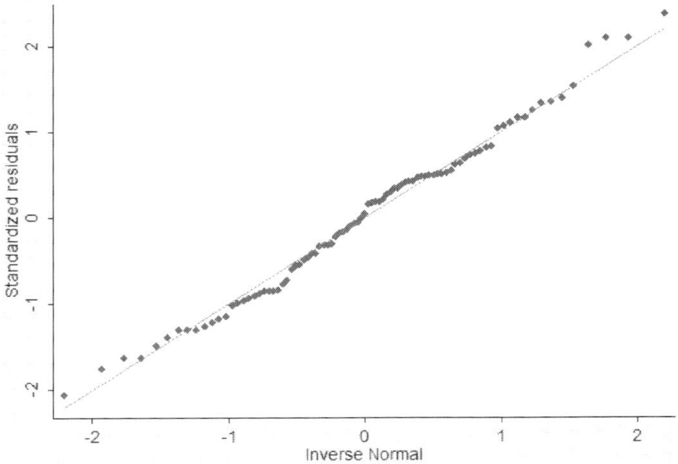

(l)

Figure S1. Q-Q plots of the standardized residuals for the models for (a) rainfall 1 month prior to sample collection, (b) rainfall 2 months prior to sample collection, (c) rainfall 3 months prior to sample collection, (d) rainfall 4 months prior to sample collection, (e) rainfall 5 months prior to sample collection, (f) rainfall 6 months prior to sample collection, (g) rainfall 7 months prior to sample collection, (h) rainfall 8 months prior to sample collection, (i) rainfall 9 months prior to sample collection, (j) rainfall 10 months prior to sample collection, (k) rainfall 11 months prior to sample collection, and (l) rainfall 12 months prior to sample collection.

ACKNOWLEDGMENTS

We are extremely grateful to all of the volunteers that have assisted with the koala surveys, with special thanks to Grace Davies, Stuart Davies, Will Goulding, Kailas Wild, Ladislas Parraud and Huiying Wu. We would like to thank all of the landholders who allowed koala surveys to be conducted on their land, their hospitality and knowledge was very much appreciated.

AUTHOR CONTRIBUTIONS

Conceived and designed the experiments: ND CM LS GB DL JR AB. Performed the experiments: ND AB. Analyzed the data: ND GG. Contributed reagents/materials/analysis tools: AB. Wrote the paper: ND GG CM LS GB DL JR AB.

REFERENCES

1. Badyaev AV (2005) Stress-induced variation in evolution: from behavioral plasticity to genetic assimilation. *Proceedings of the Royal Society of London Series B: Biological Sciences* 272: 877–886.
2. Badyaev AV (2005) Role of stress in evolution: from individual adaptability to evolutionary adaptation. In: Hallgrimsson B, Hall BK, editors. Variation: A Central Concept in Biology. Amsterdam: Elsevier Academic Press. 277–302.
3. Hoffmann AA, Hercus MJ (2000) Environmental stress as an evolutionary force. *BioScience* 50: 217–226.
4. Wingfield JC, Maney DL, Breuner CW, Jacobs JD, Lynn S, et al. (1998) Ecological bases of hormone-behavior interactions: The "emergency life history stage". *American Zoology* 38: 191–206.
5. Wingfield JC (2013) Ecological processes and the ecology of stress: the impacts of abiotic environmental factors. *Functional Ecology* 27: 37–44.
6. Jessop TS, Woodford R, Symonds MRE (2013) Macrostress: do large-scale ecological patterns exist in the glucocorticoid stress response of vertebrates? *Functional Ecology* 27: 120–130.
7. Bijlsma R, Loeschcke V (2005) Environmental stress, adaptation and evolution: an overview. *Journal of Evolutionary Biology* 18: 744–749.
8. Hoffmann AA, Parsons PA (1991) Evolutionary Genetics and Environmental Stress. Oxford, UK: Oxford University Press.
9. Hoffmann AA, Parsons PA (1997) Extreme Environmental Change and Evolution. Cambridge, UK: Cambridge University Press.
10. Seyle H (1935) A syndrome produced by diverse nocuous agents. *Nature* 138: 32–33.
11. Axelrod J, Reisine TD (1984) Stress hormones: their interaction and regulation. *Science* 224: 452–459.
12. Mostl E, Palme R (2002) Hormones as indicators of stress. *Domestic Animal Endocrinology* 23: 67–74.
13. Stewart PM (2003) The adrenal cortex. In: Larsen PR, Kronenberg HM, Melmed S, Polonsky KS, editors. Williams Textbook of Endocrinology. Philadelphia: Saunders. 491–551.
14. Wikelski M, Cooke SJ (2006) Conservation physiology. *Trends in Ecology and Evolution* 21: 38–46.

15. Turner CD, Bagnara JT (1976) General Endocrinology. 6th edition. Philadelphia: W. B. Saunders.
16. Moberg GP (1987) Problems in defining stress and distress in animals. *Journal of American Veterinary Medical Association* 191: 1207–1211.
17. McEwen BS, Schmeck HM (1994) The hostage brain. New York: Rockefeller University Press.
18. Matteri RL, Carrol JA, Dyer CJ (2000) Neuroendocrine responses to stress. In: Moberg GP, Mench JA, editors. The Biology of Animal Stress: CAB International. 43–76.
19. Keay JM, Singh J, Gaunt MC, Kaur T (2006) Fecal glucocorticoids and their metabolites as indicators of stress in various mammalian species: a literature review. *Journal of Zoo and Wildlife Medicine* 37: 234–244.
20. Sapolsky RM, Romero LM, Munck AU (2000) How do glucocorticoids influence stress responses? Integrating permissive, suppressive, stimulatory, and preparative actions. *Endocrine Reviews* 21: 55–89.
21. Dallman MF, Bhatnagar S (2001) Chronic stress and energy balance: role of the hypo-thalamo-pituitary axis. In: McEwen BS, Goodman HM, editors. Handbook of Physiology: Section 7: The Endocrine System, Coping with the Environment: Neural and Endocrine Mechanisms. New York: Oxford University Press. 179–210.
22. Sapolsky RM (2002) Endocrinology of the stress response. In: Becker JB, Breedlove SM, Crews D, McCarthy MM, editors. Behavioural Ecology. Cambridge: The MIT Press. 409–450.
23. Romero LM (2004) Physiological stress in ecology: lessons from biomedical research. *Trends in Ecology and Evolution* 19: 249–255.
24. Wingfield JC (2003) Control of behavioural strategies for capricious environments. *Animal Behaviour* 66: 807–816.
25. Wingfield JC (2005) The concept of allostasis: coping with a capricious environment. *Journal of Mammalogy* 86: 248–254.
26. Wingfield JC, Kitaysky AS (2002) Endocrine responses to unpredictable environmental events: stress or anti-stress hormones? *Integrative and Comparative Biology* 42: 600–609.
27. Romero LM, Butler LK (2007) Endocrinology of stress. *International Journal of Comparative Psychology* 20: 89–95.
28. Raynaert R, De Paepe M, Peeters G (1976) Influence of stress, age and sex on serum growth hormone and free fatty acids in cattle. *Hormone Metabolism Research* 8: 109–114.

29. Lee AK, Cockburn A (1985) Evolutionary Ecology of Marsupials. Cambridge: Cambridge University Press.
30. Kenagy GJ, Place NJ (2000) Seasonal changes in plasma glucocorticosteroids of free-living yellow pine chipmunks: effects of reproduction and capture and handling. *Endocrinology* 117: 189–199.
31. Mateo JM (2006) Developmental and geographic variation in stress hormones in wild Belding's ground squirrels (*Spermophilus beldingi*). *Hormones and Behavior* 50: 718–725.
32. McEwen BS, Wingfield JC (2003) The concept of allostasis in biology and biomedicine. *Hormones and Behavior* 43: 2–15.
33. McEwen BS, Wingfield JC (2010) What is in a name? Integrating homeostasis, allostasis and stress. *Hormones and Behavior* 57: 105–111.
34. Lane J (2006) Can non-invasive glucocorticoid measures be used as reliable indicators of stress in animals? *Animal Welfare* 15: 331–342.
35. Munck A, Guyre PM, Holbrook NJ (1984) Physiological functions of glucocorticoids in stress and their relation to pharmacological actions. *Endocrinology Review* 5: 25–44.
36. Boonstra R, Singleton GR (1993) Population declines in the snowshoe hare and the role of stress. *General and Comparative Endocrinology* 91: 126–143.
37. Creel S, Fox JE, Hardy A, Sands J, Garrott B, et al. (2002) Snowmobile activity and glucocorticoid stress responses in wolves and elk. *Conservation Biology* 16: 809–814.
38. Sheriff MJ, Bosson CO, Krebs CJ, Boonstra R (2009) A non-invasive technique for measuring fecal cortisol metabolites in snowshoe hares (*Lepus americanus*). *Journal of Comparative Physiology B* 179: 305–313.
39. Thiel D, Jenni-Eiermann S, Braunisch V, Palme R, Jenni L (2008) Ski tourism affects habitat use and evokes a physiological stress response in capercaillie *Tetrao urogallus*: a new methodological approach. *Journal of Applied Ecology* 45: 845–853.
40. Wasser SK, Bevis K, King G, Hanson E (1997) Noninvasive physiological measures of disturbance in the northern spotted owl. *Conservation Biology* 11: 1019–1022.
41. Boonstra R (2013) Reality as the leading cause of stress: rethinking the impact of chronic stress in nature. *Functional Ecology* 27: 11–23.

42. Boonstra R, Hik D, Singleton GR, Tinnikov A (1998) The impact of predator-induced stress on the snowshoe hare cycle. *Ecological Monographs* 68: 371–394.
43. Creel S, Winnie JA, Christianson D (2009) Glucocorticoid stress hormones and the effect of predation risk on elk reproduction. *Proceedings of the National Academy of Sciences USA* 106: 12388–12393.
44. Zanette LY, White AF, Allen MC, Clinchy M (2011) Perceived predation risk reduces the number of offspring songbirds produce per year. *Science* 334: 1398–1401.
45. Davenport MD, Tiefenbacher S, Lutz CK, Novak MA, Meyer JS (2006) Analysis of endogenous cortisol concentrations in the hair of rhesus macaques. *General and Comparative Endocrinology* 147: 255–261.
46. Koren L, Mokady O, Karaskov T, Klein J, Koren G, et al. (2002) A novel method using hair for determining hormonal levels in wildlife. *Animal Behaviour* 63: 403–406.
47. Palme R, Rettenbacher S, Touma C, El-Bahr S, Mostl E (2005) Stress hormones in mammals and birds: comparative aspects regarding metabolism, excretion, and noninvasive measurement in faecal samples. *Annals of the New York Academy of Sciences* 1040: 162–171.
48. Young KM, Walker SL, Lantheir C, Waddell WT, Monfort SL, et al. (2004) Noninvasive monitoring of adrenocortical activity in carnivores by fecal glucocorticoid analyses. *General and Comparative Endocrinology* 137: 148–165.
49. Oates JE, Bradshaw FJ, Bradshaw SD, Stead-Richardson EJ, Philippe DL (2007) Reproduction and embryonic diapause in a marsupial: insights from captive honey possums, *Tarsipes rostratus* (Tarsipedidae). *General and Comparative Endocrinology* 150: 445–461.
50. McKenzie S, Deane EM (2005) Faecal corticosteroid levels as an indicator of well-being in the tammar wallaby, *Macropus eugenii* . *Comparative Biochemistry and Physiology, Part A* 140: 81–87.
51. Stead-Richardson EJ, Bradshaw D, Friend T, Fletcher T (2010) Monitoring reproduction in the critically endangered marsupial, Gilberts potoroo (*Potorous gilbertii*): preliminary analysis of faecal oestradiol-17B, cortisol and progestagens. *General and Comparative Endocrinology* 165: 155–162.

52. Hogan LA, Johnston SD, Lisle AT, Keeley T, Wong P, et al. (2011) Behavioural and physiological responses of captive wombats (*Lasiorhinus latifrons*) to regular handling by humans. *Applied Animal Behavioral Science* 134: 217–228.
53. Hogan LA, Lisle AT, Johnston SD, Robertson H (2012) Non-invasive assessment of stress in captive numbats, *Myrmecobius fasciatus* (Mammalia: Marsupialia), using faecal cortisol measurement. *General and Comparative Endocrinology* 179: 376–383.
54. Davies N, Gillett A, McAlpine C, Seabrook L, Baxter G, et al. (2013) The effect of ACTH upon faecal glucocorticoid excretion in the koala. *Journal of Endocrinology* 219: 1–12.
55. Narayan EJ, Webster K, Nicolson V, Mucci A, Hero J-M (2013) Non-invasive evaluation of physiological stress in an iconic Australian marsupial: The Koala (*Phascolarctos cinereus*). *General and Comparative Endocrinology* 187: 39–47.
56. Taylor W (1971) The excretion of steroid hormone metabolites in bile and feces. *Vitamins and Hormones* 29: 201–285.
57. Goymann W (2005) Non-invasive monitoring of hormones in bird droppings: biological validations, sampling, extraction, sex differences, and the influence of diet on hormone metabolite levels. *Annals of the New York Academy of Sciences* 1046: 35–53.
58. Klasing KC (2005) Potential impact of nutritional strategy on non-invasive measurements of hormones in birds. *Annals of the New York Academy of Sciences* 1046: 5–16.
59. Palme R (2005) Measuring faecal steroids: guidelines for practical application. *Annals of the New York Academy of Sciences* 1046: 75–80.
60. Touma C, Palme R (2005) Measuring faecal glucocorticoid metabolites in mammals and birds: the importance of validation. *Annals of the New York Academy of Sciences* 1046: 54–74.
61. Sheriff MJ, Krebs CJ, Boonstra R (2010) Assessing stress in animal populations: Do fecal and plasma glucocorticoids tell the same story? *General and Comparative Endocrinology* 166: 614–619.
62. Goymann W (2012) On the use of non-invasive hormone research in uncontrolled, natural environments: the problem with sex, diet, metabolic rate and the individual. *Methods in Ecology and Evolution* 3: 757–765.

63. Sheriff MJ, Dantzer B, Delehanty B, Palme R, Boonstra R (2011) Measuring stress in wildlife: techniques for quantifying glucocorticoids. *Oecologia* 166: 869–887.

64. Goymann W, Mostl E, Van't Hof T, East ML, Hofer H (1999) Noninvasive fecal monitioring of glucocorticoids in spotted hyenas, *Crocuta crocuta* . *General and Comparative Endocrinology* 114: 340–348.

65. Harper JM, Austad SN (2000) Fecal glucocorticoids: A nonivasive method of measuring adrenal activity in wild and captive rodents. *Physiological and Biochemical Zoology* 73: 12–22.

66. Harper JM, Austad SN (2001) Effect of capture and season on fecal glucocorticoid levels in Deer Mice (*Peromyscus maniculatus*) and Red-Backed Voles (*Clethrionomys gapperi*). *General and Comparative Endocrinology* 123: 337–344.

67. Millspaugh JJ, Washburn BE (2003) Within-sample variation of fecal glucocorticoid measurements. *General and Comparative Endocrinology* 132: 21–26.

68. Monfort SL, Mashburn KL, Brewer BA, Creel S (1998) Evaluating adrenal activity in African wild dogs (*Lycaon ictus*) by fecal corticosteroid analysis. *Journal of Zoo and Wildlife Medicine* 29: 129–133.

69. Parmesan C (2006) Ecological and evolutionary responses to recent climate change. *Annual Review of Ecology Evolution and Systematics* 37: 637–669.

70. DeGabriele JL, Moore BD, Marsh KJ, Foley WJ (2010) The effect of plant secondary metabolites on the interplay between the internal and external environments of marsupial folivores. *Chemoecology* 20: 97–108.

71. Ellis W, Melzer A, Clifton ID, Carrick F (2010) Climate change and the koala *Phascolarctos cinereus*: water and energy. *Australian Zoologist* 35: 369–377.

72. Hampe A, Petit RJ (2005) Conserving biodiversity under climate change: the rear edge matters. *Ecology Letters* 8: 461–467.

73. Thomas CD, Franco AMA, Hill JK (2006) Range retractions and extinction in the face of climate warming. *Trends in Ecology and Evolution* 21: 415–416.

74. Wilson RJ, Thomas CD, Fox R, Roy DB, Kunin WE (2004) Spatial patterns in species distributions reveal biodiversity change. *Nature* 432: 393–396.

75. Sullivan BJ, Baxter GS, Lisle AT (2003) Low-density koala (*Phascolarctos cinereus*) populations in the mulgalands of south-west Queensland. III. Broad-scale patterns of habitat use. *Wildlife Research* 30: 583–591.

76. Sullivan BJ, Norris WM, Baxter GS (2003) Low-density koala (*Phascolarctos cinereus*) populations in the mulgalands of south-west Queensland. II. Distribution and diet. *Wildlife Research* 30: 331–338.

77. Strier KB, Ziegler TE, Wittwer DJ (1999) Seasonal and social correlates of fecal testosterone and cortisol levels in wild male muriquis (*Brachyteles arachnoides*). *Hormone Behaviour* 35: 125–134.

78. Beehner JC, McCann C (2008) Seasonal and altitudinal effects on glucocorticoid metabolites in a wild primate (*Theropithecus gelada*) Physiology and Behavior. 95: 508–514.

79. Foley CAH, Papageorge S, Wasser S (2001) Noninvasive stress and reproductive measures of social and ecological pressures in free-ranging African Elephants. *Conservation Biology* 15: 1134–1142.

80. Megahed G, Anwar M, Wasfy S, Hammadeh M (2008) Influence of heat stress on the cortisol and oxidant-antioxidants balance during oestrous phase in buffalo-cows (*Bubalus bubalis*): thermo-protective role of antioxidant treatment. *Reproduction in Domestic Animals* 43: 672–677.

81. Weingrill T, Gray D, Barrett L, Henzi S (2004) Fecal cortisol levels in free ranging female chacma baboons: relationship to dominance, reproductive state and environmental factors. *Hormones and Behavior* 45: 259–269.

82. Ziegler T, Scheffler G, Snowdon C (1995) The relationship of cortisol levels to social environment and reproductive functioning in female cotton-top tamarins, *Saguinus oedipus* . *Hormones and Behavior* 29: 407–424.

83. Charbonnel N, Chaval Y, Bertheir K, Deter J, Morand S, et al... (2008) Stress and demographic decline: a potential effect mediated by impairment of reproduction and immune function in cyclic vole populations. Physiological and Biochemical Zoology 81 63–73.

84. Pride ER (2004) High faecal glucocorticoid levels predict mortality in ring-tailed lemurs (*Lemur catta*). Biology Letters 1 60–63.
85. Schultner J, Kitaysky AS, Welcker J, Hatch S (2013) Fat or lean: adjustment of endogenous energy stores to predictable and unpredictable changes in allostatic load. *Functional Ecology* 27: 45–55.
86. Gordon G (1988) A koala (*Phascolarctos cinereus* Goldfuss) population crash during drought and heatwave conditions in south-western Queensland. *Australian Journal of Ecology* 13: 451–461.
87. Lunney D, Crowther MS, Wallis I, Foley WJ, Lemon J, et al... (2012) Koala populations and climate change: a case study on the Liverpool Plains, north-west NSW. In: Lunney D, Hutchings P, editors. Wildlife and Climate Change: Towards Robust Conservation Strategies for Australian Fauna. Mosman, NSW, Australia: Royal Zoological Society of NSW. 150–168.
88. Seabrook L, McAlpine C, Baxter G, Rhodes J, Bradley A, et al. (2011) Drought-driven change in wildlife distribution and numbers: a case study of koalas in south west Queensland. *Wildlife Research* 38: 509–524.
89. CSIRO (2007) Climate change in Australia - Technical Report 2007. Canberra: Commonwealth Scientific and Industrial Research Organisation.
90. Cai W, Crimp S, Jones R, McInnes KL, Durack PJ, et al... (2005) Climate change in Queensland under enhanced greenhouse conditions: Report, 2004–2005.: CSIRO Marine and Atmospheric Research report for the Queensland Government.
91. IPCC (2007) Climate Change 2007: The Physical Science Basis. Contribution of Working Group I to the Fourth Assessment Report of the Intergovernmental Panel on Climate Change - Summary for Policymakers.
92. Bureau of Meteorology (2013) Maps of average conditions. Canberra: Australian Government Bureau of Meteorology.
93. Sattler P, Williams R (1999) The conservation status of Queensland's bioregional ecosystems. Brisbane: Environmental Protection Agency, Queensland Government.
94. Thackway R, Cresswell ID (1995) An Interim Biogeographic Regionalisation for Australia: a framework for establishing the national system of reserves, Version 4.0. Canberra: Australian Nature Conservation Agency.

95. Smith AG, McAlpine C, Rhodes JR, Seabrook L, Baxter G, et al... (2012) At what spatial scales does resource selection vary? A case study of koalas in a semi-arid region.. Austral Ecology doi: 10.1111/j.1442-9993.2012.02396.x.
96. Malo AR, Nicholson SE (1990) A study of rainfall dynamics in the African Sahel using normalized difference vegetation index. *Journal of Arid Environment* 19: 1–24.
97. Davenport ML, Nicholson SE (1993) On the relation between rainfall and the normalized vegetation index for diverse vegetation types in East Africa. *International Journal of Remote Sensing* 14: 2369–2389.
98. Karabulut M (2003) An examination of relationships between vegetation and rainfall using maximum value composite AVHRR-NDVI data. *Turkish Journal of Botany* 27: 93–101.
99. StataCorp (2011) Stata Statistical Software: Release 12. College Station, TX: StataCorp LP.
100. Henderson CR (1975) Best linear unbiased estimation and prediction under a selection model. *Biometrics* 31: 423–447.
101. Moberg GP (1985) Animal Stress; Moberg GP, editor. Bethesda, Maryland: American Physiological Society.
102. Liptrap RM (1993) Stress and reproduction in domestic animals. *Annual New York Academic Science* 697: 275–284.
103. Dobson H, Smith RF (1995) Stress and reproduction in farm animals. Journal of Reproductive Fertility Supplement 49: 451–461.
104. Millspaugh JJ, Washburn BE (2004) Use of fecal glucocorticoid metabolite measures in conservation biology research: considerations for application and interpretation. *General and Comparative Endocrinology* 138: 189–199.
105. Busch SD, Hayward LS (2009) Stress in a conservation context: A discussion of glucocorticoid actions and how levels change with conservation-relevant variables. *Biological Conservation* 142: 2844–2853.
106. Stafford Smith M, McAllister RRJ (2008) Managing arid zone natural resources in Australia for spatial and temporal variability - an approach from first principles. *Rangeland Journal* 30: 15–27.
107. Munks SA, Corkrey R, Foley WJ (1996) Characteristics of Arboreal Marsupial habitat in the semi-arid woodlands of northern Queensland. *Wildlife Research* 23: 185–195.

108. Davies N, Gramotnev G, Seabrook L, Bradley A, Baxter G, et al... (2013) Movement patterns of an arboreal marsupial at the edge of its range: a case study of the koala. Movement Ecology 1.
109. Melzer A, Lamb D (1996) Habitat utilisation by a central Queensland koala colony. In: Gordon G, editor. Koalas: research for management. Brisbane, Queensland: World Koala Research. 17–22.
110. Rangel-Negrin A, Alfaro JL, Valdez RA, Romano MC, Serio-Silva JC (2009) Stress in Yucatab spider monkeys: effects of environmental condition on fecal cortisol levels in wild and captive populations. *Animal Conservation* 12: 496–502.
111. Western D, Lindsay WK (1984) Seasonal herd dynamics of a savanna elephant population. *African Journal of Ecology* 22: 229–244.
112. Spinage CA (1994) Elephants. London: T & A D Poyser.
113. Kataria N, Kataria AK (2004) Use of blood analytes in assessment of stress due to drought in camel. *Journal of Camel Practice and Research* 11: 129–133.
114. Australian Soil Resource Information System (2011) Atlas of Australian Soils. Australia: CSIRO.
115. Department of Environment and Resource Management (2012) Soils in Queensland. Queensland Government.
116. Gary JM, Murphy BW (2002) Predicting Soil Distribution, Joint Department of Land and Water Conservation (DLWC) and Australian Society for Soil Science Technical Poster. Sydney: DLWC.
117. Wu Y, McAlpine CA, Seabrook LM (2012) The dietary preferences of koalas, *Phascolarctos cinereus*, in southwest Queensland under wet winter conditions preceded by a severe drought. *Australian Zoologist* 36: 93–102.
118. Huber S, Palme R, Arnold W (2003) Effects of season, sex, and sample collection on concentrations of fecal cortisol metabolites in red deer (*Cervus elaphus*). *General and Comparative Endocrinology* 130: 48–54.
119. Kataria N, Kataria AK (2006) Endocrine and metabolic responses of Marwari sheep in arid tract. *Slovenian Veterinary Research* 43: 135–142.
120. Crowther MS, Lunney D, Lemon J, Stalenberg E, Wheeler R, et al. (In press) Climate-mediated habitat selection in an arboreal folivore. Ecography.

121. Degabriele R, Dawson TJ (1979) Metabolism and heat balance in an arboreal marsupial, the koala. *Journal of Comparative Physiology B* 134: 293–301.
122. Cavigelli SA (1999) Behavioural patterns associated with faecal cortisol levels in free-ranging female ring-tailed femurs, *Lemur catta* . *Animal Behavior* 57: 935–944.
123. Cavigelli SA, Monfort SL, Whitney TK, Mechref YS, Novotny M, et al. (2005) Frequent serial fecal corticoid measures from rats reflect circadian and ovarian corticosterone rhythms. *Journal of Endocrinology* 184: 153–163.
124. Goymann W, East ML, Wachter B, Honer OP, Mostl E, et al. (2001) Social, state-dependent and environmental modulation of faecal corticosteroid levels in free-ranging female spotted hyenas. *Proceedings of the Royal Society of London, B Biological Science* 268: 2453–2459.
125. Möstl E, Maggs JL, Schrötter G, Besenfelder U, Palme R (2002) Measurement of cortisol metabolites in faeces of ruminants. *Veterinary Research Communications* 26: 127–139.
126. Young IR, Renfree MB, Mesiano S, Shaw G, Jenkin G, et al... (2011) The Comparative Physiology of Parturition in Mammals: Hormones and Parturition in Mammals. In: Norris DO, Lopez KH, editors. Hormones and Reproduction of Vertebrates, Volume 5 - Mammals. USA: Academic Press.
127. Boonstra R (2013) The ecology of stress: a marriage of disciplines. *Functional Ecology* 27: 7–10.
128. Gordon G, McGreevy D, Lawrie B (1990) Koala populations in Queensland: major limiting factors. In: Lee AK, Handasyde KA, Sanson GD, editors. Biology of the Koala. Chipping Norton, NSW Australia: Surrey Beatty. 85–95.
129. Hanger J, Loader J (2009) Infectious disease in koalas: implications for conservation. Paper presented at the Lismore Koala Conservation Conference. Lismore.
130. Cork SJ, Braithwaite LW (1996) Resource availability, eucalypt chemical defences, and habitat quality for leaf eating marsupials. In: Gordon G, editor. Koalas: Research for Management Brisbane Koala Symposium 22–23 September 1990. Brisbane: World Koala Research. 9–16.
131. Moore BD, Foley WJ (2000) A review of feeding and diet selection in koalas (*Phascolarctos cinereus*). *Australian Journal of Zoology* 48: 317–333.

SECTION 7: WILD ANIMALS

7.6. RODENTS

CHAPTER 23

Haematology and Serum Biochemistry Parameters and Variations in the Eurasian Beaver (Castor Fiber)

Simon J. Girling[1], Roisin Campbell-Palmer[2,3], Romain Pizzi[1], Mary A. Fraser[4], Jonathan Cracknell[5], Jon Arnemo[6,7], Frank Rosell[3]

[1] Veterinary Department, Living Collection, Royal Zoological Society of Scotland, Edinburgh, United Kingdom

[2] Faculty of Arts and Sciences, Department of Environmental and Health Studies, Telemark University College, Bø i Telemark, Norway

[3] Conservation Programmes, Royal Zoological Society of Scotland, Edinburgh, United Kingdom,

[4] School of Life, Sport and Social Science, Edinburgh Napier University, Sighthill Campus, Edinburgh, United Kingdom

[5] Longleat Safari and Adventure Park, Longleat, Warminster, Wiltshire, United Kingdom

Citation: Girling SJ, Campbell-Palmer R, Pizzi R, Fraser MA, Cracknell J, Arnemo J, et al. (2015) Haematology and Serum Biochemistry Parameters and Variations in the Eurasian Beaver (Castor fiber). PLoS ONE 10(6): e0128775. https://doi.org/10.1371/journal.pone.0128775.

[6] Department of Forestry and Wildlife Management, Faculty of Applied Ecology and Agricultural Sciences, Hedmark University College, Campus Evenstad, Koppang, Norway

[7] Department of Wildlife, Fish and Environmental Studies, Faculty of Forest Sciences, Swedish University of Agricultural Sciences, Umeå, Sweden

ABSTRACT

Haematology parameters (N = 24) and serum biochemistry parameters (N = 35) were determined for wild Eurasian beavers (*Castor fiber*), between 6 months – 12 years old. Of the population tested in this study, N = 18 Eurasian beavers were from Norway and N = 17 originating from Bavaria but now living extensively in a reserve in England. All blood samples were collected from beavers via the ventral tail vein. All beavers were chemically restrained using inhalant isoflurane in 100% oxygen prior to blood sampling. Results were determined for haematological and serum biochemical parameters for the species and were compared between the two different populations with differences in means estimated and significant differences being noted. Standard blood parameters for the Eurasian beaver were determined and their ranges characterised using percentiles.

Whilst the majority of blood parameters between the two populations showed no significant variation, haemoglobin, packed cell volume, mean cell haemoglobin and white blood cell counts showed significantly greater values ($p<0.01$) in the Bavarian origin population than the Norwegian; neutrophil counts, alpha 2 globulins, cholesterol, sodium: potassium ratios and phosphorus levels showed significantly ($p<0.05$) greater values in Bavarian versus Norwegian; and potassium, bile acids, gamma globulins, urea, creatinine and total calcium values levels showed significantly ($p<0.05$) greater values in Norwegian versus Bavarian relict populations. No significant differences were noted between male and female beavers or between sexually immature (<3 years old) and sexually mature (≥3 years old) beavers in the animals sampled. With Eurasian beaver reintroduction encouraged by legislation throughout Europe, knowledge of baseline blood values for the species and any variations therein is essential when assessing their health and welfare and the success or failure of any reintroduction program. This is the first study to produce base-line blood values and their variations for the Eurasian beaver.

INTRODUCTION

Eurasian beavers (*Castor fiber*) are large (adults >20 kg), herbivorous, semi-aquatic, territorial, monogamous rodents that live in small family groups and reach sexual maturity around 3 years of age [1]. The Eurasian beaver was previously widespread throughout fresh water habitats of Europe and Asia [2]. However, by the 20th century the species had been reduced to approximately 1,200 individuals in eight isolated populations, largely through hunting [3–5].

Conservation efforts, including hunting bans and reintroduction programmes have seen beaver numbers recover throughout most of its former range [3]. Over 200 translocations have been recorded, across 25 European countries and current population figures estimate > 1.04 million animals [6]. Previous genetic studies suggest that the current Eurasian beaver population can be divided into two evolutionary significant units (ESU)—the Eastern European and Western European populations [7]. Bavarian beavers are thought to derive from a mixture of the Eastern and Western European ESU; Norwegian beavers are a relict population from the Western European ESU [8, 9]. More recent genetic analysis suggests that such variations represent the two extremes of one species [10].

Current EU legislation promotes reintroduction of beavers and other pre-existing species across their former European ranges (Council Directive 92/43/EEC, 'Habitats Directive'). With this legislation in place several European countries have started or completed the process of reintroducing Eurasian beavers. In the UK (Scottish Beaver Trial, Knapdale, Scotland, UK) a pilot project has been completed to assess the viability of reintroducing the Eurasian beaver. As part of any responsible reintroduction program or trials, pre-release health assessments, and post release health assessments are essential to ensure health and welfare legislation is complied with. One important method of assessment of health in any animal is to assess haematological and biochemical parameters [11]. It also provides one means by which the suitability of a release location and the level to which the released animals are coping in their habitat can be assessed. Blood haematology and serum chemistry have been used in health assessments for a range of wild and domestic animals and so can be an essential part of investigating the success or failure of reintroduction programs [12, 13]. To develop accurate reference values for haematological and biochemical values, baseline data must first be collected from healthy, free ranging populations [14]. Once this has been established, the health of other populations or individuals can be evaluated by comparing their blood parameters against the standard healthy, wild population.

To date, haematological and serum biochemistry parameters for Eurasian beavers have not been published. Haematological and serum biochemistry values have, however, been published for captive North American beavers (*C. canadensis*) [15–19].

This study is the first to provide baseline haematological and serum biochemistry parameters for the Eurasian beaver and to compare across two beaver sub-populations. This information may assist health monitoring and assessment of captive, reintroduced and translocated individuals and is of particular relevance at this time with the proposed reintroduction of the Eurasian beaver to Britain. It will also enable health assessment comparisons between other wild and captive populations of beaver.

MATERIAL AND METHODS

This study was approved by the Norwegian Animal Research Authority, Norwegian Environment Agency and the Royal Zoological Society of Scotland internal ethics committee. All animals were sampled as part of routine health assessments or in the process of surgical neutering, under general anaesthesia and by qualified veterinary surgeons ensuring preservation of welfare and prevention of any suffering as detailed below.

Study Area and Animals

Relict Norwegian Beavers

The relict Norwegian beavers used in this study were part of a wild population living in the Telemark region, Norway. Beavers were caught from two main rivers, the Gvarv and the Lunde in August 2010. In total 18 beavers were caught as part of a health survey undertaken by Telemark University College and the Royal Zoological Society of Scotland. This was comprised of 10 males and 8 females with an age range of 6 months to 12 years. Of these beavers, 5 were considered sexually immature (<3 years old) and 13 sexually mature (≥ 3 years old).

Bavarian Beavers

The Bavarian beavers in this study were part of a family group living in an extensive but enclosed area covering 20 hectares in southern England in September 2011. In total 17 beavers were caught as part of a health assessment and reproductive control procedure. This was comprised of 9 males and 8 females with an age range of 6 months to 12 years. Of these beavers, 9 were considered sexually immature (<3 years old) and 8 were considered sexually mature (≥3 years old).

Sampling

All beavers were live-trapped through netting from a boat as described in Rosell and Hovde [20]. After capture each beaver was fully examined for any signs of poor health or disease.

Blood samples for this study were only collected from clinically healthy individuals. Each beaver was morphometrically measured and sex was determined on basis of colour and viscosity of anal gland secretion [21]. Age class (sexually immature i.e. <3 years or sexually mature i.e. ≥3 years old) was determined through records (either from captive collection records of from data collected as part of long term-study) if known or age class estimated from body size and weight.

Chemical Restraint

The time from capture to induction of anaesthesia for both groups was between 30–90 minutes. All beavers were anaesthetised using 4% isoflurane in 100% oxygen via a face mask and maintained on 1.5–2% isoflurane in 100% oxygen for blood sample collection.

Blood sample collection

All blood samples were collected from chemically immobilised animals. Beavers were positioned in dorsal recumbancy and blood samples were obtained from the ventral coccygeal blood vessels, 2–4 cm from the base of the tail as described by Greene *et al*. [22] and Eisele *et al*. [23].

Blood for serum biochemistry analysis was collected in 10 ml plain vials then the serum separated within two hours of collection after centrifugation at 10,000 rpm for 3 minutes. Where immediate analysis was not possible, serum was then frozen at -80°C until biochemical assays were undertaken. Blood for haematology analysis was collected in 5 ml vials containing potassium EDTA as anticoagulant and a fresh blood smear made for cytological analysis. All haematology samples were analysed within 1–2 days of blood collection after refrigerated storage.

Haematology and Biochemistry

Haematological parameters were measured in 24 Eurasian beavers (leucocyte count, erythrocyte count, haemoglobin levels, mean cell volume, mean cell haemoglobin levels, mean cell haemoglobin concentrations and platelet count) using a Beckman-Coulter AcT 5 Diff. haematology analyser according to manufacturer's guidelines (Beckman-Coulter UK Ltd). Differential white cell counts were performed manually based on a blood smear assessment of 200 leucocytes.

Total proteins were measured in 35 Eurasian beavers using the Biuret method with albumin, alpha, beta and gamma globulin levels being determined using agarose gel electrophoresis on a Helena Laboratories Rapid Electrophoresis analyser according to manufacturer's guidelines (Helena Laboratories, Texas). All other biochemical parameters were analysed using a Randox Imola clinical chemistry analyser according to manufacturer's guidelines (Randox Laboratories Ltd. London) in a total of 35 Eurasian beavers.

The differences in number of beavers tested between haematological parameters (N = 24) and biochemical parameters (N = 35) was due to rapid clotting of beaver blood in potassium EDTA anticoagulant making haematological testing in 11 of the beavers sampled impossible. The makeup of the beavers sampled for haematology was for Norwegian group 7 males and 5 females, and for the Bavarian group 6 males and 6 females. Both groups covered the complete age range of 6 months to 12 years

Statistical Analysis

For each variable the mean, median, standard deviation and 90% reference range were derived to produce overall parameters for the Eurasian beaver. In addition, Welsh's two-sample unequal variance t-test was used to compare the Norwegian and Bavarian populations and results of the 95% confidence interval test for differences in means were obtained using Minitab software version 16.2.3 and NCSS version 7 for bootstrap estimation. These differences were expressed as standardised effect sizes using the pooled standard deviation of the variable. Within the population sampled, Welsh's two-sample unequal variance t-test was also used to compare male Eurasian beavers with female Eurasian beavers and to compare those under 3 years of age (sexually immature) with those over or equal to 3 years of age (sexually mature).

RESULTS

Haematology

Haematological parameters for Eurasian beavers are shown in Table 1. Comparisons of Norwegian and Bavarian beavers are shown in Table 2.

Table 1. Overall range, mean, standard deviation (SD) and median for each haematological parameter for the Eurasian beaver (combined populations)

Parameter	5th-95th percentile	Mean	SD	Median
Red cell count ($x10^{12}/L$)	3.30–4	3.69	0.25	3.7
Haemoglobin (g/dL)	11.83–14.29	12.98	0.98	12.85
Packed cell volume (L/L)	0.33–0.41	0.37	0.03	0.37
Mean cell volume (fl)	91–106	99.71	5.74	100.5
Mean cell haemoglobin (pg)	32.15–38	35.25	2.05	35
Mean cell haemoglobin concentration (g/dL)	34.15–36	35.25	0.60	35
Platelets ($x10^9/L$)	149.6–308.25	216.5	49.86	209
White cell count ($x10^9/L$)	6.87–13.43	10.45	2.09	10.80
Neutrophils segmented ($x10^9/L$)	4.15–9.56	7.5	1.72	7.75
Neutrophils banded ($x10^9/L$)	0–0.54	0.08	0.18	0
Lymphocytes ($x10^9/L$)	0.92–4.24	2.43	1.04	2.25
Monocytes ($x10^9/L$)	0.1–0.9	0.33	0.30	0.2
Eosinophils ($x10^9/L$)	0–0.2	0.08	0.09	0.1
Basophils ($x10^9/L$)	0	0	0	0

(N = 24).

Table 2. Summary of haematology values for relict Norwegian (N = 12) and Bavarian (N = 12) Eurasian beavers

Parameter	Norwegian range 5th-95th percentile	Bavarian range 5th-95th percentile	Norwegian mean (SD)	Bavarian mean (SD)	Difference in means (95% confidence interval)	Two sample T test p-value	Standardised effect size between groups
Red cell count ($x10^{12}/L$)	3.36–4.00	3.30–4.13	3.70 (0.230)	3.68 (0.298)	0.017 (-0.210, 0.243)	0.880	0.06
Haemoglobin (g/dL)	11.8–13.1	12.5–15.0	12.3 (0.473)	13.6 (0.968)	-1.30 (-1.96, -0.637)	0.001*	-1.71
Packed cell volume (L/L)	0.330–0.370	0.346–0.428	0.349 (0.0291)	0.385 (0.0144)	-0.0358 (-0.0557,-0.0160)	0.001*	-1.56
Mean cell volume (fl)	90.6–98.9	101.6–107.4	94.6 (3.18)	104.8 (2.12)	-10.2 (-12.6,-7.90)	0.000*	-3.77
Mean cell haemoglobin (pg)	32.0–35.0	35.6–38.0	33.5 (1.09)	37.0 (1.13)	-3.50 (-4.44, -2.56)	0.000*	-3.15
Mean cell haemoglobin concentration (g/dL)	35.0–36.0	34.0–36.0	35.2 (0.452)	35.2 (0.754)	0.00 (-0.53, 0.53)	1.000	0.00
Platelets ($x10^9/L$)	135.2–302.2	173.3–298.2	213.4 (58.3)	219.6 (44.8)	-6.2 (-50.4,38.1)	0.774	-0.12
White cell count ($x10^9/L$)	6.39–11.60	9.42–13.78	9.09 (1.76)	11.81 (1.55)	-2.72 (-4.12,-1.31)	0.001*	-1.64
Neutrophils segmented ($x10^9/L$)	3.83–8.94	7.32–9.66	6.50 (1.86)	8.50 (0.974)	-2.00 (-3.27,-0.73)	0.004*	-1.35
Neutrophils banded ($x10^9/L$)	0.000–0.645	0.000–0.000	0.167 (0.239)	0.000 (0.000)	-0.167 † (-0.283, -0.025)	0.034† *	-0.99
Lymphocytes ($x10^9/L$)	1.12–3.86	0.86–4.08	2.208 (0.995)	2.64 (1.12)	-0.433 (-1.334,0. 467)	0.328	-0.41
Monocytes ($x10^9/L$)	0.100–0.245	0.200–0.990	0.142 (0.067)	0.517 (0.355)	-0.375 (-0.592, -0.158)	0.003*	-1.47
Eosinophils ($x10^9/L$)	0.000–0.100	0.000–0.290	0.050 (0.0522)	0.1 (0.121)	-0.050 (-0.1314, 0.0314)	0.209	-0.54
Basophils ($x10^9/L$)	0.000–0.000	0.000–0.000	0	0	0	1	0.00

† denotes estimated by bootstrapping.

* significant at 5% level.

Significant differences ($p < 0.01$) were noted between the relict Norwegian and Bavarian beavers for haemoglobin levels, packed cell volume, mean cell volume, mean cell haemoglobin, white cell, and monocyte counts with significant differences with $p < 0.05$ for neutrophil (segmented and unsegmented) counts. In all of the above listed significant parameters the values for the Bavarian group were greater than those for the relict Norwegian group with the large standardised effect sizes (magnitude >3) being associated with mean cell volume and mean cell haemoglobin. There were no statistical significant differences in haematological parameters between different sexes ($p<0.05$) or between sexually immature (<3 years old) versus sexually mature (≥3 years old) ($p<0.05$) beavers sampled in this study.

Serum Biochemistry

Table 3 displays the serum biochemistry parameters determined for wild Eurasian beavers. Comparisons of Norwegian and Bavarian beavers are shown in Table 4.

Table 3. Overall ranges, means, standard deviations (SD) and medians for serum biochemistry parameters for the Eurasian beaver (combined populations)

Parameter	5th-95th percentile	Mean	SD	Median
Alanine transferase (IU/L)	28.7–49.9	36.83	7.28	36
Aspartate aminotransferase (IU/L)	60–93.2	74.8	12.59	73
Alkaline phosphatase (IU/L)	209.8–575.3	366.23	135.72	346
Bile acids (μmol/L)	3–6	3.54	1.13	3
Total bilirubin (μmol/L)	3.2–5.83	4.41	0.88	4.40
Total protein (g/L)	58.4–70	64.89	3.78	65
Albumin (g/L)	33.7–39	36.43	1.70	36
Alpha 1 globulin (g/L)	2.02–3.71	2.70	0.66	2.58
Alpha 2 globulin (g/L)	7.84–12.67	10.32	1.45	10.67

Beta globulin (g/L)	5.83–9.52	7.85	1.16	7.84
Gamma globulin (g/L)	4.57–10.22	7.38	1.81	7.28
Cholesterol (mmol/L)	1.87–3.86	2.71	0.65	2.70
Triglycerides (mmol/L)	0.45–1.71	0.99	0.49	0.96
NEFA (mmol/L)	740.7–1258.3	1006.89	164.67	986
Sodium (mmol/L)	129–140	133.57	3.37	133
Potassium (mmol/L)	3.87–5.53	4.44	0.58	4.20
Sodium:Potassium ratio	23.81–35.17	30.56	3.74	31.40
Chloride (mmol/L)	80.7–89	84.94	2.65	85
Urea (mmol/L)	3.14–9.06	5.99	2.02	6
Creatinine (μmol/L)	56.1–120.2	88.69	21.62	90
Total calcium (mmol/L)	2.17–2.8	2.54	0.19	2.60
Phosphorus (mmol/L)	1.20–5.22	2.88	1.65	1.50
Magnesium (mmol/L)	1.27–1.8	1.51	0.15	1.50

(N = 35).

Table 4. Summary of serum biochemistry values for relict Norwegian (N = 18) and Bavarian (N = 17) Eurasian beavers

Parameter	**Relict Norwegian Eurasian beaver range 5th-95th percentile**	**Bavarian Eurasian beaver range 5th-95th percentile**	**Relict Norwegian Eurasian beaver mean (SD)**	**Bavarian Eurasian beaver mean (SD)**	**Difference in means (95% confidence interval)**	**Two sample T test p-value**	**Standardised effect size between groups**
Alanine transferase (IU/L)	28.8-52.6	29.4–45.8	37.5 (8.30)	36.12 (6.45)	1.38 (-3.73,-6.49)	0.585	0.19
Aspartate aminotransferase (IU/L)	60.8–93.1	58.0–92.8	74.7 (12.8)	74.9 (13.2)	-0.27 (-9.21,8.66)	0.951	-0.02
Alkaline phosphatase (IU/L)	215.7–599.8	204.8–573.2	352.6 (156.0)	380.7 (118.0)	-28.1 (-123.3,67.0)	0.550	-0.20
Bile acids (μmol/L)	3.00–6.15	3.00–4.00	3.94 (1.47)	3.12 (0.332)	0.827 (0.077,1.576)	0.032*	0.78
Total bilirubin (μmol/L)	2.99–5.95	3.44–4.94	4.56 (1.14)	4.24 (0.52)	0.320 (-0.293, 0.933)	0.292	0.36
Total protein (g/L)	59.8–70.0	56.6–69.6	64.9 (3.61)	64.82 (4.17)	0.12 (-2.57, 2.82)	0.928	0.03

Albumin (g/L)	33.8–39.0	33.8–38.2	36.7 (1.87)	36.1 (1.54)	0.605 (-0.572,1.781)	0.303	0.35
Alpha 1 globulin (g/L)	2.05–4.49	1.83–3.06	2.86 (0.80)	2.53 (0.45)	0.325 (-0.122,0.773)	0.148	0.50
Alpha 2 globulin (g/L)	7.65–11.14	10.1–12.8	9.35 (1.24)	11.34 (0.90)	-1.98 (-2.73,-1.24)	0.000*	-1.83
Beta globulin (g/L)	6.17–9.42	5.61–9.30	7.96 (1.27)	7.74 (1.10)	0.214 (-0.601,1.030)	0.596	0.18
Gamma globulin (g/L)	5.86–10.51	3.19–8.96	7.99 (1.75)	6.74 (1.74)	1.25(0.053,2.458)	0.041*	0.72
Cholesterol (mmol/L)	1.87–3.45	2.04–4.04	2.48 (0.56)	2.96 (0.68)	-0.487 (-0.016,-0.058)	0.027*	-0.78
Triglycerides (mmol/L)	0.51–1.55	0.44–1.51	1.03 (0.43)	0.94 (0.58)	0.083 (-0.268,0.434)	0.633	0.16
NEFA (mmol/L)	673.9–1353.1	956.0–1134.0	995.0 (25.0)	1019.2 (72.5)	-24 (-140.5, 92.5)	0.672	-0.44
Sodium (mmol/L)	129.0–138.4	129.0–140.0	132.9 (3.16)	134.3 (3.64)	-1.41(-3.76, 0.95)	0.233	-0.41
Potassium (mmol/L)	3.90–5.69	3.80–4.42	4.75 (0.67)	4.112 (0.21)	0.638 (0.291,0.986)	0.001*	1.29
Sodium:Potassium ratio	23.3–34.4	29.2–35.1	28.5 (4.02)	32.7 (1.87)	-4.24 (-6.41,-2.06)	0.000*	-1.35
Chloride (mmol/L)	79.8–88.1	81.0–89.0	84.8 (2.65)	85.1 (2.80)	-0.340 (-2.220,1.540)	0.715	-0.12
Urea (mmol/L)	5.47–10.01	2.88–7.02	7.35 (1.63)	4.56 (1.35)	2.791 (1.762,3.820)	0.000*	1.86
Creatinine (µmol/L)	89.5–123.4	51.8–86.2	106.3 (12.1)	70.0 (12.0)	36.33 (28.01, 44.66)	0.000*	3.01
Total calcium (mmol/L)	2.57–2.80	2.10–2.62	2.68 (0.11)	2.39 (0.16)	0.284 (0.187,0.380)	0.000*	2.07
Phosphorus (mmol/L)	1.20–1.50	3.28–5.54	1.35 (0.10)	4.51 (0.71)	-3.156 (-3.523,-2.789)	0.000*	6.22
Magnesium (mmol/L)	1.29–1.72	1.28–1.80	1.506 (0.147)	1.506 (0.171)	0.000 (-0.111, 0.110)	0.996	0.00

* denotes significant at 5% level.

Significant differences ($p<0.05$) were noted between the relict Norwegian and Bavarian Eurasian beaver groups for bile acids, alpha 2 globulins, gamma globulin, cholesterol, potassium, sodium:potassium ratio, urea, creatinine, phosphorus and total calcium. The values were significantly greater for the Bavarian group versus the relict Norwegian group for alpha 2 globulins, cholesterol, sodium:potassium ratio and phosphorus. The reverse was true for bile acids, gamma globulin, potassium, urea, creatinine and total calcium, in which Norwegian beavers displayed significantly greater values than Bavarian. Large standardised effect sizes (magnitude >3) were associated with phosphorus and creatinine. There were no statistical significant differences in biochemical parameters tested based on differences in the sex ($p<0.05$) or between sexually immature (<3 years old) versus sexually mature (≥3 years old) ($p<0.05$) beavers sampled in this study.

DISCUSSION

This study is the first to our knowledge to provide baseline information on the haematology and serum biochemistry parameters and their variations in the Eurasian beaver.

Kitts *et al.* [15] investigated haematology of the North American beaver and concluded their results were of comparative values to other rodents. However, North American beavers did possess a lower red cell count than other rodents. The average diameter of the North American beaver erythrocyte is slightly larger than expected from other mammalian species, thought to be due to the semi-aquatic nature of this genus [15]. In the case of this study, the Eurasian beaver mean cell volumes are nearly double that of non-aquatic rodents such as rats (*Rattus norvegicus*) (upper range 64fl) and mice (*Mus musculus*) (upper range 57fl) [24] and are comparable with previously published volumes in North American beavers (83.2–118.8fl [15,19]) and in nutria (*Myocastor coypus*) another semi-aquatic rodent (84–102.5fl [25]). Red cell counts may be elevated in dehydrated animals, and reduced in those demonstrating blood loss or other causes of anaemia.

Whilst many parameters were found to be similar for both groups of beavers it is interesting to note that there are significantly different haematological results ($p<0.05$) between the relict Norwegian and Bavarian Eurasian beavers for some parameters. Haemoglobin levels have been shown to vary with the time of year in North American beavers, being higher in the autumn than the summer [15]. In this study, blood samples were collected at the same time of year ruling this out as a possibility for differences between the two location groups. Values for haemoglobin from both sets of Eurasian beavers in this study overlapped those of Kitts *et al.* [15] and were comparable to ranges for most other rodent species [26]. Elevated haemoglobin levels are often associated with dehydration and reduced levels with anaemia and liver disease.

Packed cell volumes (PCV) were slightly lower in the Eurasian beavers in this study compared with published ranges for the North American beaver (overall range 0.33–0.43 versus North American beavers of 0.4–0.46 [15]) or nutria (0.39–0.45 [25]). This indicates a lower cellularity of the blood although the total number of red cells for both populations of Eurasian beavers were within the lower end of the reported North American species range (3.7-5x10^{12}/L [15]) and reported nutria range (3.9-6x10^{12}/L [25]) suggesting this is related to the differences in mean cell volume seen. PCV, mean cell volumes (MCV) and mean cell haemoglobin concentration (MCHC) were also significantly greater in the Bavarian versus the relict Norwegian Eurasian beavers. It has long been known that these parameters vary between different species so it is interesting to note these differences within a species and may suggest that there are population differences between the relict Norwegian and Bavarian Eurasian beavers [27]. The age of the animal can also increase MCV in other species with very young animals often possessing larger MCVs [27]. However the age distribution for both populations was similar with no neonates included in the study and no evidence of significant differences were noted in this parameter between sexually immature and sexually mature animals. Iron deficiency anaemia can lead to a decrease in MCV, MCHC and haemoglobin levels [27]; but no clinical anaemia was noted in any individual included in this study suggesting that this also could not account for the differences.

White blood cell counts were again comparable to that previously published in North American beavers (7–18.2x10^{9}/L [15]) and nutria (9.9–13.1x10^{9}/L [25]). Between the two location groups in this study there was again a significant difference with the Bavarian group demonstrating a greater white cell count than the relict Norwegian. It should be noted that in some rodents such as rats and mice, significant diurnal variation is seen in leukocyte levels (increases during the light phase and decrease during the dark phase) but it is not known whether the same occurs in beavers [28]. Other causes for variation in white cell counts include infection, different handling techniques, neoplasia and stress [27]. Bacterial and fungal disease usually result in an overall increase in leukocyte levels and a left shift towards band forms of neutrophils with an increase in the neutrophil and in chronic cases of disease, monocyte counts. Immune stimulation and viral diseases may result in an increase in lymphocyte counts, although chronic antigenic depletion may also be seen with a reduction in lymphocyte numbers. Stress may result in an overall increase in the leucocyte, neutrophil and monocyte count but a relative reduction in the lymphocyte count (lymphopaenia). The process for collection was the same in both groups, so it is unlikely that stress or handling techniques could account for the differences. Blood borne neoplasia could also be ruled out as no evidence of neoplastic cells could be seen. Infection however is more difficult to rule out based on purely a clinical examination and it is highly possible that the 'left shift' seen with band forms of neutrophils in individuals in the Norwegian beavers may indicate a low level of inflammatory disease. This is however contrary to the overall white cell count which was actually lower in the relict Norwegian group. The population of monocytes was also significantly greater in the Bavarian group. Monocytosis may be seen in chronic active inflammatory disease, in conjunction with a leucocytosis and often neutrophilia [27]. Clinically, this was not apparent in individuals of this group but could account for this variation as subclinical disease, or it may again be a genuine variation between the two location groups.

Total protein levels in Eurasian beavers in this study fall within the ranges published for North American beavers, although as with albumin levels seasonal variations were noted in the North American beaver study which cannot be assessed in this study [16]. In comparison with nutria they were at the lower end of their range (nutria total protein ranges 63–89g/L [25]). Blood proteins play an important role in maintaining overall oncotic potential preventing significant extravasation of fluid from the circulation and so again it is not surprising that there is no significant variation between the two location groups for total proteins or the overall levels of the globulins or albumin. However within the different subset of globulins analysed, one significant difference was observed in the alpha 2 globulins with the Bavarian Eurasian beaver group demonstrating higher levels than the relict Norwegian Eurasian beaver group. Alpha 2 globulins are manufactured in the liver and include: alpha 2 macroglobulin whose levels may be raised with the nephrotic syndrome, nephritis, acute inflammatory disease and chronic active hepatitis; and very low density lipoproteins which may also be elevated with the nephrotic syndrome, diabetes mellitus and hypothyroidism [12]. These acute phase proteins found in the alpha and beta globulin sections of the protein electrophoretogram can be used therefore to assess a non-specific inflammatory response to disease or trauma or organ failure and so are useful in assessing animal health. Gamma globulins contain the main immunoglobulin IgG in mammals and may be used to assess humeral immunity stimulation by an antigen such as a virus and so can be useful in assessing disease exposure [12]. Values here were similar between the two subgroups.

Creatinine and urea are often used as an indicator of renal function in mammals with elevations in both parameters being seen in the latter stages of renal failure [12, 29]. It is interesting to note that there is a statistically significant difference between the two location groups studied here with the relict Norwegian group being greater than the Bavarian for both parameters. This is an important feature to note when attempting to interpret results in any renal health assessment. Values are similar to published values for nutria for creatinine but lower than nutria for urea [25].

Stevenson *et al.* [16] study demonstrated high levels of serum cholesterol in North American beavers. The beavers in that study were wild caught North American beavers but then retained in captivity for the study and so these high cholesterol levels may reflect the captive diet rather than the true cholesterol readings of wild North American beavers. In this study, cholesterol levels for Eurasian beavers were largely within the range published for many rodents including nutria [25, 26, 29]. There is a difference between the two location groups studied here with the Bavarian group being greater than the relict Norwegian at $p<0.05$, but not at $p<0.01$. This could be a reflection of the different herbage present in the two locations as cholesterol levels can be influenced by diet in animals with access to high levels of saponins (e.g. as found in clovers and members of the pea family) these have been shown to reduce cholesterol absorption levels. Genetic differences may also alter cholesterol levels as is seen in humans and other species [30]. Interestingly, triglyceride and NEFA levels were not significantly different at $p<0.05$ or $p<0.01$. NEFA and triglyceride levels are similar to those reported in other rodents and are directly associated with dietary levels maybe suggesting again a genetic difference in the cholesterol levels seen between the two groups [26, 29, 30].

Elevations in liver leakage enzymes alanine transferase (ALT) and aspartate transferase (AST) indicate hepatocellular damage in many mammals [12]. In rodents, AST is more liver specific but is also found in skeletal muscle [29]. Values for both enzymes were similar to those published for a wide range of rodent species [31] and slightly lower than that published for nutria [25]. Bile acid levels are often used as a liver function test with fasting values of <20umol/L being quoted as normal for most mammals [32]. Elevations in fasting animals indicate liver dysfunction [32]. Values here were clearly unlikely to be fasting results, but were nonetheless low with no difference between the two location groups tested. In rodents bilirubin is the end product of haemoglobin metabolism, and so when elevated can be used to indicate liver damage or haemolysis [29]. Values determined here were similar to those reported for other rodents [31].

Alkaline phosphatase (ALKP) is a leakage enzyme found in bone, the intestines and the liver predominantly [32]. Elevations can therefore occur due to bone, intestinal or liver disease, or due to bone growth in young animals. Values for the Eurasian beaver in this study were similar to those reported for other rodents [31] including nutria [25]. Electrolytes sodium and chloride values are also similar to that reported for nutria and other rodents however potassium levels varied significantly between the two location groups studied with the relict Norwegian group being greater than the Bavarian Eurasian beaver group. In all mammals, physiologically normal values for potassium do not exceed 7.5mmol/L as cardiac dysfunction occurs, or become less than 2.5mmol/L due to muscular dysfunction [33]. The values in this study fit within this range. It is possible with haemolysis for elevation in potassium levels to occur, and serum has higher levels of potassium than plasma. However handling and processing of the samples was the same for both groups making these less likely reasons. Potassium loss due to renal or gastrointestinal disease is unlikely as no other signs of disease were noted in the animals included in this study. Variations in the intake of potassium due to differing diets however is a more likely possibility.

Diet may also account for the significantly increased total calcium and lower phosphate levels in the relict Norwegian versus the Bavarian Eurasian beavers. Calcium:phosphorus ratios often follow an inversely proportionate relationship and again this may account for the results seen here. Calcium and phosphate levels reported here are within the range for other rodents and mammals [31]. Elevated phosphorus levels may also be seen in cases of advanced renal failure [29]. Lowered calcium levels may be seen in diets deficient in calcium, high in phosphorus or containing compounds such as oxalates (e.g. beet leaves) or phytates (e.g. grains) which bind ionised calcium in the gut and prevent its absorption.

Magnesium is an important micromineral aiding in the homeostatic mechanisms which maintain serum calcium and potassium levels with deficits often resulting in seizures, weakness, and unresolving hypocalcaemia and hypokalaemia [30]. Results here tally with the narrow physiological range from a variety of other mammals [30].

The data from this study could be used to assess and monitor the health of captive animals and other wild populations. This will have significant effects when assessing the compatibility of reintroduced beavers to new habitats. Currently in the UK this is of even greater significance as not only has a recent trial reintroduction been completed in Knapdale, Scotland, but wild-living populations of beavers have now been found throughout in other parts of Britain, most notably in South-West England and Scotland, numbering in their hundreds in some cases. The beaver has been shown to be a potential intermediate host for significant pathogens such as *Giardia* spp., *Franciscella tularensis* and the tapeworm *Echinococcus multilocularis* and as such assessing the health of any wild, captive or beavers of unknown origin is important for human health as well as for reintroduction assessments [34, 35]. Changes in blood chemistry values can be used in the diagnosis of diseases [29]. However, when establishing blood parameters for wild populations, capture methods must be taken into account as these can cause significant differences between reference values [36, 37] as will diet and other environmental factors [29].

Further studies should sample a larger number of individuals, from both wild populations and captive collections across Europe. The influence of habitat condition (e.g. free-living vs. captive) and habitat quality is another area requiring further research. Correlating blood parameters with known survival outcomes could generate fitness predictors for demographic class or populations.

ACKNOWLEDGMENTS

We would like to acknowledge the Royal Zoological Society of Scotland, the Scottish Beaver Trial and Telemark University College for technical support of this research. Phoebe Carter for her help arranging the Bavarian beaver sampling and Fiona Howie MRCVS of the Scottish Agricultural College for help with the blood analysis. Christian Robstad, Robert Needham and Helga Veronica Tinnesand for their assistance in the field. We would also like to thank Professor George Gettinby for his help in preparing the statistical analysis. We would also like to thank the editors and reviewers of PLOS One for their help in preparing this manuscript.

AUTHOR CONTRIBUTIONS

Conceived and designed the experiments: SG RCP RP. Performed the experiments: SG RCP RP MF JC JA FR. Analyzed the data: SG RCP MF. Contributed reagents/materials/analysis tools: SG RCP RP JC JA FR. Wrote the paper: SG RCP MF.

REFERENCES

1. Novak M. Beaver In: Novak M, Baker JA, Obbard ME, Malloch B,editors. *Wild furbearer management and conservation in North America*. Ontario: Ministry of Natural Resources; 1987. p. 283–312.
2. Zharkov IV, Sokolov VE. The European beaver (*Castor fiber* Linnaeus,1758) in the Soviet Union. *Acta Theriol*. 1967;12:27–46.
3. Nolet BA, Rosell F. Comeback of the beaver *Castor fiber*: An overview of old and new conservation problems. *Biol Conserv*. 1998;83(2):165–173.
4. Halley DJ, Rosell F. The beaver's reconquest of Eurasia: status, population development and management of a conservation success. *Mamm Rev*. 2002;32:153–178.
5. Halley DJ, Rosell F. Population and distribution of European beavers(*Castor fiber*). *Lutra*. 2003;46:91–101.
6. Halley DJ, Rosell F, Saveljev A. Population and distribution of Eurasian beavers (*Castor fiber*). *Baltic Forestry*. 2012;18:168–175.
7. Durka W, Babik W, Ducroz JF, Heidecke D, Rosell F, Samjaa R, et al. Mitochondrial phylogeography of the Eurasian beaver *Castor fiber* L. *Mol Ecol*. 2005; 14(12):3843–3856.
8. Rosell F, Campell-Palmer R, Parker H. More genetic data are needed before populations are mixed: response to 'Sourcing Eurasian beaver *Castor fiber* stock for reintroductions in Great Britain and Western Europe' *Mamm Rev*. 2012;42(4):319–324.
9. Halley DJ. Sourcing Eurasian beaver *Castor fiber* stock for reintroductions in Great Britain and Western Europe. *Mamm Rev*. 2011;41(1):40–53.
10. Senn H., Ogden R, Frosch C, Syruckova A, Campbell-Palmer R, Munclinger P, et al. Nuclear and mitochondrial genetic structure in the Eurasian beaver (*Castor* fiber)–implications for future reintroductions *Evol Appl*. 2014;7(6):645–662. 10.1111/eva.12162
11. Milner JM, Stien A, Justin Irvine R, Albon SD, Langvatin R, Ropstad E. Body condition in Svalbard reindeer and the use of blood parameters as indicators of condition and fitness. *Canad J Zool*. 2003;81:1566–1578.
12. Kaneko JJ, Harvey JW, Bruss ML. Appendices In: Kaneko JJ, Harvey JW, Bruss ML, editors. *Clinical biochemistry of domestic animals*. 6th edition San Diego California: Academic Press; 2008. p. 889–895

13. Thoresen SI, Arnemo JM, Liberg O. Hematology and serum clinical chemistry reference intervals for free-ranging Scandinavian gray wolves (*Canis lupus*). *Vet Clin Path*. 2009;38:224–229. 10.1111/j.1939-165X.2008.00105.x
14. Mašek T, Konjevic D, Severin K, Janicki Z, Grubešić M, Krapinec K, et al. Hematology and serum biochemistry of European mouflon (*Ovis orientalis musimon*) in Croatia. *Europ J Wildl Res*. 2009;55(6):561–566.
15. Kitts WD, Robertson MC, Stephenson B, Cowan IMcT. The normal blood chemistry of the beaver (*Castor canadensis*). A. Packed-cell volume sedimentation rate, haemoglobin, erythrocyte diameter, and blood cell counts. *Canad J Zool*. 1958;36:279–283.
16. Stevenson AB, Kitts WD, Wood AJ, Cowan IMcT. The normal blood chemistry of the beaver (*Castor canadensis*). B. Blood glucose, total protein, albumin, globulin, fibrinogen, non-protein nitrogen, amino acid nitrogen, creatine, creatinine, cholesterol, and volatile fatty acids. *Canad J Zool*. 1959;37:9–14.
17. Clark JD, Olfert OD. Rodents (Rodentia) In: Fowler ME, editor. *Zoo and wild animal medicine*. 2nd edition Philadelphia: WB Saunders; 1986. p. 727–748.
18. Bennet PM, Gascoyne SC, Hart MG, Kirkwood JK, Hawkey CM. Development of LYNX: a computer application for disease diagnosis and health monitoring in wild mammals, birds and reptiles. *Vet Rec*. 1991;128:496–499.
19. ISIS Reference values for captive North American beaver, *Castor canadensis*, both sexes and all ages combined (CD-ROM) St Paul Minnesota: ISIS; 2002.
20. Rosell F, Hovde B. Methods of aquatic and terrestrial netting to capture Eurasian beavers. *Wildl Soc Bull*. 2001;29:269–274.
21. Rosell F, Sun L. Use of anal gland secretion to distinguish the two beaver species *Castor canadensis* and *C.fiber* . *Wildl Biol*. 1999;5:119–123.
22. Greene SA, Keegan RD, Gallagher LRV, Alexander JE, Harari J.Cardiovascular effects of halothane anesthesia after diazepam and ketamine administration in beavers (*Castor canadensis*) during spontaneous or controlled ventilation. *Am J Vet Res*. 1991;52:665–668.

23. Eisele PH, Faith TL, Menth PM, Parker JC, Van Vuren DH. Ketamine isoflurane combination anesthesia for surgical implantation of intraperitoneal radio transmitters in the beaver. *Contemp Top Lab Anim Sci.* 1997;36:97–99.
24. Ness RD. Rodents In: Carpenter JW editor. *Exotic animal formulary*, 3rd Edition St Louis Missouri: Elsevier Saunders; 2005. p. 375–408.
25. Martino PE, Arauz SM, Anselmino F, Cisterna CC, Silvestrini MP, Corva S, Hozbor FA. Hematology and serum biochemistry of free ranging nutria (*Myocastor coypus*). *J Zoo and Wildl Med.* 2012;43:240–247.
26. Wesche P. Rodents: clinical pathology In: Keeble E, Meredith A, editors. *BSAVA Manual of rodents and ferrets* 4th edition Quedgeley Gloucester: BSAVA; 2009. p. 42–51.
27. Swenson MJ. Blood circulation and the cardiovascular system In: Swenson MJ editor. *Duke's physiology of domestic animals*, 10th Edition London: Cornell University Press; 1984. p. 15–40.
28. Pilny A. Clinical hematology of rodent species. *Vet Clin North Am: Exot Anim Pract.* 2008;11(3):523–533. 10.1016/j.cvex.2008.04.001
29. Jenkins JR. Rodent diagnostic testing. *J Exotic Pet Med.* 2008;17:16–25.
30. Nelson RW, Turnwald GH, Willard MD. Endocrine, metabolic and lipid disorders In: Willard MD, Tvedten H editors, *Small animal clinical diagnosis by laboratory methods* 4th Edition St Louis Missouri: Elsevier Saunders; 2004. p.165–207.
31. Mayer J (2013) Rodents In: Carpenter JW, Marion CJ editors. *Exotic animal formulary* 4th Edition St Louis Missouri: Elsevier Saunders; 2008. p. 476–516.
32. Willard MD, Twedt DC. Gastrointestinal, pancreatic and hepatic disorders In: Willard MD, Tvedten H, editors. *Small animal clinical diagnosis by laboratory methods* 4th Edition St Louis Missouri: Elsevier Saunders; 2004. p. 208–246.
33. DiBartola SP, Green RA, Autran de Morais HS, Willard MD. electrolyte and acid-base disorders In: Willard MD, Tvedten H, editors. *Small animal clinical diagnosis by laboratory methods* 4th Edition St Louis Missouri: Elsevier Saunders; 2004. p.117–134.
34. Barlow AM, Gottstein B, Mueller N. *Echinococcus multilocularis* in an imported captive European beaver (*Castor fiber*) in Great Britain. *Vet Rec.* 2011;169:339 10.1136/vr.d4673

35. Goodman G, Girling S, Pizzi R, Rosell F, Campbell-Palmer R. Establishment of a health surveillance program for the reintroduction of the Eurasian beaver (*Castor fiber*) into Scotland. *J Wildl Dis*. 2012;48:971–978. 10.7589/2011-06-153
36. Marco I, Lavín S. Effect of the method of capture on the haematology and blood chemistry of red deer (*Cervus elaphus*). *Res Vet Sci*. 1999;66:81–84.
37. Rosef O, Nystøyl HL, Solenes T, Arnemo JM. Haematological and serum biochemical reference values in free-ranging red deer (*Cervus elaphus atlanticus*). *Rangifer*. 2004;24:79–85.

CHAPTER 24

Parasitism and Physiological Trade-Offs in Stressed Capybaras

Ayelen T. Eberhardt[1], Sebastián A. Costa[1], M. Rocío Marini[2], Andrea Racca[1], Cecilia J. Baldi[1], M. Rosario Robles[3], Pablo G. Moreno[1], Pablo M. Beldomenico[1]

[1] Laboratorio de Ecología de Enfermedades, Instituto de Ciencias Veterinarias del Litoral, Universidad Nacional del Litoral - Consejo de Investigaciones Científicas y Técnicas (UNL - CONICET). Esperanza, Santa Fe, Argentina

[2] Cátedra de Patología Básica, Facultad de Ciencias Veterinarias, Universidad Nacional del Litoral. Esperanza, Santa Fe, Argentina

[3] Centro de Estudios Parasitológicos y de Vectores (CONICET) La Plata, Buenos Aires, Argentina

ABSTRACT

Parasites play a key role in regulating wildlife population dynamics, but their impact on the host appears to be context-dependent. Evidence

Citation: Eberhardt AT, Costa SA, Marini MR, Racca A, Baldi CJ, Robles MR, et al. (2013) Parasitism and Physiological Trade-Offs in Stressed Capybaras. PLoS ONE 8(7): e70382. https://doi.org/10.1371/journal.pone.0070382.

indicates that a synergistic interaction between stress, host condition and parasites is implicated in this phenomenon, but more studies are needed to better understand this context-dependency. With the goal to assess the net effect of two types of chronic stress on various host-parasite interactions, we conducted an experiment in capybaras to evaluate the impact of food restriction and physical restraint on the infection intensity of specific gastrointestinal nematodes and coccidia, and how these stressors affected the growth, body condition, and some immuno-physiological parameters. Our hypothesis was that both forms of stress would result in an alteration in the host-parasite interactions, with deteriorated condition and reduced immunological investment leading to high parasite burdens and vice versa. Stressed capybaras had significantly higher coccidia infection intensities; but among individuals that were smaller, those stressed consistently showed lower helminth burdens than controls. Both stress treatments had a marked negative impact on growth and body condition, but concomitantly they had a significant positive effect on some components of the immune system. Our results suggest, on the one hand, that during prolonged periods of stress capybaras preventatively invest in some components of their immunity, such as innate humoural defenses and cells that combat helminths, which could be considered a stress-dependent prophylaxis. On the other hand, stress was found to cause greater infection intensities of protozoans but lower burdens of nematodes, indicating that the relationship between stress, physiological trade-offs and infection depends on the type of parasite in question. Moreover, both findings might be related in a causal way, as one of the immunological parameters enhanced in stressed capybaras is associated with the immune response to control helminths.

INTRODUCTION

Mounting empirical evidence supports the notion that parasites play a key role in wildlife population dynamics [1]–[3]. Parasites are detrimental to the health of their hosts and can cause a decrease in their probability of survival and/or reproduction by exerting a specific pathogenic effect, extracting host's resources and/or inducing a nutritionally demanding immune response [4]–[6]. Due to this, parasites have considerable potential to regulate the growth of host populations in nature [7], which has been experimentally demonstrated for some host-parasite systems [2], [3]. However, the impact of a parasite on its host appears to be context-dependent [8].

Although the factors that determine a context that enhances parasite virulence are yet to be elucidated, some hypotheses have been suggested. Lochmiller and Deerenberg posited that nutrient limitation is amongst the most important environmental stressors that influence immunocompetence and, subsequently, population regulation [5]. Some endemic parasites may be only rarely pathogenic, but they could become important population regulators when hosts are stressed [8]–[10]. A few studies show that certain components of the immune system are enhanced by stress [11]. However, it is widely accepted that prolonged (chronic) stress decreases immune function through mechanisms involving the hypothalamic-pituitary-adrenal (HPA) axis and glucocorticoids, leaving individuals more susceptible to infection [12], [13]. It has been suggested that this stress-dependent vulnerability of the host is the mechanism by which parasites exert a control on their host populations, as hosts tend to be stressed and in poor condition (thus becoming more vulnerable to their parasites) when their densities are high [14]. In a recent study, Pedersen and Greives [3] demonstrated that the interaction between food availability, stress and parasites could drive oscillations in wild mouse populations. Furthermore, their study revealed a synergy between nutrient availability and parasitic infection. Beldomenico and Begon [14] suggested that such synergy might be caused by reciprocal effects between parasites and physiological condition. Individuals in poor condition are more susceptible to infections, which further weakens their condition and predispose them to even more severe infections, and so on, creating a vicious circle in which infection intensity is both a cause and consequence of the immune condition of the host. However, empirical data from multiple host-parasite systems are needed to enhance our understanding of the mechanisms by which parasites regulate host populations.

The capybara, *Hydrochoerus hydrochaeris* Linnaeus (Rodentia: Caviidae), the largest living rodent on earth, is one of the most intensely utilized wildlife species in South America [15]. Capybaras are hosts to a very rich parasite community, including several specific helminths and protozoans that show high prevalence and ubiquity [16], [17]. The nematodes most frequently reported in capybaras are *Strongyloides chapini* Sandground (Rhabditoidea, Strongyloididea), *Hydrochoerisnema anomalobursata* Arantes & Artigas (Trichostrongyloidea, Viannaiidae), *Viannella hydrochoeri* Travassos (Trichostrongyloidea, Viannaiidae), *Trychostrongylus axei* Cobbold (Trichostrongyloidea, Trychostrongylidae), *Protozoophaga obesa* Diesing (Oxyuroidea, Oxyuridae), *Trichuris* sp. Roederer (Trichinelloidea, Trichuridae) and *Echinocholeus hydrochaeri* Travassos (Trichinelloidea,

Trichinellidae). Among the protozoans, the most common coccidia are *Eimeira hydrochoeri* Carini, *E. trinidadensis* Casas, Duszynski and Zalles, *E. ichiloensis* Casas, Duszynski and Zalles, *E. boliviensis* Casas, Duszynski and Zalles and *E. araside* Gurgel, Sartori and Araújo [18]. Capybara population dynamics studies showed density-dependent effects on body mass gain, fecundity, survival of newborn and mortality of adults [19], but the involvement of parasites in these effects have not been investigated. Despite the large number of gastrointestinal parasites found in capybaras, no associated pathology has been described. Nevertheless, there are reports of negative associations between body condition and helminth intensity for *V. hydrochoeri* and the cestode *Monoecocestus macrobursatum* [20].

Previous studies have examined separately the impact of nutritional restriction and other forms of stress on growth [21], immunity [22], and parasites [9]. Here we assess these outcomes altogether to establish the net effect of two types of stress on host-parasite interactions, producing evidence that might shed some light on the underlying interactions and trade-offs. We evaluated the impact of food restriction and stress induced by capture and restraint (physical stress) on the parasitism intensity of specific gastrointestinal nematodes and coccidia, and measured how these stressors affected the growth, body condition, immunological investment and other physiological parameters of capybaras. Our hypothesis was that both forms of chronic stress alter host-parasite interactions, with deteriorated condition and reduced immunological investment leading to high parasite burdens and vice versa in a circular process, as in the vicious circle hypothesis mentioned above.

MATERIALS AND METHODS

Ethics Statement

The animal care and treatments used in this experiment were approved by the Bioethical Committee of the School of Veterinary Medicine of Universidad Nacional del Litoral (Permit Number: 36/09).

Animals and Enclosures

Capybaras are gregarious rodents in the suborder Histricognathi (cavy-like rodents) that inhabit South American wetlands. Adult weight, which is around 60 Kg of body mass, is reached after they are 2 years old, and their lifespan range from 10 to 14 years [23]. Females reach physiologic puberty between 12 and 18 months of age.

The experiment was carried out in specially designed enclosures built at the experimental zoological station "Granja la Esmeralda" in Santa Fe, Argentina. All enclosures were identical, measuring 7×3.5 m, with soil ground, each including a 1.75×1.50 m shelter, a tank for water, and half of its surface was covered by a cloth shade to provide protection from direct sunlight. The enclosures were built contiguously, creating a big rectangle, and a cloth shade covered the subdividing fences to prevent capybaras from seeing other enclosures. The capybaras that were used in the experiment were acquired from the commercial farm "Ayuí" (owned by heirs of Julio Cesar Storti), Santo Tomé, Corrientes, where they are bred and kept in semi-captivity until sold for their meat and pelt, and parasitism occurs naturally and is not controlled by antiparasitic drugs. To reduce the variability induced by sex and age on the studied variables, the individuals used in the study were all females between 6 and 12 months old at the beginning of the experiment. At the outset, the study began with 30 individuals (5 animals per enclosure), but soon three needed to be excluded due to different causes. One was mis-sexed, one died immediately after the translocation and another one escaped and could not be found for several days. The remaining 27 were allocated to the six enclosures through stratified random sampling to ensure an even distribution of initial body mass (which also determines that ages are distributed homogeneously, as body mass correlates very strongly with age during the first two years of age [24]). Half of the enclosures received five animals and the other half, four. Water was available *ad libitum* for the whole duration of the experiment. Veterinarians inspected the capybaras daily making sure that no clinical signs of disease were apparent. The only medical treatment that was administered during the experiment was the use of insecticide spray on a wound to prevent infestation by maggots (one individual, single application).

Acclimation and Baseline Comparisons

Before beginning the treatments, the animals were left to acclimate in their new environment for four weeks, during which they were fed *ad libitum* and were not subject to capture and physical restraint. The meals were administered twice a day and consisted of fresh alfalfa, hay (sorghum or maize) and a mixture of rice bran and rice meal. The food was provided on the ground, one lump per individual to avoid differential access to food due to social hierarchy. During these weeks the total daily consumption per capybara was determined to be 800 gr. of mixture of rice bran+meal, 300 gr. of sorghum or maize hay, and 500 gr. of fresh alfalfa.

Also, this 4-week acclimating period was used to carry out baseline comparisons that assured that the treatment groups were not different at the beginning of the experiment in terms of body mass and size, body mass index, and faecal parasite egg and oocyst counts. These comparisons were made using Kruskal-Wallis tests. For these baseline comparisons only, α was set at 0.1 to reduce the probability of a type II error. A statistically significant test indicated that a re-allocation into enclosures via a new stratified random sampling was needed.

Treatments

Three treatments were established: animals in two of the enclosures were fed a restricted diet (hereafter, food restricted group), others in two different enclosures were captured and physically restrained three times a week (physically stressed group), and the individuals in the remaining two enclosures served as control groups. The treatments were spatially distributed in a way that ensured that enclosures with controls and food-restricted groups were adjacent to a physically stressed group.

The treatments were implemented for twelve consecutive weeks, and commenced immediately after the four acclimation weeks were completed. Each treatment was applied to 9 capybaras (in two enclosures). The food restricted group was provided with a diet of 50% less rice bran+meal (400 gr. per capybara) and of 40% less hay and fresh alfalfa (150 gr. and 300 gr., respectively) than that consumed when fed *ad libitum* during the acclimation period. Dietary restriction while avoiding malnutrition can be accomplished by a 20 to 60% reduction from average unrestricted food intake, including balanced decrease in calories, protein, vitamins, and minerals [25]. Three times a week (on Mondays, Wednesdays and Fridays), animals in the physically stressed group were chased, captured using a net and then physically restrained by tying their limbs for 10 minutes. They were fed 800 gr. of mixture of rice bran+meal, 300 gr. of sorghum or maize hay, and 500 gr. of fresh alfalfa. The control group was fed the same diet as the physically stressed groups but they were not stressed by capture and restraint.

After twelve weeks of treatment, all animals were anesthetized with 10 $mg.kg^{-1}$ ketamine and 0.5 $mg.kg^{-1}$ xylazine [26] and euthanized by exsanguination immediately after collecting a blood sample from the cava vein and recording the body mass and morphometric measures. Later, they were necropsied to obtain samples from different organs.

Stress Assessment

In order to evaluate if the treatments were inducing measurable stress on the capybaras, we compared the proportion of the adrenal cortex that corresponded to the fascicular portion, which is the part of the gland that synthesizes glucocorticoids in response to stress [27]. The adrenal glands fixed in 10% buffered formalin were dehydrated in a graded series of ethanol, cleared in xylene and embedded in paraffin. Serial sections were cut at 5 µm, mounted on glass slides and stained with hematoxylin and eosin. Stained sections were examined under a light microscope at 200× and measured at 1000× using an eyepiece graticule.

The behaviour of the animals along the experiment indicated that habituation of individuals or stressing of the wrong group was not occurring. For the whole duration of the experiment, individuals from the physically stressed group acted with conspicuous evasive behaviour as soon as a person approached the enclosure, while this was not observed in capybaras of the other two treatments.

Measures of Growth and Body Condition

Before and after the experiment, all individuals were weighed (initial body mass, *IBM*; final body mass, *FBM*) using a mechanic scale (precision 0.2 kg.). The morphometric measure used in this study was total length (*TL*). The measures of growth and body condition used were, body mass gain (*FBM - IBM*), body mass index (*FBM/TL*) [24], and body condition score. The latter was estimated by palpating the fat and muscle cover over the thoracic vertebrae and pelvic bones [28]. Each area was scored on a scale between 1 and 5 and both scores were summed. Body condition scoring was always carried out by the same person (ATE).

Measures of the Immune System and Other Physiological Parameters

We evaluated the effect of the treatments on physiological parameters associated with the immune system and general health of mammals. Once animals were anesthetized, a 15–20 ml blood sample was taken from the cava vein and stored in tubes with and without anticoagulant (EDTA 10%). Samples with anticoagulant were kept refrigerated and processed within 8 h of collection. The other sample was centrifuged and the sera obtained were kept frozen at −20°C until further processing. At necropsy, the spleen was weighed using a digital scale and the adrenal glands were stored in 10% buffered formalin.

The sample with anticoagulant was used to produce blood cell counts. They provided an indication of aerobic capacity and energetic balance (red blood cells, RBC), and immunological investment (white blood cells, WBC) [29]. Whole blood was diluted 1:10 in saline solution, and then used to produce two final dilutions: 1:20 in 3% acetic acid and 0.5% of methylene blue (Türk solution) and 1:200 in saline solution to count WBC and RBC, respectively, using Improved Neubauer's chambers [26]. The remaining blood was used to produce blood smears for differential WBC counts to estimate the levels of Lymphocytes, Neutrophils, Monocytes, Eosinophils and Basophils. Smears were fixed and stained with May Grunwald-Giemsa. At least 200 WBCs were counted.

Serum samples were used for the determination of total plasma protein (TPP), albumin (A) and globulins (Gb) [30] using a colorimetric assay (Proti2, Wiener lab., Rosario, Argentina). The A/Gb ratio was also analysed. Variations in the A/Gb ratio may indicate underproduction (high ratio) or overproduction (low ratio) of antibodies, or underproduction of albumin (low ratio) [31]. Other measures of immunological investment used were the spleen mass index, calculated by dividing spleen mass by initial body mass (spleen mass/*IBM*) [9], [22] and natural antibodies (NAb) titers [32]. The determination of NAb titers was done by a hemagglutination assay described by Matson and collaborators [32], with some modifications. Briefly, serum samples (25 μl) were added to columns 1 and 2 of 96-well round bottom plates (Corning Costar) and 25 μl of PBS were added to columns 2 to 10, so that the second well contained a 1:2 dilution. Doubling dilutions were made by transferring 25 μl to from one well to the next, up to well 10, leaving well 11 as positive control (IgM monoclonal antibody specific to α-galactosamine of glucoproteins and glucolipids; Alexis Biochemicals) and well 12 as negative control (PBS only). A stabilized rabbit red blood cells suspension (1.5%) was added to each well and the plates were mechanically shaken for several seconds followed by incubation for 2 h at room temperature. Titers were recorded as the column number of the last serum dilution showing clear evidence of agglutination.

Parasites

We chose to study direct-cycle parasites because burdens of those with an indirect cycle (i.e. cestodes and digeneans) depend directly on the intake of intermediary hosts, which would be very difficult to recreate under our experimental conditions. At necropsy, the entire gastrointestinal tract was removed. The stomach, small and large intestines, and caecum were isolated by ligature. Each section was then opened and the contents washed into separate measuring containers. Mucous membranes were extensively washed and scraped under running water to remove adhered worms [33]. Due to the large volume of caecal and small intestine contents, we estimated helminth intensity at these sections by taking aliquots, which have shown to be good estimators of the entire gut section in other mammals [33]. The contents were put into measuring pails, stirred to homogenize, and a 10% aliquot was collected. The large intestine was opened and examined macroscopically during the necropsy. These aliquots were fixed in 5% buffered formalin and then examined under microscope to classify and count helminths. Species identification was done using specific keys and descriptions [34]–[47].

For baseline comparisons in the 4-week acclimation period, we used faecal egg/oocyst counts. Fresh faecal samples from each individual were evaluated for coccidian oocysts and helminth eggs using a modified Wisconsin Sugar Flotation quantitative method [48] at Laboratorio de Estudios Parasitológicos, Facultad de Ciencias Veterinarias, Universidad Nacional del Litoral. We also used this technique to estimate the coccidian infection intensity at the end of the experiment.

Statistical Analysis

Given the nature of the treatments implemented, capybaras that received a given treatment needed to be confined in the same enclosure (e.g. chasing one capybara stresses out the whole lot in the enclosure). Although the enclosures were identical and contiguous, an individual receiving a given treatment shared the enclosure with other 3 or 4 individuals receiving the same treatment. Because observations from the same enclosure were not independent, an effect attributed to a treatment might in fact be an "enclosure" effect. To account for this lack of independence of observations from individuals in the same enclosure we used linear mixed models (LMM) or generalized linear mixed models (GLMM) with 'enclosure ID' as a random intercept, which takes into account that groups of observations belong to the same enclosure [49], [50]. The type of model used (LMM or GLMM) was decided on the basis of the distribution of the response. When the distribution of the response was approximately normally distributed or it could be transformed to approximate normality by exponentiation, a LMM was used. All parasite counts showed an aggregated distributional pattern, for which GLMM with a negative binomial response was the most appropriate model type [51].

To assess the impact of induced stress on body mass gain, body condition and immuno-physiological parameters, we used LMM, transforming by exponentiation the response variable to approach normality, where appropriate. We used the *lme4* [52] and *languageR* [53] packages of the statistical software R (functions *lmer* and *pvals.fnc*, respectively). The initial models had the main effects *Treatment* and *IBM,* and the interaction term *Treatment*IBM*, to take into account the potential effect modification of differential *IBM*, as well as its potential confounding effect. A likelihood ratio test was used to assess the significance of the interaction term. When the interaction term was not-significant, it was removed from the model and only the main effects were retained (whether significant or not).

To evaluate the impact of induced stress on the intensity of infection of selected parasite species, we used GLMM with negative binomial responses (*glmmADMB* [54] package of R). The initial models included the same main effects and interaction as the LMM above. Removal of unimportant interaction terms from the models was done by likelihood ratio test as described above. The data will be made freely available upon request.

RESULTS

The descriptive statistics of the parameters measured in the capybaras are shown in Table 1. Baseline comparisons carried out during the four weeks prior to establishing the treatments showed that experimental groups were not significantly different in terms of body mass and body mass index, and with respect to faecal parasite egg/oocyst counts of all parasite species investigated (p>0.1 for all comparisons).

Table 1. Parameters of health (including physiological values, immunological parameters, growth and body condition measures) of captive capybaras under three different feeding and physical stress regimes

Health parameters	Control group		Food restricted group		Manipulation group	
	LI	HI	LI	HI	LI	HI
	Mean (Range)	Mean (Range)	Mean (Range)	Mean (Range)	Mean (Range)	Mean (Range)
Adrenal Fascicularproportion (%)	73.7 (64.0,86.7)	75.6 (72.7,77.4)	76.9 (68.9,82.3)	78.8 (77.4, 81.2)	82.2 (78.2,85.7)	74.8 (66.7,81.1)
Body mass gain (kg)	4.2 (3.0,5.5)	4.4 (3.0,6.5)	1.0 (–0.5,3.0)	1.5 (1.5, 1.5)	3.6 (1.5,6.9)	1.2 (–1.0,4.0)
Body condition score	6.5 (6.0,7.0)	7.6 (7.0,8.0)	4.6 (3.0,5.5)	6.0 (6.0, 6.0)	5.8 (5.5,6.5)	7.3 (6.0,8.0)
Body mass index (bodymass/total length)	0.198 (0.19,0.21)	0.262 (0.23,0.30)	0.158 (0.10,0.18)	0.240 (0.22,0.27)	0.193 (0.14,0.25)	0.246 (0.22,0.29)
RBC (millons of cells/µl)	2.97 (2.06,3.75)	3.43 (2.30,5.05)	4.09 (3.00,4.95)	3.80 (3.27, 4.34)	3.49 (3.41,3.64)	4.08 (3.25,5.30)
WBC (thousand of cells/µl)	6.58 (9.20,4.80)	6.58 (7.50,5.40)	6.37 (9.85,3.55)	8.57 (6.45, 11.40)	6.57 (4.20,9.35)	6.59 (5.42,7.55)
L (thousand of cells/µl)	3.10 (1.84,4.44)	3.24 (2.51,4.29)	3.05 (2.16,4.44)	4.20 (3.20, 5.03)	3.17 (1.99,3.93)	3.20 (2.33,4.28)
N (thousand of cells/µl)	2.97 (2.46,3.79)	2.75 (2.08,3.73)	2.25 (1.09,3.41)	3.50 (2.59,5.24)	2.59 (1.83,4.12)	2.62 (1.97,3.22)
E (cells/µl)	229 (170,391)	315 (70,716)	730 (121,2036)	702 (428, 975)	580 (304,971)	477 (181,587)
B (cells/µl)	124 (0,222)	158 (0,422)	151 (0,415)	72 (0, 185)	120 (60,185)	128 (26,219)
M (cells/µl)	126 (50,261)	108 (0,214)	168 (69,258)	83 (0, 153)	101.23 (0,145)	166 (20,330)
Log2 NAb titer	7.0 (6,9)	7.8 (6,9)	8.8 (7,12)	11.3 (10, 13)	6.7 (6,7)	5.8 (4,7)
Spleen mass index	3.61 (2.84,4.37)	3.07 (2.27,3.75)	2.99 (1.97,4.69)	2.81 (2.50, 3.28)	3.23 (3.02,3.37)	2.74(2.22,3.81)
TPP (g/dl)	4.26 (3.14,5.26)	5.14 (4.54,6.22)	5.10 (4.46,5.59)	5.44 (3.90,7.02)	4.72 (4.09,5.59)	5.58 (4.85,7.01)
A (g/dl)	2.81 (1.64,3.60)	2.99 (2.80,3.60)	3.17 (2.58,3.58)	3.15 (2.88,3.42)	3.06 (2.69,3.28)	3.20 (2.52,3.84)
A/Gb (g/dl)	2.04 (1.10,3.18)	1.46 (1.02,1.84)	1.72 (1.04,2.36)	1.78 (0.81,2.82)	1.99 (1.35,2.73)	1.59 (0.56,2.18)

In this table, for reporting purposes, the median of initial body mass (17 kg) was used as the criterion to divide into body mass classes (lighter and heavier individuals).

Abbreviations: LI: lighter individuals at the beginning of the experiment; HI: heavier individuals at the beginning of the experiment; RBC: red blood cells; WBC: white blood cell; L:Lymphocytes; N: Neutrophils; E: Eosinophils; B:Basophils; M: Monocytes; NAb: natural antibodies; PTT: total proteins; A: albumin; Gb: globulins.

Stress Assessment

An analysis of the histo-architecture of the adrenal cortex after the finalization of the experiment revealed that individuals belonging to both treatments had significantly greater fascicular portions than controls, indicating that treatments induced measurable stress on the capybaras (Table 2). The difference between controls and physically stressed individuals was greater among individuals that were initially lighter than among heavier individuals. This trend was also observed in food-restricted individuals, but the interaction term was not quite significant (p=0.089) (Table 2).

Table 2. Linear mixed model describing the effect of treatments on the fascicular portion of the adrenal gland of capybaras (*N*=27)

Response=fascicular proportion (μm)				
Term	Coefficients	Standard error	*P*-value[b]	F-value[c]
Intercept	0.582	0.062	<0.001	
$Treat._{(Food\ restricted)}$[a]	0.175	0.073	0.026	2.74
$Treat._{(Physically\ stressed)}$[a]	0.327	0.076	<0.001	2.74
IBM	0.008	0.003	0.019	0.73
IBM: $Treat._{(Food\ restricted)}$	−0.007	0.004	0.089	8.33
IBM: $Treat._{(Physically\ stressed)}$	−0.015	0.004	0.001	8.33

[a]Simple contrasts – reference level: control (the coefficients reflect comparison with control groups).

[b]P-values obtained from Markov chain Monte Carlo samples (*pvals.fnc* function in R).

[c]For the factor 'Treatment': numerator degrees of freedom=2; denominator degrees of freedom=24.

Measures of Body Mass Gain and Body Condition

The effect of both treatments was largely evident on body mass gain, and it was also significant on body condition score and body mass index, although for these two the effect was only strong for food restricted individuals (Table

3; Figure 1). While control individuals gained on average 4.3 kg during the experiment, food restricted and physically stressed individuals grew only 25% (p=0.0005) and 50% (p=0.0178) that value, respectively.

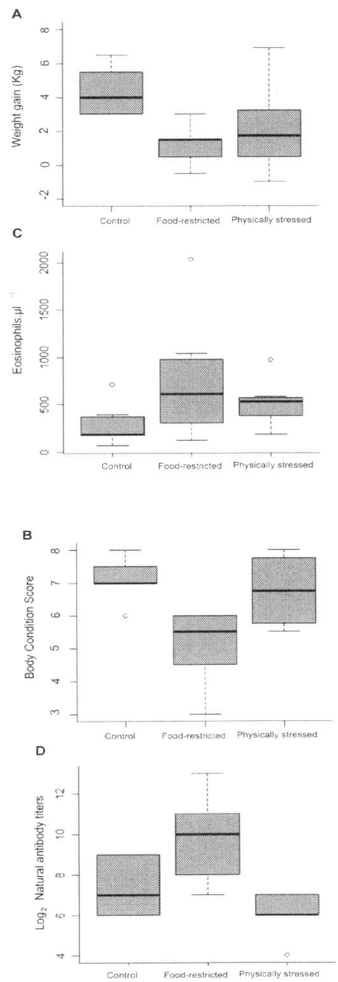

Figure 1. The effect of treatments on body mass gain, body condition score and immunological parameters. Boxplots showing the effect of three different feeding and physical stress regimes on, (A) body mass gain during the duration of the experiment; (B) body condition score; (C) eosinophil concentration in blood; and (D) natural antibodies titers. Boxplots depict the median (bold bar), 25–75% quartiles (box), 10–90% quantiles (whiskers) and outliers (points).

Table 3. Linear mixed models describing the effect of treatments on body mass gain, body condition score and body mass index (*N*=27)

Response=body mass gain				
Term	Coefficients	Standard error	P-value[b]	F-value[c]
Intercept	4.333	0.564	<0.001	
Treat. (Food restricted) [a]	−3.294	0.798	<0.001	8.29
Treat. (Physically stressed) [a]	−2.119	0.823	0.017	8.29
IBM	−0.060	0.058	0.314	1.06
Response=body condition score				
Intercept	4.803	0.396	<0.001	
Treat. (Food restricted) [a]	−1.785	0.267	<0.001	34.24
Treat. (Physically stressed) [a]	−0.578	0.274	0.0467	34.24
IBM	0.128	0.019	<0.001	43.73
Response=body mass index				
Intercept	0.109	0.012	<0.001	
Treat. (Food restricted) [a]	−0.032	0.008	<0.001	19.98
Treat. (Physically stressed) [a]	−0.018	0.008	0.040	19.98
IBM	0.007	<0.001	<0.001	138.72

Effects of Treatments on the Immune System and Other Physiological Parameters

For blood cells, no differences between treatments were observed for red blood cells, lymphocytes, neutrophils and monocytes. The only significant differences between treatments were found for eosinophils (Figure 1, Table 4). Individuals in the food restricted and physically stressed groups had eosinophil counts significantly higher than control individuals (p=0.0113 and 0.0494, respectively) (Figure 1, Table 4).

Table 4. Linear mixed models describing the effect of treatments on eosinophils and natural antibody titers (N=27)

Term	Coefficients	Standard error	P-value[b]	F-value[c]
Intercept	1.650	0.042	<0.001	
Treat. (Food restricted)[a]	0.165	0.059	0.011	3.915
Treat. (Physically stressed)[a]	0.122	0.061	0.049	3.915
IBM	0.004	0.004	0.348	0.917
Response=log2(NAb)$^{0.01}$				
Intercept	1.020	0.001	<0.001	
Treat. (Food restricted)[a]	0.003	0.001	0.011	10.84
Treat. (Physically stressed)[a]	−0.002	0.001	0.051	10.84
IBM	0.000	<0.001	0.498	0.47

[a]Simple contrasts – reference level: control (the coefficients reflect comparison with control groups).

[b]P-values obtained from Markov chain Monte Carlo samples (*pvals.fnc* function in R).

[c]For the factor 'Treatment': numerator degrees of freedom=2; denominator degrees of freedom=24.

Individuals from the food-restriction groups showed significantly higher NAb titers than controls and physically stressed animals (p=0.011 and p=0.002, respectively; Table 4; Figure 1). Physically stressed individuals showed a nearly significant (p=0.051) trend to have lower NAb titers than controls.

Neither food restriction nor physical stress had a significant effect on TPP, A, Gb, A/Gb ratio or spleen mass. However, there was an almost significant trend of food restricted individuals to have lower spleen mass than controls (p=0.053).

Impact of Induced Stress on the Intensity of Infection of Selected Parasite Species

The nematodes found in necropsied capybaras were S. chapini, Trichostrongylus sp., Trichuris sp., E. hydrochaeri, H. anomalobursata and V. hydrochoeri (the latter two were counted together as 'Viannaiidae', because both females are undistinguishable). Although low counts of P. obesa eggs were occasionally observed at the beginning of the experiment, adults were not present at necropsy. The effects of treatments on nematode intensity depended on the capybara body mass at the beginning of the experiment for all nematode species found, except for Trichuris sp. (Table 5; Figure 2).

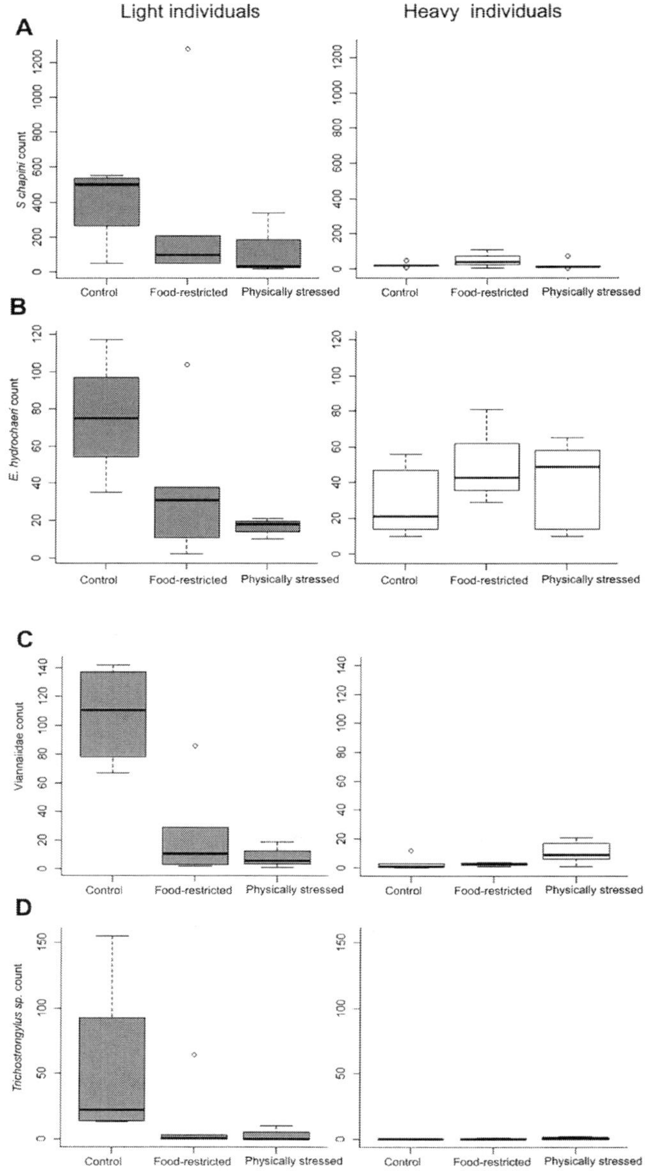

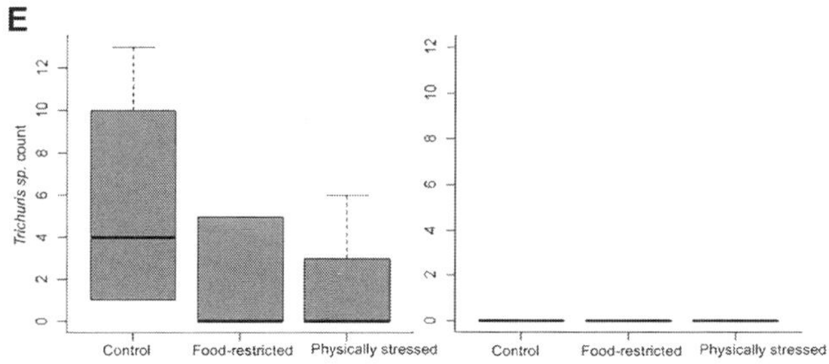

Figure 2. Intensity of different parasite species by treatments. In this boxplot, the median of initial body mass (17 kg) was used as the criterion to divide into body mass classes (lighter and heavier individuals). (A) *Strongyloides chapini*; (B) *Echinocholeus hydrochaeri*; (C) Family Viannaidae; (D) *Trichostrongylus* sp.; and (E) *Trichuris* sp. Boxplots depict the median (bold bar), 25–75% quartiles (box), 10–90% quantiles (whiskers) and outliers (points).

Table 5. Generalized linear model with a negative binomial response describing the effect of treatments on nematode and coccidian intensity (N=27)

Response=*S. chapini*				
Term	Coefficients	Standard error	P-value	Deviance/Residual dev.
Intercept	10.288	1.188	<0.001	
Treat. (Food restricted) [a]	−3.466	1.429	0.015	7.62/60.92
Treat. (Physically stressed) [a]	−4.200	1.513	0.005	7.62/60.92
IBM	−0.325	0.062	<0.001	24.62/36.29
IBM*Treat. (Food restricted)	0.214	0.075	0.004	7.41/28.89
IBM*Treat (Physically stressed)	0.189	0.074	0.011	7.41/28.89
Response= *E. hydrochaeris*				
Intercept	5.249	0.875	<0.001	
Treat. (Food restricted) [a]	−2.087	1.061	0.049	2.24/34.91
Treat. (Physically stressed) [a]	−3.155	1.138	0.005	2.24/34.91
IBM	−0.078	0.047	0.099	1.12/33.79
IBM*Treat. (Food restricted)	0.111	0.058	0.058	5.81/27.99

IBM* Treat (Physically stressed)	0.140	0.058	0.016	5.81/27.99
Response= ***Trichostrongylus*** **sp.**				
Intercept	18.066	5.606	0.001	
Treat. (Food restricted) [a]	−11.607	6.007	0.053	8.95/44.91
Treat. (Physically stressed) [a]	−16.538	5.811	0.004	8.95/44.91
IBM	−1.057	0.383	0.006	9.60/35.31
IBM*Treat. (Food restricted)	0.674	0.412	0.102	15.05/20.26
IBM* Treat (Physically stressed)	0.993	0.389	0.010	15.05/20.26
Response=Viannaiidae				
Intercept	8.722	1.439	<0.001	
Treat. (Food restricted) [a]	−3.558	1.739	0.040	10.59/45.68
Treat. (Physically stressed) [a]	−6.707	1.774	<0.001	10.59/45.68
IBM	−0.311	0.074	<0.001	4.60/41.07
IBM* Treat. (Food restricted)	0.139	0.101	0.166	11.98/29.09
IBM* Treat (Physically stressed)	0.325	0.095	<0.001	11.98/29.09
Response= ***Trichuris*** **sp.**				
Intercept	10.452	3.012	<0.001	
Treat. (Food restricted) [a]	−2.998	1.045	0.004	3.99/50.28
Treat. (Physically stressed) [a]	−3.021	1.378	0.028	3.99/50.28
IBM	−0.648	0.209	0.001	39.91/10.37
Response=Coccidian				
Intercept	4.908	0.761	<0.001	
Treat. (Food restricted) [a]	2.646	0.494	<0.001	26.27/33.40
Treat. (Physically stressed) [a]	1.565	0.511	0.002	26.27/33.40
IBM	−0.074	0.038	0.051	3.42/29.97

When considering the initial body mass, light individuals in control groups at the end of the experiment had significantly greater *S. chapini*, *E.*

hydrochaeri, Viannaiidae and *Trychostrongylus* sp. burdens than physically stressed groups (Table 5), while no significant difference was observed within heavier animals (Table 5; Figure 2). Food restricted capybaras showed the same trend (Figure 2), but the interaction was only statistically significant for *S. chapini* (Table 5).

In the case of *Trichuris* sp., both treatments showed lower parasite counts than controls (Figure 2, Table 5). The magnitude of the difference was much greater between controls and food restricted animals.

Regarding coccidians, oocyst counts were much higher in both treatments than in control groups (Table 5). The magnitude and significance of this difference was much greater between controls and food restricted individuals (Figure 3). Food restricted individuals had significantly higher oocyst counts than physically stressed ones (p=0.035).

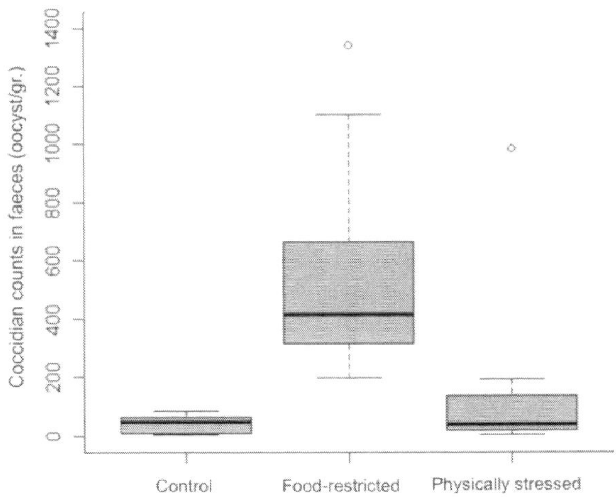

Figure 3. Coccidian infection intensity in capybaras under different treatments. Boxplots showing coccidian infection intensity (faecal oocyst count) at the end of the experiment in capybaras under three different feeding and physical stress regimes. Boxplots depict the median (bold bar), 25–75% quartiles (box), 10–90% quantiles (whiskers) and outliers (points).

DISCUSSION

The most ecologically significant results of this experiment were, on the one hand, that stress had a marked negative impact on the body mass gain and body condition of capybaras, but a positive effect on nonspecific components of the immune system: eosinophil counts were greater in capybaras from both treatment groups and NAb titers were greater in food restricted animals, compared to controls. On the other hand, the consequence of both treatments on the gastrointestinal parasitism dynamics depended on the type of parasite and on the size of the individuals at the beginning of the experiment. Stressed capybaras had significantly higher coccidian infection intensities than controls (this difference was much stronger in the food restricted group), but controls had, in general, higher helminth intensity than treated individuals among those that were lighter, and therefore younger, at the beginning of the experiment (especially in physically stressed capybaras). Our prediction that stressed capybaras would be in poorer condition, immunosuppressed and with high parasite burdens at the end of the experiment was only partially supported. Only for coccidian infection did we observe our expected scenario, although we did not find a significant decrease in any of the immunological parameters studied.

Stress, Body Mass Gain and Body Condition

Large context-dependent differences in growth and body condition have been observed in some wild rodent populations. Voles and lemmings, for example, are larger (20–30% heavier) during the increase and peak phases of the population cycle than in the declining and low phases of their multi-annual fluctuations [55]. This phenomenon, referred to as the 'Chitty' effect, is believed to be a consequence of a dynamic allocation of energy among physiological functions [56]. In good years, rodents have a surplus energy that will allow continuous growth and deposition of additional body mass, while when the context is not favourable they suppress growth. Here we experimentally reproduced this context-dependency of somatic effort, showing that body mass gain and body condition are very sensitive to both nutritional and physical stress in capybaras.

Previous reports found that, under nutritional stress, rats and chicks reared under germ-free conditions had greater rate of growth than conventionally reared animals [21], [57]. This suggests a trade-off between growth and immunity, with immunity having priority over body growth when nutrients are limited [5]. The results of the present study are in agreement with the above, as animals under nutritional stress showed different priorities in their use of resources than non-stressed individuals, reducing their investment in somatic effort while enhancing some compartments of the immune system.

Stress and Immunological Investment: Stress-induced Prophylaxis

The physiological parameters measured did not show the expected response to treatments. There was only a nearly significant trend of food restricted animals to have lower spleen mass index than controls (25% reduction in the spleen mass). A relationship between food restricted animals and reduced spleen size was found in an experiment with chicken [22]. It might be that this effect was not strong enough to be detected by the limited statistical power of our experiment.

Of the parameters measured, one showed to be sensitive to the influence of both types of stress, and a second one to nutritional stress, but in the opposite direction than predicted. While stressed individuals were significantly reducing their body mass gain and body condition, they were diverting part of their limited available energy to some nonspecific compartments of the immune system, as they had greater eosinophil counts and much higher Nab titers (the latter only in food restricted capybaras) than controls.

NAb are unique among the immunoglobulins, as their production does not require previous exposure to antigen [58]. Levels of NAb have been demonstrated in animals bred in pathogen-free environments [59]. Given that they confer nonspecific humoural immunity independent of pathogen exposure, NAb have the potential to be good indicators of the humoural immunocompetence of wild animals [60]. While it is known that chronic stress determines a reduction in specific antibody levels [61], how it affects Nab has been completely ignored. Our results suggest that nutritional stress enhances their production. Eosinophils, on the other hand, are key protagonists in the helminth-induced immune response, known as the T helper 2 cell (Th2) response, being the main leukocyte that increases their levels in the presence of helminth infections [62]. Contrary to our finding, chronic stress is known to cause a reduction in circulating eosinophil levels, as glucocorticoids suppress eosinophil proliferation and diminishes its survival [62].

This previous knowledge on NAb and eosinophils led us to anticipate that their levels would be lower in treated groups, as a result of immunosuppression caused by chronic stress and lack of resources to invest in immunity [12]. It is sensible to expect that individuals fed *ad libitum* and undisturbed should be able to invest more in immune function than stressed individuals. In this experiment, however, the only indication that stressed individuals were investing less in immune function was a not quite statistically significant reduction in the spleen size (25%). Moreover, investment in eosinophils was greater in stressed individuals, and NAb titers were much higher in food restricted animals. It is noteworthy that both TPP and the A/Gb ratio were not different between treatment groups, indicating that it was unlikely that immunoglobulins other than NAb were also elevated in food restricted animals. Similarly, leukocytes other than eosinophils were not affected by treatments.

The stress-immunity interaction depends on coping styles (the behavioural and physiological efforts to master the stressing situation) [63]. Prey species often show a passive coping style, which is associated with high HPA reactivity and a Th2 biased immune response. The latter is because prey species are more likely to be exposed to macroparasites, as they show exploratory nature and greater intake of novel resources [64]. Increased eosinophils in stressed capybaras are reflecting an enhanced Th2 immune response in these individuals. It is unlikely that this difference in eosinophil counts was caused by differential helminth parasitism, as treated individuals had lower or similar nematode burdens as controls (faecal egg counts throughout the duration of the experiment were never higher in the treated groups; data not shown). This anti-helminthic immune response in the absence of increased helminth infection might be indicating preparedness in anticipation of greater risk of parasite exposure.

In rodents, the effect of stressors on the immune function depends on their duration [11]. They enhance immunity in the short term (hours-days) and cause immunosuppression when it is sustained for longer (weeks-months) [11]. Our results suggest that in capybaras chronic stress produces a diversion of resources to some nonspecific compartments of the immune system. Similar results were obtained by Hangalapura and collaborators [22] in chicken, who conducted an experiment and concluded that sustained food restriction did not have a significant effect on specific antibody responses, but rather suppressed parts of cell-mediated immunity and enhanced others of the innate immunity (increased production of reactive oxygen intermediates by phagocytes in whole blood). This suggests that the organism stressed over long periods of time invests in nonspecific immunity rather than in acquired cell mediated or humoural specific immunity, which makes biological sense. Firstly, because it makes little logic that organisms under chronic stress would suppress the whole immune system, as it could be critical for recovery from stressors [65]. Secondly, considering the wide diversity of parasites that infect a host under natural conditions, an optimal strategy may be to invest preemptively in defences against an array of pathogens rather than invest in a specific immune response to antigens of every member of the parasite community. This could be interpreted as a sort of stress-induced prophylaxis. A similar preventative strategy is used by some insects. When some lepidopteran species are reared under crowded conditions, they prepare for higher pathogen circulation by enhancing their immune system, resulting in butterflies that are significantly more resistant to infections than those reared solitarily, a phenomenon termed 'density-dependent prophylaxis' [66].

This stress-dependent nonspecific immune enhancement might explain some contradictory results observed in eco-immunological studies. For example, Vinkler and collaborators [67] found a negative association between the response to the phytohaemagglutinin skin-swelling test and ornament saturation of rosefinch males, indicating stronger immune responsiveness in inferior males. In that study, a major involvement of basophils in the swelling response was shown, which indicates the nonspecific nature of the reaction. The authors concluded that poor ornamentation was associated with greater immune response because both were indicative of low quality, as the immune response could be detrimental to the birds (inflammation is a destructive process). However an alternative interpretation of their findings is that, like in our experiment, stressed individuals invested more in nonspecific immunity at the expense of other physiological functions such as ornamentation. In such case, it would not be a matter of quality (genotypes) but rather of differential life-histories (phenotypes).

The physiological trade-offs observed in our experiment arise under stressing circumstances, thus allowing animals to adjust to changing environmental conditions. In favourable contexts, resources are readily available and the organism should be able to maintain its core body temperature, fight an infection, and produce and rear viable offspring concomitantly [68]. However, if resources are limited, the same organism will face severe challenges and will need to establish priorities in order to overcome the difficult moment. Its success will depend on the degree of stress faced by the animal. For example, an experimental study on tree lizards [69] showed that when food was unlimited, females were able to invest both in reproduction and heal their wounds. When food was restricted, they could still heal a cutaneous wound, but they needed to decrease reproductive investment, and under extreme food limitation both reproductive investment and wound healing were suppressed. We did not investigate different levels of stress in our experiment. It is to be expected, however, that capybaras would be unable to invest on eosinophils or NAb, should they be subject to a more severe stress. Of course the differential costs of the various immune responses (Th1 v. Th2, specific v. generic, etc.) should be taken into account carefully in order to comprehend the trade-offs underlying changes in host-parasite interactions during stress. Further studies should focus on determining if the influence of stress on a specific compartment of the immune system depends on its relative cost compared to other compartments (e.g. are inexpensive immune responses enhanced and costly ones suppressed?).

Stress and Parasite Intensity

Infection with species of *Eimeria* causes coccidiosis, a common disease of domestic animals. It is characterised by catarrhal to haemorrhagic enteritis that results in diarrhoea, anemia, dehydration, anorexia, weight loss and eventually death [70], [71]. We did not observe these signs, but as expected, both food restricted and physically stressed treatments had significantly greater oocyst counts than control groups, which could result in subclinical effects, as the severity of the enteritis is positively correlated with the infection intensity [72]. This result is in accordance with the notion of a vicious circle, as stressed individuals were in poorer body condition, and were also suffering more severe coccidian infections, which in turn would lead to an even more deteriorated condition [14]. However, no associated significant decrease in immune function was observed.

Although we expected similar results in the infection intensity of gastrointestinal nematodes, we observed the opposite. In individuals that were lighter at the beginning of the experiment, the pattern was consistent for all parasite species, controls had much higher parasite burdens than physically restrained individuals. The pattern held for food-restricted capybaras, although the difference was significant only for *Trichuris* sp. and *S. chapini.* We offer two not mutually exclusive hypotheses to explain this observation, and argue that the evidence gathered supports one over the other.

The first explanation relates to differential consumption rates of food resources. Unlike coccidians, which can multiply within the hosts like all microparasites, the intensity of nematode infection is strictly dependent on the amount of immature stages (eggs) that are ingested by a host. Individuals that were fed *ad libitum* would have greater intake of parasite eggs than those that had a lower consumption rate (food restricted). In addition, decreased host nutrition may negatively impact parasite populations through a simple reduction in resource availability [73]. However, the fact that the pattern of reduced helminth burdens in smaller capybaras was stronger or as strong in physically stressed capybaras than in food restricted ones argues against this hypothesis.

The other explanation would be that higher nematode intensity in controls might arise from a reduced parasite resistance (the ability to limit parasite burden) or increased tolerance (the ability to limit harm caused by a given burden). The lower levels of eosinophils in control individuals is an indication that they had less allocation of resources to immune response against macroparasites than stressed ones, which in turn could also lower parasite resistance [65], [30]. The measure of tolerance is the slope of the regression of host fitness on parasite burden [30]. In a supplementary analysis, we observed that the slopes of the relationship between body condition and parasite intensity were similar in all treatment groups. This lack of significant differences in slopes suggests that tolerance was not affected by the treatments, but the difference in eosinophil counts indicates that resistance could have been affected. Therefore, this lower investment in Th2 response in control individuals was very likely the cause of the higher helminth infection intensity they suffered. What remains to be explained is why this was mainly observed in individuals that were smaller at the beginning of the experiment. This could be related to the immunological experience (acquired immunity), which would be greater in older individuals, but it could also be attributed to differential effect of stress on smaller and larger individuals. The latter was reflected in the adrenal glands, as the signs of stress were greater among physically restrained individuals that were initially lighter than among heavier individuals. Food restricted showed a similar trend, although the interaction term was not significant. In addition, even though all capybaras in our experiment were still growing, younger animals (i.e. initially lighter) were at a more demanding phase of body growth than older individuals [24], which constitutes an additional source of stress.

Further studies should explore the involvement of stress on the interaction between the immune response to typical intracellular parasites (e.g. coccidians) mediated by Th1, and that against extracellular parasites (e.g. helminths) primarily mediated by Th2, as cytokines produced by Th1 cells may suppress Th2 immune function and vice versa [74]. The immune response of capybaras might be Th2 biased, as explained above for passive coping species [63], which is in agreement with our findings; stress resulted in higher infection intensities in a parasite that elicits a Th1 response, while the infection intensities of those that are controlled by a Th2 response were indeed lower. It would be important to know what consequences in the life-history of capybaras result from these changes in the susceptibility to different parasite types induced by stress. Our results suggest that during stressful periods capybaras might be more vulnerable (although not necessarily more exposed) to microparasites like coccidians than to macroparasites.

Conclusions

Although short to medium duration stressors elicit an enhancement of immunity, it is widely accepted that chronic stress results in immunosuppression through neuroendocrine mechanisms involving the HPA axis and glucocorticoids, in order to redirect resources towards physiological processes that are important to overcome the adverse situation [11], [13]. Although we have only assessed a limited set of components of the complex vertebrate immune system, we were able to show that this chronic stress-mediated immunosuppression does not hold for the whole immune system in capybaras. Our results suggest that parts of the nonspecific arm of the immune system are favored over specific components during stress, preparing the organism to face its very diverse parasite community with generic defences instead of investing in specific immunity to every single parasite species and opportunistic infectious agents they are exposed to. Animals use many environmental signals to modify their phenotypes in preparation for recurrent environmental challenges [11]. We named the immunological preparedness observed here 'stress-dependent prophylaxis'.

We also showed that food restriction and physical stress had opposite influences on gastrointestinal microparasites and macroparasites. This highlights that special attention should be placed on the parasite type or species involved when attempting to interpret how the synergistic interaction between food availability, stress and parasites affects individual fitness and population dynamics.

ACKNOWLEDGMENTS

Ayelen Eberhardt is a fellow of the Argentine Council for Science and Technology (CONICET). We thank Molino Arrocero Del Litoral SA and Sergio Bassega for kindly donating the rice bran and fresh alfalfa, respectively. This work was made possible thanks to the invaluable help provided by Federico "Fama" Miranda, Evelina Tarragona, Nicolás Welscen, Ana Inés Correa, Mateo Rochi, Carolina Magni, Juan Pablo Picasso, Juan Micheloud,

Matías Belotti, Amorina Sánchez, Natalia Pezzone, Hugo Ortega, Carolina Gaggiamo, Carolina Panzani, Matías Stangaferro, Guido Bordesse, Florencia Bono, Dana Plaza, Viviana Orcellet, Marcelo Ruiz, Cristina Cerra, Eduardo Eberhardt, Rubén Miranda, Rubén D. Quintana, María Luisa Bolkovic, Guillermo Principe, Pablo Sirosky, Alejandro Larriera, Daniel Hunziker, and the experimental zoological station "Granja la Esmeralda" staff.

AUTHOR CONTRIBUTIONS

Conceived and designed the experiments: ATE PMB. Performed the experiments: ATE PMB SAC PM. Analyzed the data: PMB ATE MRM AR CJB MRR. Contributed reagents/materials/analysis tools: PMB ATE MRM AR CJB MRR. Wrote the paper: ATE PMB MRR AR PM MRM.

REFERENCES

1. Grenfell BT, Wilson K, Isham VS, Boyd HEG, Dietz K (1995) Modelling patterns of parasite aggregation in natural populations: trichostrongylid nematode-ruminant interactions as a case study. *Parasitol* 111: S135–S151.
2. Hudson PJ, Dobson AP, Newborn D (1998) Prevention of Population Cycles by Parasite Removal. *Science* 282: 2256–2258.
3. Pedersen AB, Greives TJ (2008) The interaction of parasites and resources cause crashes in a wild mouse population. *J Anim Ecol* 77: 370–377.
4. Gulland FMD (1995) Impact of infectious diseases on wild animal population: a review. In: Grenfell BT, Dobson AP, editors. Ecology of Infectious Diseases in Natural Population. Cambridge: Cambridge University Press. 20–51.
5. Lochmiller RL, Deerenberg C (2000) Trade-offs in evolutionary immunology: just what is the cost of immunity? *Oikos* 88: 87–98.
6. Frank SA, Schmid-Hempel P (2008) Mechanisms of pathogenesis and the evolution of parasite virulence. *J Evol Biol* 21: 396–404.
7. Tompkins DM, Begon M (1999) Parasites can regulate wildlife populations. *Parasitol Today* 15: 311–313.
8. Brown MJF, Schmid-Hempel R, Schmid-Hempel P (2003) Strong context-dependent virulence in a host-parasite system: reconciling genetic evidence with theory. *J Anim Ecol* 72: 994–1002.
9. Ould P, Welch HE (1980) The effect of stress on the parasitism of mallard ducklings by *Echinuria uncinata* (Nematoda: Spirurida). *Can J Zool* 58: 228–234.
10. Scott ME (1988) The Impact of Infection and Disease on Animal Populations: Implications for Conservation Biology. Conserv Biol 2, N°1.
11. Martin LB (2009) Stress and immunity in wild vertebrates: Timing is everything. *Gen Comp Endocrinol* 163: 70–76.
12. Webster JI, Tonelli L, Sternberg EM (2002) Neuroendocrine Regulation of Immunity. *Annu Rev Immunol* 20: 125–163.
13. Sapolsky RM, Romero LM, Munck AU (2000) How Do Glucocorticoids Influence Stress Responses? Integrating Permissive, Suppressive, Stimulatory, and Preparative Actions. *Endocrine Rev* 21: 55–89.

14. Beldomenico PM, Begon M (2010) Disease spread, susceptibility and infection intensity: vicious circles? *Trends Ecol Evol* 25: 21–27.
15. Bolkovic ML, Quintana RD, Ramadori D, Elisech M, Rabinovich J (2006) Proyecto Carpincho. Propuesta para el uso sustentable del carpincho (*Hydrochaeris hydrochaeris*) en la Argentina. In: Bolkovic ML, Ramadori D, editors. Manejo de fauna silvestre en la Argentina. Programas de uso sustentable. Dirección de Fauna Silvestre, Secretaría de Ambiente y Desarrollo Sustentable de la Nación. 105–119.
16. Salas V, Herrera EA (2004) Intestinal helminths of capybaras, *Hydrochoerus hydrochaeris*, from Venezuela. *Mem Inst Oswaldo Cruz* 99: 563–566.
17. Sinkoc AL, Brum JGW, Muller G (2009) Gastrintestinal helminths of capybara (*Hydrochoerus hydrochaeris*, Linnaeus, 1766) in Cattle Breeding Farm in the area of the Ecological Reserve of Taim, Rio Grande. *Braz Arch Biol Technol* 52: 327–333.
18. Albuquerque GR, Berto BP, Catenacci LS, Nogueira SSd, Nogueira-Filho SL, et al. (2008) Eimerid coccidia from capybaras (*Hydrochoerus hydrochaeris*) in southern Bahia, Brazil. *Pesqui Vet Bras* 28: 323–328.
19. Ojasti J, Sosa Burgos LM (1985) Density regulation in population of capybara. *Acta Zool Fen* 173: 81–83.
20. Salas V, Herrera EA (2004) Intestinal helminths of capybaras, *Hydrochoerus hydrochaeris*, from Venezuela. *Mem Inst Oswaldo Cruz* 99: 563–566.
21. Snyder DL, Wostmann BS (1987) Growth rate of male germfree Wistar rats fed ad libitum or restricted natural ingredient diet. *Lab Anim Sci* 37: 320–325.
22. Hangalapura BN, Nieuwland MG, Vries Reilingh G, Buyse J, Van Den Brand H, et al (2005) Severe feed restriction enhances innate immunity but suppresses cellular immunity in chicken lines divergently selected for antibody responses. *Poult Sci* 84: 1520–1529.
23. Herrera E, Macdonald DW (1984) The capybara. In: Macdonald DW, editor, Encyclopaedia of Mammals. George Allen & Unwin, 696–699.
24. Ojasti J (2011) Estudio Biológico del chigüire o capibara. 2 nd Edition. Caracas: Equinoccio, University of Simón Bolívar. 353 p.
25. Weindruch R, Walford RL (1988). The Retardation of Aging and Disease by Dietary Restriction. Springfield, Illinois, USA: Thomas Press.

26. Madella DA, Neto EJ, Felisberto ME, de Souza CE (2006) Valores hematológicos de capivaras (*Hydrochoerus hydrochaeris*) (Rodentia:Hydrochoeridae) de vida livre na região de Campinas-SP. *Ciência Rural* 36: 1321–1324.
27. Jubb K, Kennedy P, Palmer N (2007) Pathology of domestic animals. Sauders Ltd.
28. Burthe S, Telfer S, Lambin X, Bennett M, Carslake D, et al. (2006) Cowpox virus infection in natural field vole *Microtus agrestis* populations: delayed density dependence and individual risk. *J Anim Ecol* 75: 1416–1425.
29. Beldomenico PM, Telfer S, Gebert S, Lukomski L, Bennett M, et al. (2008) The dynamics of health in wild field vole populations: a haematological perspective. *J Anim Ecol* 77: 984–997.
30. Råberg L, Graham AL, Read AF (2009) Decomposing health: tolerance and resistance to parasites in animals. *Philos Trans R Soc Lond B Biol Sci* 364: 37–49.
31. Latimer KS, Mahaffey EA, Prasse KW (2003) Duncan & Prasse's veterinary laboratory medicine: clinical pathology. Ames, IA: Iowa Satate Press.
32. Matson K, Ricklefs R, Klasing K (2005) A hemolysis-hemagglutination assay for characterizing constitutuve innate humoral immunity in wild and domestic birds. *Dev Comp Immunol* 29: 275–286.
33. Ueno H, Gutierres V (1988) Manual para diagnóstico das helmintoses de rumiantes. Tokyo, Japan: 176 p.
34. Travassos L (1916) Informações sobre a fauna helmintológica Sul-Fluminense. II. *Brazil Méd* 30: 313–314.
35. Freitas Texeira JF, Lent H (1936) Estudo sobre os Capillarinae parasitos de mamíferos. *Mem Inst Oswaldo Cruz* 3: 123–156.
36. Travassos L (1937) Revisão da família Trichostrongylidae Leiper 1912. Monogr. Inst. Oswaldo Cruz 512 pp.
37. Boero JJ, de Boehringer I (1967) Los parasitos del Carpincho (*Hydrochoerus hydrochaeris*) y del quiyá (Myocastor coypus). *Rev Fac Cien Vet La Plata* 9: 161–172.
38. Anderson RC, Chabaud AG, Willmont S (1974) CHI. Keys to the Nematode Parasites of Vertebrates. Commonwealth Agricultural Bureaux. Farnham Royal, Bucks, UK. 170 pp.

39. Barus V, Madjumdar G, Mikailov TK (1975) Morphology and taxonomy of *Trichocephalus myocastoris* Enigk 1933. *Folia Parasitologica* 22: 207–213.
40. Diaz-Ungria C, Quentin JC (1976) Morphologie et position systematique de l'oxyure *Protozoophaga obesa* (Diesing, 1851) Travassos, 1923. Exc Parasitol en Mem del Dr Eduardo Caballero y Caballero.
41. Moravec F (1982) Proposal of a new systematic arrangement of nematodes of the family Capillaridae. *Folia Parasitol* 29: 119–132.
42. Vicente JJ, Olivera Rodrígues H, Correa Gomes D, Pinto RM (1997) Nematoides do Brasil. Parte V: Nematoides de mamiferos. *Rev Brasil Zool* 14: 1–452.
43. Anderson RC (2000) Nematode parasites of vertebrates: Their development and transmission, 2nd ed. CAB International, Wallingford, U.K. 650 p.
44. Ribeiro SMB, Amato SB (2003a) Descrição de estructuras associadas com a bolsa copuladora e cone genital de *Hydrochoerisnema anomalobursata* ARANTES & ARTIGAS, 1980 (TRICHOSTRONGYLOIDEA, VIANAIIDAE). Arquiv do Instit Biol 70, 145–147.
45. Ribeiro SMB, Amato SB (2003b) Redescrição da bolsa copuladora de *Vianella hydrochoeri* (TRAVASSOS, 1914) TRAVASSOS, 1918 com ênfase no cone genital (TRICHOSTRONGYLOIDEA, VIANAIIDAE). Arquiv do Instit Biol 70, 155–157.
46. Gibbons LM (2010) Keys to the nematode parasites of vertebrates. Supplementary Volume. CAB International, Wallingford, U.K., 416 p.
47. Robles MR (2011) New species of *Trichuris* (Nematoda: Trichuridae) from *Akodon montensis* Thomas, 1913 of the Paranaense forest in Argentina. *J of Parasitol* 97: 319–327.
48. Cox D, Todd A (1962) Survey of gastrointestinal parasitism in Wisconsin dairy cattle. *J Amer Vet Med Assoc* 141: 706–709.
49. Chavez LF (2010) An Entomologist Guide to Demystify Pseudoreplication: Data Analysis of Field Studies with Design Constraints. *J Med Entomol* 47: 291–298.
50. Paterson S, Lello J (2003) Mixed models: getting the best use of parasitological data Trends Parasitol. 19: 370–375.
51. Wilson K, Bjørnstad ON, Dobson AP, Merler S, Poglayen G, et al. (2002) Heterogeneities in macroparasite infections: patterns and processes. In Hudson PJ, Rizzoli A, Grenfell BT, Heesterbeek H, Dobson AP, editors. The Ecology of Wildlife Diseases. 6–44.

52. Douglas B, Martin M, Ben B (2013) lme4: Linear mixed-effects models using S4 classes. R package version 0.999999-2. Available: http://CRAN.R-project.org/package=lme4. Accessed 14 June 2013.
53. Baayen RH (2011) LanguageR: Data sets and functions with "Analyzing Linguistic Data: A practical introduction to statistics". R package version 1.4. Available: http://CRAN.R-project.org/package=languageR. Accessed 14 June 2013.
54. Fournier DA, Skaug HJ, Ancheta J, Ianelli J, Magnusson A, et al. (2012) AD Model Builder: using automatic differentiation for statistical inference of highly parameterized complex nonlinear models. Optim Methods Softw 233–249.
55. Chitty D (1952) Mortality among Voles (*Microtus agrestis*) at Lake Vyrnwy, Montgomeryshire in 1936–9. *Philos Trans R Soc Lond B Biol Sci* 236: 505–552.
56. Oli MK (1999) The Chitty Effect: A Consequence of Dynamic Energy Allocation in a Fluctuating Environment. *Theor Popul Biol* 56: 293–300.
57. Furuse M, Yokota H (1984) Protein and energy utilization in germ-free and conventional chicks given diets containing different levels of dietary protein. *Br J Nutr* 51: 255–264.
58. Casali P, Schettino EW (1996) Structure and function of natural antibodies. *Curr Top Microbiol Immunol* 210: 167–179.
59. Pereira P, Forni L, Larsson EL, Cooper M, Heusser C, et al. (1986) Autonomous activation of B and T cells in antigen-free mice. *Eur J Immunol* 16: 685–688.
60. Townsend AK, Clark AB, McGowan KJ, Miller AD, Buckles EL (2010) Condition, innate immunity and disease mortality of inbred crows. *Proc Biol Sci* 277: 2875–2883.
61. Merino S, Moreno J, Tomás G, Martínez J, Morales J, et al. (2006) Effects of Parental Effort on Blood Stress Protein HSP60 and Immunoglobulins in Female Blue Tits: A Brood Size Manipulation Experiment. *J Anim Ecol* 75: 1147–1153.
62. Rothenberg MV, Hogan SP (2006) The Eosinophil. *Annu Rev Immunol* 24: 147–174.
63. Koolhaas JM (2008) Coping style and immunity in animals: Making sense of individual variation. *Brain Behav Immun* 22: 662–667.

64. Korte SM, Koolhaas JM, Wingfield JC, McEwen BS (2005) The Darwinian concept of stress: benefits of allostasis and costs of allostatic load and the trade-offs in health and disease. *Neurosci Biobehav Rev* 29: 3–38.
65. Martin LB, Weil ZM, Nelson RJ (2008) Seasonal changes in vertebrate immune activity: mediation by physiological trade-offs. *Philos Trans R Soc Lond B Biol Sci* 363: 321–339.
66. Wilson KENN, Reeson AF (1998) Density-dependent prophylaxis: evidence from Lepidoptera-baculovirus interactions? *Ecol Entomol* 23: 100–101.
67. Vinkler M, Schnitzer J, Munclinger P, Albrecht T (2012) Phytohaemagglutinin skin-swelling test in scarlet rosefinch males: low-quality birds respond more strongly. *Anim Behav* 83: 17–23.
68. French SS, Moore MC, Demas GE (2009) Ecological immunology: The organism in context. *Integr Comp Biol* 49: 246–253.
69. French SS, Johnston GIH, Moore MC (2007) Immune activity suppresses reproduction in food-limited female tree lizards *Urosaurus ornatus* . *Func Ecol* 21: 1115–1122.
70. Nogueira, MF Fukuta da Cruz T (2007) Doenças da Capivara. Corumbá, MS: Embrapa Pantanal. 74 p.
71. Busato A, Lentze T, Hofer D, Burnens A, Hentrich B, et al. (1998) A case control study of potential enteric pathogens for calves raised in cow-calf herds. *J Vet Med B* 45: 519–528.
72. Daugschies A, Najdrowski M (2005) Eimeriosis in Cattle: Current Understanding. *J Vet Med B* 52: 417–427.
73. Bundy DAP, Golden MHN (1987) The impact of host nutrition on gastrointestinal helminth populations. *Parasitol* 95: 623–635.
74. Yazdanbakhsh M, Kremsner PG, van RR (2002) Allergy, parasites, and the hygiene hypothesis. *Science* 296: 490–494.

INDEX

C

D

I

J

K

L

Q

R

S

T